18th International Conference on Textures of Materials (ICOTOM-18)

IOP Conference Series: Materials Science and Engineering
Volume 375

St. George, Utah, USA
6 - 10 November 2017

Editors:

Stuart I. Wright
David T. Fullwood
Matthew M. Nowell

ISBN: 978-1-5108-6488-7
ISSN: 1757-8981

Printed from e-media with permission by:

Curran Associates, Inc.
57 Morehouse Lane
Red Hook, NY 12571

Printed by Curran Associates, Inc. (2018)

For permission requests, please contact the Institute of Physics at the address below.

Institute of Physics
Dirac House, Temple Back
Bristol BS1 6BE UK

Phone: 44 1 17 929 7481
Fax: 44 1 17 920 0979

techtracking@iop.org

Additional copies of this publication are available from:

Curran Associates, Inc.
57 Morehouse Lane
Red Hook, NY 12571 USA
Phone: 845-758-0400
Fax: 845-758-2633
Email: curran@proceedings.com
Web: www.proceedings.com

Table of contents

Volume 375

18th International Conference on Textures of Materials (ICOTOM-18)

6–10 November 2017, St George, Utah, USA

Accepted papers received: 30 May 2018
Published online: 18 June 2018

Preface

Papers

ICOTOM Pioneers

Texture in Advanced Processes, Materials & Engineering Applications

Transformation Textures

Crystallization, recrystallization and growth textures

Deformation textures

Mathematical, Numerical and Statistical Methods of Texture Analysis

Texture and Microstructure Characterization

IOP Conf. Series: Materials Science and Engineering **375** (2018) 011001 doi:10.1088/1757-899X/375/1/011001

Preface

The 18th International Conference on Textures of Materials (ICOTOM 18) took place in St George, Utah, USA, November 5-10, 2017. ICOTOM-18 follows in the long "triennial" tradition of ICOTOM meetings initiated in 1969 in Clausthal, Germany and followed with: Cracow (1971), Pont-à-Mousson (1973), Cambridge (1975), Aachen (1978), Tokyo (1981), Noordwijkerhout (1984), Santa Fe (1987), Avignon (1990), Clausthal (1993), Xian (1996), Montreal (1999), Seoul (2002), Leuven (2005), Pittsburgh (2008), Mumbai (2011), Dresden (2014).

ICOTOM 18

18th International Conference on Textures of Materials
5-10 November 2017 • St George, Utah, USA

ICOTOM-18 continued the fine scientific tradition of previous ICOTOMs fostering the fundamental understanding of the basic processes underlying the formation of texture and its relation to the anisotropic properties of polycrystalline materials. One of the motivating factors for holding ICOTOM in Utah was the long-standing contributions of Professor Brent L. Adams to the texture community. Professor Adam's spent much of his life and professional career in Utah. Brent's work in linking crystallographic orientation to microstructure led to the automation of Electron Backscatter Diffraction or EBSD, with seminal journal papers appearing 25 years before ICOTOM 18.

The entire texture community has been instrumental in the incubation of this technology and the continued cultivation of both its development and application. With EBSD and other techniques providing textural information from the nano to the macro scale, our understanding of the links between the local orientation and texture evolution and material properties continues to grow. ICOTOM-18 provided a forum for both materials scientists and geologists to engage in the discussion of recent progress in texture and anisotropy research.

These proceedings offer a snapshot of the strong research presented in plenary, invited and regular presentations and posters at ICOTOM. We are particularly pleased to have contributions from four pioneers in the texture community.

- Paul Van Houtte's contribution entitled "Statistical models for deformation texture prediction using vortex-type accommodation of local strain misfits" continues to show Professor Van Houtte's forward thinking in texture modelling. Professor Van Houtte received the *Bunge Award* at ICOTOM 18. This award is given in honor of Professor Hans-Joachim Bunge (1929-2004), to whom the texture community is very much indebted for his distinguished work on the mathematic theory of quantitative texture analysis as well as his lifelong promotion of the field of textures of polycrystalline materials.
- Hans-Rudolf Wenk's contribution presents new work on the simulation of texture and anisotropy development in the lowermost region of the earth's mantle. A symposium was held at ICOTOM to celebrate the contributions of Professor Wenk to the texture community. Rudy is one of those rare scientists whose work in fabric (as texture is termed in the earth sciences) has bridged the divide between geology and materials science.

- Professor David J. Dingley contributed both a look back as well as a look forward on EBSD. David first introduced EBSD to the ICOTOM community 30 years ago at ICOTOM 1987 in Santa Fe which eventually led to the modern automated EBSD systems that have become integral to linking texture with microstructure.
- Ranjit K. Ray has contributed a paper of electrodeposition of copper thin films on textured substrates. Professor Ray has been a key member of the texture community for many years and is singularly responsible for the rapid growth of texture research in India.

Bunge Award recipient Paul Van Houtte with the Chair of the ICOTOM International Committee, Anthony Rollett.

Nearly 300 scientists from 26 countries gathered together for ICOTOM-18 and our sincere thanks go to all those who made the effort to come and participate. We are indebted to the members of the local organizing committee for making the conference a success as well as the International ICOTOM Committee for their valuable guidance. We extend our gratitude to the commercial exhibitors for their financial support. We gratefully acknowledge a grant from the National Science Foundation (Program Manager, Alexis Lewis) which partially supported around 40 students to attend the event. We express our gratitude for the assistance of Southwest Adventure Tours, the staff at the Dixie Convention Center, and the staff of the Mechanical Engineering Department of Brigham Young University.

At the conference, the International ICOTOM Committee announced that the next ICOTOM will be held in Osaka, Japan in 2020. We are certain ICOTOM-19 will be a success and look forward to gathering the texture community together once again in Japan.

Stuart I. Wright (EDAX)
David T. Fullwood (Brigham Young University)
Matthew M. Nowell (EDAX)

Guest Editors

IOP Conf. Series: Materials Science and Engineering **375** (2018) 011002 doi:10.1088/1757-899X/375/1/011002

Peer review statement

All papers published in this volume of *IOP Conference Series: Materials Science and Engineering* have been peer reviewed through processes administered by the proceedings Editors. Reviews were conducted by expert referees to the professional and scientific standards expected of a proceedings journal published by IOP Publishing.

IOP Conf. Series: Materials Science and Engineering **375** (2018) 012001 doi:10.1088/1757-899X/375/1/012001

Article I. Statistical models for deformation texture prediction using vortex-type accommodation of local strain misfits

P. Van Houtte[1], B. Van Bael[1], Q. Xie[1] and M. Seefeldt[1]

[1] Department of Materials Engineering, K.U.Leuven, Kasteelpark Arenberg 44, B3000 Leuven, Belgium
paul.vanhoutte@kuleuven.be

Abstract. Many advanced models for deformation texture prediction make use of crystal plasticity finite element or fast Fourier methods. They manage to deal with strain heterogeneities at very small length scales. However, they demand a lot of calculation power, too much for daily use as tools for computer aided design/manufacturing of forming operations of steel or aluminium sheet parts. So-called statistical models for deformation texture and plastic anisotropy are still of interest in such cases, as they are fast and can indeed be incorporated as constitutive models in finite element simulations of metal forming processes. The present work discusses the existing statistical models. Predicted ODFs of rolling textures of a few steel and aluminium alloys will be compared with those of older models as well as with experimental results. A currently undertaken attempt to further improve the quality of the predictions of one of the statistical models (ALAMEL) will be proposed. Not only the effect of local deviations from homogeneous strain ('relaxations') on the plastic work in the grains will be considered, but also the increase of plastic work due to the plastic accommodation of these misfits. This is done by assuming the existence of certain flow patterns in the matrix surrounding the grains, some of which resembling the 'vortices' sometimes observed in liqsuids.

1. Introduction

Sheet metal (aluminum or steel, used in the metal forming industry for the production of car bodies) usually features a strong crystallographic texture which causes a strong plastic anisotropy in the material. For various applications (product design, quality monitoring) it is then desirable to model the deformation process including its effect on the change of the material properties. This sets the following problem: to properly model the forming process, one needs to know the texture in a given location in the workpiece, because it makes the constitutive law anisotropic. But this anisotropy itself is constantly changing due to the progress of the forming process. So one needs a modelling procedure that simulates the intricate coupling between the mechanical deformation, the texture evolution being caused by it, and the evolution of the plastic anisotropy (itself caused by the evolution of the texture). In principle this could be dealt with by a crystal-plasticity finite element code (CP-FEM), see e.g. Roters et al. [1], or by a Crystal-Plasticity Fast Fourier (CP-FFT) model, see e.g. Lebensohn et al. [2]. Using CP-FEM for the entire polycrystal would require an enormous computational power in case the objective would be to optimize an industrial production process; even the much more efficient CP-FFT method would (at the time being) require a calculation power beyond reach of SMEs or single research groups at universities. So we have recently proposed the "Hierarchical Multi-Scale method" (HMS) [3]. It would not treat the entire workpiece as a single polycrystalline material in which each grain has a different velocity gradient, but rather try to capture its over-all behavior by introducing "integration points" at three different length scales:

- largest: "engineering length scale", looking at the entire work piece, modelling its elastic-plastic or visco-plastic behavior by an standard engineering FE model such as Abaqus. The smallest "items" considered at this length scale are the integration points used by this software.
- these integration points are then considered as polycrystals with typically 5000 grains. They are the representative volume elements (RVEs) of the 'macroscopic' length sale. A 'polycrystal deformation model' will be used to obtain the anisotropic stress-strain relationship at the macroscopic length scale, to be used by the constitutive model at the engineering length scale.

Now, consider that all this is done for a single time increment. Abacus will then provide a new velocity gradient for each of its own integration points (at the 'engineering length scale'). This means that the effect of the deformation step can now be implemented at the smaller length scales. Indeed, a strain increment tensor is obtained from the velocity gradient (and for a given time increment) at each

Abacus-integration point, leading to a typical polycrystal problem: given the average strain increment of the polycrystal, what is the strain increment of each of its grains? The problem can in principle be solved by means of a **polycrystal deformation model**, leading to new textures at the integration points of the 'engineering length scale', i.e. the textures of the RVEs of the macro-scale ('macro-RVEs'). However, it should be noted that other updates must also be done, namely of the constitutive laws at each integration point (the anisotropic stress-strain relationship at the macroscopic length scale). Once this has been done, all is ready for the simulation of a next deformation step by Abacus (at the 'engineering length scale').

2. Polycrystal deformation models

So it is clear that a good polycrystal deformation model is needed. Just how good it needs to be is an other matter. The present authors are dedicated to texture research, so they have decided to work with macro-RVEs consisting of 2000 or 5000 grains, as in earlier work it had been found that a texture represented by a discrete set of 2000 grains was about good enough to make acceptable calculations of r-values of fcc metals, the standard deviation of the error on the r-values being of the order of 0.002 (Tòth and Van Houtte [4]). Another issue is the choice of the polycrystal model. Excellent quality is expected from CPFEM (see e.g. [1]) and CP-FFT methods (see e.g. [2]), but also from VPSC methods see e.g. Lebensohn and Tomé [5]. Of these methods, CPFEM requires most computer power followed by CP-FFT, in any case too much for the purpose set in the present work. Note that these two models try to explicitly model the heterogeneity of stresses and strain rates inside the individual grains as well as the mechanical interactions between them. In contrast to this the so called 'statistical methods' only adopt certain assumed relationships between for example the plastic velocity gradient tensors of individual grains and certain statistical quantities of the entire polycrystal (such as its average velocity gradient tensor). This is even true for the currently most complex statistical method: the visco-plastic self-consistent method (VPSC). Predictions (deformation textures, flow stresses etc.) made by statistical models are less precise as those made by CPFEM and CP-FFT models but nevertheless deserve a discussion (see e.g. Van Houtte et al. [6]) as some of these achieve good quality of predictions combined with high calculation speeds, which makes their application in the context of the HMS method [3] possible. Some well known statistical models have been described in detail by Van Houtte et al. [6]. The simplest is the Taylor-Bishop-Hill model which requires that the macroscopic displacement gradient applies to each of the individual grains.

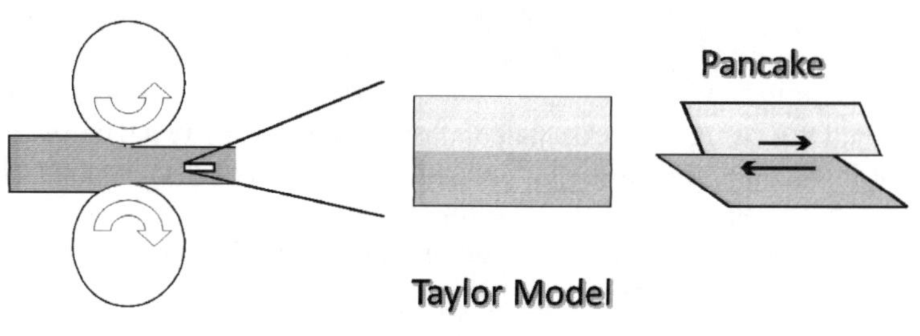

(a) (b) (c)

Fig. 1 (a) Schematic representation of a rolling process applied on a stack of 2 grains, both of which originally had a square cross section. (b) After rolling (compression + elongation), their cross section has become rectangular, at least according to the Taylor model. (c) according to one of the relaxed constraints models (Pancake), each grain may have undergone an additional simple shear called 'relaxation'. These relaxations are chosen as to minimize the plastic work in the grain. There is no relationship between the relaxation shears of the two grains.

Another class of models are the conventional relaxed constraints models, which differ from the Taylor-Bishop-Hill model by the fact that some of the 'simple shear' components of the macroscopic displacement gradient tensor are NOT enforced on the individual grains, but instead chosen in such

IOP Conf. Series: Materials Science and Engineering **375** (2018) 012001 doi:10.1088/1757-899X/375/1/012001

way that that the plastic work required to deform the grain is minimized (Fig. 1c). This deviation from the prescribed macroscopic velocity gradient is described by a tensor called 'relaxation tensor' which is the product of a user-defined velocity gradient tensor describing a unit shear rate with a scalar $\dot{\gamma}_r$. This scalar is a sort of 'additional slip rate' to be estimated by the model in order to minimize the plastic work required to deform the grain.

3. Cluster models

Such models subdivide the grains of a polycrystal in 'clusters' of *n* grains, neighboring each other or not (Fig. 2d-e). It is requested that the average velocity displacement gradient tensor of each cluster is equal to the average at the macroscopic length scale (i.e. overall average of the polycrystal). This consistency requirement is easily achieved in case of the ALAMEL model [6] (Fig. 2e).

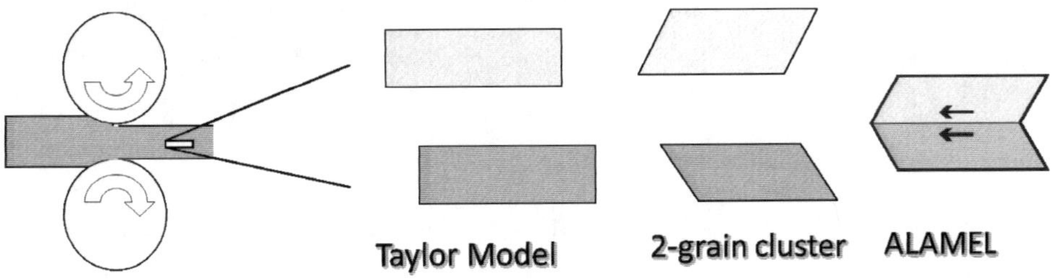

(d) (e)

Fig. 2 As fig. 1, except: (d) is a cluster of 2 grains. The relaxations come in pairs of opposite shears so that the average relaxation is zero. Usually a second pair of relaxations is also applied, which has the same shear plane but a shear direction perpendicular to the plane of the drawing. (e) ALAMEL model: similar to (d), but in addition it is assumed that the two grains have a common grain boundary, which is parallel to the shear plane of the relaxation.

More general cluster model have been proposed by other authors: the GIA model by Crumbach et al. [7] and the RGC model by Eisenlohr et al. [8] operate on 8 grains organized in a 2x2x2 cluster. The grains touch each other (Fig. 3a). Initially the 8 grains have the same brick shape. The cluster has 3 mutually orthogonal symmetry planes. Their intersections form an orthogonal reference frame. Each relaxation consists of a set of 2 shears with opposite signs at both sides of one of the symmetry planes. Note that this preserves the consistency, the average relaxation of the cluster being kept equal to zero. Fig. 3 illustrates a configuration that would be obtained in case of rolling. The plane of the figure is one of the symmetry planes of the process (the plane formed by ND and RD). The effects of 2 out of the 6 relaxations can be seen in this figure. Please note that each of the 8 grains will feel the overlapping effects of 3 of the 6 relaxations, themselves organized in 3 pairs of 2 relaxations with opposite sign. So three independent values (called $\dot{\gamma}_I$,$\dot{\gamma}_{II}$, and $\dot{\gamma}_{III}$) characterize the 6 relaxations. Their effect on the strain rate in each of the 8 grains can easily be mathematically described, and so can their effect on the strain rate tensor of each of the 8 grains. The authors of this model have used the latter in a mathematical model based on crystal plasticity to calculate the rate of plastic work in each of the grains, as a function of the three relaxations. Note however that the authors of the GIA model [7] have realized that such shear relaxations would create additional misfits between the 8 grains of the cluster and the matrix surrounding the cluster. This would require additional plastic accommodation strains in the matrix and/or the grains. In some variants of their model this is taken into account by estimating the density of geometrically necessary dislocations (GNDs) which would

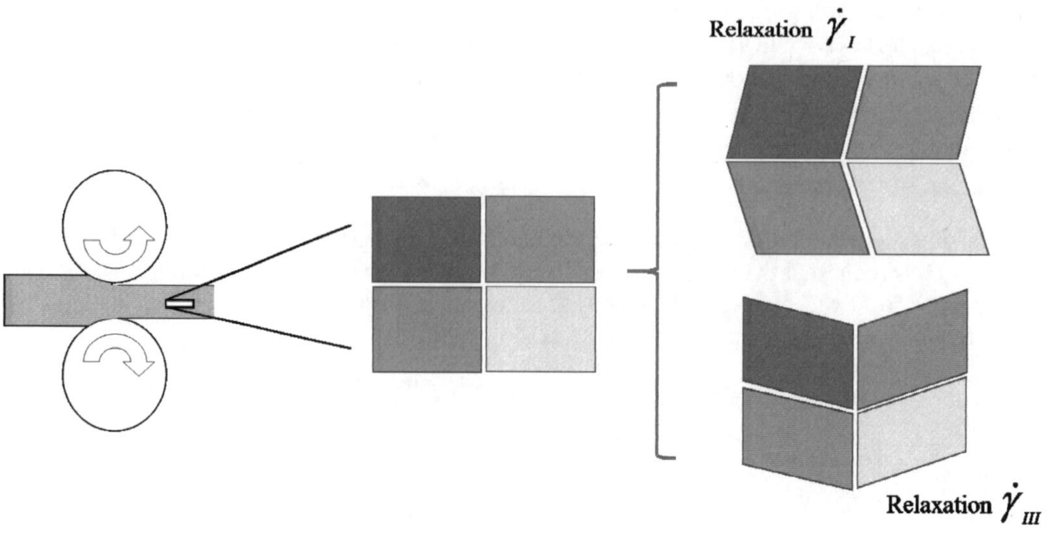

(a) without relaxations (b) with 2 relaxations (overlapping)

Fig. 3. Cluster model: sideview on a cluster of 8 brick-shaped grains in the microstructure of a rolled polycrystal. The deformation of the grains is supposed to be 'relaxed' by 6 pairs of opposite shear relaxation systems. Pair I (with $\dot{\gamma}_I$) operates along the rolling plane in RD. Pair II (with $\dot{\gamma}_{II}$, not visible in the figure) also operates along the rolling plane, but in TD. In a similar way, pairs III and IV operate along the plane normal to RD (in ND with $\dot{\gamma}_{III}$ as well as in TD with $\dot{\gamma}_{IV}$, not visible). Finally, pairs V and VI operate along the plane normal to TD (in RD and ND, $\dot{\gamma}_V$ and $\dot{\gamma}_{VI}$ both invisible in the figure).

be required .to accommodate the misfit, calculate the extra plastic work associated to them, and include it in the global minimization procedure of the dissipated plastic work.
Note that this description

- focuses on deviations from the Taylor theory (uniform plastic strain) in the form of simple shears; in the case of grains or clusters of two or more grains, these shears are chosen in such way that the resulting shape changes do not create regions inside the cluster with increased or decreased material density;
- so far, most papers on statistical method do hardly discuss stress equilibrium issues and neither effects due to strain heterogeneity created in the surroundings of grains/clusters. Exception are the cluster methods [7-8] which introduce GND density calculations.

Some remarks can be given about the stress equilibrium issue. Van Houtte [9] published a short summary of work done mainly by other authors demonstrating that the introduction of a shear-type relaxation in a Taylor-Bishop-Hill type model is equivalent to requesting that the corresponding shear stress is zero. Doing this for all grains would means that the model sets the corresponding shear stress component to zero in all grains. Introducing a relaxation in the form of opposite shears in adjacent grains (as is done by cluster models) creates equilibrium of the corresponding shear stress components at the grain boundary [6]. However, it remains true that the requirement for stress equilibrium at all grain boundaries is far from being satisfied by these statistical models; this remains the privilege of CPFEM and CP-FFT models. Benchmarks can be done by comparing the results of cluster models with those of CPFEM models. This was done for ALAMEL [10], giving insight in the degree of approximation of CPFEM: not perfect, but acceptable given the difference in calculation time.

4. VPSC model and associated cluster model

The visco-plastic model for metallic crystals uses a linear relationship between the deviatoric stress and the plastic strain rate. The effect is rather weak: multiplying the strain rate by 100 might only slightly increase the stress. The effect can nevertheless be used to establish a formula giving a linear relationship between stress deviators and plastic strain rates. That has allowed several authors to develop the 'Visco-Plastic Self Consistent (SC) model' [5, 11-12], inspired by the self-consistent method from elasticity theory. It uses the Eshelby relations which derive the relationship between stress concentrations acting on an 'inclusion' which for some reason undergoes a different strain as the nominal average strain of the material. In the VPSC method, the linear relationship between stresses and elastic strains is replaced by one between stress and plastic strain rate deviators. The model assures consistency between the macroscopic and the mesoscopic levels: the macroscopic stress and plastic strain rate deviators are made equal to the volume averages of the corresponding values in each grain. Note that:

- as to the main strain mode: in contrast to the Taylor-Bishop-Hill theory and its 'relaxed' versions, the model sees to it that 'hard' grains are less deformed than 'soft' grains;
- note however that the VPSC model has not been designed to assure stress equilibrium at grain boundaries, as CPFEM does.

Xie et al. [13] have proposed a cluster model called 'cluster 2' (2-grain clusters) or 'cluster 5' (5-grain clusters). It makes use of the VPSC method. Note that

- the average strain rate of each of these clusters needs to be equal to the macroscopic strain rate.
- the grains in the clusters need not be direct neighbors. Each cluster is treated independently as a small polycrystal by the Neffect-VPSC model (Tomé [12]) which delivers the strain rates and deviatoric stresses of each of the grains. Some results will be given in Table 2.

Comment: as far as the interaction between grains is concerned, an intermediate length scale is introduced between the macroscopic and the mesoscopic (=grain) length scales.

5. Some results of the existing models for cold rolling

Reduction	**40%**	**63%**	**86%**	**95%**	**98%**
FC	0.208	0.584	0.636	0.272	0.388
VPSC	0.429	0.824	0.501	0.409	0.465
GIA	0.208	0.355	0.257	0.190	0.253
LAMEL	0.344	0.653	0.346	0.183	0.196
ALAMEL	0.189	0.433	0.306	0.156	0.171
$\boldsymbol{I_{exper}}$	*2.59*	*2.45*	*4.51*	*6.87*	*9.23*

Table 1 Normalised texture index (ID_N) of difference ODFs for AA1200. FC= FC Taylor theory; VPSC: Visco-Plastic Self-Consistent Model (Lebensohn et al.[5]); GIA (Crumbach et al. [7]); LAMEL: as ALAMEL starting with elongated grains [14]; ALAMEL: standard version [6], starting with equiaxed grains.

Table 1 (from [6]) gives an idea about ID_N after cold rolling of an AA1200 aluminum alloy. The ODF of the initial texture had been measured and used as initial texture in the simulations. The quality of agreement between predicted deformation and observed textures was then assessed by means of the following parameters:

$$ID = \int [f_1(g) - f_2(g)]^2 \mathrm{d}g \qquad \text{and} \qquad ID_N = \frac{ID}{I_{exper}}$$

in which $f_1(g)$ is the experimental ODF and $f_2(g)$ the predicted one. The integral ID is taken over entire Euler space. ID_N is the value of ID normalized by the texture index I_{exper} (integral of the square of the ODF) of the experimental texture. ID_N can only be 0 when both ODFs (experimental and predicted) are exactly equal to each other in all 6859 (=19^3) points of the discrete representation of both ODFs . Table 2 compares ID_N results obtained by the new cluster model described above with those of other models.

	Taylor	ALAMEL	Cluster2	Cluster5	VPSC
DC06_compression_ND_40%	0.04	0.05	0.04	0.04	0.09
DC06_simple shear_RD_40%	0.18	0.14	0.09	0.07	0.07
DC06_simple shear_RD_100%	0.45	0.25	0.12	0.13	0.21
AA6016_cold rolling_86% reduction	0.51	0.24	0.25	0.22	0.37
T61_30% cold reduction	0.20	0.10	0.05	0.06	0.10
T61_60% cold reduction	0.67	0.20	0.15	0.14	0.23

Table 2 Normalized texture index of the difference ODFs (ID_N) for Taylor FC, ALAMEL, VPSC, Cluster2 with 2500 clusters and Cluster5 with 1000 clusters[13].

6. Current work

6.1. Concept

CPFEM models applied on polycrystalline materials (e.g. 1000 grains or more) are expected to be able to capture effects due to the heterogeneity of plastic deformation in a real polycrystal. The average plastic deformation differs from one grain to another, which in fact can only be achieved when the internal deformation of each crystal is heterogeneous, otherwise the integrity of the material is lost. Only in this way material integrity can be achieved everywhere in the material while simultaneously guaranteeing stress equilibrium everywhere. Now, suppose that one needs at least 30 elements to capture the heterogeneity of stress and strain rate inside a grain, while at least 1000 grains at least are needed to predict the overall anisotropy of a polycrystal in a satisfactory way. That means that CPFEM models with 30000 elements or more would be needed to make satisfactory texture and anisotropy predictions. The present authors then consider it a challenge to try to develop new statistical models which would offer a good compromise: still sufficiently accurate for industrial applications however less of course as CPFEM or CP-FFT models. When looking at their own models (especially ALAMEL), the present authors feel that what is mainly lacking is a way to deal with the misfits that are created between the grains of the polycrystal. CPFEM does offer solutions without such misfits - but at a price. So a 2-grain cluster model (Fig. 2 d) is currently being developed, which

- tries to describe the reduction of plastic work in the grains due to shear relaxations as an analytical function f_r^- of these shear relaxations;
- assumes that each grain is surrounded by a "matrix" which has the average plastic properties as the polycrystal in the current condition;
- uses a kinematically admissible velocity field in the matrix that would indeed accommodate the misfit created in the matrix;
- calculates the increase of plastic work that this would cause in the matrix, also in the form of an analytical function: f_r^+ ;
- minimizes the sum $f_r^- + f_r^+$ by choosing the best values for the shear relaxations.

The authors believe that this can be done, although they do not yet know whether the resulting model would deserve the effort.

6.2. Constitutive model of the matrix

The direct surroundings of each grain of such 'cluster' model will be treated as if they have the average plastic properties of the entire polycrystal. In the first version of the model a von Mises model will be used for these. At the beginning of each deformation step, the Taylor model will be used to estimate the flow stress of the polycrystal for the nominal strain mode, however taking the current texture into account. This particular flow stress/strain rate combination must then be reproduced by the 'matrix' model, i.e. the von Mises model mentioned above.

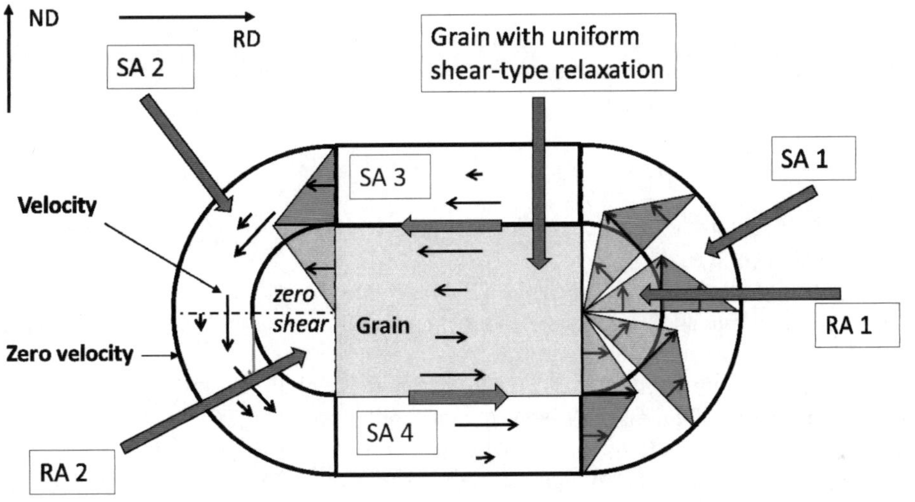

Fig. 4 Kinematical model for a grain with a shear relaxation. The grain is in this example is already elongated. It is supposed to undergo plane strain deformation as in cold rolling. The shear relaxation in the grain is indicated by velocity vectors. The matrix regions RA1 and RA2 right and left of the grains are supposed to locally accommodate the grain relaxation by a rigid body rotation overlapping the deformation of the matrix itself. This rotation does not require additional plastic work in RA1 and RA2. The regions in the matrix above and below the grain (SA3 and SA4) accommodate the grain relaxation by an opposite shear (overlapped on the nominal strain rate in the matrix). Note that the velocity caused by the relaxation at the top of SA3 and the bottom of SA4 is zero. This is not the case at the half circular outer boundaries of RA1 and RA2. The tangential velocities there are continued in SA1 and SA2 where they decrease linearly when moving in outward direction, to become zero when arriving at the outward boundary with the matrix; this resembles a vortex-type pattern.

6.3. Kinematic model for a shear-type relaxation and its accommodation by the matrix

Fig. 4 shows a 2-D kinematically admissible model (i.e. no density changes anywhere in the material) for a grain undergoing shear relaxation, including the accommodation of this shear relaxation by the matrix around the grain. That means that at the outer boundary of the accommodation zone the change of the material velocity due to the relaxation is exactly zero in all points. As a result this hypothetical accommodation field does not violate the integrity of the material. It has been inspired by one which has been proposed many years ago by Brown [16] in a paper called '*Transition from laminar to rotational motion in plasticity*' in which he proposes a model for the hardening of an aluminium alloy by hard particles. It is possible to generalize the velocity field of fig. 4 in order to also achieve such compatibility at the interfaces not shown in the figure (those which are parallel to the plane of the drawing.). Note also that before the era of powerful computers, scientists and engineers already used kinematically admissible velocity fields to estimate the metal flow and the forces required for it, using the 'upper bound theorem' [17] from plasticity theory to get an upper limit of the plastic work required to achieve the forming operation. The application of this theorem would in the present case mean that the relaxations must be chosen in such way that the total dissipated plastic work in the grain and in the matrix must be minimized (relaxations decreasing the plastic work in the grain, but increasing it in the matrix). In the present work, it will be tried to use only analytical equations for the estimation of the dissipated plastic work in order to get a good compromise between

IOP Conf. Series: Materials Science and Engineering **375** (2018) 012001 doi:10.1088/1757-899X/375/1/012001

accuracy and calculation time. All equations had been already developed at the time that the present paper was written. The software tests have however not yet been completed.

Acknowledgements

The authors wish to thank P. Eyckens, J. Gawad, L. Kestens and S. Saimoto for their help in the validation work and for many useful discussions.

References

[1] Roters F, Eisenlohr P, Hantcherli L, Tjahjanto D D, Bieler TR and Raabe D (2010) Acta Materialia 58, 1152.
[2] Lebensohn RA, Rollet A and Suquet P (2011) JOM 63 Nr. 3, 13.
[3] Van Houtte P, Gawad J, Eyckens P, Van Bael A, Samaey G and Roose D (2011) JOM 63 Nr. 11, 37.
[4] Tòth LS and P. Van Houtte P (1992) Textures and Microstructures 19, 229.
[5] Lebensohn RA and Tomé CN (1993) Acta Metall. Mater. 41, 2611.
[6] Van Houtte P, Li S, Seefeldt M and Delannay L (2005) Int. J. Plasticty 21, 589.
[7] Crumbach M, Pomana G, Wagner P and Gottstein G (2001). A Taylor Type Deformation Texture Model Considering Grain Interaction and Material Properties. Part I - Fundamentals. *Recrystallisation and Grain Growth, Proc. First Joint Conference,* ed G Gottstein and D.A. Molodov (Berlin: Springer) pp 1053-1060.
[8] Eisenlohr P, Tjahjanto D D, Hochrainer T, Roters F and Raabe, D (2009) Int. J. Mat. Res. 100, 500.
[9] Van Houtte P (1982) Mater. Sci. Eng. 55, 69.
[10] Kanjarla A K, Van Houtte P and Delannay L (2010) Int. J. Plasticty 26, 1220.
[11] Molinari A, Canova G R and Ahzi S (1987). Acta Metall. 35, 2983.
[12] Tomé CN (1999) Modell. Simul. Mater. Sci. Eng. 7, 723.
[13] Xie Q, Van Bael A, Sidor J, Moerman J and Van Houtte P (2014) Acta Materialia, 69 175.
[14] Van Houtte P, Delannay L, Samajdar I (1999) Textures and Microstructures 31, 109
[15] Brown L M (1997) Phil. Trans. R. Soc. Lond. A 355, 1979.
[16] Hill R (1950). *Mathematical Theory of Plasticity*. (Oxford: Clarendon Press).

IOP Conf. Series: Materials Science and Engineering **375** (2018) 012002 doi:10.1088/1757-899X/375/1/012002

A Refined Approach to Model Anisotropy in the Lowermost Mantle

B C Chandler[1], K Yuan[1], M Li[2], S Cottaar[3], B Romanowicz[1,4,5], C N Tomé[6] and H R Wenk[1]

[1] Department of Earth and Planetary Science, U.C. Berkeley, Berkeley CA 94720, USA
[2] School of Earth and Space Exploration, Arizona State University, Tempe, AZ 85287, USA
[3] University of Cambridge, UK
[4] Collège de France, Paris, France
[5] Institut de Physique du Globe, Paris, France
[6] MST Division, Los Alamos National Laboratory, Los Alamos, NM 87545, USA

E-mail: wenk@berkeley.edu

Abstract.
Seismic anisotropy in the lowermost mantle has been attributed to texture development during mantle convection. This study models texture evolution in a subducting slab impinging on the core-mantle boundary. Using a 3-dimensional geodynamic model with tracers recording the deformation history, a visco-plastic self-consistent (VPSC) model for polycrystal deformation, and relying on experimentally determined slip systems, aggregate grains with volume fractions of 60% orthorhombic silicate perovskite ($MgSiO_3$), 20% cubic calcium perovskite ($CaSiO_3$), and 20% cubic ferropericlase ((Mg,Fe)O) were deformed plastically and developed crystal preferred orientation. Forward and reverse perovskite (Pv)-postperovskite (PPv) phase transitions were included by allowing for likely orientation variant selections. Grain orientations, P (compression) and S (shear) wave velocity pole figures were calculated for each phase as well as the aggregate. The results show that dominant (001) slip on PPv can produce strong texture and shear wave anisotropy of 0.01-3.07% with $V_{SH} > V_{SV}$ which agrees with seismological observations in selected areas of the D$''$ layer.

1. Introduction

The presence of seismic anisotropy in the upper mantle has been attributed to preferred orientation of olivine [1-4]. With increasing depth in the mantle, uncertainties and speculations increase due to lack of seismic sampling as well as difficulties in performing experiments at relevant pressure and temperature conditions. Some reviews highlight the current state-of-the-art [5-7]. One specific area in the lower mantle, which Bullen (1950) called the D$''$ region [8], has intrigued scientists for quite some time. The D$''$ region acts as a thermomechanical boundary layer [9] between the Earth's liquid outer core and the solid mantle. While the bulk of the lower mantle is mostly isotropic [10,11], seismic observations have revealed strong seismic anisotropy in D$''$ where upwelling and subduction occur [10,12-15]. With the discovery of the Mg-perovskite (Pv, also known as bridgmanite [16]) Mg-post-perovskite (PPv) phase transition [17,18] at deep mantle conditions, highly anisotropic PPv emerged as a likely candidate to explain the observed anisotropy in D$''$. It is expected that, for a pyrolytic mantle composition, Pv and PPv dominate in the lower mantle, with ferropericlase ((Mg,Fe)O) and calcium perovskite ($CaSiO_3$) contributing the rest in varying amounts (Fig. 1.a). Figure 1.b shows an enlarged depth/temperature phase diagram of the lowermost mantle with geotherms for a cold subducting slab and a warm upwelling plume.

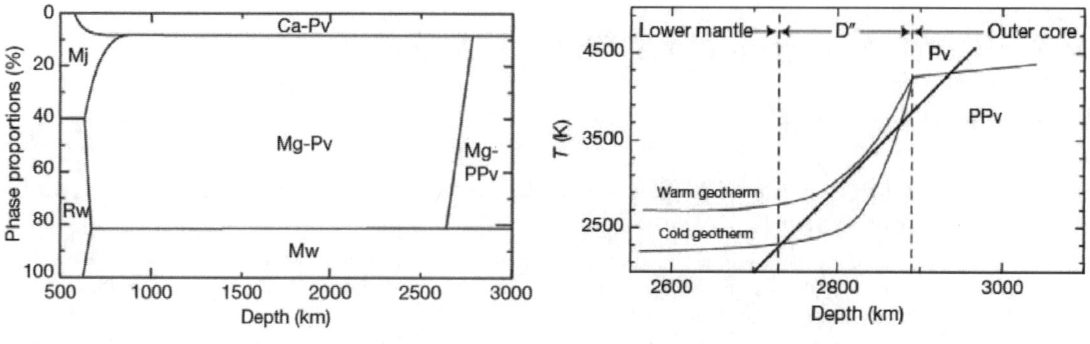

Figure 1: (a) Phase diagram of the lower mantle assuming a pyrolytic composition [19]. (b) Depth/temperature phase diagram, plotting the Pv-PPV phase boundary and cold and warm geotherms. Black line separates Pv and PPv stability fields [20].

Modern geodynamic models predict significant plastic strain in areas of slab subduction into D$''$ as well as plume upwelling [21,22], with large variations in density, temperature and viscosity. Most geodynamic models assume an isotropic medium. In order to predict texture development, tracers are introduced into the convecting medium and record strain. From the strain history, texture evolution can be predicted based on deformation mechanisms. It is generally assumed that plastic deformation in the lower mantle occurs dominantly by dislocation creep leading to crystal rotations and resulting in systematic preferred orientation.

Since Mg-Pv, Mg-PPv and Ca-Pv are unstable at ambient conditions, potential deformation mechanisms must be studied *in situ* at high pressure and diamond anvil cell (DAC) experiments have become a favorite method, using diamonds not only to produce pressure but also as pistons to axially compress samples. With X-rays at synchrotrons the texture evolution in powders such as PPv can then be studied *in situ* by analyzing diffraction patterns [23-25]. Assuming a strain history from a geodynamic model and deformation mechanisms from ultrahigh pressure experiments, the evolution of texture in the lower mantle can then be modeled with polycrystal plasticity theory. This has been done before for 2D [26,27] and 3D geodynamic models [25,28,29]. Here we present an improved 3D modeling, including phase and texture transformations. We start with Pv at a relatively shallow depth. At greater depth Pv transforms to PPv and corresponding variant selection rules are applied. Slip systems are updated according to latest information and other mechanisms such as climb and grain boundary sliding are accounted for. Also, single crystal elastic properties are adjusted for temperature-pressure conditions. Results are shown for a single tracer that is representative for the system.

2. Methods

2.1. Geodynamic model

As slabs subduct to the deep mantle, large-strain deformation occurs. In this study, the convection-induced deformation within subducted slabs is calculated in a 3D regional spherical geodynamical model. The governing equations for conservation of mass, momentum, and energy are solved using a modified version of code CitcomCU [31] under Boussinesq approximation. The geodynamic model includes tracers that follow the flow, defining location, pressure, temperature and velocity gradient tensors for each step to define the incremental deformation. Figure 2 shows a tracer as function of depth projected onto the core-mantle boundary, indicating top and bottom of D″. The depth, temperature, and pressure along the streamline at selected steps are included for reference. Grey colors along the streamline represent the presence of perovskite while red represents the presence of post-perovskite. The forward transition from Pv to PPv occurs at a shallower depth than the reverse transition during upwelling. This can be explained by increasing temperatures in the upwelling section decreasing the PPv stability field (Fig. 1.b).

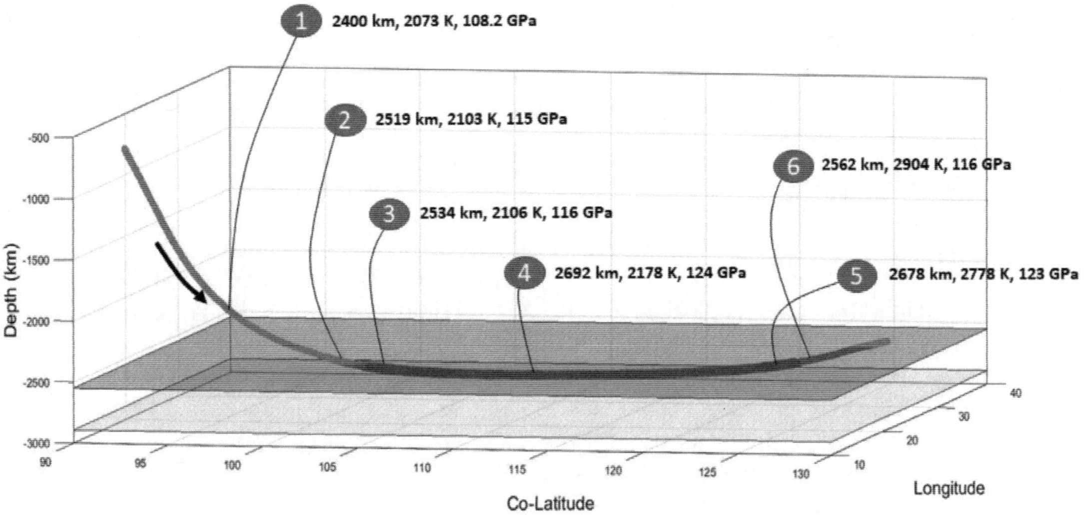

Figure 2: Spatial representation of selected tracer projected on the core-mantle boundary. Grey represents Pv and red represents presence of PPv. Blue layer represents D″ at 2550 km and green the core mantle boundary.

2.2. Deformation mechanisms

Information about deformation mechanisms were largely derived from DAC experiments, observing texture development at ultrahigh pressure (>30 GPa) and interpreting patterns with plasticity theory, e.g. for Pv [32,33], PPv [23-25], Ca-Pv [34], ferropericlase [35-38]. Additional information may be obtained from low-pressure analogs, although bonding characteristics are different [39,40], and from atomic force calculations over a wide range of pressure-temperature conditions [41-44]. Slip systems used in the simulations are shown in Table 1. Since other deformation mechanisms that do not produce texture may be active, such as climb and grain boundary sliding [45,46,47], it is assumed that only 50% of the deformation is accommodated by dislocation glide.

IOP Conf. Series: Materials Science and Engineering **375** (2018) 012002 doi:10.1088/1757-899X/375/1/012002

Table 1: Proposed dominant slip systems (hkl)[uvw] and critical resolved shear stress coefficients of deep Earth minerals.

PPv	(100)⟨001⟩	(100)⟨010⟩	(110)⟨110⟩	(010)⟨100⟩	(001)⟨100⟩	(001)⟨010⟩	(010)⟨001⟩
[23,24]	2	2	3	4	1	1.5	4
Pv	(100)⟨010⟩	(100)⟨011⟩	(010)⟨100⟩	(010)⟨101⟩	(001)⟨100⟩	(001)⟨010⟩	(001)⟨110⟩
[32,33]	2	4	3	4	1	1	1.5
Ca-Pv	(111)⟨10-1⟩	(110)⟨-110⟩	(100)⟨011⟩	-	-	-	-
[34]	3	1	1.5	-	-	-	-
Periclase	{111}⟨10-1⟩	{110}⟨-110⟩	{100}⟨011⟩	-	-	-	-
[35,36,37]	1.5	0.5	0.5	-	-	-	-

2.3. Visco-plastic self consistent model

Based on velocity gradients from the geodynamic model and slip systems derived from diamond anvil cell (DAC) experiments, a polycrystal plasticity model can simulate texture evolution. Here the Visco-Plastic Self-Consistent (VPSC6) method [48] is used, starting with 500 grains for each phase that are randomly oriented. In this method, known as the Eshelby inclusion approach, each grain is treated as a finite inclusion that is under constant strain. The inclusion is interacting with a homogenous but anisotropic medium with the averaged properties over all single crystals. Grain reorientation within this medium is a result of deformation by slip along specified lattice planes resulting in rotations. At each step the properties of the effective medium are updated by fitting the constitutive equations between stress and strain within the grain to that of the aggregate. The behavior of the individual grains in the anisotropic medium is obtained by iteratively fitting the behavior of indivudual grains as well as the averaged response of the medium, leading to a self-consistent solution. Each grain is initialized as spherical in shape but becomes elongated into an ellipsoid due to deformation.

Three phases are deformed simultaneously with slip systems for each phase defined by slip planes, slip directions and critical resolved shear stresses (Table 1). The stress exponent n, given by ($\varepsilon \sim \sigma^n$), was set to 3 that appears appropriate for high temperature/slow strain rate creep conditions. Four phases are included in accordance with a pyrolite composition (Fig. 1.a), Mg-perovskite (Pv), Ca-perovskite (Ca-Pv), ferropericlase, and at high pressure Mg-post-perovskite (PPv). Each streamline follows a trajectory beginning with downward flow followed by transversal flow parallel to the core-mantle boundary, and ending with near vertical upwelling in the back-arc region. CitcomCU 3D uses a coordinate system colatitude, longitude and depth to define the location of a tracer and velocity gradients. This is converted to Cartesian coordinates with Z=N/S and X=E/W to convert velocity gradients as input for VPSC. After computation, orientation distribution functions (ODF) as well as velocity surfaces of the aggregate are calculated using BEARTEX [49]. ODF's as well as elastic tensors for each step are rotated such that the core-mantle boundary is the horizontal (x,y) plane in each image.

2.4. Accounting for phase transitions

During subduction Pv transforms to PPv and during upwelling PPv reverts to Pv. This transition from the Pv Pbnm structure to the PPv Cmcm structure has been studied on the analog system $NaNiF_3$ [40]. Each Pv grain transforms to two PPv orientations, keeping the c-axis unchanged (Fig. 3). In terms of Euler angles it corresponds to a ±16.75°deg rotation of Φ around [001] keeping (Ψ, Θ) constant.

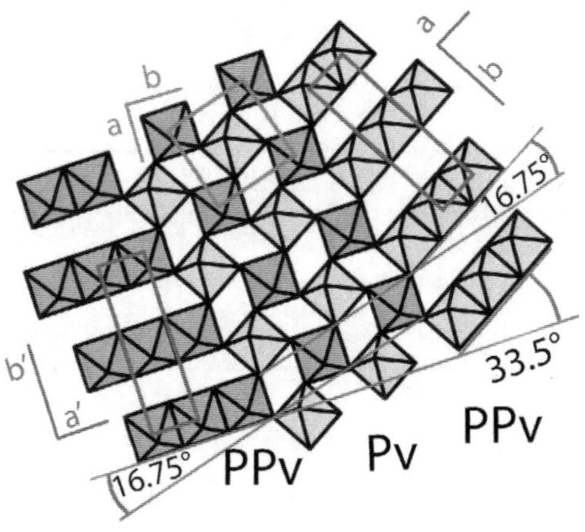

Figure 3: Orientation relation between Pv (center) and PPv (left and right) during phase transition [41].

During the simulation, at the forward phase transition the number of Pv grains is doubled from 500 to 1000 to account for both possible orientations of each PPv grain. The new 1000 grains have PPv slip system information and elastic properties. To maintain the correct volume fraction, the other phases are also doubled from 500 to 1000 grains. VPSC is reinitiated and continues deforming the aggregate until the reverse phase transition point is reached, again doubling grains (1000 to 2000). The PPv grain then transitions back to Pv crystals by the reversal of the same symmetry operation and the deformation continues through the upwelling section of the streamline.

2.5. Elastic properties and seismic velocities
We computed single crystal elastic properties at different pressure and temperature based on a linear approximation to the second derivative. Pressure information was determined by the Preliminary Reference Earth Model (PREM) [50]. Temperature was recorded by the geodynamic model at each timestep. Table 2 provides the densities and single crystal stiffness coefficients (GPa) for each phase based on first principle calculations and figure 4 shows images for P and dS velocity surfaces for single crystals of Pv and PPv. Single crystal elastic tensors for each phase were averaged over the crystal orientation distributions in VPSC and from aggregate elastic properties corresponding P and dS wave velocities were calculated.

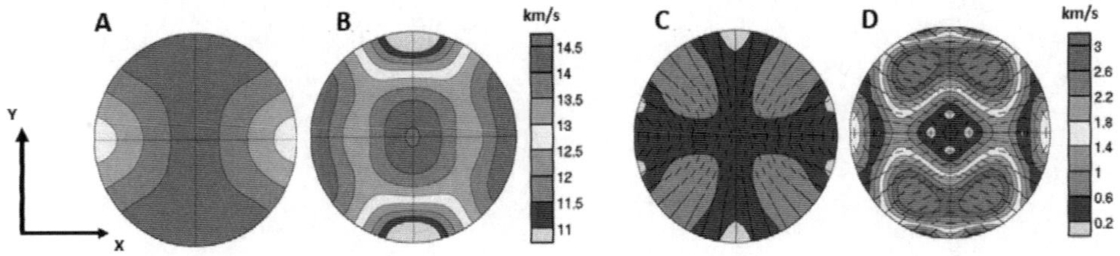

Figure 4: Equal area projection P-wave and dS wave velocities (where dS-wave velocites (S_1-S_2 and ticks are orientation of fast S-wave polarizations) of perovskite (A,C) and post-perovskite (B,D) single crystals from values in Table 2.

IOP Conf. Series: Materials Science and Engineering **375** (2018) 012002 doi:10.1088/1757-899X/375/1/012002

Table 2: Phase density (gm/cm^3) and elastic stiffness (GPa) of deep Earth materials at 3000 K and 125 GPa.

Phase	density	C_{11}	C_{12}	C_{13}	C_{22}	C_{23}	C_{33}	C_{44}	C_{55}	C_{66}
Post-perovskite [51]	5.35	1107.0	429.0	318.0	847.0	441.0	1131.0	251.0	221.0	361.0
Perovskite [52]	5.25	860.0	535.5	437.0	1067.5	467.5	1053.0	294.0	249.5	284.5
Calcium perovskite [53]	5.6	970.0	505.0	505.0	970.0	505.0	970.0	305.0	305.0	305.0
Periclase [54]	5.0	1008.6	233.4	233.4	1008.6	233.4	1008.6	191.5	191.5	191.5

3. Results

Figure 5 shows (001) pole figures for texture, P-wave velocities and dS-wave (S_1-S_2) in km/s for each phase at step increments marked in Figure 2 while Figure 6 displays velocities for the aggregate. Depth, pressure, and temperature for each row is indicated in Figure 2. All projections are plotted with the core-mantle boundary horizontal for reference. PPv is present from rows 3-5, while Pv is present in rows 1, 2, and 6.

The strongest texture developed in PPv reaching a maximum of 15 multiples of random (m.r.d.) shortly after the PPv-Pv phase transition during upwelling (step 5 in Fig. 5.A). Textures of Pv are weaker since Pv is much stronger than Ca-Pv and ferropericlase. Minimal texturing is seen in cubic Ca-Pv (Fig 5.B) and ferropericlase (Fig. 5.C) throughout the entire streamline with maxima of 3.5 m.r.d. at 2500 km and 115 GPa. During high temperature upwelling, the PPv and Pv textures remained stable while both Ca-Pv and ferropericlase textures were reduced. (001) planes of PPv preferentially align parallel to the core mantle boundary. Rotation of the aggregate material is observed in the final steps of the simulation, during upwelling of slab material.

P-waves are fast perpendicular to the core-mantle boundary (Fig. 5.D and Fig. 6). Anisotropy of P-wave velocities remains fairly stable throughout the streamline with intensities that align with the texture developed in both Pv and PPv. S-wave anisotropy, however, shows strong variability due to phase transitions. The forward Pv-PPv transition (steps 2-3, Fig. 5.E) leads to an immediate change in S-wave polarization, becoming dominantly horizontal and a S-wave anisotropy of 4.48% with $V_{SH} > V_{SV}$. With further deformation, anisotropy in PPv decreases to 4% with $V_{SH} > V_{SV}$ in sections where the slab is traveling parallel to the core mantle boundary at a depth of 2690 km. Velocity pole figures for the aggregate (Fig. 6) show that anisotropy from PPv texture development dominate at deep mantle conditions. At the Pv-PPv phase transition (steps 2-3) S-wave anisotropy increases immediately by over 1% and reaches a maximum of 3.07% (step 4). After the PPv-Pv transition fast S-wave polarization is no longer predominantly horizontal but maintains shear wave anisotropy of 3.0% just after the reverse phase transition. P wave anisotropy also reaches a maximum of 2% at step 4.

IOP Conf. Series: Materials Science and Engineering **375** (2018) 012002 doi:10.1088/1757-899X/375/1/012002

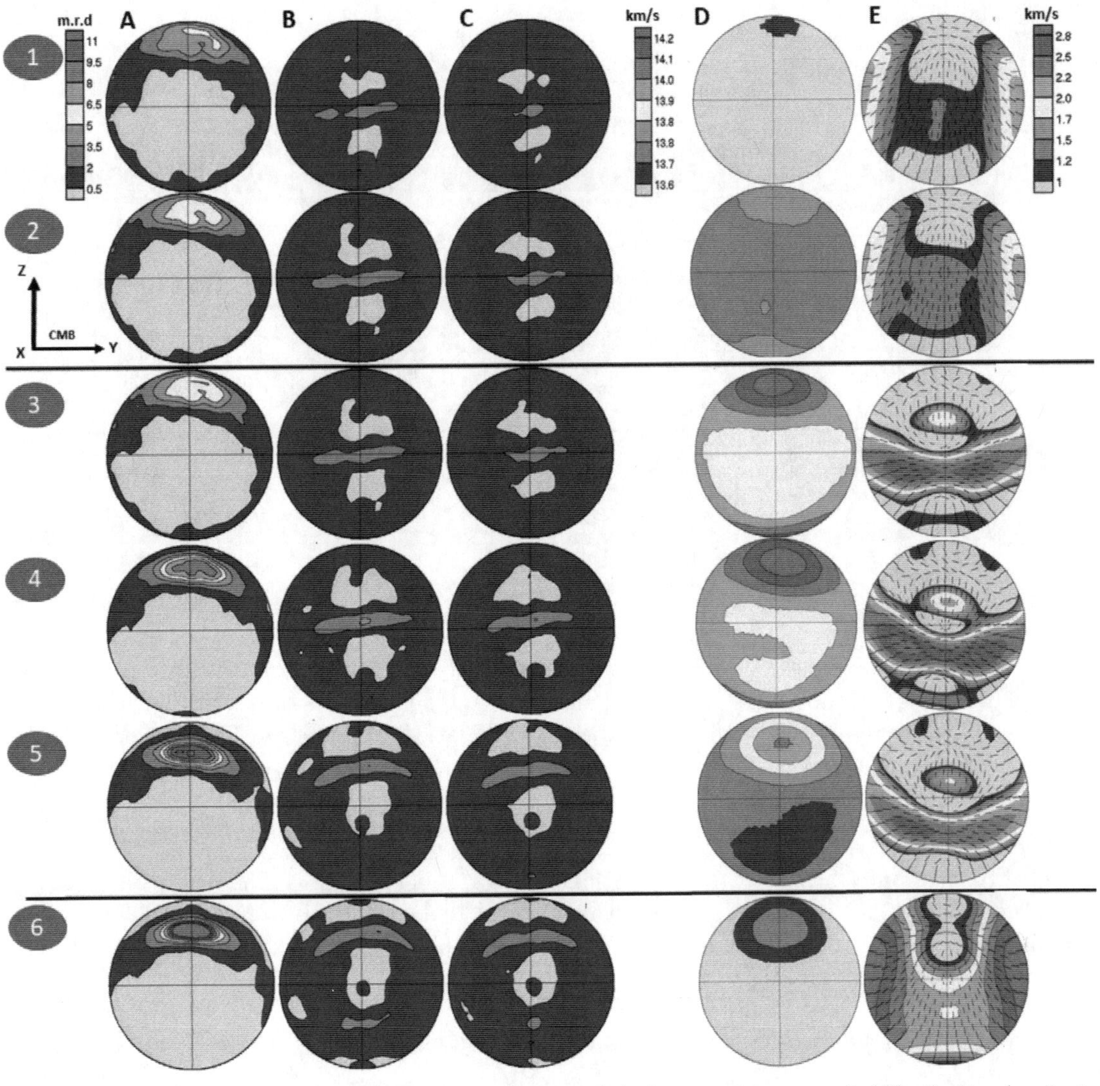

Figure 5: (001) Pole figures (A) Pv and PPV, (B) ferropericlase, and (C) Ca-Pv (scale in multiples of random distribution), (D) P-wave, (E) dS wave velocites (S_1-S_2)(ticks are orientation of fast S-wave polariation) along streamline intervals shown in Figure 2. Equal area projection. Core-mantle boundary is horizontal. Black lines indicate Pv-PPv and PPv-Pv phase transitions, respectively.

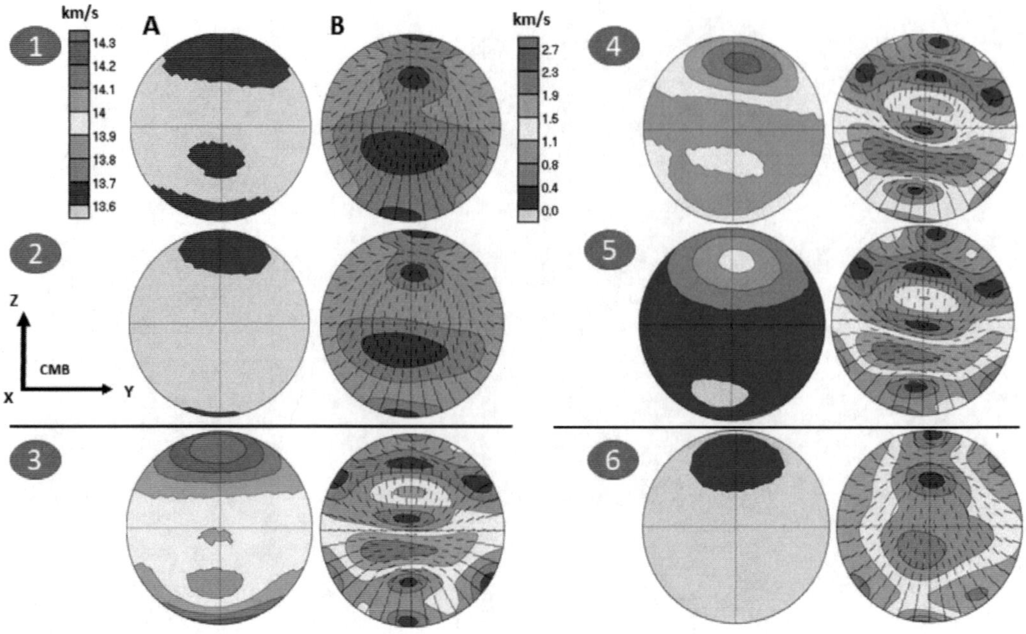

Figure 6: P-wave velocites (A), dS-wave velocites (S_1-S_2 where ticks are orientation of fast S-wave polarizations) (B) of 3 phase aggregate along streamline intervals shown in Figure 2. Equal area projection. Core-mantle boundary is horizontal. Black lines indicate Pv-PPv and PPv-Pv phase transitions, respectively.

4. Discussion

The goal of this study has been to link anisotropy in the deep mantle that has been observed by seismologists [10,12-16] to the evolution of mineral texture during convective deformations by combining geodynamic flow models with polycrystal plasticity. Earlier models [24-27] were refined by including variant selections during phase transitions and adjusting elastic properties to actual PT conditions. The model predicts strong texture patterns in Pv and PPv that develop rapidly during slab subduction. Texture is inherited across the Pv-PPv phase transition and remains stable as the slab travels along the core-mantle boundary. The deformation texture is further inherited and strengthened across the PPv-Pv phase transition. VPSC simulations also provide information about activity of slip systems as displayed in Figure 7. Ferropericlase is assumed to be the weakest phse (Table 1) and particularly 100<011> slip accounts for most of the strain, even though this phase only contributes 20% of the volume. (001) slip dominates in both Pv and PPv and is responsible for the (001) pole figure maximum. The c-axis orientations do not change during the phase transition, assuming a texture memory (Fig. 3). During strong texture development in PPv in Leg 2 there are significant changes in relative activities. Deformation of Ca-Pv is minimal. In this model only half of the deformation is achieved by dislocation glide with the rest accommodated by non-texture contributing mechanisms such as climb and grain boundary sliding.

The shear wave anisotropy of 0.01 to 3.07% predicted by the aggregate grain with $V_{SH} > V_{SV}$ is consistent with seismic observations in the D layer in areas such as the circum-Pacific rim, Caribbean and Antarctic Ocean [10,13,14] and further corroborates that dominant (001) slip on PPv is a plausible cause for seismic anisotropy observed in the lower mantle. The low anisotropy in the upper part of the lower mantle can be attributed to relatively low single crystal anisotropy of Pv.

IOP Conf. Series: Materials Science and Engineering **375** (2018) 012002 doi:10.1088/1757-899X/375/1/012002

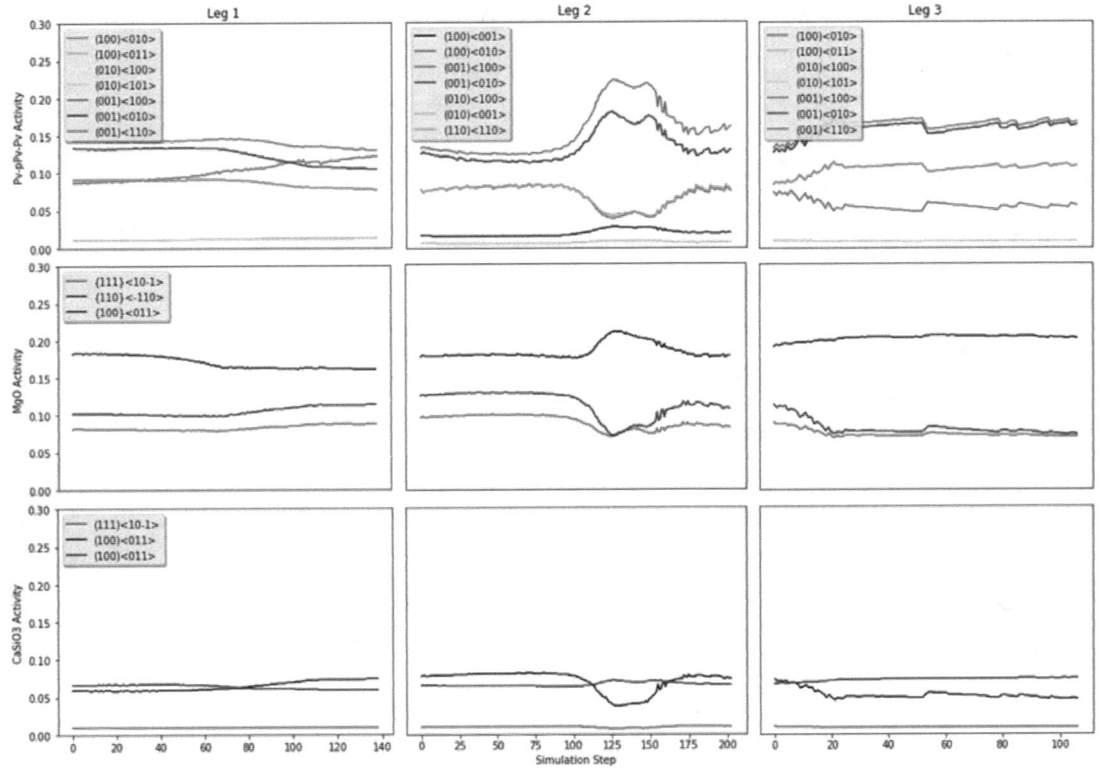

Figure 7: Slip system activity plotted against simulation time step along the tracer streamline. Legs 1 and 3 include Pv,MgO, CaPv (vertical columns 1 and 3) while Leg 2 includes PPV, MgO, CaPv (vertical column 2).

5. Conclusions

This study models texture development in a subducting slab traveling through the lower mantle of the Earth, combining a 3D geodynamic model with the visco-plastic self-consistent approach to polycrystal plasticity. The results portray a characteristic seismological pattern of anisotropy that is consistent with observations demonstrating the importance of an interdisciplinary approach using material science, mineral physics, and seismology to better understand deep Earth processes. Here a single tracer was used. Moving forward, the model will be extended to many tracers traversing the lower mantle and obtaining a topological map with seismic properties of areas consisting of subducted slabs as well as initiation of deep mantle plumes.

Acknowledgments

HRW is grateful to the organizers of ICOTOM18 for arranging this fascinating overview of texture research, from history to state-of-the-art. We acknowledge discussions with Pamela Kaercher, Lowell Miyagi and D. Dobson as well as support from NSF (EAR 1343908, 1417229, CSEDI-106751), DOE (DE-FG02-05ER15637) and CDAC.

References

[1] Blackman D K, Kendall J M, Dawson P R, Wenk H R Boyce D and Morgan J P 1996 *Geoph. J. Int.* **127** 415-426
[2] Dawson P R and Wenk H R 2000 *Phil. Mag. A* **80** 537-598
[3] Karato S, Jung H, Katayama I and Skemer P 2008 *Ann. Rev. Earth Planet. Sci.* **36** 59-95

[4] Long M D and Becker T W 2010 *Earth Planet. Sci. Lett.* **297** 341-354
[5] Long M D 2013 *Rev. Geophysics* **51** 76-112
[6] Weidner D J and Li L 2015 *Treatise on Geophysics 2nd Edtn.* 351-368
[7] Romanowicz B and Wenk H R 2017 *Phys. Earth Planet. Int.* **269** 58-90
[8] Bullen K E 1950 *Geophys. J. Int.* **6** 50-59
[9] Helmberger D, Lay T, Ni S and Gurnis M 2005 *PNAS* **102** 17257-17263
[10] Panning M P and Romanowicz B 2004 *Phys. Earth Planet. Int.* **269** 58-90
[11] Moulik P and Ekstrom G 2014 *Geophys. J. Int.* **199** 1713-38
[12] Vinnik L P, Breger L and Romanowicz B 1998 *Nature* **393** 564-567
[13] Nowacki A, Wookey J and Kendal J M 2011 *J. Geodyn.* **52** 205 - 228
[14] Cottaar S and Romanowicz B 2013 *Geophys. J. Int.* **195** 1184-1195
[15] Lay T 2015 *Treatise on Geophysics 2nd Edtn.* **1** 684-723
[16] Tschauner O, Ma C, Beckett JR, Prescher C, Prakapenka V and Rossman G 2014 *Science* **346** 1100-1102
[17] Murkami M, Hirose K, Kawamura K, Sata N and Ohishi Y 2004 *Science* **304** 855-858
[18] Oganov A and Ono S 2004 *Nature* **303** 351-353
[19] Irifune T and Tsuchiya T 2015 *Treatise on Geophysics 2nd Edtn.* **2nd** 33-60
[20] Herlund J W, Thomas C and Tackley P I 2005 *Nature* **434** 882-886
[21] McNamara A K, Van Keken P E and Karato S I 2003 *J. Geophys. Res.* **108** 2230
[22] Li M, McNamara A K, Garnero E J and Yu S 2014 *Nature Geoscience* **7** 336-370
[23] Merkel S, McNamara A K, Kubo A, Speziale S, Miyagi L, Meng Y, Duffy T S and Wenk H R 2007 *Science* **316** 17291732
[24] Miyagi L, Kanitpanyacharoen W, Kaercher P, Lee K K M and Wenk H R 2010 *Science* **329** 1639-1641
[25] Wu X, Lin J F, Kaercher P, Mao Z, Liu J, Prakapenka V and Wenk H R 2017 *Nature Communications* **8** 14669
[26] Wenk H R, Speziale S, McNamara A K and Garnero E J 2006 *Earth Planet. Sci. Lett.* **245** 302-314
[27] Wenk H R, Cottaar S, Tome C, McNamara A and Romanowicz B 2011 *Earth Planet. Sci. Lett.* **306** 33-45
[28] Cottaar S, Li M, McNamara A, Romanowicz B and Wenk H R 2014 *Geophys. J. Intern.* **199** 164-177
[29] Walker A M, Dobson D P, Wookey J, Nowacki A and Forte A M 2018 *Phys. Earth Planet. Int.* **276** 159-171
[30] Zhong S 2006 *J. Geophs. Res. Solid Earth* **111** B04409
[31] Li M, McNamara A K, Garnero E J and Yu S 2017 *Nature Communications* **8** 177
[32] Wenk HR, Lonardelli I, Pehl J, Devine J, Prakapenka V, Shen G and Mao H K 2004 *Earth Planet. Sci. Lett.* **226** 507-519
[33] Miyagi L and Wenk H R 2016 *Phys. Chem. Minerals* **43** 597-613
[34] Miyagi L, Merkel S, Yagi T, Oshishi Y and Wenk HR 2009 *Phys. Earth. Planet. Inter.* **174** 159-164
[35] Stretton I, Heidelbach F, Mackwell S and Langenhorst F 2001 *Earth Planet. Sci. Lett* **194** 229-240
[36] Merkel S, Wenk H R, Shu J, Shen G, Gillet P, Mao H K and Hemley RJ 2002 *J. Geophys. Res.* **107** 2271
[37] Immoor J, Marquardt H, Miyagi L, Lin F, Speziale S, Merkel S, Buchen J, Kurnosov A and Liermann H P 2018 *Earth Planet. Sci. Lett.* **489** 251-257
[38] Lin J F, Wenk H R, Voltolini M, Speziale S, Shu J and Duffy T 2009 *Phys. Chem. Minerals* **37** 585-592
[39] Merkel S, Kubo A, Miyagi L, Speziale S, Duffy S, Mao H K and Wenk H R 2006 *Science* **311** 644-646
[40] Dobson D P, Miyajima N, Nestola F, Alvaro M, Casati N, Liebske C, Wood I G and Walker A M 2013 *Nature Geoscience* **6** 575-578
[41] Carrez P, Ferre D and Cordier P 2007 *Nature* **481** 177-180
[42] Amodeo J, Dancette S and Delannay L 2016 *Int. J. Plasticity* **82** 177-191
[43] Cordier P, Amodeo J and Carrez P 2007 *Nature* **481** 177-180
[44] Goryaeva A M, Carrez P and Cordier P 2015 *Phys. Chem. Minerals* **42** 793-803
[45] Kraych A, Carrex P, Hirel P, Clouet E and Cordier P 2016 *Earth Planet. Sci. Lett.* **452** 60-88
[46] Ammann M, Brodholt J, Wookey J and Dobson 2010 *Nature* **465** 462-465
[47] Boioli F, Carrez P, Cordier P, Devincre B, Gouriet K, Hirel P, Kraych A and Ritterbex S 2017 *Science Adv.* **3** e1601958
[48] Lebensohn R A and Tomé C N 1993 *Acta Metall. Mater.* **41** 2611-2644
[49] Wenk HR, Matthies S, Donovan J and Chateigner D 1998 *J. Appl. Cryst.* **31** 262-269
[50] Dziewonski A M and Anderson D L 1981 *Phys. Earth and Planet. Int.* **25** 297-356
[51] Stackhouse S and Brodholt J 2007 *Geophysical Monograph Series* **174** 99-114
[52] Wentzcovitch R, Karki B, Cococcioni M and de Gironcoli S 2004 *Phys. Rev. Lett.* **92** 1-4
[53] Kawai K and Tsuchiya T 2015 *Geophys. Res. Lett.* **42** 2718-2736
[54] Karki B B, Wentzcovitch R M, De Gironcoli S and Baroni S 2000 *Phys. Rev. B* **61** 8793-8800

IOP Conf. Series: Materials Science and Engineering **375** (2018) 012003 doi:10.1088/1757-899X/375/1/012003

A review of EBSD: from rudimentary on line orientation measurements to high resolution elastic strain measurements over the past 30 years.

David.J.Dingley[1] Graham Meaden[2] Damian. J. Dingley[3] and Austin P Day[4]
[1]H.H.Wills Physics Laboratory University of Bristol, UK
[2]BLG Productions Ltd., Bristol, UK.
[3]Vantage Software Inc., Salt Lake City, Utah, USA
[4]Aunt Daisy Scientific Ltd, Monmouth, UK.

Abstract. Considering that the Electron Back Scatter Diffraction technique, EBSD, features in more than 60% of the papers published in the current conference proceedings, this review concentrates on the most recent development in the last ten years, that is, High Resolution EBSD. An outline of the theory is presented and four-point bend test results are shown proving the sensitivity of the technique for measuring elastic strain is 1 part in 10000. Other examples of its use included here are strains surrounding indents in silicon, mapping the stress concentration at grain boundaries ahead of dislocation pile ups and dislocation generation at grain boundaries due to strain ahead of nano indentations. An application is presented to distinguish the c axis direction in slightly tetragonal PZT crystals and developments in obtaining absolute strain measurement as opposed to the relative strain measurement currently the norm are discussed.

1. Historical Note

The paper, 'On Line Microtexture Determination Using Backscatter Electron Kikuchi Diffraction in a Scanning Electron Microscope' was the first in which the EBSD technique was described at an ICOTOM meeting, Dingley 1987 [1]. In it the concept of direct crystal orientation measurement using a scanning electron microscope rather than the then universally used x-ray diffraction methods was presented. Now 30 years later, over 60% of the papers at the ICTOM meetings are applications of the EBSD technique and in excess of 1000 papers are published annually in which it is a major tool used in the investigation. The plenary lecture at the 1987 meeting by Professor Sir Charles Frank [2] was concerned entirely with orientation representation. He wrote "Thus, *because of the EBSD technique,* we now have a practical need for comprehensible displays of complete orientation statistics. To degrade that information to pole figures, showing the statistics of orientation of crystal planes, and not of crystals, is a criminal disregard of significant information." He would have been pleased then that although true orientation maps are not the normal method of orientation representation the alternative representation of the data in the form of inverse pole figure maps, Adams et al. 1993 [3] is so demonstrative that it has made EBSD a universally adopted tool in materials science.

The ICOTOM 1987 paper was one in a series starting with those of Prof. John Venables et al published between 1972 and 1977, [4]. On-line indexing evolved over the period 1983 to 1987 in the Bristol group lead by David Dingley [5]. Early enthusiasts using the Dingley system were Rolls Royce Bristol, UK, (M Parsley), National Physical Laboratory UK, (P Quested), SINTEF Norway (J Hjelen), Canadian Aluminium Company, Canada (E Balls), Universite Paris Sud, France, (R Pennelle), Nippon Steel, Japan (J Harase), Alcoa USA (H Weiland), Riso Denmark (D Juul Jensen and N-H Schmidt). However, the ICOTOM 1987 paper was particularly significant in that it lead directly to the development of what is known as Orientation Imaging Microscopy under the direction of Prof. Brent Adams and his team at

Yale University, Stuart Wright, David Field and Karsten Kunze with the author David Dingley as visiting professor, (1994), [3],[6] .

The main theme of the current presentation however, is not a review of the development of OIM over the succeeding 30 years as the standard EBSD technique is very well known, but to concentrate on the most recent advance, High Angular Resolution Electron Backscatter Diffraction, HREBSD. It is a tool developed principally at Oxford University in a materials group headed by Prof. Angus Wilkinson and provides quantitative data on internal strain, both the elastic strains and the dislocation distribution following plastic deformation. It is worth emphasising that these parameters, especially when mapped to show their spatial distribution, are of equal importance in materials characterisation as those of grain size, grain orientation, grain misorientation, grain boundary type and phase. A complete understanding of the origin of texture and its influence on mechanical properties and on behaviour during subsequent thermal mechanical treatment relies on knowledge of the entire microstructural description and the way it influences and is influenced by the strain and dislocation state. Some may have wished to include the energy state and its microstructural distribution in the list of essential microstructural features, but the energy state can be calculated form the stress strain data obtained from HR EBSD and as in one of the examples below it is implied in calculation of the von Mises parameter.

Standard EBSD does provide some semi quantitative data on strain. An early example is the image quality factor map, IQ map, as illustrated in figure 1, [7].

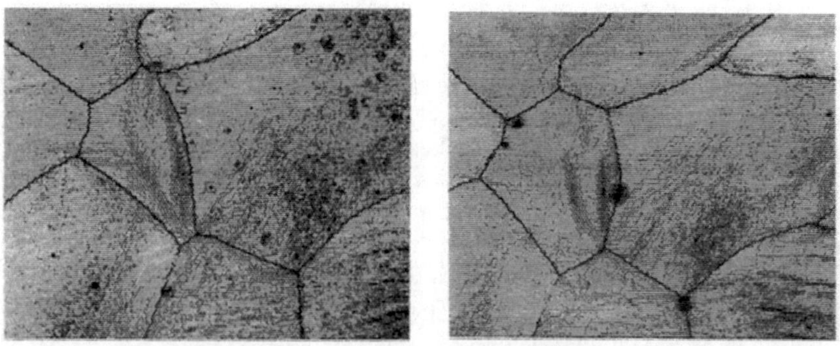

Figure 1. *Image quality maps of bcc iron adjacent to micro-hardness indentation. The second was recorded after polishing away a few microns of material.*

A more quantitative map is that of the kernel average map, KAM, originally included in the TSL commercial software.

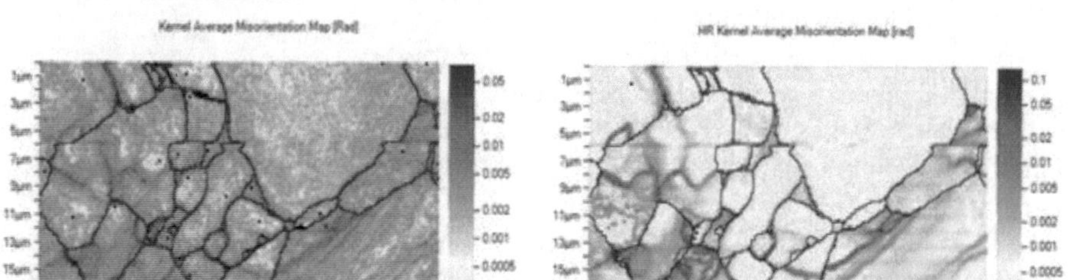

Figure 2, *kernel average maps a) standard EBSD, b) HR EBSD*

IOP Conf. Series: Materials Science and Engineering **375** (2018) 012003 doi:10.1088/1757-899X/375/1/012003

The KAM parameter is the value of the average misorientation between an element in a map with its nearest neighbours. An example map is shown in figure 2. The colour key is show on the right. For comparison a High Resolution KAM is shown alongside to demonstrate the improved angular sensitivity possible using the advanced technique.

2. The basis of HR EBSD

The basic point is that any crystal distortion caused by internal or insitu applied stress results in a corresponding distortion of the EBSD pattern. Zone axes shift in different directions and by different amounts according to the strain. At 0.2% strain, the strain at which yield stress values are often quoted, zone axes shifts in the direction of the applied tensile strain would be 2 parts in 1000. An elastic strain measurement tools needs to be ten times more sensitive. Although there had been a number of early publications exploring methods of determining this distortion, notably those by Wilkinson, reference [8] for example, the major step forward was published in 2006 by Wilkinson Meaden and Dingley [9] titled "High-resolution elastic strain measurement from electron backscatter diffraction patterns: New levels of sensitivity." The procedure is well detailed there and in many subsequent papers, e.g. [10] so that apart from some essential points most of the detail need not repeated here.

The Wilkinson method achieves the required level of sensitivity by using an image cross correlation procedure in which an EBSD pattern 'the test pattern' is correlated with a reference in the same grain. The reference pattern is nominally from a stress free region or one in minimal stress. The respective values for pattern centre coordinates PCx PCy and specimen to screen distance Z, for the two patterns have to be known with the same precision as that of the sort after strain sensitivity, though the absolute values can be in error by up to 1% of the screen diameter, [11]. Finding a suitable reference is straight forward, for example, in an indent in a silicon single crystal where the strain can be considered zero far from the indent and the relative positions of the PCx PCy Z values between reference and test points are known from the electron beam movement between the two patterns and the sample tilt. Finding a suitable reference point is less certain in other cases, in a deformed polycrystalline sample for example. Whatever the situation, the results for test points are always relative to the actual stress state of the point from which from which the reference pattern was taken.

Figure 3 illustrates the relevant geometry. On the left is a schematic of the distortion of an unstrained unit cell from cubic to skew. The vector **r** in the undistorted cell extends and rotates to form the vector **r'**. The displacement is the vector **Q**. To the right is shown the projection of this distortion onto the phosphor screen that images the EBSD pattern. The vector **Q** is projected as the vector **q**.

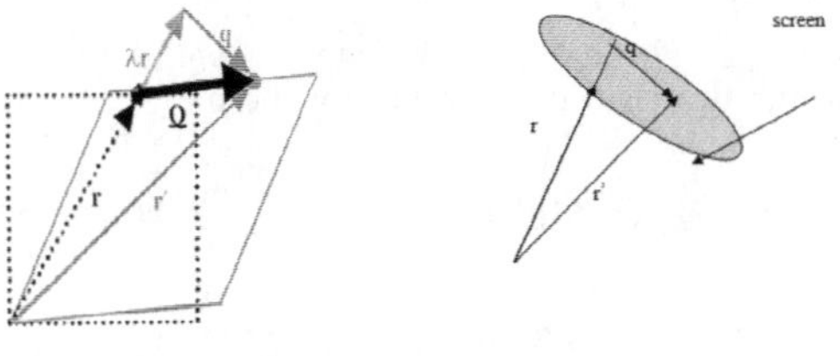

Figure 3. *Geometry relating lattice distortion vector **Q** to EBSD pattern distortion vector **q***

IOP Conf. Series: Materials Science and Engineering **375** (2018) 012003 doi:10.1088/1757-899X/375/1/012003

The two vectors are related by the equation

$$q = Q - \lambda r \qquad 1)$$

in which λ is a scalar of unknown magnitude.

The critical equations that describe the displacement are

$$zq_x - yq_z = (yz[a_{xx} - a_{zz}] + xza_{yx} + z^2 a_{yz} - xya_{zx} - y^2 a_{xy}) \qquad 2)$$

$$zq_y - xq_z = (xy[a_{xx} - a_{zz}] + yza_{xy} + z^2 a_{xz} - x^2 a_{zx} - xza_{zy}) \qquad 3)$$

In these equations x y z are the coordinates of the point r in the reference pattern, with respect to a coordinate system with origin at the pattern source point in the specimen. The quantities q_x q_y q_z are the respective vector displacements of the point r in the test pattern as measured in the phosphor screen. The crystal distortion components are a_{xx}, a_{yy}, a_{zz}. There are 8 unknowns between the two equations. We thus need to make measurements on a minimum of four points in the pattern. In practice, instead of measuring the shift of individual points, a small region of interest, ROI, is drawn around each point and the average shift of all image elements within the ROI are determined. A cross correlation operation is used to do this and the average value found is assigned to the central point of the respective ROI. To reduce error and sampling problems many more than 4 ROI are selected and to improve the sensitivity to a fraction of a pixel a cubic spline procedure is included in the analysis.

If a non-symmetric deformation tensor is found from experiment this indicates rigid body rotation as well as strains are present. The deformation tensor is divided into its symmetric and asymmetric parts. The symmetric part is given by ½ $(A+A^T)$ which equals the strain tensor**.** The anti-symmetric part is given by ½$(A-A^T)$ which equals the rotation tensor. This provides the shear components e_{12} e_{13} e_{23} which because the strain tensor is symmetric are respectively equal to e_{21} e_{31} e_{31} . (Notation has changed here from subscripts xy etc, to subscripts 12 and the terms a_{xy} etc. to e_{12} to conform to accepted forms distinguishing a strain tensor from a distortion tensor). The rotation components are w_{12} w_{13} w_{23} which are respectively equal to $-w_{21}$ $-w_{31}$-w_{32}. The diagonal components of the strain tensor are only known at this stage as the differences $[e_{11} - e_{33}]$ and $[e_{22} - e_{33}]$ and 0. The term e_{33} can not be determined directly as the strain it represents is normal to the phosphor screen. However, noting that the free surface has zero traction across it, i.e. $\sigma_{33}= 0$ and using the generalized Hooke's law

$$\sigma_{ij} = \sum_{k=1}^{3} \sum_{l=1}^{3} C_{ijkl} \cdot e_{kl} \qquad 4)$$

where C is the fourth rank tensor defining the elastic constants for the material, we can write

$$C_{33}\, e_{33} + C_{13}\,(e_{11} + e_{22}) = \sigma_{33} = 0 \qquad 5)$$

Using the appropriate substitutions we can solve separately for the three normal strains e_{11} e_{22} e_{33}.

Like all other metrics it is important to have an indication of the quality of the result. Two are used, geometric peak height of the cross correlation factor and the mean angular error. The former measures to what extent, after beam shift and image zoom between reference and test patterns have been taken into account, the two EBSD patterns match. If the two patterns were both unstrained and there was no rotation between them then the correlation factor would be unity. If one were strained with only a small rotation between them the factor will be less but still close to unity. Factors as low as 0.6 might indicate a problem, e.g. a large rotation between the two patterns; one of the patterns had been degraded because of dirt on or damage to the sample surface, or because the beam shift and zoom correction values were in error. The latter is a likely occurrence if the distance between reference and test points on the sample exceed 300µm. This is because the SEM beam shift calibration can differ on day to day basis. The second cross correlation quality factor, the mean angular error, is calculated by taking the measured strain tensor and back calculating the pattern distortion that such a strain would produce in each of the ROI. The mean angular error is the angular difference between the back calculated distortion values and the experimental values.

3. Sensitivity.

It was shown in the 2006 WMD [9] paper that using an imaging camera with a sensor containing 1k x 1k pixels the cross correlation procedure could reliably detect a beam shift over the surface of a single silicon crystal of ±4µm, equivalent to ±0.1 of the size of one camera pixel or 0.001% of the screen width. The standard deviation was ±0.02pixels. This test has been repeated many times using different cameras with the ±0.1pixel level being the same no matter what the camera type. This means that a camera with a 2kx2k pixels sensor can detect a shift in the EBSD pattern twice as small and is clearly an advantage in HR EBSD work. The phosphor and aberrations of the imaging lens used in the system will of course also have an effect [12], but the values quoted are the best the author has measured. It follows that a map of the strain level for a scan over the surface of an unstrained region of a silicon crystal obtained using a 1kx1k camera should show no strain above the 1 to 2 parts in 10000 level. Figure 4, shows one such map, the shear strains e_{12}, demonstrating this point. The silicon crystal chosen had been deformed locally using a nano indenter. Whereas the strain surrounding the indent as seen in figures 4 a & b exceeds 0.005 it falls rapidly within a few microns. The reference EBSD pattern was recorded from a point 12µm from the indentation. It is at the top right of the map. The graph shows the e_{12} strain along the entire top line of the map, drawn as a blue line and from right to left. The red line shows the e_{12} strain down the last vertical column. The vertical column shows the strain is within 0.0002 of that of the reference pattern for 8 microns after which it rises as the strain from the indentation takes effect.

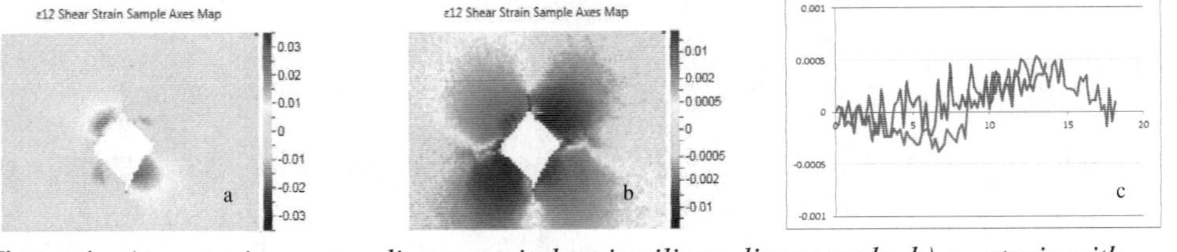

Figure 4, *a) e_{12} strain surrounding nano indent in silicon, linear scale, b) e_{12} strain with strain colouration according to a log scale, c) plots of e_{12} strains for the first row (right to left, coloured blue) and the last column (top to bottom, coloured red). (Data supplied by T Jaepel GmbH Dortmand).*

The horizontal row remains within 0.0002 strain of the reference point for 3 microns followed by a small negative shear and then a small positive shear. This proves that the reference point was truly in a region of zero strain and that the noise of the measurements, for the pixels close to the reference point, was within the 0.0002 limit claimed for the method.

4. Testing accuracy

The simplest experiment to verify that the technique can both accurately and precisely detect strain at a sensitivity comparable to that demonstrated above is the three point or four point bend test. The simplifying factor here is that the axis midway between the convex side and concave side of the bent sample is a neutral axis of zero strain and that the strain normal to the neutral axis increases linearly with distance from the axis. Thus selecting a reference point on this axis can confidently be assumed to be at zero strain. Figure 5a shows results from the work of Plancher et al [13][14]. The sample was stainless steel. The results shown are those Plancher obtained using a 1k x 1k EBSD camera, EDAX Digiview. EBSD patterns were recorded at 0.1µm intervals along the x axis as drawn in the figure. The data analysis was carried out using Crosscourt 4 software. The measurements were repeated by D. Loisnard and D. Dingley at EDF Moret-sur-Loing, using a 2k x 2k camera, Nordif HR4M, and the resulting strain variation is shown in figure 5b. Figure 5c shows an overlay of the results from the 2kx2k camera with those using micro Laue x-ray diffraction. All three data sets show the expected linearity of strain with distance from the neutral axis and the overlay shows exact correspondence of the data between the 2k x 2k camera result and the x-ray data. The noise level in the data from the 2k x 2k camera is half that from the 1k x 1k camera as also expected. The line profiles in figure 5c include the data collected outside the linear range in which the sample had deformed plastically. The EBSD data and x-ray data do not correspond perfectly in these regions as the line traces could not be set exactly in the same places for the two experiments.

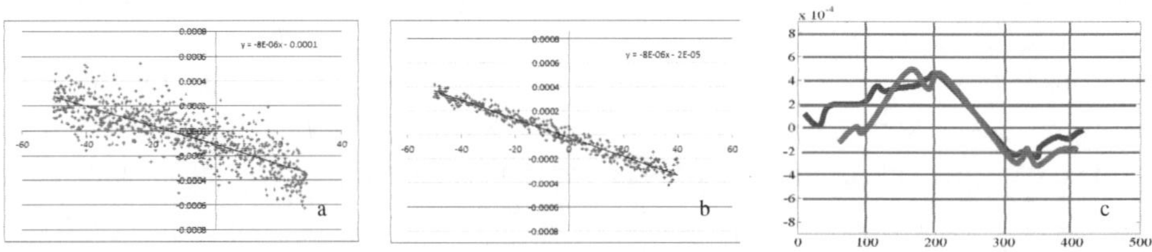

Figure 5, *a, Plancher data, b, Loisnard data and c, overlays of Loisnard data and micro x-ray data in which blue line is micro Laue and red HR EBSD.*

5. Relative rotations and geometrically necessary dislocations.

As explained above the ant-symmetric components of the off axis distortion tensor are a measure of the rotation of the point sampled in the experiment with respect to the orientation of the reference point. The relative rotations obtained have a precision of 0.006°. This means IPF maps can be drawn with 100 times the sensitivity of those using the standard Hough measurements. The relative orientations as measured using cross correlation are transformed to absolute orientation measurements by adding to all the relative measurements the absolute orientation of the reference point as obtained using a conventional Hough transform or as outlined below using an improved high resolution Hough based measurement. Although the so determined absolute measurements have the accuracy of the Hough measurement the relative orientations retain the 1 part in 10000 precision. This is important because the well-established Kernel Average Map used as a qualitative representation of plastic deformation retains the 100 times increased sensitivity available when calculated using cross correlation. The two Kernel Average maps shown in figure 1 demonstrate this difference and show that much of the data obtained using conventional Hough technique reflects its poor relative orientation measurement sensitivity.

The 0.006° precision however, also means that the strain gradient between points in a map can also be obtained at sufficient precision for a meaningful generation of maps displaying the distribution of the Geometric Necessary Dislocation (GND) density. A gradient precision of 1 part in 10000 allows for a GND density of 10^{12} to be detected, i.e. a dislocation spacing of 1 micron. The mathematical procedures involved can be found in the papers by Arsenlis and Parks, [15] and in the paper by Wilkinson and Randman [16]. The data shown here in figure 6b was obtained using the Wilkinson Randman approach in software supplied by BLG Vantage (CC4). In figures 6 a, b and c are shown a correlation between the maximum principal strain value, the GND density distribution and the stress value normal to the grain boundary between the points marked x-x. The sample was tantalum with nanoindents at regular distances from the grain boundaries. Data supplied by courtesy of B Dunlop, Mitchigan State University.

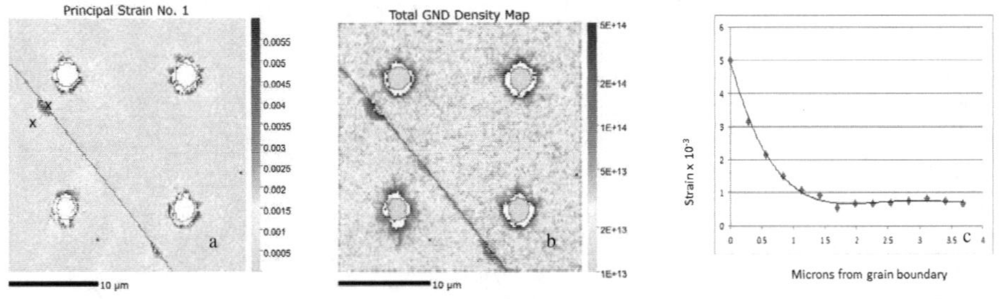

Figure 6. *a, Principal strain map surrounding nano indentations, b corresponding total GND density map, c, strain versus distance from grain boundary between points x x . (With acknowledgements to Bret Dunlap Michigan State University).*

6. Antiphase boundaries.

The high precision achieved by the method also permits a novel method for determining the presence of antiphase boundaries. Consider for example a PZT sample in the unpolarised state. The crystal is naturally tetragonal but the c to a ratio is often too small,·1.02, for conventional EBSD to distinguish the <100> axis from the <001> axis. However, this is well within the detection limit of HR EBSD. It is not possible though, to use the strain algorithm as described above as the tetragonality is not caused by strain but is simply because of the crystal class. Hence, to distinguish the c from the a b axes (a=b in this case) the procedure differs slightly. A reference point as selected as normal within a grain and the distortion between the reference pattern and a test pattern carried out as described above. We know from the crystal system that the crystal extension is along one of the 100 type axes. As the tetragonality is small we can at first treat the crystal as though it were cubic. We can thus determine its orientation using the Hough transform and find the real space directions of the 001 type axes. Figure 7 shows two domains in the PZT microstructure oriented at 90° to each

IOP Conf. Series: Materials Science and Engineering **375** (2018) 012003 doi:10.1088/1757-899X/375/1/012003

other. The extension of the c axis is exaggerated to clarify the argument. The real space directions for these axes are labelled x y and z in the figure. The problem is to determine which of the crystal axes parallel to x y z is the longest. We next need to consider the initial output of the distortion tensor. As discussed following equations 2 & 3 the output from the cross correlation procedures gives us x distortion = $[e_{11}-e_{33}]$, y distortion = $[e_{22}-e_{33}]$ and z= 0. If we view the example shown in figure 7, which is one of the three possible orientation relationships between two disorientated domains in PZT, we see that the pattern from the test point compared with that from the reference point will show a shortening only in the x reference direction. From the figure the shortening is a-c. This is equivalent to a strain e_{11}. Along the z reference direction the change is c-a= e_{33}. Hence $[e_{11}-e_{33}]$ which is equal to the x component of the distortion tensor =2(a-c). Similar arguments for the changes $[e_{22}-e_{33}]$ show that this difference =a-c, half that for the former. Thus, as shown in the inset to figure 7, for a tetragonality of 2.7%, the x axis will show a distortion of -0.054, the z axis, a contraction of -0.027 with no change in the y direction. Thus the c axis of the reference pattern is parallel to x and that of the test pattern parallel to z. The other two possible domain orientation relationships provide different values along the x y and z reference directions. Hence it is possible to distinguish for all three domain orientation relationships along which of the x y z real space directions the c axis lies. Figures 7b and c are two HR EBSD maps showing these measurements on a selected grain. The black arrows are the surface projections of the measured real space 100 type directions.

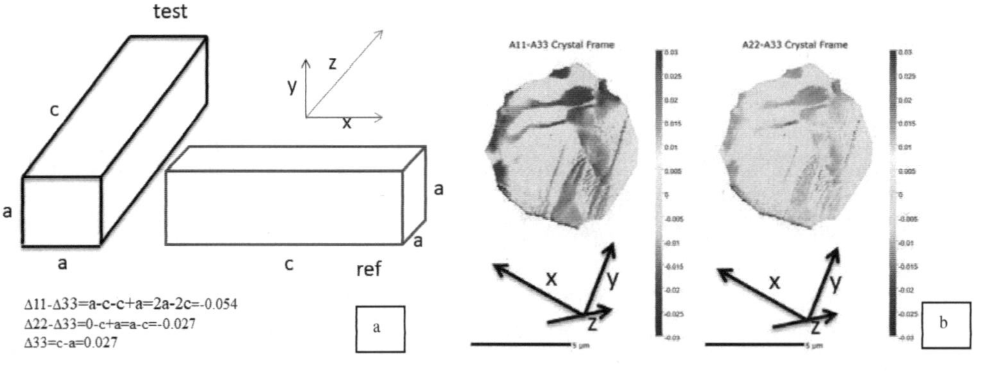

Figures 7 *a) schematic of domain orientations, b), the dark blue regions are the [001] domains, the light blue and green regions are the {001} domains.*

From the schematic and the attached analysis, which is for test point minus reference, the darker blue pixels in 7b correspond to the lattice parameter parallel to the x axis being shorter than that in the reference crystal. The [001] axis of the reference point, shown as a black dot in the green central grain is parallel to the X axis in the map. Similarly it can be deduced that the [001] axis in the dark blue regions is parallel to the Y axis in the map. In the particular example shown the domains are also under elastic strain so that the actual measurements for a particular orientation will show a variation in the values $[e_{11}-e_{33}]$ and $[e_{22}-e_{33}]$ hence the small deviations in blue and green colours shown in the map within domains in the same orientation.

7. Absolute strain measurement

The Wilkinson cross correlation procedure for measurement of residual strain provides values relative to the strain state of the point in the sample used to obtain the reference pattern.

IOP Conf. Series: Materials Science and Engineering **375** (2018) 012003 doi:10.1088/1757-899X/375/1/012003

Though the measurement of strain distribution and relative strain between points remains a valid and powerful addition to microstructural characterisation as illustrated in figure 5; obviously it fails in providing absolute, i.e. true strain, stress and if required energy values. For example the maximum strain ahead of the grain boundary as shown in figure 5c is given as 0.005 but this is relative to the measured background strain truly being at the 0.0008 level. There are two ways the uncertainty in the strain of the reference point might be overcome; by the use of simulated EBSD pattern as the reference pattern in place of a real pattern, or by improving the sensitivity and accuracy of Hough based analysis. The use of simulated patterns has been explored over the past 8 years beginning with Kacher et al [17] who used kinematical diffraction theory to calculate the intensity distribution for different reflections. Maurice et al [18] showed soon after that such simulations based on kinematical theory lead to false results. Dynamical theory calculations produce much better correspondence with real patterns,. These simulations, beginning with the work of Winkelmann [19], have improved the situation and recently Alkorta et al [20] have reported that further improvements in the correspondence can be achieved by using the differential of both the real and simulated patterns as input to the cross correlation procedure. These latter papers have concentrated on the problem that it is imperative for the pattern centre and specimen to screen values used to generate the simulation have to be the same as the values in the test pattern. As the experimental values are not know better than 1% to 2% of the screen diameter, then an iterative procedure needs to be followed to find exactly identical values. This might be possible if the sample pattern was from an unstrained point in the crystal for example. Doing the same for a test point in the crystal that is of unknown strain and uncertain orientation considerably magnifies the problem. Figures 8a b & c illustrate the matching problem. The real EBSD pattern in 8a is from germanium. The simulations 8b 8c, were provided by Aunt Daisy Scientific. The Hough based values for PCx PCy and z and crystal orientation were those used for computing the simulations

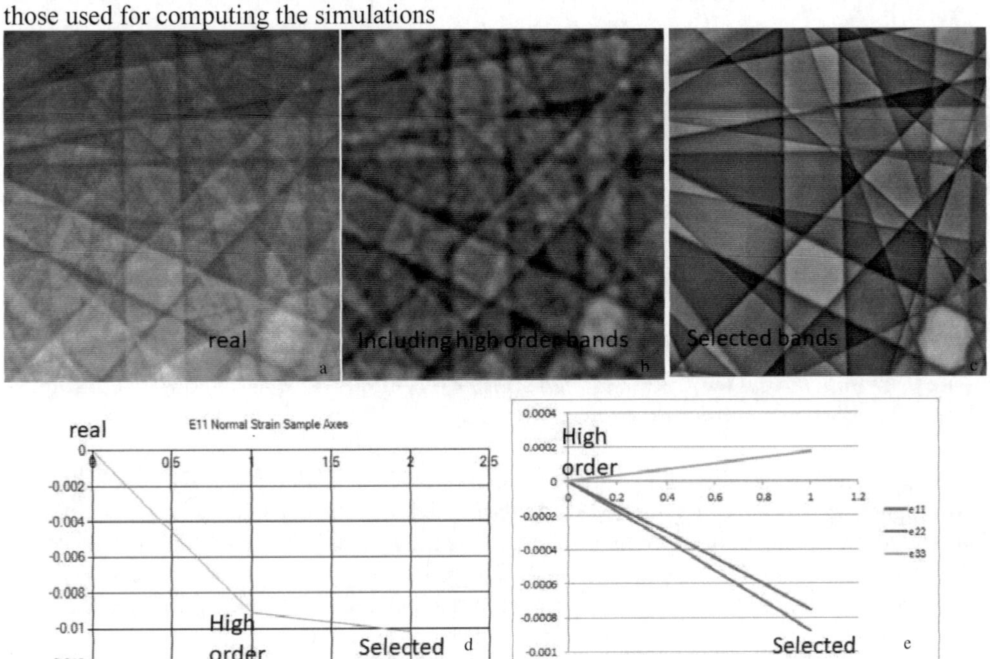

Figures 8 a)EBSD pattern from Ge at 2kx2k resolution, b) dynamic simulation at same resolution with large number of reflectors used in the computation, c) dynamic simulation

using limited number of reflectors, d) e_{11} strain values using real pattern as reference, e) normal strain values for pattern c using simulated pattern, b, as reference pattern.

Figure 8a is the real pattern; 8b is a simulated pattern using a large number of reflectors and 8c a simulated pattern using only bands seen in the real pattern. The two simulated patterns were cross correlated first against the reference pattern and then pattern 8b was used as reference for cross correlation with pattern 8c. Figure 8d shows the e_{11} strain as measured for the 3 patterns. Figure 8e is a plot of the e_{11} e_{22} e_{33} strains of pattern 8c with respect to pattern 8b used as reference. The data in figure 8e indicates that even when the PC x y & z coordinates (as well as the orientation) are the same, the cross correlation can fail to give the correct answer, zero strain in this case. The error in e_{22} and e_{33} show strains of almost 0.0008. They can only be due to differences in the number of bands used in the simulations. The large difference in apparent strain values (figure 8d), when the real pattern 8a is cross correlated against the simulated patterns, is a combination of insufficient correspondence between real and simulated patterns as well as the PC x y z values of the simulated patterns not being the same as the real pattern. Which of the two simulated patterns best fits the quality of the real pattern and whether either are sufficiently similar to the real pattern to provide reliable cross correlation data is open to question. We leave the arguments of using simulations to determine the strain state as still a very open ended problem.

The advent of a Nordiff HR4M high resolution camera of 2050x2050 pixels with 12 bit deep images as currently used at the EDF laboratory, Moret-sur-Loing, has encouraged a re-examination of using Hough type analysis of the EBSD pattern to improve its sensitivity. Figure 9 shows a detail of an HR EBSD Hough peak obtained from analysis of one such EBSD pattern a germanium crystal.

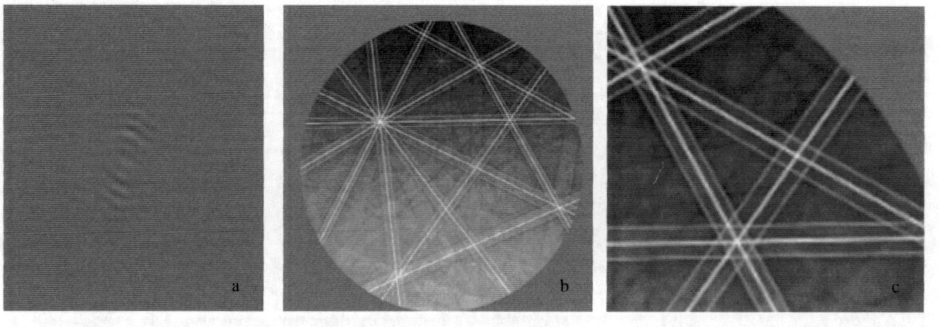

Figure 9, *a) Hough peak for [220] and higher order reflections, b) superimposed band edges as found from Hough analysis, c) detail of part of overlay.*

In Hough space (9a) the vertical axis is the normal distance ρ, from the EBSD pattern centre to a straight line drawn through the EBSD pattern. The horizontal axis is the inclination, θ, of that line. At high resolution, the transform highlights the Kikuchi band edges in the pattern rather than the band itself. The detail is for the 220 reflection and shows three orders of reflections. The straight lines that are tangential to the hyperbola give rise to the dark and bright regions in Hough space. The angular change between tangents at the highest resolution is 5×10^{-4} radians. When the tangent point in the EBSD pattern passes through the point where the vector defined by ρ, θ is shortest possible for each Kikuchi line bordering a Kikuchi band, the shape of the Hough peak shows an inflection. This is clearly seen in the figure. The Kikuchi band edges on the opposite sides of the bands produce intensity profiles in the

IOP Conf. Series: Materials Science and Engineering **375** (2018) 012003 doi:10.1088/1757-899X/375/1/012003

Hough plot which are the inverse of each other. This is due to an asymmetry in the diffracted intensities which is a function of the tilt of the sample with respect to the incident beam. The intensity difference is zero for reflections that are normal to the tilt axis and maximum for those parallel to it. The geometric mid-point in Hough space between the Hough inversion points corresponding to either side of the Kikuchi band, defines the rho value corresponding to the centre line of the Kikuchi band. The normal to this line passing through the mid-point also passes through the true pattern centre. (It leads to a better determination of the PC position as will be explained in a coming publication). The centre line of the Kikuchi band is 7the line defining the intersection of the projection of the diffracting plane onto the phosphor screen. Thus, careful analysis of all Hough peaks permits the rho theta values of the centre lines of all Kikuchi bands to be determined. In addition because of the higher resolution used in the analysis and also the method of interpreting the Hough peak, the accuracy and precision is greater than in conventional Hough applications. Hence, measurement of interplane angles and consequent orientation determination is also improved. Figure 9b shows an overlay of the found band edges on the original EBSD pattern for all pairs of bands. The important features to observe are the intersection points of the centre bands for three intersecting bands. In the detail shown in figure 9c, two such intersections are shown. If the result were perfect the centre lines would intersect at a point. In one case this is true but in the other the splitting is calculated as just 2 pixels in 2000.

8. Conclusions

Thirty years of EBSD at ICOTOM has seen a transition from a rudimentary procedure for on-line determination of crystal orientation in the SEM, to a technique that can now acquire data at a rate approaching one thousand patterns a second and produce maps relating all the major factors affecting the physical properties of the material. The growth of applications has been remarkable, but yet there are still areas that require further development and enhancement. In this paper the most challenging application, that concerned with the measurement of elastic strain and local rotation gradients, both at the 1 part in 10000 level, has been reviewed in detail. The claimed precision has been proven for relative measurements with respect to a pattern obtained from a reference point in the sample. Absolute measurement remains a challenge with two possible approaches; that of using a computer simulation as the reference pattern and that of improving the accuracy of measurement of interplanar angles using a high resolution Hough procedure, are part of an ongoing investigation.

9. Acknowledgements

The author would like to acknowledge that the advancement of EBSD in materials research following the first publication at an ICOTOM meeting would not have happened without the contributions in particular from, Prof Brent Adams, Prof. David Field, Dr. Stuart Wright, Dr. Karsten Kunze, Matt Nowell Dr. Krieger Larsen and more recently by Prof. Angus Wilkinson and Dr. Ben Britton.

10. References

[1] Dingley D J 1987 On line microtexture determination using backscatter Kikuchi diffraction in a scanning electron microscope, Eighth International Conference on Textures of Materials (ICOTOM 8) (The Metallurgical Society) ed. J S Kallend and G Gottstein 189-194
[2] Frank C 1987 Orientation Mapping, Ibid

[3] Adams B L Wright S I and Kunze K 1993 Orientation imaging: The emergence of a new microscopy. *Metallurgical Transactions A* **24**(4): 819-831
[4] Venables J A and Bin-Jaya A 1977 Accurate micro-crystallography using electron backscatter diffraction. *Phil Mag.* **35** (5) 1317-1332
[5] Dingley D J, Alderman J, Longden M and Weinbren J, 1987 Online analysis of electron back scatter diffraction patterns 1. Texture Analysis of Zone Refined Polysilicon." *Scanning Microscopy* **1**(2) 451-456
[6] Adams B L, Dingley D J, Kunze K and Wright S I' (1994). "Orientation Imaging Microscopy: New Possibilities for Microstructural Investigations using Automated BKD Analysis." *Materials Science Forum* **157-62**: 31-42
[7] Dingley D J and Kunze K (1994) Orientation imaging Microscopy and serial sectioning, ICEM Paris *Les editions de physique* editing B Jouffrey and C Colliex **2B** 1313-1314
[8] Wilkinson A J, 1996 "Measurement of elastic strains and small lattice rotations using electron back-scatter diffraction." *Ultramicroscopy* **62**(4): 237-247
[9] Wilkinson A J, Meaden G and Dingley D J' 2006 "High-resolution elastic strain measurement from electron backscatter diffraction patterns: New levels of sensitivity." *Ultramicroscopy* **106** (4-5): 307-313
[10] Dingley D J, 2016 High angular resolution EBSD for measurement of strain and residual dislocation density. *EBSD: Analyse par diffraction des electrons retrodiffuses,* Groupement National de Microscopie Elecronique, edit. Francois Brisset
[11] Villert S, Maurice C, Wyon C and Fortunier R. (2009) Accurate assessment of elastic strain measurement by EBSD, *Jounal of Microscopy* **233** 290-301.
[12] Mingard K, Day A., Maurice C, and Quested P N, (2011) Towards high accuracy calibration of EBSD systens. *Ultramicroscopy* **111** 320 329
[13] Plancher E, 2015 Mesures de champs de deformations elastique et totalepour la determination du comportement mecanique local de materiaux cristallins. Doctorat ParisTech l'École Nationale Supérieure d'Arts et Métiers
[14] Plancher E, Favier V, Maurice C, Bosso E, Rupin N, Stodolna J, Loisnard D, Marijon JB, Petit , Micha J-S, Robache O and Castelnaua O, (2017) Measurement of local constitutive relations at the micrometre scale, in bulk metallic alloys, *J. Appl. Cryst.* **50**, 940–948
[15] Arsenlis A, Parks D M, Becker R and Bulatov, V, (2004) "On the evolution of crystallographic dislocation density in non-homogeneously deforming crystals." *Journal of the Mechanics and Physics of Solids* **52**(6): 1213-1246
[16] Wilkinson A J and Randman D, (2010). "Determination of elastic strain fields and geometrically necessary dislocation distributions near nanoindents using electron back scatter diffraction." *Philos. Mag.* **90**(9)
[17] Kacher J H, Landon C, Adams B L and Fullwood D, (2009) Bragg's law diffraction simulations for electron backscatter diffraction analysis, *Ultramicroscopy* 109 (9) (2009) 1148–1156.
[18] Maurice C, Fortunie R, Driver J, Day A, Mingard J A K and Meaden G, (2010) Comments on the paper Bragg's law diffraction simulations for electron backscatter diffraction analysis by Josh Kacher, Colin Landon, Brent L. Adams & David Fullwood, *Ultramicroscopy* **110** 7.
[19] Winkelmann A, Trager-Cowan C, Sweeney F, Day A P, Parbrook P, (2007) Many-beam dynamical simulation of electron backscatter diffraction patterns, *Ultramicroscopy* 107 (4–5) 414–421.
[20] Akorta J, Marteleur M, Jacques P J, (2017) Improved simulation based HR EBSD procedure using gradient-based DIC techniques, *Ultramicroscopy* **182** 17-27

IOP Conf. Series: Materials Science and Engineering **375** (2018) 012004 doi:10.1088/1757-899X/375/1/012004

Characteristics of Thin Cu Films Electrodeposited On Textured Ni-Co Substrates

B Panda[1] R K Ray[2] and Rakesh Kumar[3]
[1] Associate Professor, CMR Institute of Technology, Bangalore, India
[2] Visiting Professor, Dr. MN Dastur School of Materials Science and Engineering, Indian Institute of Engineering Science and Technology Shibpur, Kolkata, India
[3] Assistant Professor, RV College of Engineering, Bangalore, India

Email: rkray@iitk.ac.in

Abstract. An attempt has been made to study the effects of current density and the texture of the underlying substrate on electrodeposited copper. For this purpose five Ni-Co alloys, Ni-10 Co, Ni-20 Co, Ni-30 Co, Ni-40 Co and Ni-60 Co, in the cold rolled as well as in the annealed conditions, were used as substrates. Acid copper sulfate solution was used to deposit thin Cu layers on the substrates, using a Cu plate as anode, at four different current densities, 1, 10, 30 and 50 mA/cm^2. The thickness of the electrodeposited Cu layer increased nearly linearly with current density, but was independent of the texture and composition of the underlying substrate. In general, the X-ray diffraction line intensities for the deposited Cu layer sharpened and those for the substrate Ni alloys weakened with increase in current density. The textural developments in the Cu deposits appeared to be quite independent of the textures as well as the compositions of the substrate materials. The deposited Cu layers did not inherit the textures of the substrates at the lower current densities, and also developed their own textures at higher current densities. The Cu (220) peak ultimately became the strongest XRD peak for the electrodeposited layer. The surface roughness of the deposited layers was distinctly smoother for the annealed substrates, as compared to the cold rolled Ni-Co alloys.

1. Introduction

In the present work, five Ni-Co alloys (Ni-10Co, Ni-20Co, Ni-30Co, Ni-40Co and Ni-60Co) in cold rolled and annealed condition have been used as substrates for electrodeposition of Cu. The textures of Ni-Co alloys have been studied previously **[1-3]**. As the Co concentration increases, the SFE decreases drastically (for pure Ni, the SFE is 130mJ/m^2). Pure Ni shows a rolling texture of copper or pure metal type, in which the Cu {112}<111>, S {123}<634> and Bs {110}<112> are equally strong. In the Ni-Co alloys, there is a texture transition from the copper type to the α-brass type at a cobalt content of around 40%. When the cold worked Ni-Co alloys having low SFE are annealed, the Bs texture transforms to {332}<113>. The compositions having higher magnitude of SFE (10, 20, 30%Co) yield {100}<001> cube texture. The intermediate composition Ni-40Co exhibits some amount of recrystallized cube texture along with some amount of twin of cube texture {122}<212>.

Electroplating on articles is widely carried out for purposes such as appearance, protection, special surface properties and engineering or mechanical properties. Electrodeposited films have textures, which influence their electronic, magnetic, optical and other properties and are closely linked with their microstructures and surface morphologies. Copper is considered as the material for semiconductor metallization because of its high conductivity and corrosion resistance in high temperature. However, the nature of the electrodeposited layer will depend on the structure of the base metal and the electrodeposition parameters. If the interatomic distances in the lattice plane of the substrate match those of the lattice plane of the deposit, the structure of the substrate may be continued into that of the deposit. Therefore, in the present work, an attempt has been made to study the effect of deposition parameter (as current density) and the texture of the underlying Ni-Co alloy substrate on characteristics of the electrodeposited copper film.

2. Experimental Procedure

2.1. *Substrate Material*

The chemical compositions of the Ni-Co alloys (in weight percent) taken for the experimental work are given in **Table 1**. Both the nominal and detailed compositions of each alloy are shown in the table. A total of five Ni-Co alloys, Ni-10Co, Ni-20Co, Ni-30Co, Ni-40Co and Ni-60Co were melted under vacuum. Segregation was avoided by magnetic stirring during melting. The ingots were cold rolled 50% to a thickness of 10mm and then homogenization annealed in vacuum at 1150^{0}C for 24 hours. These were then again cold rolled 50% and annealed at 1100^{0}C for a period of 3 hours to yield the starting materials of almost random texture and a grain size of 0.1 mm.

Table 1 Chemical compositions of Ni-Co alloys (wt%)

Alloy nominal composition		**Detailed composition (wt%)**					
		Co	**C**	**S**	**Si**	**Cu**	**Ni**
A	Ni-10Co	11.15	0.006	0.003	0.03	0.03	Bal.
B	Ni-20Co	22.85	0.007	0.003	0.06	0.03	Bal.
C	Ni-30Co	30.90	0.006	0.003	0.03	0.03	Bal.
D	Ni-40Co	41.05	0.006	0.004	0.03	0.03	Bal.
E	Ni-60Co	60.50	0.006	0.004	0.06	0.03	Bal.

2.1.1. *Cold Rolling*

The starting materials were then cold-rolled by an amount of 95% reduction, using a laboratory rolling mill having 250 mm diameter rolls. The directions of the strips were reversed end to end after each pass. In between any two successive passes, the strips were dipped into cold water bath to minimize any unwanted rise in temperature. The change in thickness after each pass was noted. In order that homogeneous deformation throughout the specimens might be achieved, the ratio of the length of contact with the rolls to specimen thickness was maintained at a value greater than unity. Reproducible texture was found over almost the whole of the thickness except in a very narrow zone near the outermost surface. From each specimen, material was etched off from one side up to the mid-thickness section on which all texture measurements were carried out. These cold rolled alloys were made available for the purpose of this work along with the relevant texture plots.

2.1.2. Recrystallization Anneal

From the cold rolled strips of the five alloys, samples of dimensions 14 mm x 25 mm were cut. Out of the total number of cut pieces from any alloy, one half were subjected to annealing heat treatment at 800°C for 1 hour in a Horizontal Tube furnace using argon atmosphere for complete recrystallization. After heat treatment, the samples were cooled in the furnace. The textures measured from the recrystallized samples were also made available for the purpose of this study. All the textures, for both cold worked and recrystallized samples were in the form of pole figures.

2.2 Electroplating

High purity copper metal plates of dimension 14 mm x 25 mm were used as anode while Ni-Co alloy plates of same dimensions were used as cathode. Prior to electroplating, the cold-rolled and annealed samples were polished with emery papers of 240, 400, 600, and, finally, 800 grit. Following which, the samples were cleaned with an organic solvent (C_2HCl_3) and then with an alkali solvent. The temperature of the alkali solvent solution was kept at 70-90°C and degreasing was carried out for 10-30 minutes.

Additive-free acid copper plating solution was used for plating, which was composed of $CuSO_4.5H_2O$ (188 g/l) and H_2SO_4 (75 g/l). The distance between cathode and anode was maintained at 6 cm. Four different current densities of 1, 10, 30 and 50 A/cm^2 were used for plating and the bath temperature was kept at room temperature. The bath was stirred at constant speed and controlled to maintain a pH of 3.0. Time of electroplating was kept constant at 10 minutes for each case. Each of the five Ni-Co alloys (both cold-rolled and annealed) was electroplated at the four different current densities as stated above. After deposition, the Ni-Co alloy plate was rinsed thoroughly with water and then blow dried with air.

2.3. Measurement of Thickness of Deposit

The thickness of the electrodeposit was measured from the weight increase of the cathode after deposition, using the following relationship

$$t=W/ \rho A$$

where t is the thickness of the deposit, W is the increase in weight of the cathode plate after electrodeposition, ρ is the density of copper and A is the surface area of the cathode.

2.4. X-ray Diffraction

X-ray diffraction of the Cu electrodeposited Ni-Co alloy was carried out using a SIEFERT ISO-DEBYEFLEX 2002 X-ray diffractometer equipped with a Cu-tube. The samples were scanned in a 2θ range of 35° to 150° at a sweep rate of 3°/ minute.

2.5. Surface Roughness Measurement

The surface roughness measurement of the electrodeposited Cu layer was carried out using a FEDRAL SURFANALYZER 5000. The cut-off for all the measurements was 0.8 mm. Drive speed of probe was 0.25 mm/sec and the probe range was +/- 50 microns. The average surface roughness value (R_a) was used for comparing the relative roughness of the electrodeposited layer as a function of the current density as well as the composition of the Ni-Co substrates.

3. Results and Discussion

3.1. Thickness of Cu Layer

The thickness of copper deposit (in μm) on the Ni-Co alloys as a function of the current density, in both cold rolled and annealed conditions are shown in **Fig. 1(a) and (b)**, respectively. As expected, the thickness of the electroplated copper layer increases with the increase in current density for all the alloys. The thickness had almost linear relationship with current density. There was hardly any difference in the thickness of the electrodeposited layer, for either the cold rolled or the annealed substrates. Therefore, the thickness of the electrodeposited layer and the course of its development were found to be more or less constant, irrespective of the texture of the substrate and the chemical compositions of the substrate.

One observation was that, at lower current densities, the thickness of the deposited copper is almost same for all alloys both in cold rolled and annealed condition. However, as the current density increases, the scatter in the thickness value increases for the alloys.

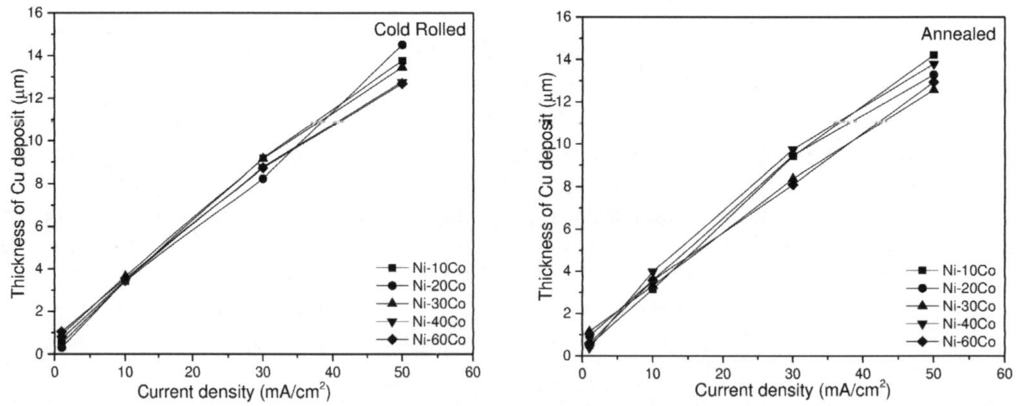

Fig. 1 Thickness of Cu deposit on (a) cold rolled and (b) annealed Ni-Co alloys

3.2 X-Ray Diffraction

3.2.1. Cu Electrodeposited on Cold Rolled Ni-Co Alloys

The variation of the intensity of the most prominent XRD peak from Ni-Co alloy and Cu for cold rolled substrate is shown in **Figs. 2 (a) and 2 (b)**, respectively. In general, it was found that out of the Cu peaks, the (220) line of copper is most intense followed by (111) and (200) lines at all current densities except at 1 mA/cm^2. At the lowest current density (1 mA/cm^2), the Cu (111) peak is most intense. Therefore, Cu(111)/Cu(220) ratio decreases with current density. Out of the Ni-Co peaks, the (220) peak is most intense at all current densities. But there is a decrease in the peak intensities of all the three lines from Ni alloy as we move from lower to higher current density while the Cu pattern sharpens. This is quite natural, since with increase in current density the thickness of the Cu layer increases.

The same pattern was found for cold rolled Ni-10Co, Ni-20Co, Ni-30Co, Ni-40Co and Ni-60Co substrate. From these results it is understandable that with increases in current density there is a definite change in the texture of the electrodeposited layer. Usually in

IOP Conf. Series: Materials Science and Engineering **375** (2018) 012004 doi:10.1088/1757-899X/375/1/012004

untextured Cu, (111) is the strongest peak, while the present investigation shows a progressive increase of the Cu(220) peak with the increment in current density. However, strangely enough these textural changes in the Cu deposit seem to be quite independent of the texture as well as the composition of the cold rolled substrate materials.

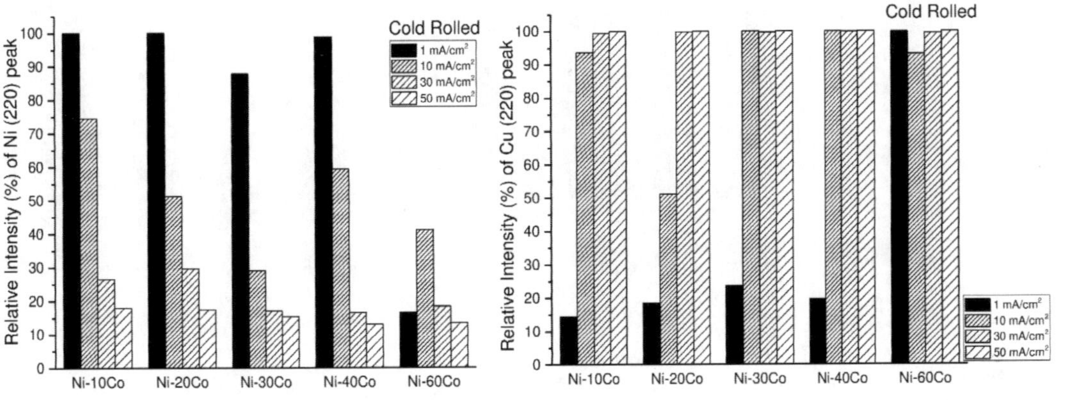

Fig. 2 Relative intensity (%) of (a) Ni(220) and (b) Cu(220) for cold rolled Ni-Co alloys

3.2.2. Cu Electrodeposited on Annealed Ni-Co Alloys

The variation of the intensity of the most prominent XRD peak from Ni-Co alloy and Cu for annealed substrate is shown in **Figs. 3 (a) and 3 (b)**, respectively. At lower current densities, Cu (111) has the highest intensity while at higher current densities, Cu (220) has highest intensity out of the other two copper lines. The variation of intensity of the copper lines with respect to current density, are thus, similar to that obtained for the cold rolled alloys. However, the Ni-Co lines show a pattern different from that of the cold rolled alloys. Here, the (200) peaks of Ni-Co are more intense in contrast to the (220) peaks in the cold rolled alloy. For all current densities starting from $1mA/cm^2$ to $50\ mA/cm^2$, the same pattern is obtained. Similar to cold rolled alloys, for the annealed substrates also, the Ni pattern gradually weakens with increase in current density and Cu pattern sharpens. These results again indicate that the texture of electrodeposited layer is more or less independent of the texture of the annealed substrate material also. The most interesting result is that whether the substrates are in the cold rolled or in annealed conditions (and therefore vary widely in their textures), the Cu(220) peak ultimately becomes the strongest peak of the electrodeposited layer. Thus the electrodeposited layer exhibits a definite strong texture when it finally develops. In all the above cases the deposited Cu layer did not inherit the texture of the substrate at the lower current densities, and also developed its own texture at higher current densities.

IOP Conf. Series: Materials Science and Engineering **375** (2018) 012004 doi:10.1088/1757-899X/375/1/012004

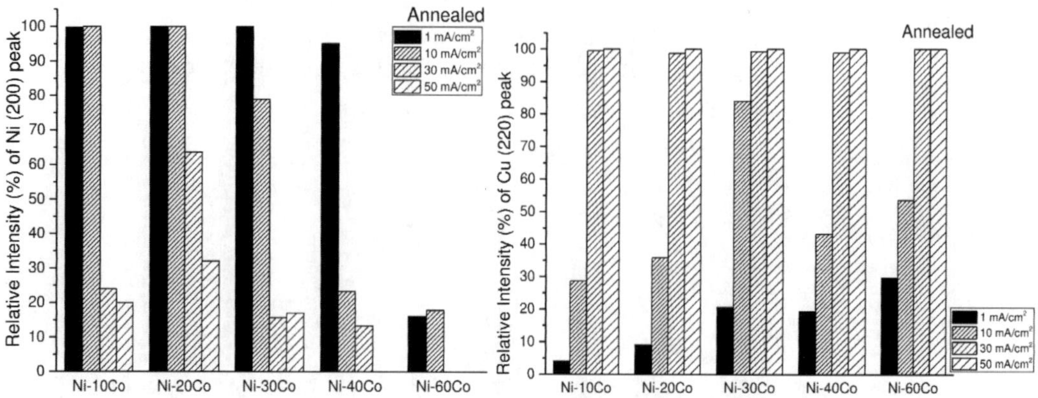

Fig. 3 Relative intensity (%) of (a) Ni(200) and (b) Cu(220) for annealed Ni-Co alloys

3.3. *Surface Roughness*

The roughness parameters R_a (average value) for the cold rolled and annealed Ni-Co alloys are shown in **Figs. 4 (a) and 4 (b),** respectively. It can be seen that within each group, variation of R_a with current density does not follow a general trend, it is rather erratic. However, for different alloy compositions and different current densities, the average roughness value of Cu on annealed Ni-Co alloys was found to be lower as compared to cold rolled alloys (Table 2 and 3, respectively).

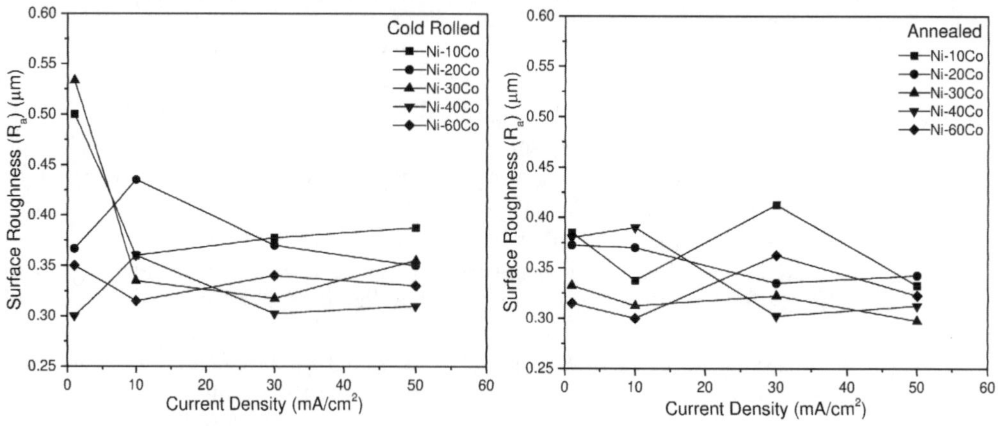

Fig. 4 Variation of average surface roughness (R_a) in µm with current density for (a) cold rolled and (b) annealed Ni-Co alloys

Table 2 Variation of R_a for different compositions

Composition of substrate	R_a(μm)	
	Cold rolled alloy substrate	Annealed alloy substrate
Ni-10Co	0.14	0.08
Ni-20Co	0.16	0.04
Ni-30Co	0.10	0.03
Ni-40Co	0.13	0.08
Ni-60Co	0.04	0.06

Table 3 Variation of R_a at different current densities

Current density	R_a(μm)	
	Cold rolled alloy substrate	Annealed alloy substrate
1mA/cm^2	0.18	0.07
10A/cm^2	0.19	0.09
30mA/cm^2	0.07	0.11
50mA/cm^2	0.08	0.05

These results seem to indicate that the substrate texture could have some effect on the roughness of the electrodeposited layer. It should be noted that the cold rolled as well as annealed samples were polished to the same surface roughness prior to electroplating. Even then, the deposited layers are definitely much smoother in case of the annealed than in the cold rolled Ni-Co alloys. The exact reason for this is not very clear though.

Overall, the crystallographic textures of Ni-Co alloy substrates do not seem to have any tangible effect on the texture and grain size of the electrodeposited Cu layer, whereas the roughness properties of the layers seem to be affected somewhat by the substrate texture.

4. Conclusions

a. The thickness of the electrodeposited layer and the course of its development was found to be more or less constant, irrespective of the texture of the substrate.

b. The substrate composition also does not seem to have any effect on the thickness of the electrodeposited layer. The thickness has nearly linear relationship with current density.

c. From the X-ray diffraction patterns of Cu electroplated on cold rolled and annealed Ni-Co alloys it is evident that in general, the Cu pattern sharpens and Ni pattern weakens as current density increases. For annealed substrate, initially the Ni(200) is the sharpest (at 1mA/cm^2), contrary to that obtained for the cold rolled substrates(where Ni(220) is the sharpest at the lowest current density).

d. In the present investigation a progressive increase of the Cu{220} peak with the increment in current density has been found. However, these textural changes in the Cu deposit seem to be quite independent of the texture as well as the composition of the cold rolled substrate materials. The deposited Cu layer did not inherit the texture of the substrate at the lower current densities, and also developed its own texture at higher

IOP Conf. Series: Materials Science and Engineering **375** (2018) 012004 doi:10.1088/1757-899X/375/1/012004

current densities. The Cu (220) peak ultimately becomes the strongest peak of the electrodeposited layer.

e. The deposited layers are definitely much smoother in case of the annealed than in the cold rolled Ni-Co alloys.

References:

[1] Carstensen J V, Ray R K and Leffers T 2002 *Materials Science Forum* **408-412** 607-12
[2] Ray R K *Acta Metall. Mater.* 1995 **43** 3861-72
[3] Mukherjee K 2002 *Correlation of Textures and Microstructures in a Series of Ni-Co Alloys*, Masters Thesis

IOP Conf. Series: Materials Science and Engineering **375** (2018) 012005 doi:10.1088/1757-899X/375/1/012005

About the combined role of texture and grain size on hardening behavior of cp titanium sheets

F. Wagner[1,2], T. Richeton[1,2], C. Chen[1,3] and L.S. Toth[1,2]

[1] Labex DAMAS, Université de Loraine, France
[2] LEM3 CNRS (UMR 7239) - Université de Lorraine, Metz, France
[3] now at Nanjing University of Science and Technology, China
E-mail: francis.wagner@univ-lorraine.fr

Abstract. The goal of the study is to analyze the role of both the grain size distribution and the crystallographic texture on the plastic deformation. Data consist, on one hand, in tensile curves obtained for commercially pure titanium samples elaborated to correspond at several mean grain sizes and textures. Data consist, on the other hand, in calculated tensile curves using a crystal plasticity EVPSC model modified to include grain size effects. The trend for the yield stresses and the initial plateaus are satisfying showing that the effects of grain size are captured in the model. Under the conditions used in the model the hardening is however not enough grain size dependent when compared with experiments.

1. Introduction

Since decades the plastic deformation of polycrystalline materials has been studied from a very large number of experiments as well as from many model calculations [1, 2] in order to understand in details the mechanisms which govern this deformation and to have predictive tools. It was recognized very soon that the texture is a key factor for the plastic behavior. Although known at the macroscopic level (Hall-Petch relationship for example), the role of the mean grain size and the grain size distribution have attracted less attention. Only few papers consider the effects of both texture and grain size distribution on the mechanical behavior (e.g., [3-8]). The goal of the present paper is to contribute to a better understanding of the combined role of crystallographic texture and grain size during plastic deformation. For that purpose a set of specimens showing various microstructures was prepared, characterized using a FEG-SEM (with EBSD) and then submitted to tensile tests. The used material was cp (commercially pure) titanium (grade 2). On the other hand a crystal plasticity model, EVPSC (elasto-visco-plastic self-consistent) type [9], was modified to include grain size effects. The comparison of the experimental and simulated results was used to clarify the respective role of the texture and grain size.

2. Materials and experiments

2.1 Preparation of the specimens

The as-received material was a 2 mm-thick plate of commercially pure titanium (Ti grade 2 or T40 according to TIMET*) in a fully recrystallized state with a mean grain size of about 10 μm. Specimens cut off from this plate were cold rolled, most of the time along the previous RD (Rolling Direction) and in a few cases along the previous TD (Transverse Direction). After the deformation, the specimens were annealed under various conditions in a furnace working under secondary vacuum. The conditions of preparation for the seven selected specimens used in this study are reported in Table 1. The specimens had a dog bone shape (total length 43 mm) and could be put entirely in a FEG-SEM.

IOP Conf. Series: Materials Science and Engineering **375** (2018) 012005 doi:10.1088/1757-899X/375/1/012005

2.2 Microstructural characterization of the specimens

The microstructural characterization of these specimens was made by using EBSD maps. An example of such maps is presented at Figure 1. From each EBSD map a set of grains was extracted. The size and the orientation of each grain were determined using a post-processing software (Chanel 5* or JTEX*). D_m, in Table 1, is the mean grain size. In all the cases the specimens were fully recrystallized.

Table 1: List of the seven specimens, their preparation conditions, some metallurgical data and some mechanical characteristics (the quantities related to mechanical properties are defined in section 2.4)

Specimen name	Cold rolling reduction	Rolling direction	Annealing conditions	Number of grains	D_m (µm)	σ_e (MPa)	LP1 (%)	σ_{max} MPa	ε* (%)
R_2.8	75%	// RD	500°C–40mn	6254	2.8	390	2.1	485	21.7
T_2.8	75%	// TD	500°C–40mn	7328	2.8	383	4.7	436	22.3
R_4.8	75%	// RD	650°C – 1 h	8793	4.8	331	1.0	442	16.4
RR_9.8	75%	// RD	730°C – 2 h	3262	9.8	354	-	494	14.6
RT_9.8	75%	// RD	730°C – 2 h	3262	9.8	375	2.0	440	9.7
R_11.7	75%	// RD	740°C – 2 h	4075	11.7	353	-	466	18.4
R_21.3	30%	// RD	840°C – 4 h	1273	21.3	282	-	484	18.8

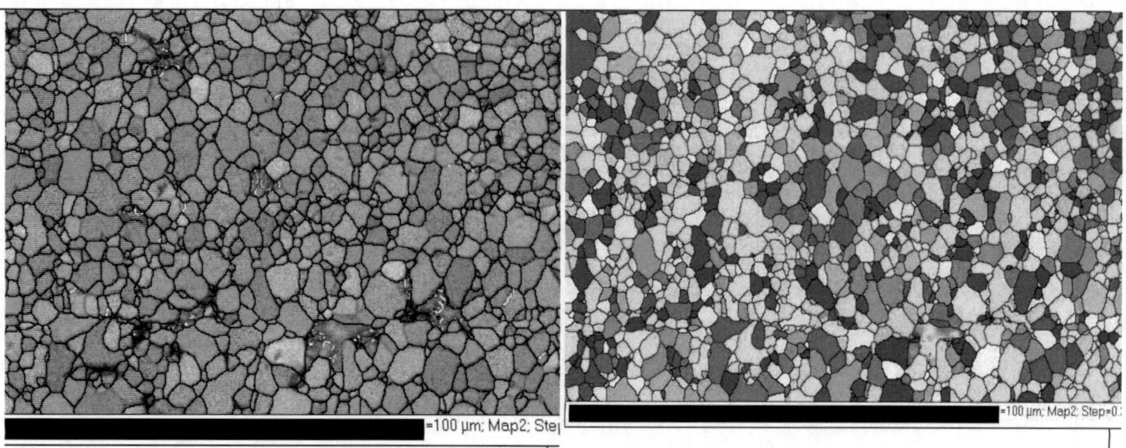

a) BC (Band Contrast) map	**b)** IP (Inverse Pole) map

Figure 1: Part of an EBSD map for the specimen R_2.8 (mean grain size D_m=2.8 µm)
a) BC map b) IP map for the direction RD

The specimens with the smallest grains were close to the end of the primary recrystallization. Note that the number in the names of the specimens (Table 1) refers to the mean grain size in microns. For the two specimens with a mean grain size 9.8 µm, the notations used in Table 1 distinguish the case rolling // RD and extension // RD (sample RR_9.8) from the case rolling // RD and extension // TD (sample RT_9.8).

2.3 Texture

For each specimen the data set of the grains (size and orientation) was used to calculate the ODF. An example is given at figure 2 where the section Phi1=0° is presented for two specimens. The ODF's correspond to well known textures for this material [10, 11]. When the grains are small, which means that the annealing was stopped close to the end of the primary recrystallization, the texture resembles the deformation one whereas the 'grain growth texture' is obtained after heat treatments made at higher temperatures (see [12] for further information).

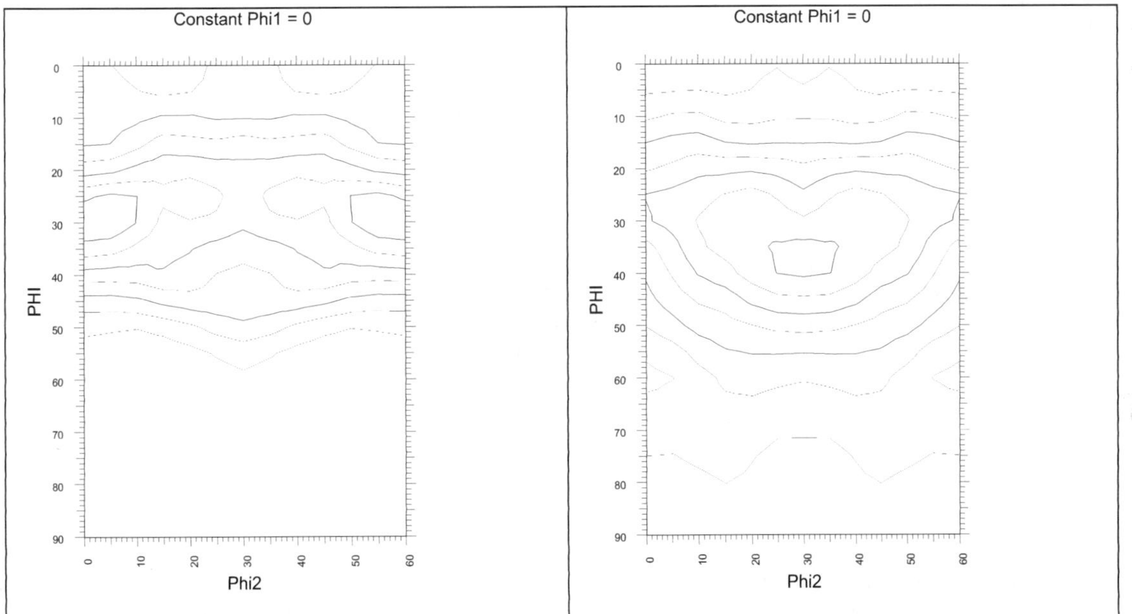

a) Sample R_2.8	**b)** Sample R_11.7
Contour levels: 1., 2., 3., 4., 5., 6., 7.	Contour levels: 1., 2., 3., 4., 5., 6., 7.

Figure 2: Section Phi1=0° of the ODF for two specimens a) sample R_2.8 b) sample R_11.7

For the simulations which are shown later on, the data consist in a list of grains with, for each one, its orientation (Euler angles) and its grain diameter *Dg*.

2.4 Mechanical characterization

The samples were small dog-bone-shaped specimens (total length L_0=43 mm, gauge length L_g=11.0 mm, width w_0=3.0 mm, thickness t_0=0.5 mm). Tensile tests were conducted at ambient temperature on a Deben Microtest MT10331-1kN tensile machine at a constant displacement speed of 0.5mm/mn which corresponds to an initial strain rate of 7.5 10^{-4} s^{-1}. The extension direction was parallel to the last rolling direction except for the sample RT_9.7 (extension direction // TD). An example of the tensile curves is shown at Figure 3.

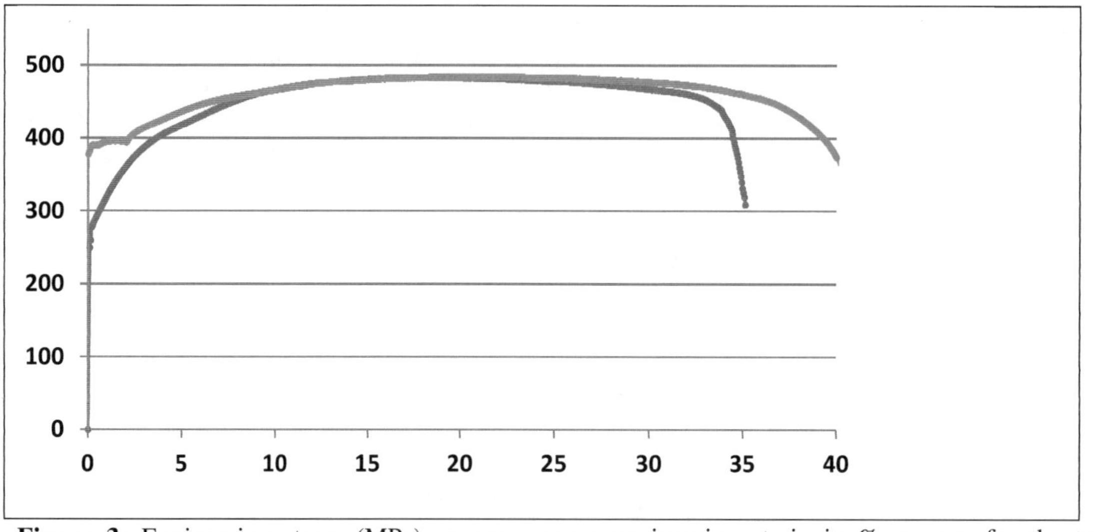

Figure 3: Engineering stress (MPa), σ_eng, versus engineering strain in %, ε_eng, for the two specimens R_21.3 (red curve) and R_2.8 (green curve).

Several quantities were extracted from the tensile curves to characterize (and discuss in the followings) the mechanical behavior: σ_e is the yield stress, LP1 stands for the length (in %) of the first plateau which appears sometimes in the beginning of the plastic deformation, σ_{max} is the maximal engineering stress, ε* is the corresponding engineering strain (in %). These values are reported in Table 1. We will use also further on the quantity $\Delta\sigma = \sigma_{max} - \sigma_e$ to characterize the 'hardening amplitude'.

From these experiments two trends were remarked:

i) the yield stress increases with decreasing grain size and follows rather well a Hall-Petch relationship [9, 10],
ii) the 'hardening amplitude', Δσ, tends to decrease with decreasing grain size [9].

3. Used model

3.1 Characteristics of the model

The model used to simulate the tensile behavior is the one that was developed in reference [13] to study hardening mechanisms in cp Ti. It is an EVPSC model based on an affine linearization of the viscoplastic flow rule and the translated fields technique to formulate the 1-site self-consistent

IOP Conf. Series: Materials Science and Engineering **375** (2018) 012005 doi:10.1088/1757-899X/375/1/012005

approximation [9]. At the level of the single crystal, glide on prismatic, basal, pyramidal <a>, 1st order pyramidal <c+a> and 2nd order pyramidal <c+a> is considered but not twinning. The slip rates $\dot{\gamma}^{(s)}$ result from the Orowan relation where the evolution of mobile dislocation densities and dislocation velocities are treated separately on each slip system. This allows to account for an initial fast multiplication of mobile dislocations. It was shown that this setting, together with the consideration of lower strain-rate sensitivity for prismatic systems, succeeded to reproduce the three-stage hardening behavior of cp Ti in tensile conditions.

3.2 Taking into account the effects of grain size

In order to take into account grain size effects in the model, two modifications were made in the constitutive laws. Following references [3-8], the assumption was made that the critical resolved shear stresses (crss) $\tau_c^{(s)}$ depends on the *individual grain size* D_g in a Hall-Petch type relationship:

$$\tau_c^{(s)} = \tau_0^{(s)} + \mu^{(s)} b^{(s)} \sqrt{\sum_l a^{(sl)} \rho_f^{(l)}} + \frac{k_1^{(s)}}{\sqrt{D_g}} \qquad (1)$$

In this formula, s or l designates the slip system, $\tau_0^{(s)}$ the lattice friction stress, $\mu^{(s)}$ the directional shear modulus, $b^{(s)}$ the Burgers vector magnitude, $a^{(sl)}$ interaction coefficients between systems, $\rho_f^{(l)}$ forest (or sessile) dislocation density that increases with plastic deformation and $k_1^{(s)}$ that is thus a Hall-Petch type slope resolved on the slip system. Note that with this supplementary Hall-Petch -type term, the crss are different *from grain to grain* according to their size, the smallest grains being harder that the largest ones.

The second modification made was based on the idea that, with decreasing grain size, i.e. with increasing ratio grain boundary area over grain volume, the balance between dislocation absorption and emission at grains boundaries is changed in favor of absorption. In order to account for this effect, the dynamic recovery term, which is present within the evolution equation of forest densities [13], was made grain size dependent as follows:

$$\dot{\rho}_f^{(s)} = \frac{1}{b^{(s)}}\left(\frac{1}{L^{(s)}} - 2k_c^{(s)} b^{(s)}\left(\rho_f^{(s)} + \frac{k_2^{(s)}}{D_g^2}\right)\right)|\dot{\gamma}^{(s)}| \qquad (2)$$

Through equations (1) and (2), it is seen that $k_2^{(s)}$, by influencing the evolution of the forest dislocation densities, will modify the hardening behavior. k_1 and k_2 were taken the same for all the slip systems. Their values are reported in Table 2 along with the values of $\tau_0^{(s)}$ and ρ_{f_0} (initial forest density). The other model's parameters are the same as in [13].

Table 2: List of the model's parameters.

k_1 (MPa.m$^{0.5}$)	k_2	ρ_{f_0}(m^{-2})	$\tau_0^{(s)}$(MPa)				
			Prismatic	Basal	Pyr. <a>	1st order pyr. <c+a>	2nd order pyr. <c+a>
0.114	40	10^{12}	100	145	145	130	290

These crss values seem in accordance with some of the data reported in [14].

4. Results and discussion

Two examples of simulated tensile curve, compared to the experimental one, are shown at Figure4.

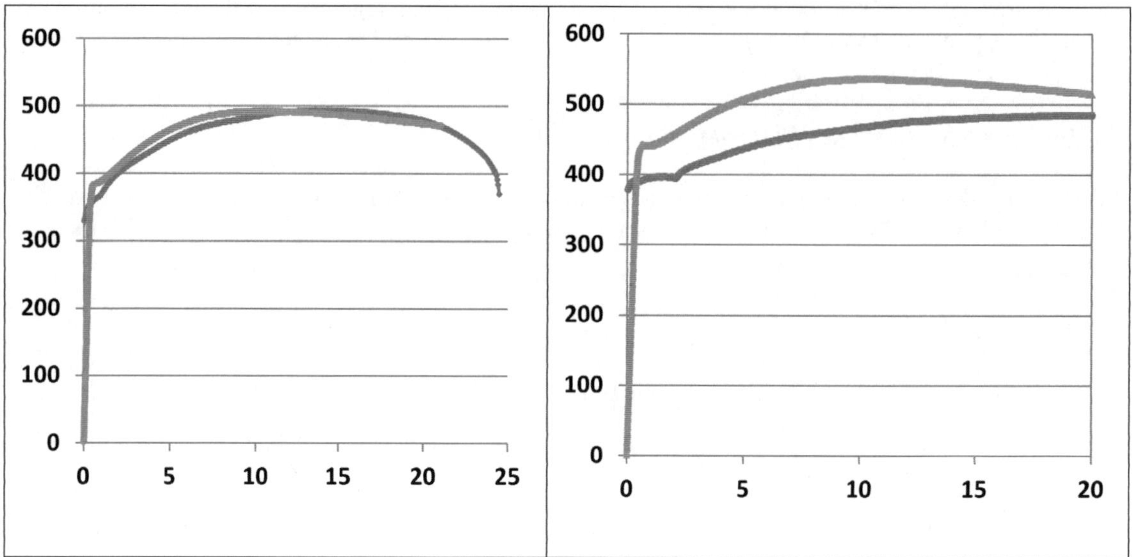

a) exp. R_11.7 (red) and simu. R_11.7 (green)	**b)** exp. R_2.8 (red) and simu. R_2.8 (green)

Figure 4: Experimental and simulated tensile curves (σ_eng (MPa) versus ε_eng (%))
a) sample R_11.7 b) sample R_2.8

4.1 Yield stress

The values of the yield stress for experiments and for simulations as function of the mean grain size are shown in Figure 5. The yield stresses calculated from the Hall-Petch relationship [9] are also plotted.

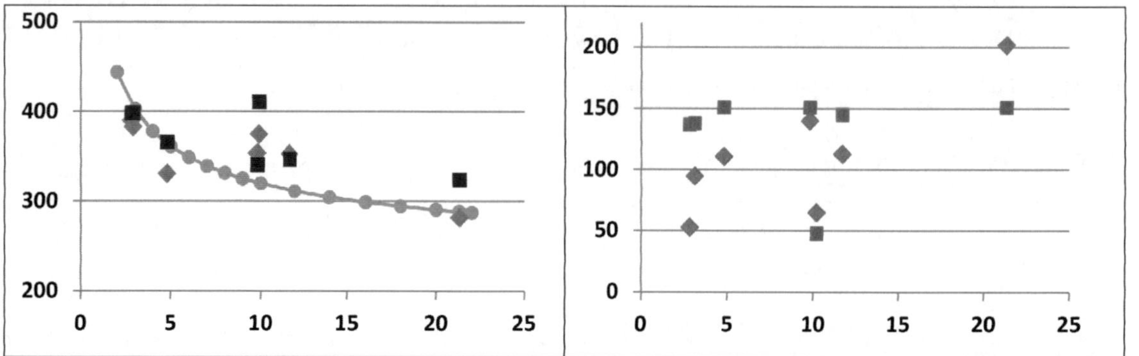

a) Yield stress (MPa) versus the grain size (μm); exp. red, simu. blue, HP green	**b)** Δσ, the 'hardening amplitude', (MPa) versus the grain size (μm); exp. red, simu. blue square

Figure 5: Comparison of experimental and calculated values of a) the yield stress b) Δσ ($=\sigma_{max}-\sigma_e$)

The calculated yield stresses follow a correct trend which means that the assumption made for the dependence of the crss on the grain size (eq. 1) is relevant. The main discrepancy arises for the case where the tensile direction is parallel to TD (sample RT_9.8) due likely to crss which are somewhat too high for the pyramidal slips.

4.2 First plateau

When an initial plateau exists (see the values of LP1 in Table 1), the model reproduces it but somewhat shorter (see a typical example in Figure 4b). From the simulations, it is possible to analyze the activity of the several slip systems in the beginning of the plastic deformation. No plateau appears when the prismatic slip dominates. A plateau appears when there is a more balanced activity between prismatic, pyramidal <c+a> and pyramidal <a> slip systems. Under such conditions plastic strain incompatibilities and the induced internal stresses are strongly decreased. Along with the fast multiplication of mobile dislocations, this gives rise to an initial state with nearly no hardening.

4.3 Hardening

In all the cases the simulations reproduce correctly the three stages of the hardening rate $d\sigma/d\varepsilon$ which have been described in the literature for tensile experiments [13, 15, 16]. When considering the 'hardening amplitude', $\Delta\sigma$, it can be seen in Figure 5b that this quantity is nearly constant in the simulations whereas it increases with increasing grain size (except for the particular case of extension along TD – sample RT_9.8). This indicates that the term added in the model, eq. (2), by using a coefficient k_2=40, does not change enough the forest dislocation density rate. In order to produce a significant effect, one should use a much higher value of this coefficient. A simulation performed with k_2=1000, in a case with small grains (sample T_2.8), shows that $\Delta\sigma$ decreases strongly (from 137 MPa, with k_2=40, to 30 MPa). In the same time ε^*, the limit for homogeneous deformation, decreases somewhat and is smaller than the experimental value. This way to account for grain size effect in the hardening behavior should clearly be further calibrated.

5. Conclusion

By considering the tensile curves of a set of cp titanium specimens with several microstructures, especially with several mean grain sizes, it was recognized that the yield stress depends on the grain size following a Hall-Petch relationship. The 'hardening amplitude', defined as $\Delta\sigma = \sigma_{max} - \sigma_e$, shows also some dependence with the grain size. If a crystal plasticity model should account for such effects, it is necessary to include some 'internal length' in it. In this study, where a EVPSC model was used, it was made by adding a term that depends on the individual grain size, in a Hall-Petch manner, to the crss. The yield stresses obtained in the simulations reproduce rather correctly the experimental values, which validates up to some extent the assumption for the crss expression. The simulations lead to interpret the initial plateau, which sometimes appear in the beginning of the plastic deformation, in terms of combined slip activity compared to the dominant prismatic slip activity when no plateau arises. To account for the dependence of hardening on the grain size, a supplementary term was added in the evolution equation of the forest density dislocation. It is expected to amplify the dislocation absorption when the grain size decreases. The simulations show that the three stages for the hardening rate are correctly reproduces but that the 'hardening amplitude' is not enough sensitive to this supplementary term. A short test with a higher value of this term highlights that the parameter which governs the contribution of this term should be significantly increased to obtain more realistic results for the 'hardening amplitude'.

Acknowledgments: The authors thank Mr. Yvon Millet from the company Timet for providing cp Ti sheets and Dr. Laurent Peltier for performing various heat treatments. This work was supported by the French State through the program "Investment in the future" operated by the National Research Agency (ANR) and referenced by ANR-11-LABX-0008-01 (LabEx DAMAS).

References

[1] Kocks U F Tomé C N and Wenk H R **1998** Texture and Anisotropy, Cambridge University Press

[2] Roters F. Eisenlohr P Hantcherli L Tjahjanto D D Bieler T R and Raabe D **2010** *Acta Mat.* **58** p. 1152-1211

[3] Bunge H J Wagner Welch P I and Van-Houtte P **1985** *J. Phys. Lett.* **46** p.L1109-L1113.

[4] Sinclair C W Poole W J and Brechet Y **2006** *Scripta Mat,* **55** p.732-742

[5] Berbenni S Favier V and Berveiller M **2007** *Int. J. of Plasticty* **23** p.114-142

[6] Raeisinia B Sinclair C W Poole W J and Tomé C N **2008** *Mater. Sci. Eng.*, **16,** p.1–15

[7] Fromm B S Adams B L Ahmadi S and Knezevic M X **2009** *Acta Mater* **57** p.2339-2348

[8] Nicaise N Berbenni S Wagner F Berveiller M and Lemoine X **2011** *Int. J. of Plasticity*, **27**, p.232-249

[9] Mareau C and Berbenni S **2015** *Int. J. of Plasticty* **64** p. 134-150

[10] Wagner F Bozzolo N Van Landuyt O and Grosdidier T **2002** *Acta Mat.* **50** p.1245-1259.

[11] Singh A K and Schwarzer R A **2008**, *Trans Indian Inst. Met.* **61** n°5 p.371-387.

[12] Wagner F Ouarem A Richeton T and Toth L S **2018**, to appear in *Adv. Eng. Mat* DOI: 10.1002/adem.201700237

[13] Amouzou K E K Richeton T Roth A Lebyodkin M A Lebedkina T A **2016**, *Int. J. of Plasticity* **80** p. 222-240

[14] Wang L Zheng Z Phuka H Kenesei P Park J-S Lind J Suter R M and Bieler T R **2017** *Acta Mat.* **132** p.598-610

[15] H. Becker H and Pantleon W **2013** *Comput. Mater. Sci.* **76** p.52–59.

[16] Roth A Lebyodkin M A Lebedkina T A Lecomte J-S Richeton T and Amouzou K E K **2014** *Materials Science and Engineering A* **A596** p.236–243.

IOP Conf. Series: Materials Science and Engineering **375** (2018) 012006 doi:10.1088/1757-899X/375/1/012006

Texture and microstructure of HPT-processed Fe-based shape memory alloys

R Chulist[1], M Czerny[1], A Panigrahi[2], M Zehetbauer[2] N Schell[3] and W Skrotzki[4]

[1] Instytut Metalurgii i Inżynierii Materiałowej Polskiej Akademii Nauk, PL-30-059 Krakow, Poland

[2] Physik Nanostrukturierter Materialien, Universität Wien, A-1090 Wien, Austria

[3] Institut für Werkstoffforschung, Helmholtz-Zentrum Geesthacht, D-21502 Geesthacht, Germany

[4] Institut für Festkörper- und Materialphysik, Fakultät für Physik, Technische Universität Dresden, D-01062 Dresden, Germany

E-mail: r.chulist@imim.pl

Abstract. Texture and microstructure evolution of Fe-based shape memory alloys subjected to high pressure torsion (HPT) at room temperature was investigated using synchrotron radiation and electron backscatter diffraction in the scanning electron microscope. As starting material for HPT a directionally solidified Fe-based alloy characterized by a <100> fiber texture and coarse columnar grains with a mean size of about 300 µm was used. Prior to HPT the samples were quenched and heat treated for precipitation hardening for 24 h at 700°C. HPT-processing changes the initial texture to a typical shear texture with dominating C component. As deformation proceeds an intensive stress-induced precipitation of a brittle NiAl-B2 phase takes place. The precipitates form a typical shear texture of *bcc* metals with F component dominating. The C and F components may lead to specific interphase orientation relationships.

1. Introduction

Fe-based alloys belong to a new family of shape memory alloys (SMAs) showing a large superelastic strain at room temperature [1, 2]. The superelastic effect is due to thermoelastic transformation of γ to α'. The alloy with a composition of Fe-28Ni-17Co-11.5Al-0.5Ta-0.05B (in at.%, abbreviated NCATB) with a strong {035}<100> recrystallization texture obtained by rolling and a special thermomechanical treatment shows a superelastic strain up to 13.5% in tensile deformation [1]. The essential prerequisite for achieving large superelastic strains in this system is precipitation hardening with size of precipitates less than 15 nm [1-16]. The main strengthening component of these alloys is the *fcc* phase with composition given by the formula $(Ni, Fe, Co)_3(Al, Ta)$ [10-16]. The boron addition promotes formation of the coherent Ni_3Al-type phase. On the other hand, a small amount of B is added in order to suppress grain boundary precipitation of the brittle and incoherent NiAl-B2 type phase [17].

The superelastic strain has also been observed for NCATB and FeNiCoAlTa (NCAT) single crystals, however, with smaller strain values than those reported for their polycrystalline counterparts [3]. Most recently, an abnormal high reversible strain of 15.3% for FeNiCoAlNb single crystal was observed at 77 K in tension [15]. This value was higher than the theoretical one resulting from the theory of martensite formation. The authors explain it with an unusual reversible twinning mode. Nevertheless, both FeNiCoAl-based single crystals and polycrystals are very promising for the alternative refrigeration concept based on a giant entropy change associated with the diffusionless martensitic transformation.

As a unique property of SMA, superelastic strain, can be considered as a deformation mode activated only along certain crystallographic planes and directions. Therefore, it is strongly anisotropic and directly related with the crystallographic orientation or texture. Depending on various deformation processes different textures can be obtained. Therefore, in order to change the texture, which can drastically change the superelastic properties, high pressure torsion (HPT) processing was applied. In torsion the deformation mode is simple shear, with the amount of shear strain changing along the sample radius from zero at the center to a maximum at the edge of the disc. This allows investigating the influence of different shear strains and strain rates even on one and the same sample. As this kind of severe plastic deformation produces a submicro/nanocrystalline microstructure with a pronounced texture its application leads to properties differing from those of conventional materials. Therefore, to study the effect of HPT on texture and microstructure formation as well as precipitation hardening, the NCATB alloys were investigated using synchrotron radiation and electron backscatter diffraction in the scanning electron microscope.

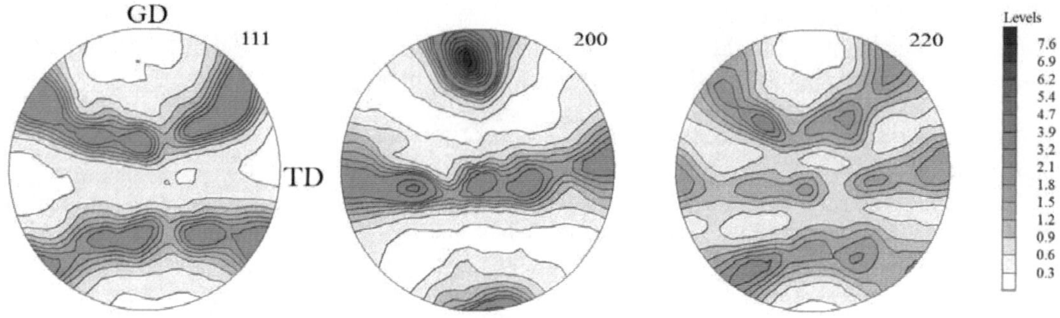

Fig. 1: {111}, {200} and {220} pole figures showing an initial <100> fiber texture. (GD = growth direction).

2. Experimental

The starting material for HPT-processing was as-cast material with Fe-28Ni-17Co-11.5Al-2.5Ta-0.05B composition [1, 4]. Before deformation the alloy was subjected to thermal hardening with rapid cooling and subsequent annealing for 24 h at 973 K [1]. From the initial ingot 3 small discs with a diameter of 8 mm and a height of 0.8 mm were cut by spark erosion. Cutting alignment was done in such a way that the growth direction (GD) was parallel to the normal direction Y of the disc. Subsequently, the discs were deformed at room temperature by HPT up to 0.25, 1 and 2 rotations at a constant pressure of 8 GPa. The texture of the HPT-processed material was measured by diffraction of high-energy synchrotron radiation (87.1 keV) using the HZG materials science beamline P07B at DESY in Hamburg, Germany [18, 19]. A pin of size 1.0 mm × 0.8 mm × 8 mm was cut from each sample along the radial direction for local synchrotron measurements. Such prepared samples were transmitted with synchrotron radiation with a beam size of 0.7 × 0.7 mm along the radial direction starting at the middle of the pin. The shear strain value given corresponds to the middle of the respective analyzed volume. The experimental pole figures were calculated using StressTexCalculator software [20]. The orientation distribution function (ODF) was determined from the measured pole figures {111}, {200}, {220} for austenite and {110}, {200}, {211}, {224} for NiAl-B2 type precipitates using LABOTEX software [21]. To account best for the monoclinic symmetry of simple shear, the crystal and sample reference systems are defined as X ǁ shear direction (SD), Y ǁ shear plane normal (SPN), and Z ǁ transverse direction (TD) [22]. The textures are represented by $\varphi_2 = 0°$ and $\varphi_2 = 45°$ ODF sections, which for face- and body-centered cubic (*fcc* and *bcc*) metals contain all major shear components. After deformation the samples were investigated by scanning electron microscopy and EBSD in order to reveal the microstructure evolution of the NCATB alloys.

3. Results and discussion

The initial cast sample was characterized by long columnar grains with an average size of about 300 μm. The matrix (austenite) had a strong <100> fiber texture along GD, Fig. 1 [23]. Fig. 2 shows EBSD maps taken from the initial material and samples after HPT achieving shear strains of 1.96, 3.92, 7.84

and 15.68. As can be seen, with increasing shear strain the grains become more elongated. Shear deformation changes the shape of the austenitic grains from about equiaxed to elliptical with the long axis oriented towards SD.

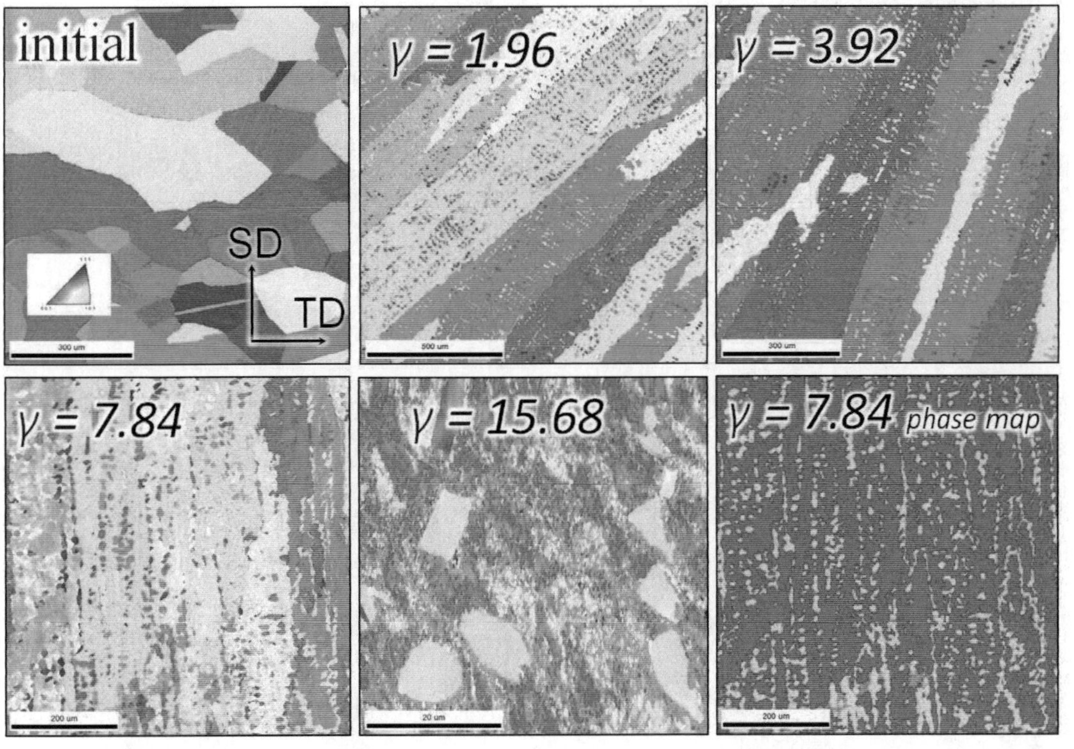

Fig. 2: EBSD maps of the initial NCATB material and subjected to HPT at different shear strains. The phase map shows austenite (red) and NiAl-B2 phase (green) for γ = 7.84.

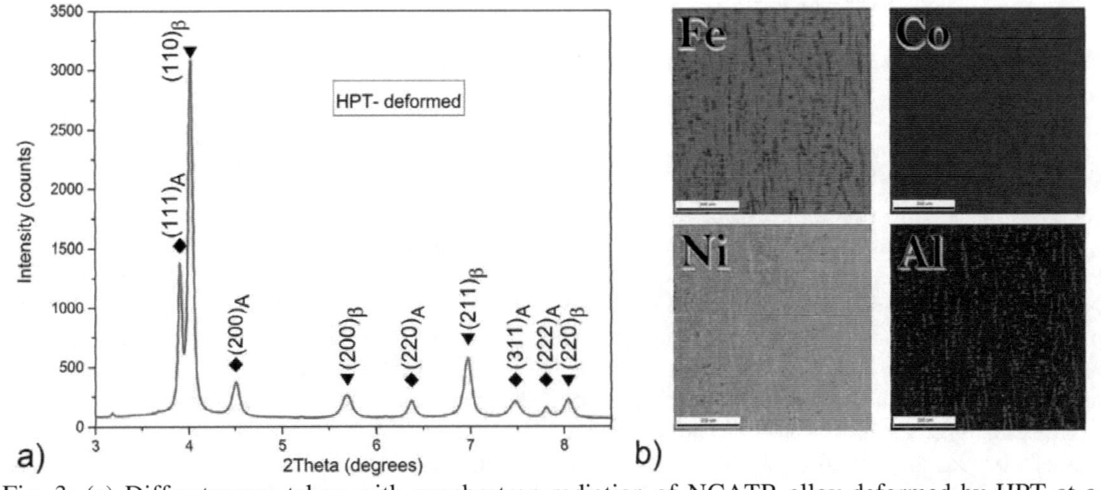

Fig. 3: (a) Diffractogram taken with synchrotron radiation of NCATB alloy deformed by HPT at a shear strain of 15.68 (A – austenite, β – NiAl-B2 phase). (b) EDS maps show the distribution of four main elements Fe, Co, Ni, Al.

Additionally, HPT leads to an intensive precipitation process promoting the NiAl-B2 type intermetallic phase. The high amount of NiAl-B2 phase can be clearly seen in one of the X-ray diffractograms given in Fig. 3a which was used for texture measurements. The same holds for the EDS maps given in Fig. 3b where almost an equal distribution of Ni and Al is unveiled in all precipitates.

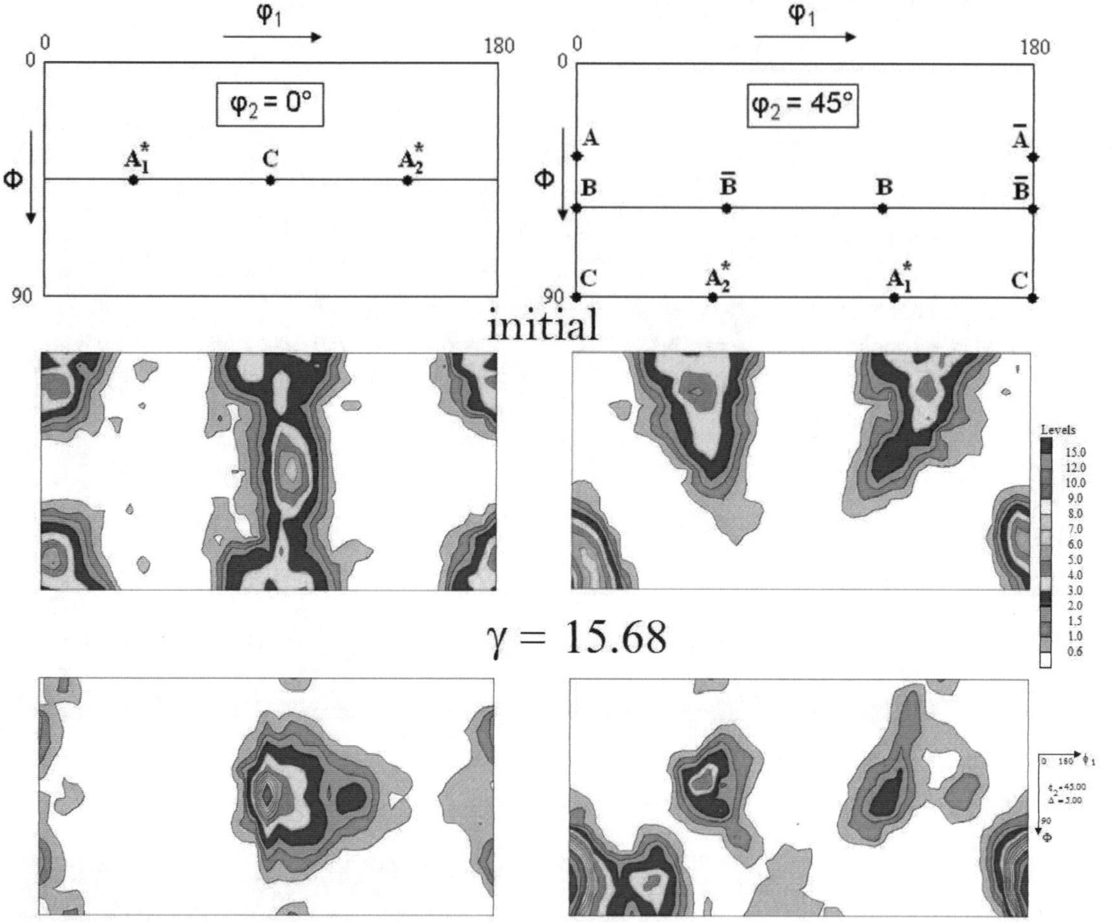

Fig. 4: ODF sections at φ_2 = 0° and 45° of the initial cast sample and after HPT at a shear strain of 15.68 for austenite phase. Key figures give the ideal components of simple shear of *fcc* metals. Intensities are given in multiples of random distribution (mrd).

The HPT process changes the initial fiber texture of the austenitic phase (denoted as A) to a shear texture with typical simple shear components for *fcc* metals with strong C = {100}<110> ({plane parallel shear plane}<direction parallel SD>), Fig. 4. The C component is typical for simple shear of *fcc* metals with high stacking fault energy and may also be promoted by the initial <100> fiber texture. Moreover, HPT performed at room temperature generates a massive precipitation of the NiAl-B2 phase. With increasing shear the volume fraction of this phase increases up to about 31%. Fig. 5 exhibits the ODF sections at φ_2 = 0° and 45° which clearly show that the shear deformation of particles causes a typical shear texture for *bcc* metals. The main component detected is F = {110}<100>, typical for ordered metals with B2 structure, like NiAl [24]. Taking both texture components possible interphase orientation relationships are:

$$\{100\}_{fcc} \parallel \{110\}_{bcc}, <110>_{fcc} \parallel <100>_{bcc}$$

$$\{110\}_{fcc} \parallel \{110\}_{bcc}, <110>_{fcc} \parallel <100>_{bcc}$$

$$\{110\}_{fcc} \parallel \{100\}_{bcc}, <100>_{fcc} \parallel <110>_{bcc}.$$

The orientation relationship between two main shear components for *fcc* and *bcc* metals (C and F) is shown in Fig. 6. The obtained results also show that the addition of 0.05 at.% B was not sufficient to suppress the stress-induced formation of the incoherent NiAl-B2 phase.

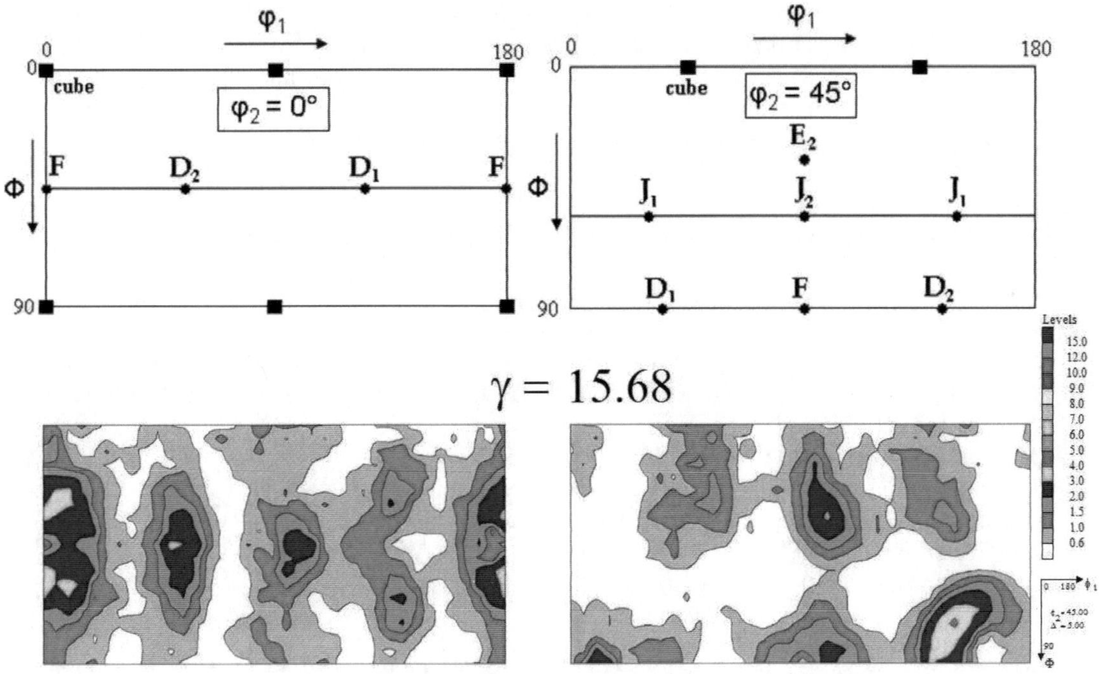

Fig. 5: ODF sections at $\varphi_2 = 0°$ and 45° after HPT of the NiAl-B2 phase at a shear strain of 15.68. Key figures give the ideal components of simple shear of *bcc* metals.

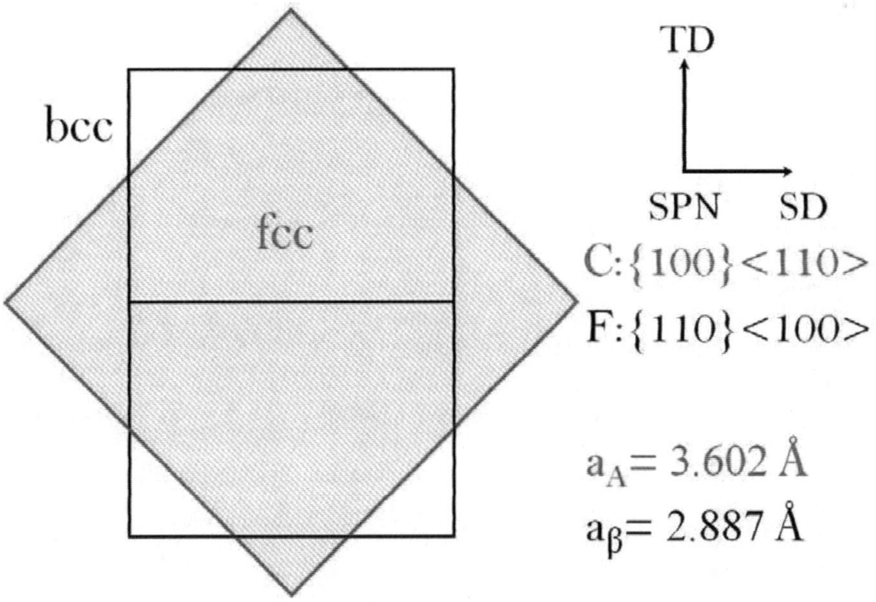

Fig. 6: Orientation relationship between two main texture components (C and F) for *fcc* (red) and *bcc* (black) metals. Lattice parameters determined by synchrotron X-ray diffraction.

4. Conclusions

The results of HPT-processed Fe-based alloys can be summarized as follows:

1. HPT-processing at room temperature leads to an intensive precipitation of NiAl-B2 phase.
2. The texture of HPT-processed Fe-based alloys is characterized by a strong C component for the austenitic phase and a strong F component for NiAl-B2 precipitates. These may give rise to certain interphase orientation relationships.
3. The texture components observed are typical for *fcc* metals with high stacking fault energy and for *bcc* metals of ordered B2 structure.
4. The addition of boron by 0.05 at.% does not suppress the formation of NiAl-B2 phase.

Acknowledgment

The financial support of the National Science Centre of Poland through project 2014/13/D/ST8/03108 is acknowledged.

References

[1] Tanaka Y, Himuro Y, Kainuma R, Sutou Y, Omori T and Ishida K 2010 *Science* **327** 1488
[2] Omori T, Ando K, Okano M, Xu X, Tanaka Y, Ohnuma I, Kainuma R and Ishida K 2011 *Science* **333** 68
[3] Roy AK and Venkatesh A 2010 *J. Alloys Comp.* **577** 393
[4] Chumlyakov YI, Kireeva I, Kretinina I, Kuts I, Kejnih OA, Kirillov KS, , Karaman I, , Maier I and Karaka H Thermoelastic martensitic transformations into new high-strength FeNiCoAlX (X= Ta, Ti, Nb, B) single crystals. *Abstract Book*, ICOMAT (2014) Bilbao Spain
[5] Hornbogen E, Meyer W 1967 Acta Metall. **15** 584
[6] Maki T, in: Otsuka K, Wayman CW (Eds.) 1999 Shape Memory Materials, Cambridge University Press; Revised edition
[7] Sehitoglu H, Karaman I, Zhang XY, Chumlyakov Y and Maier HJ 2001, *Scripta Mater* **44** 779
[8] Kireeva IV, Chumlyakov YI, Kirillov VA, Karaman I and Cesari E 2011 *Tech. Phys. Lett.* **37** 86
[10] Ma J, Kockar B, Evirgen A, Karaman I, Luo ZP and Chumlyakov YI 2012 *Acta Mater.* **60** 2186
[11] Chumlyakov YI, Kireeva IV, Panchenko EY, Kirillov VA, Timofeeva EE, Kretinina IV, Danilson YN, Karaman I, Maier H and Cesari E 2012 *Russ. Phys. J.* **54** 937
[12] Ma J, Hornbuckle BC, Karaman I, Thompson GB, Luo ZP and Chumlyakov YI, 2013 *Acta Mater.* **61** 344
[13] Lee D, Omori T and Kainuma R 2014 *J. Alloys and Compd.* **617** 120
[14] Tseng LW, Ma J, Karaman I, Wang SJ and Chumlyakov YI 2015 *Scripta Mater.* **101** 1
[15] Chumlyakov YI, Kireeva IV, Kutz OA, Karaca HE and Karaman I 2016 *Scripta Mater.* **119** 43
[16] Krooß P, Holzweissig MJ, Niendorf T, Somsen C, Schaper M, Chumlyakov YI and Maier HJ 2014 *Scripta Mater.* **81** 28
[17] Tanaka Y, Kainuma R, Omori T and Ishida K 2015 *Mat. Today Proc.* **2S** 485
[18 Pagounis E, Chulist R, Lippmann T, Laufenberg M and Skrotzki W 2013 *Appl. Phy. Lett.* **103** 111911
[19] Chulist R, Straka L Sozinov A, Lippmann T and Skrotzki W, 2013 *Scripta Mater.* **68** 671
[20] Randau C, Garbe U and Brokmeier H-G 2011 *J. Appl. Crystallogr.* **44** 641
[21] Pawlik K and Ozga P *Gottinger Arb. Geol. Palaont.* SB4 (1999) 146
[22] Toth LS and Molinari A 1994 *Acta Metall.* **42** 2459
[23] Chulist R, Skrotzki W, Oertel C-G, Bohm A, Brokmeier H-G and Lippmann T 2012 *Int. J. Mater. Res.* **103** 575
[24] Tränkner C, Chulist R, Skrotzki W, Beausir B, Lippmann T, Horky J and Zehetbauer M 2014 *IOP Conf. Series Mat. Sci. Eng.* **63** 012154

IOP Conf. Series: Materials Science and Engineering **375** (2018) 012007 doi:10.1088/1757-899X/375/1/012007

Microstructure and Texture Evolution of Magnesium alloy after Shear Assisted Processing and Extrusion (ShAPE™)

M Jamalian[1], V. V. Joshi[2], Scott Whalen[2], Curt Lavender[2], D.P. Field[1]

[1]*School of Mechanical and Materials Engineering, Washington State University, Pullman, USA.* [2]*Pacific Northwest National Laboratory, Richland, USA.*

maryam.jamalian@wsu.edu

Abstract. Shear Assisted Processing and Extrusion (ShAPE™) is a novel procedure used for fabricating thin-walled magnesium tubing. The technique involves pressing a billet through a rotating sleeve, thus causing severe deformation in the metal as the tube forms. This study analyzes the microstructure and crystallographic texture of the alloy in three different states by using electron backscatter diffraction. Notable differences in grain size and texture were observed depending on the location examined within the extrusion. Wall thicknesses of 1.5 and 3.0 mm (0.06 and 0.12 in) were characterized showing that both the grain size and basal plane orientation changed significantly as a function of position through the wall thickness, with abrupt changes near the outside wall. This type of microstructural information can be used to tune the parameters of the forming process to obtain the desired properties.

1. Introduction

With a view to fuel efficiency, the demand for materials with a high ratio of strength to weight increases significantly [1-3]. To this end, magnesium alloys are excellent candidates to be substituted for steel and other metal alloys due to their exclusive mechanical properties such as the lowest density as a metal, noticeable strength, suitable for casting and machining, great damping characteristics, high thermal stability, conductivity, and recyclability [3-4]. However, the lack of sufficient slip systems because of the HCP crystal structure causes restricted ductility and toughness at room temperature [5]. Thus, one of the main obstacles to achieve this goal is forming while optimizing the microstructure to obtain strength and ductility [6]. Recently, severe plastic deformation (SPD) has been used to alter grain size and texture to gain superior mechanical properties [7]. SPD can be applied by different techniques, for instance, shear force in high-pressure torsion [8], compression force in rolling [9], or equal channel angular pressing (ECAP) [10], among others. Size and cost of the product are two important challenges required to be addressed to turn these into appropriate metal forming processes in industry [11-12]. Indirect shear extrusion has been utilized to overcome tension-compression yield strength asymmetry in ZK60 and other magnesium alloys [13,3]. Recently, Shear Assisted Processing and Extrusion (ShAPE), a new direct shear extrusion procedure, has been applied on ZK60 for manufacturing large pipes [12]. It is important to investigate the microstructure along with its temporal and spatial variability in such processes. Electron backscatter diffraction (EBSD) is a robust technique that can be leveraged to quantify grain size, crystallographic texture, grain boundary structure, and deformation character in such materials [14].

IOP Conf. Series: Materials Science and Engineering **375** (2018) 012007 doi:10.1088/1757-899X/375/1/012007

In this paper, the gap between the die and the mandrel (wall thickness) and the number of spiral scrolls (starts) on the die shoulder were varied to determine their effect on the grain size, crystallite lattice orientation, and defect structure for specimens processed using the ShAPE procedure. Hence, three states were investigated, 1) 4 starts and 3 mm (120 mil) wall thickness, 2) 4 starts and 1.5 mm (60 mil) wall thickness, 3) 16 starts and 1.5 mm wall thickness. For each state, at least two samples were examined in different locations along the tube length to ensure sample uniformity.

2. Experimental procedure

2.1. ShAPE procedure

ShAPE is a direct extrusion procedure that applies shear loading for optimizing the grain size and texture. Figure 1 contains a schematic picture of the ShAPE system, in which the extrusion die is fixed and other parts rotate at 200 RPM. The actuator pushes the rotating billet and applies axial force and torque to the extrusion die shoulder while the applied torque, the rotational speed, and the chilled water flow rate through the mandrel control the generated heat. At the critical point, the area of billet touching the extrusion die softens and plastically flows through the spiral scrolls on the extrusion die. According to the combination of applied forces and thermal conditions, a strong gradient in crystallographic texture has been obtained. The texture depends on the thermal conditions, rotational speed, extrusion rate, and the number of spiral scrolls (starts) on the extrusion die. These quantities are expressed in Table 1[12]. In this work, eight specimens from different areas of the tube for three distinct states have been investigated. To study the effect of the wall thickness and number of starts as the two main parameters that control the microstructure in ShAPE, three states were defined. These are shown in Table 2.

Table 1. ShAPE procedure details

Processing Details	
Ram force	40 kN (9000 lbs)
Torque	550 N-m (405 lb-ft)
Power consumption	11.5 kW
Temperature 1 mm from die surface	450°C
k factor	3.33 MPa
Outside diameter of pipe	50.8 mm
Feed rate	3.81 mm/min

Table 2. The three defined states

State's number	Number of starts	Thickness
1	4	3 mm
2	4	1.5 mm
3	16	1.5 mm

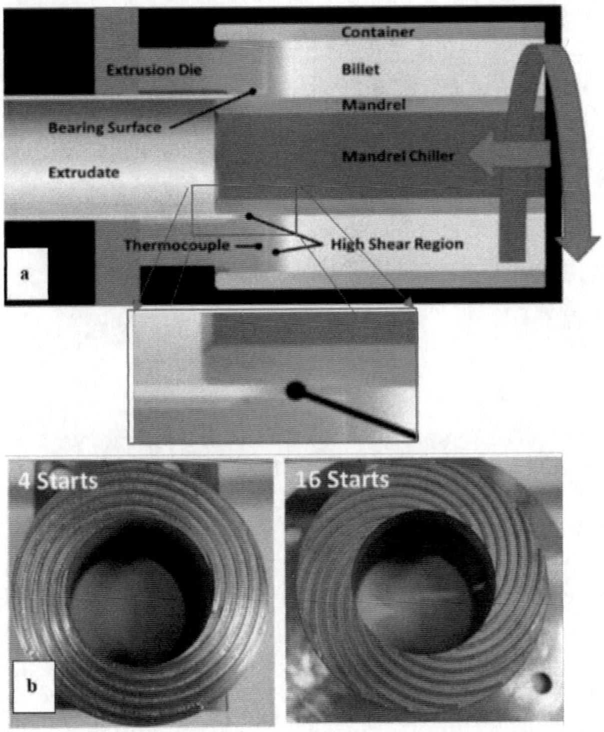

Figure 1 a) Schematic image of ShAPE procedure b) shape and number of starts at shoulder of Die

2.2. Microstructural investigation

The specimens were cut and mounted in the longitudinal direction for the three states. They were mechanically ground and polished to a final step of 0.05 μm colloidal silica on a low nap cloth until the quality of Kikuchi patterns was acceptable. EBSD with an accelerating voltage of 20 KeV and a probe current of ~10 nA was used. For analyzing EBSD data, Orientation Imaging Microscopy (OIM) was leveraged to define grain size, misorientation angles, crystallite lattice orientation distributions, and recrystallization fraction [15]. Grain orientation spread (GOS) was used to obtain a measure of deformation in the grains and to estimate the fraction of recrystallization. GOS of less than 1.3 degrees was used to determine that the grain was recrystallized [16]. In this paper, the blue color represents the recrystallized grains in the GOS images.

3. Result and discussion

3.1. Microstructure analysis

As shown in Figure 2-a, the mass flow lines were created as a function of axial and shear force. These lines effectively show the material flow direction at each position in the structure. The change in grain shape orientations and crystallographic plane direction follows these flow lines. As shown Figure 2-a, the slope of these curves changes smoothly from the inside to the outside of the tube wall, while it abruptly comes back to the initial slope near to the outside edge. Although grains were refined to less than 10 μm, large grains were detected that created the bimodal structure (Figure 2-b). In the orientation map (Figure 2-a), the mass flow line can be observed by the orientation of the major axis of the elongated larger grains and by the path traced by the contiguous fine grains.

IOP Conf. Series: Materials Science and Engineering **375** (2018) 012007 doi:10.1088/1757-899X/375/1/012007

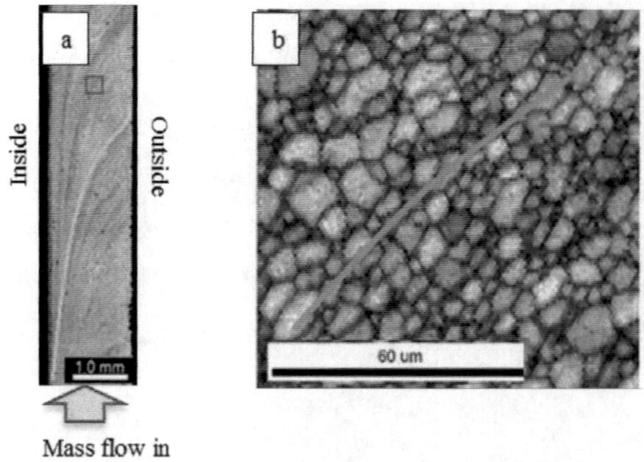

Figure 2 a) Optical micrograph of sample after final polishing step shows the mass flow line from the inside to the outside, b) Image quality map of EBSD near to the center.

In Figure 3, grain structures through the thickness of the three states are shown. As demonstrated, by increasing the number of starts and reducing the wall thickness, the number of large grains decreased. As a result, a more homogeneous structure was obtained except near the outside edge. The average grain size for the three states were 7.3 +/- 3.71 μm, 6.06 +/- 3.06 μm, and 5.94 +/- 3.0 μm, respectively. The error bars are determined by the standard deviation of the distribution. Large grains were detected in two different locations in all the states. The size of these grains was near 20 μm, approximately three times the average grain size, while there were very fine grains (1-3 μm) near these larger grains. Larger grains near the outside edge are created as a result of greater stored energy and higher temperatures in the structure due to frictional forces, allowing for more rapid grain growth. On the other hand, larger grains in the mid-thickness region occurred in conditions where lower stored energy was imposed, including tubes with thicker walls and for processing with lower expected mass flow (smaller number of spiral scroll starts). In tubes with the 1.5 mm wall thickness, the grain size distribution was generally more homogeneous.

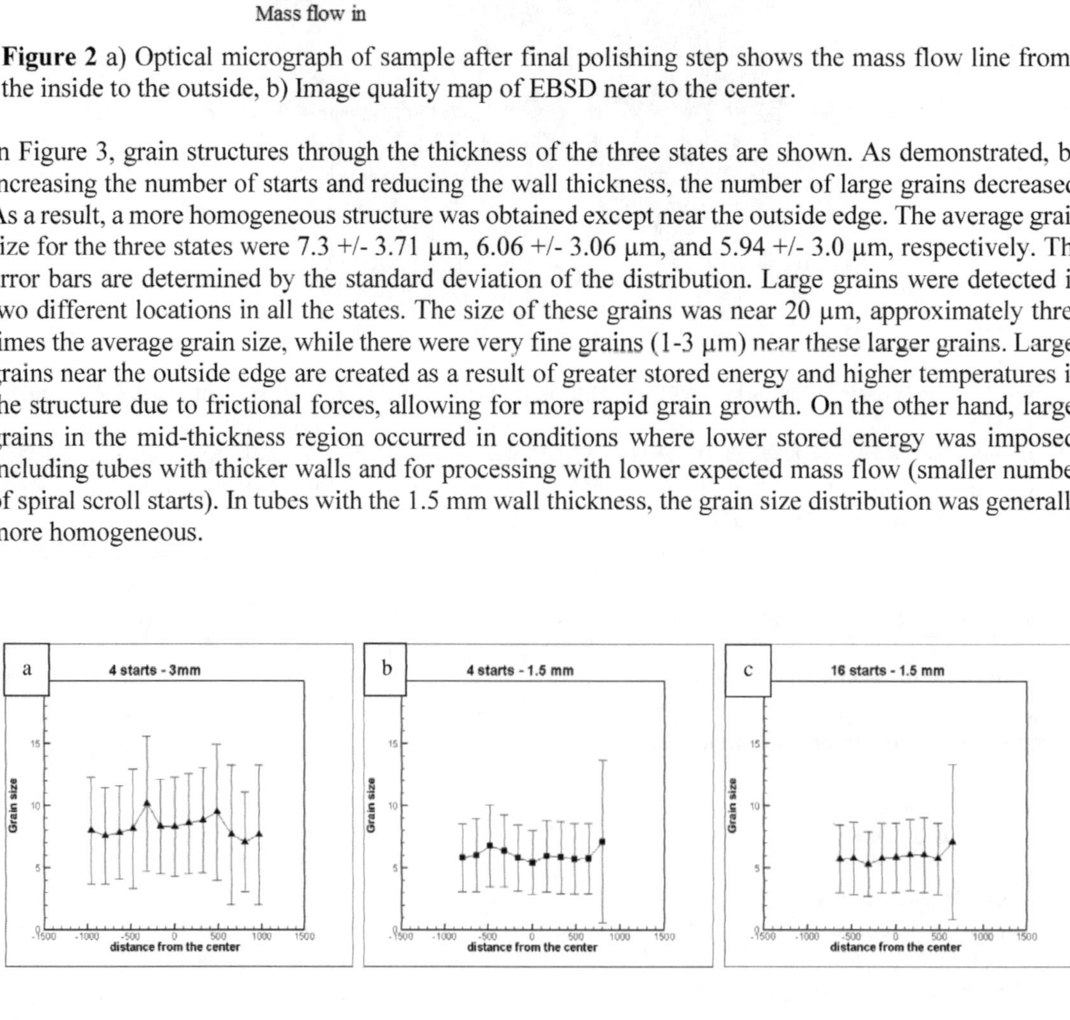

Figure 3 Grain size histogram through the thickness a) 3 mm 4 starts b) 1.5 mm 4 starts c) 1.5 mm 16 starts

3.2. Texture Analysis

All calculations were done with respect to the radial direction. As shown in Figure 4-a, Basal poles were always perpendicular to the material flow line that gradually rotated due to the change of flow line slope. Thus, according to the majority of grain orientations, the thickness could be divided into five different regions each demonstrating a given texture (Figure 4-b). Although the first region demonstrated weak texture at the inner edge, the second region had a strong texture with basal poles lying along the radial direction. On the other hand, region 3 was a transition between regions 2 and 4 that had a c-axis fiber texture at 45 degrees while region 4 had a strong texture approximately along the extrusion direction. Finally, in region 5, a strong texture in a small area near the outside edge was observed where the crystallite lattices were rotated near 90 degrees with respect to those seen in region 4. The misorientation angle histogram for a randomly textured HCP metal is expected to follow that described by Mackenzie [17-18]. The ideal histogram for a perfect basal texture in this material has a constant frequency from 0-30 degrees and nothing outside of that range. The misorientation angle histograms in this structure were consistent with a random texture overlaid on a c-axis fiber texture in the region 1. In contrast, the disorientation distributions were similar and consistent with c-axis fiber textures in the rest of the structure. This pattern was revealed in all three states as observed in Figure 5-a except that in the 1.5 mm pipe, the grains in the extruded direction were eliminated and the region 2 grains grew by increasing the number of starts that led to the stronger texture seen in Figure 5-b.

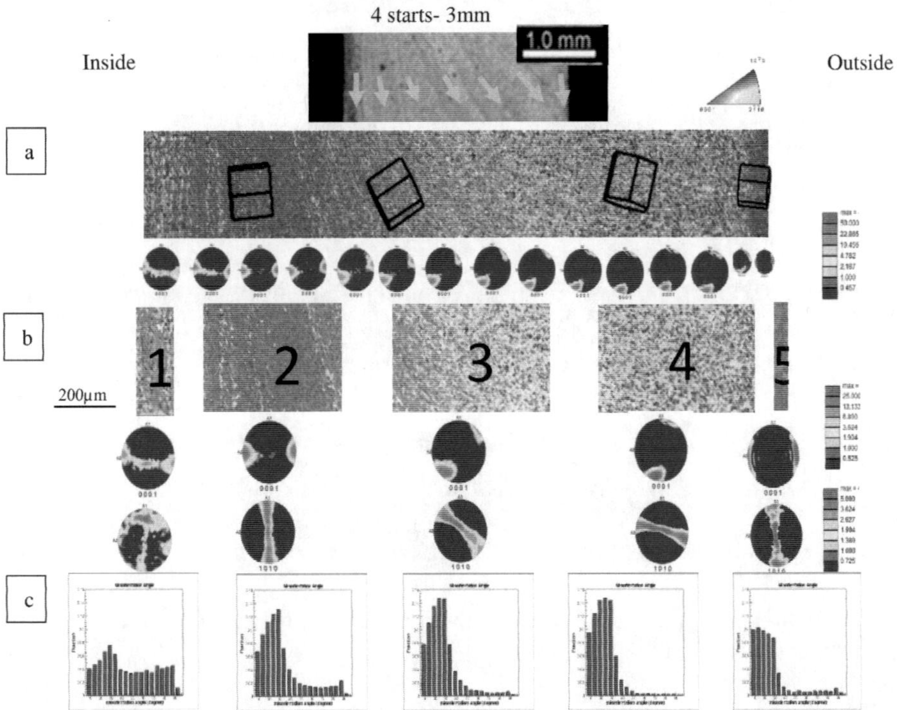

Figure 4 a) Orientation map through the thickness from inside to outside, b) five different texture conditions according to basal plane orientation, and c) misorientation histogram for defining the regions

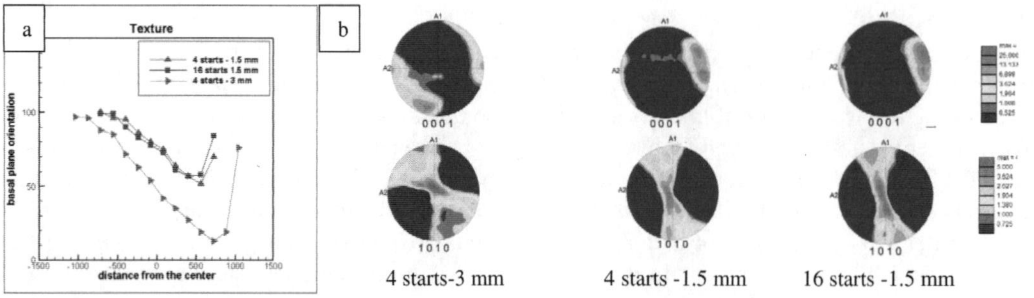

Figure 5 a) Changing basal plane angle along the thickness for the three states, b) (0001) pole figure for 3 mm - 4 starts, 1.5 mm - 4 starts, and 1.5 mm - 16 starts, respectively

EBSD scans were also taken at two positions separated by 20 cm along the tube length. The results showed that as opposed to the radial direction, no significant change was observed, in crystallographic texture or grain size distribution, as shown in Figure 6.

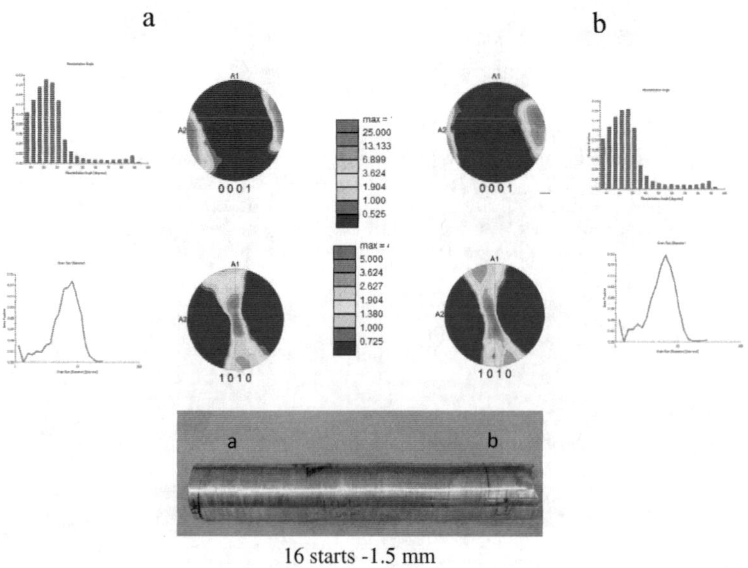

Figure 6 Misorientation histogram, grain size distribution and pole figures ((0001) & (10-10)) at two different locations in the longitudinal direction

3.3. Recrystallization

Although more than 90% of grains were recrystallized as a result of dynamic recrystallization, the number of un-recrystallized grains increased near both edges. According to Figure 7-a all pipes thickness and number of starts did not have a considerable effect on recrystallization. Thus, no significant difference was observed in the three states (Figure 7-b). A closer look reveals that the un-recrystallized grains were larger grains that had been dynamically recrystallized and deformed again as the process progressed. This process was likely repeated in these regions through the entirety of the process. As mentioned, large grains were in two locations including those near the outside wall and those near the middle of the wall thickness. For the recrystallized larger grains near the outside wall, static grain growth likely occurred; therefore, they do not have in-grain orientation gradients.

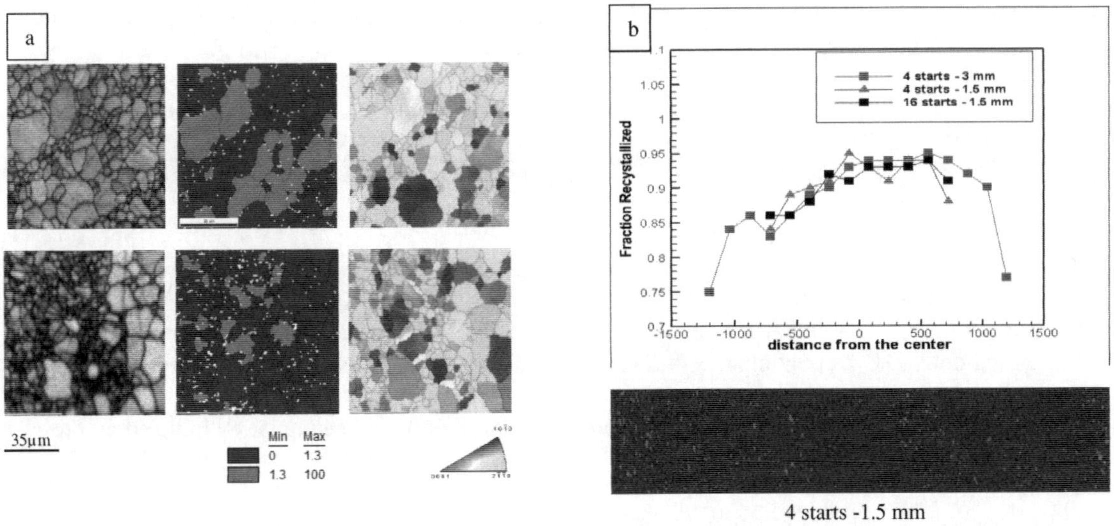

Figure 7 a) Different recrystallization on large grains of two different locations, b) recrystallization through thickness in the three different states

4. Conclusion

Shear Assisted Processing and Extrusion (ShAPE) is a procedure that can easily be scaled up for cost reductions in such processing. ShAPE processing leads to grain refinement that gives a somewhat bimodal structure depending on processing parameters. The basal plane orientations change from 90 degrees to near zero with respect to the extrusion direction, through the wall thickness. However, raising the number of spiral scrolls and decreasing the wall thickness led to more uniform microstructures. Tuning the mentioned parameters successfully addressed the issue of the larger grains in the middle of the wall thickness. However, those near the outer wall thickness persisted, due to the increased strain energy and higher temperatures in these locations.

Acknowledgements

The authors thank Dr. William Joost of the U.S. Department of Energy Vehicles Technologies Office (DOE/VTO) for financial support of this work. The Pacific Northwest National Laboratory is operated by the Battelle Memorial Institute for the United States Department of Energy under contract DE-AC06-76LO1830.

References

[1] M. K. Kulekci, "Magnesium and its alloys applications in automotive industry," *Int. J. Adv. Manuf. Technol.*, vol. 39, no. 9–10, pp. 851–865, Nov. 2008.
[2] J. González, S. Bagherifard, M. Guagliano, and I. Fernández Pariente, "Influence of different shot peening treatments on surface state and fatigue behaviour of Al 6063 alloy," *Eng. Fract. Mech.*, Mar. 2017.
[3] S. Zheng, Q. Yu, and Y. Jiang, "An experimental study of fatigue crack propagation in extruded AZ31B magnesium alloy," *Int. J. Fatigue*, vol. 47, pp. 174–183, Feb. 2013.
[4] V. Manakari, G. Parande, and M. Gupta, "Selective Laser Melting of Magnesium and Magnesium Alloy Powders: A Review," *Metals*, vol. 7, no. 1, p. 2, Dec. 2016.
[5] M. R. Barnett, "Twinning and the ductility of magnesium alloys," *Mater. Sci. Eng. A*, vol. 464, no. 1–2, pp. 1–7, Aug. 2007.

[6] H. Wang *et al.*, "Achieving high strength and high ductility in magnesium alloy using hard-plate rolling (HPR) process," *Sci. Rep.*, vol. 5, no. 1, Dec. 2015.
[7] R. Z. Valiev, Y. Estrin, Z. Horita, T. G. Langdon, M. J. Zehetbauer, and Y. T. Zhu, "Fundamentals of Superior Properties in Bulk NanoSPD Materials," *Mater. Res. Lett.*, vol. 4, no. 1, pp. 1–21, Jan. 2016.
[8] H.-J. Lee, B. Ahn, M. Kawasaki, and T. G. Langdon, "Evolution in hardness and microstructure of ZK60A magnesium alloy processed by high-pressure torsion," *J. Mater. Res. Technol.*, vol. 4, no. 1, pp. 18–25, Jan. 2015.
[9] A. Styczynski, C. Hartig, J. Bohlen, and D. Letzig, "Cold rolling textures in AZ31 wrought magnesium alloy," *Scr. Mater.*, vol. 50, no. 7, pp. 943–947, Apr. 2004.
[10] A. Muralidhar, S. Narendranath, and H. Shivananda Nayaka, "Effect of equal channel angular pressing on AZ31 wrought magnesium alloys," *J. Magnes. Alloys*, vol. 1, no. 4, pp. 336–340, Dec. 2013.
[11] X. Chen, L. Liu, and F. Pan, "Strength improvement in ZK60 magnesium alloy induced by pre-deformation and heat treatment," *J. Wuhan Univ. Technol.-Mater Sci Ed*, vol. 31, no. 2, pp. 393–398, Apr. 2016.
[12] S. Whalen, V. Joshi, N. Overman, D. Caldwell, C. Lavender, and T. Skszek, "Scaled-Up Fabrication of Thin-Walled ZK60 Tubing Using Shear Assisted Processing and Extrusion (ShAPE)," in *Magnesium Technology 2017*, K. N. Solanki, D. Orlov, A. Singh, and N. R. Neelameggham, Eds. Cham: Springer International Publishing, 2017, pp. 315–321.
[13] V. V. Joshi, S. Jana, D. Li, H. Garmestani, E. Nyberg, and C. Lavender, "High Shear Deformation to Produce High Strength and Energy Absorption in Mg Alloys," in *Magnesium Technology 2014*, M. Alderman, M. V. Manuel, N. Hort, and N. R. Neelameggham, Eds. Cham: Springer International Publishing, 2014, pp. 83–88.
[14] A. J. Schwartz, Ed., *Electron backscatter diffraction in materials science*, 2nd ed. New York: Springer, 2009.
[15] J. P. Young, H. Askari, Y. Hovanski, M. J. Heiden, and D. P. Field, "Thermal microstructural stability of AZ31 magnesium after severe plastic deformation," *Mater. Charact.*, vol. 101, pp. 9–19, Mar. 2015.
[16] F. Zhang, A. Ruimi, P. C. Wo, and D. P. Field, "Morphology and distribution of martensite in dual phase (DP980) steel and its relation to the multiscale mechanical behavior," *Mater. Sci. Eng. A*, vol. 659, pp. 93–103, Apr. 2016.
[17] A. Morawiec, *Orientations and Rotations*. Berlin, Heidelberg: Springer Berlin Heidelberg, 2004.
[18] J. K. Mackenzie, "Second paper on statistics associated with the random disorientation of cubes," *Biometrika*, vol. 45, no. 1–2, pp. 229–240, 1958.

Synergistic effect of Mg addition and hydrostatic extrusion on microstructure and texture of biodegradable low-alloyed zinc

A.Jarzębska[1*], M.Bieda[1], J.Kawałko[1,3], P.Koprowski[1], Ł.Rogal[1], R.Chulist[1], B.Kania[1], K.Sztwiertnia[1], W.Pachla[2], M.Kulczyk[2]
[1]Institute of Metallurgy and Materials Science, Polish Academy of Sciences, Krakow, Poland
[2]Institute of High Pressure Physics, Polish Academy of Sciences, Warszawa, Poland
[3]AGH University of Science and Technology Academic Centre for Materials and Nanotechnology, Krakow, Poland

*a.jarzebska@imim.pl

Abstract. The influence of synergic effect of different content of Mg and plastic deformation in the room temperature on microstructure and texture of low-alloyed zinc (ZnMg) was investigated in this work. Cast ZnMg alloys with three different Mg contents were hot extruded at 250 °C and subsequently subjected to hydrostatic extrusion (HE). Obtained alloys were compared to pure zinc. Detailed analysis of microstructure and local texture of ZnMg was performed on the longitudinal cross section of each sample, using scanning electron microscope equipped with electron backscatter diffraction. The macroscopic texture was investigated on the transverse cross section of each sample by means of the X-ray diffraction. Results showed that HE leads to grain refinement level unattainable with classic methods of deformation for both pure Zn and ZnMg alloys. Microstructural observation both with local and macroscopic texture analysis suggests the occurrence of dynamic recrystallization during plastic deformation. Addition of Mg caused major changes in microstructure of deformed materials and enhanced grain refinement. The bimodal structure consisted of grains elongated in the extrusion direction, as well as equiaxed ones was observed on the longitudinal cross section of ZnMg samples. Mg alloying combined with hydrostatic extrusion caused modification of the microstructure and texture in the zinc alloy samples which resulted in significant improvement of mechanical properties.

1. Introduction

In the last decade, zinc has gained a lot of interest due to the discovery of its new application as biodegradable metal suitable for production of temporary implants. According to definition biodegradable metals corrode slowly in human body in such a way that released corrosion products do not cause any harm to the host. After completing its mission to support tissue healing, the biodegradable metal-based implant dissolves completely [1]. Among currently studied candidates for such an application, zinc appears to be the most promising due to its adequate corrosion (below 20 μm/year) rate, placed between magnesium and iron which are respectively too fast and too slow [2,3]. However, zinc possesses mechanical properties too weak to be used for stents. Therefore, it is important to increase the strength of Zn without deteriorating other crucial features. One of the strengthening methods is alloying with essential elements, for example

IOP Conf. Series: Materials Science and Engineering **375** (2018) 012008 doi:10.1088/1757-899X/375/1/012008

magnesium, calcium, or strontium [4,5]. The second method is plastic deformation, which usually is implemented in a form of hot extrusion or hot rolling [6,7]. Such treatments yield improved strength and plasticity approaching required values for stent application. Nevertheless, even the combination of these two methods leading to improved mechanical properties does not ensure meeting the requirements [8]. During deformation conducted in elevated temperatures like hot rolling or extrusion, recrystallization occurs extensively and inhibits strengthening of pure and low-alloyed zinc. It should be mentioned that zinc exhibits a low melting temperature and hence a low recrystallization temperature. Furthermore, as hexagonal close packed material, zinc has only 3 independent slip systems, which imposes difficulties in grain refinement.

In this paper, the hydrostatic extrusion (HE) method was applied to perform deformation at ambient temperature. During the HE the billet is forced through an extrusion die by pressure of surrounding liquid, which limits friction and allows processing of hard-deformable materials. The aim of the proposed work was to investigate the synergistic effect of alloying and plastic deformation on microstructure and texture changes, and to find the relationship between microstructure and mechanical properties.

2. Experimental procedure

ZnMg alloys were prepared by gravity casting of pure zinc (99.99%) and magnesium (99.99%) in metal molds. 27 mm diameter ingots with compositions: ZnMg0.3, ZnMg1 and ZnMg1.5 were conventionally hot extruded (CE) at 250 °C to 10 mm diameter rods afterwards. Before CE, ZnMg1 ingot was annealed at 300 °C for 24 h and subsequently quenched in the water. Pure zinc rod provided by Goodfellow was used as a reference material. Rods after CE were subjected to hydrostatic extrusion (HE) in a single step with reduction R = 4.11, to a final diameter of 4.9 mm. In addition pure zinc and ZnMg1 alloy samples were hydrostatically extruded in a multi-step process from 27 mm to 4.9 mm, using 5 and 4 extrusion passes respectively, in an attempt to investigate the effect of cumulative deformation on texture evolution and grain refinement mechanisms. Within this paper, samples after CE are assumed as the initial state while after HE as the deformed state.

Microstructural investigations by means of SEM/EBSD measurements were carried out on an FEI Quanta 3D FEG scanning electron microscope equipped with an EDAX OIM TSL EBSD system. Observations were performed on longitudinal cross section (LS), parallel to the extrusion direction (ED). 100 x 100 μm^2 maps with step size of 100 or 200 nm were gathered. An analysis of collected orientation data was performed with a grain defined as a set of at least five measurement points, surrounded by a continuous grain boundary segment with a misorientation of at least 15 °. Grain size, misorientation angle, IPF (Inverse Pole Figure) maps and local texture were studied. Macroscopic texture information was obtained from XRD pole figure measurements conducted on a Bruker diffractometer D8 Discovery on transverse cross section (TS), using Co K_α radiation operated at 45 kV and 40 mA.

The mechanical properties were measured in a tensile test carried out on a Zwick/Roell Z250 kN Static Tensile Machine.

3. Results and discussion

The results obtained from SEM/EBSD measurements revealed the influence of both HE and magnesium addition on microstructure changes. The microstructure observation indicated that HE induced grain refinement in all investigated samples. In the case of pure Zn, more than 15 fold reduction in grain size was noticed (fig.1). Moreover, after plastic deformation, elongation of grains in the extrusion direction was observed. This may suggest that dynamic recrystallization did not occur. The addition of magnesium has a positive impact on reduction of grain size, even in

case of the initial material as shown on fig.1. An increase of Mg content lead to a decrease in grain size due to the formation of the intermetallic phase Mg_2Zn_{11}, which strongly inhibits grain growth. Initial samples are characterized by inhomogeneous structure. Grains elongated in the ED, as well as equiaxed ones can be observed. Two groups may be distinguished from equiaxed grains according to their size, small ones (<1 μm) and large ones (>>1 μm). The small ones were located in the vicinity of the intermetallic phase. The presence of the equiaxed grains on the longitudinal cross section after the deformation indicated the occurrence of dynamic recrystallization. The ZnMg1 varied slightly from the other low-alloyed zinc samples in that only large equiaxed grains were observed. Such difference results from different heat treatment after casting in this particular case. This produced large particles of hard-to-deform Mg_2Zn_{11} phase, and affected the plasticity of this material. In general the increase of Mg amount results in an increased number of smaller grains.
Microstructure of alloys after HE is more homogeneous than in their initial state. After the HE process, grains with diameters larger than 10 μm were not found in low-alloyed Zn (Fig. 2).

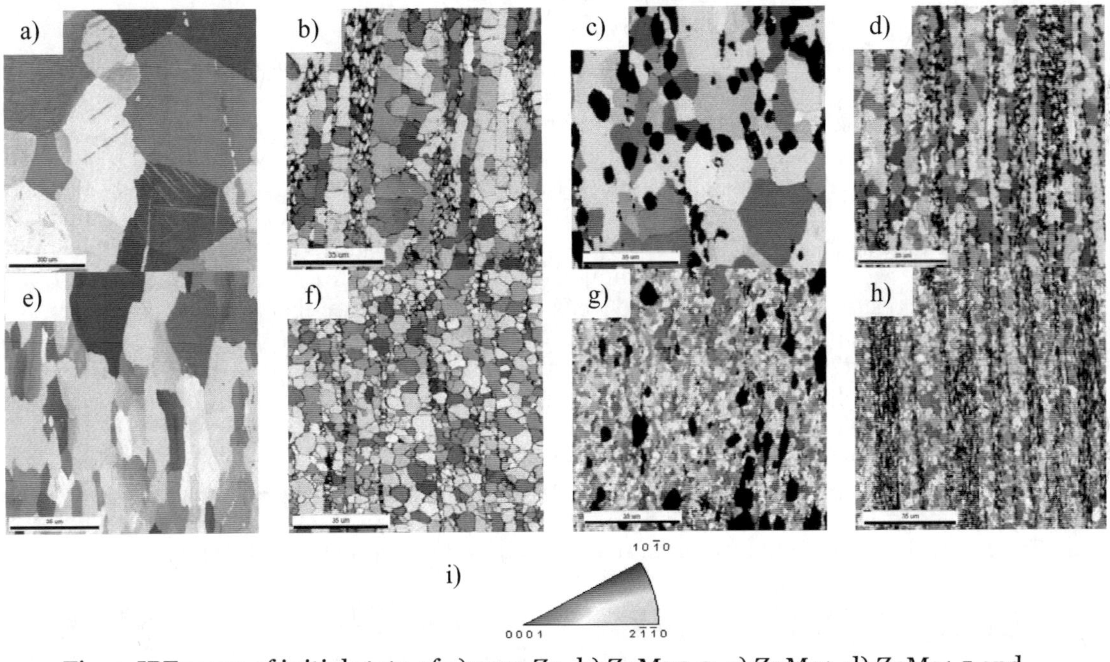

Fig. 1. IPF maps of initial state of a) pure Zn, b) ZnMg0.3, c) ZnMg1, d) ZnMg1.5 and corresponding microstructures after HE: e), f), g), h) respectively. Dark areas on ZnMg alloys maps corresponds to intermetallic phase Mg_2Zn_{11}. All IPF maps were colored based on depicted unit triangle i).

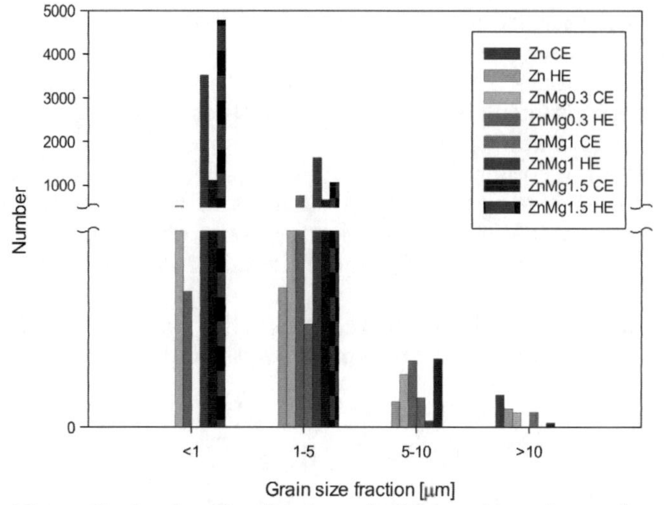

Fig. 2. Grain size distribution of all investigated samples.

Implementation of cumulative HE induced further changes in microstructure. In the case of pure zinc, a bimodal structure composed of two differently sized equiaxed grain groups was observed. This may suggest that additional passes of extrusion caused dynamic recrystallization leading to an average grain size in this sample bigger than that for pure Zn undergoing fewer passes. However, results for ZnMg1 showed a positive effect from cumulative deformation on grain size reduction. A bimodal structure with elongated and equiaxed grains was also noticed. Such microstructure can indicate that dynamic recrystallization occurred only partially. Differences in the microstructure of deformed low-alloyed Zn in one or multiple passes can be referred to the generation of heat during the process. Cumulative extrusion does not require such large pressure as in the case of single pass extrusion, and thus generates less heat. In case of ZnMg1, dynamic recrystallization processes are controlled by obstacles in the form of second phase particles Mg_2Zn_{11} so the cumulative extrusion led to greater refinement of the microstructure contrary to pure Zn. Fig. 3 shows IPF maps of pure Zn and ZnMg1 after cumulative extrusion.

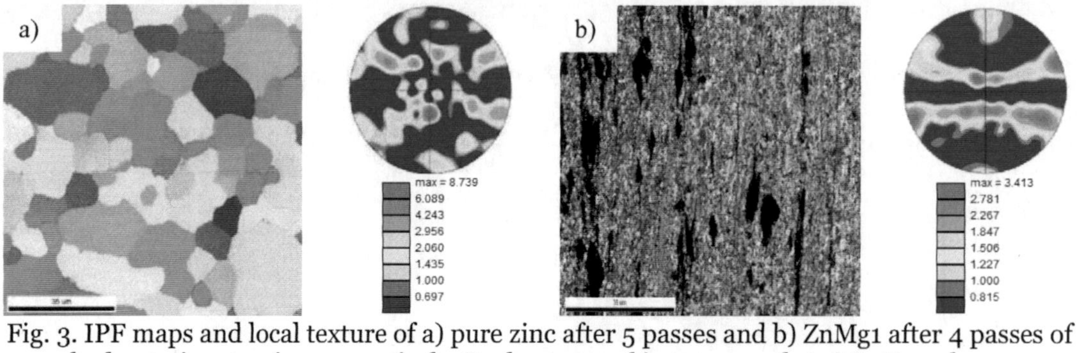

Fig. 3. IPF maps and local texture of a) pure zinc after 5 passes and b) ZnMg1 after 4 passes of hydrostatic extrusion respectively. Dark areas on b) corresponds to Mg_2Zn_{11} phase.

The results of macroscopic texture investigation are presented in fig. 4. In the initial state, a typical cylindrical texture (in which the basal pole is distributed evenly in the plane perpendicular to extrusion direction) for extruded materials was observed [e.g. 9]. However, what was found to be different than in available literature, an additional maximum corresponding to the extrusion direction was also noticed. After HE substantial evolution of texture was noted. The deformation caused formation of a new c-axis component. The cylindrical texture shifted from 90° to 75-85°. Moreover, directions <0001> orientated parallel to the extrusion direction were observed after all. This may suggest that the twinning, besides the slip, is an important mechanism of deformation during HE. During the deformation, twins can undergo fragmentation, so that their typical shapes did not occur in the microstructure, but their typical misorientation angle can be observed (fig. 5). The number fraction of twins increases after HE, especially {11-22} twins. Similar results were obtained for rolled pure titanium [10]. Other studies have revealed that an increasing strain increases the number of twins [10].

The texture characterization results indicate that the addition of Mg did not affect texture. Differences between pure and low-alloyed zinc in both cases of CE and HE are related to grain size, therefore coarse grains provide less information about macroscopic texture.

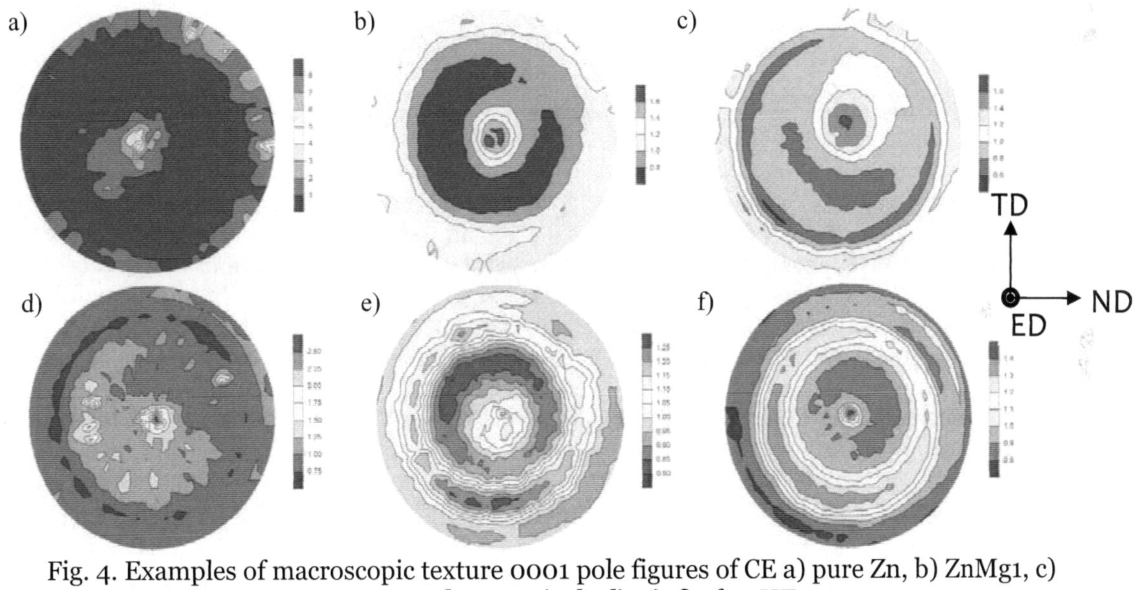

Fig. 4. Examples of macroscopic texture 0001 pole figures of CE a) pure Zn, b) ZnMg1, c) ZnMg1.5 and respectively d), e), f) after HE.

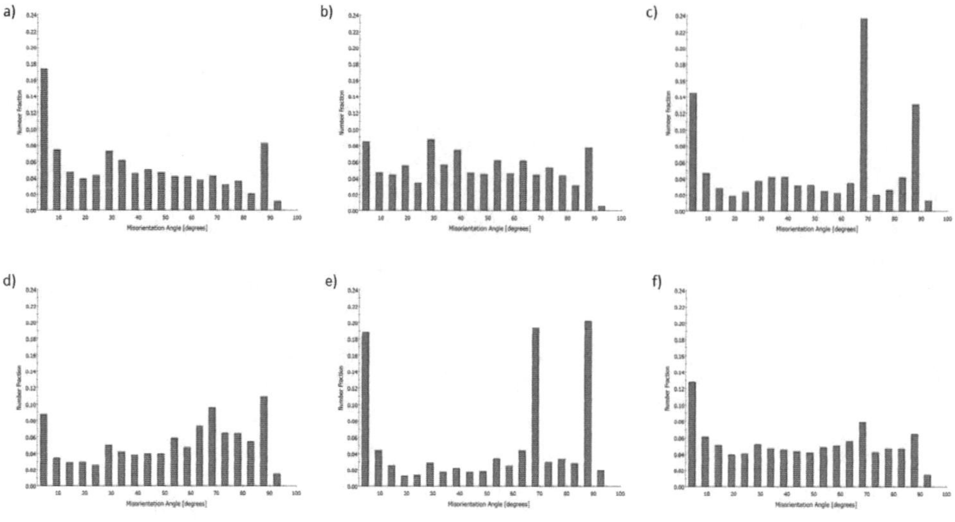

Fig. 5. Misorientation angle distribution of CE a) ZnMg0.3, b) ZnMg1, c) ZnMg1.5 and respectively d), e), f) after HE.

Mechanical testing results showed that HE improved strength and plasticity of pure Zn. However, only synergic effect of Mg addition and HE lead to mechanical properties required for biodegradable stents application. The only instance of decreased elongation was noticed for ZnMg1. The reason of such behavior could be the intermetallic phase, which in the form of large secondary phase precipitates is harder to deform and can propagate cracking. The general improvement in mechanical properties is related to the microstructure. It was observed that a decrease in grain size caused an increase in mechanical properties. This indicates that grain refinement is the main strengthening mechanism of ZnMg alloy. The observed relationship between microstructure and mechanical properties is presented in table 1.

Table 1. Microstructure – mechanical properties relation for all investigated materials.

Sample	Average grain size [μm]	YS [MPa]	UTS [MPa]	E [%]
Design requirements for stent application [3]	< 30	> 200	> 300	> 15
Zn HE	11.1	113	185	17
ZnMg0.3 HE	5.0	350	366	20
ZnMg1HE	2.4	408	431	4
ZnMg1.5HE	1.8	392	433	28
ZnMg1 cumHE [11]	0.7	316	435	35

4. Conclusions

The synergistic effect of hydrostatic extrusion and magnesium addition causes grain refinement, which is the main strengthening mechanism of low-alloyed zinc. Such treatment enabled the mechanical property requirements for stent application to be satisfied. Applying cumulative extrusion to low-alloyed zinc can lead to an even greater reduction in grain size.

The differences in texture of the pure and low alloyed zinc after different deformation processes can be recognized. At the same time no influence of further Mg additions on texture was observed. The misorientation associated with twins was observed but no typical twins shapes were observed in the microstructure. Fragmented twinning may be an additional mechanisms contributing to grain refinement and improved mechanical properties.

Acknowledgement

This work was supported by the National Science Center (Poland) UMO-2016/23/B/ST8/00724.

References

[1] Li H, Zheng Y, Qin L 2014 Prog. Nat. Sci: Mater, Int. **24** pp. 414–422
[2] Bowen P K, Drelich J, Goldman J 2013 Adv. Mater. **25** pp. 2577–2582
[3] Bowen P K, Shearier E R, Zhao S, Guillory R J, Zhao F, Goldman J, Drelich J W 2015 Adv. Healthcare Mater. **5** pp. 1121–1140
[4] Li H F, Xie H, Zheng Y F, Cong Y, Zhou F Y, Qiu K J, Wang X, Chen S H, Huang L, Tian L, Qin L 2015 Sci. Rep. **5** pp. 10719
[5] Niu J, Tang Z, Huang H, Pei J, Zhang H, Yuan G, Ding W 2016 Mater. Sci. Eng. C **69** pp. 407–413
[6] Mostaed E, Sikora-Jasinska M, Mostaed A, Loffredo S, Demir A G, Previtali B, Mantovani D, Beanland R, Vedani M 2016, J. Mech. Behav. Biomed. **60** pp. 581–602
[7] Liu X, Sun J, Qiu K, Yang Y, Pu Z, Zheng Y 2015 J. Alloy Compd. **664** pp. 444-452
[8] Levy G K, Goldman J, Aghion E 2017 Metals **7**, pp. 402
[9] Wang Y N, Huang J C 2003 Mater. Chem. Phys. **81** pp. 11-26
[10] Won J W, Park C H, Hong S G, Lee C S 2015 J. Alloy Compd. **651** pp. 245-254
[11] Jarzębska A, Bieda M, Kawałko J, Rogal Ł, Koprowski P, Sztwiertnia K, Pachla W, Kulczyk M 2018 Mater. Lett. **211**, pp. 58-61

Transmission Kikuchi diffraction from nano-crystalline Ti and TiN thin-films

T Mungole[1], J Zhang[1], B Mansoor[2], G Ayoub[2, 3], D P Field[1]

[1] School of Mechanical and Materials Engineering, Washington State University, Pullman, Washington, 99164-2920, USA, PO Box 642920

[2] Mechanical Engineering Program, Texas A & M University, Doha, Qatar, PO Box 23874.

[3] Industrial and Manufacturing Systems Engineering Department, University of Michigan, Dearborn, Michigan USA 48128-2406

E-mail: tarang.mungole@wsu.edu

Abstract. Nano-crystalline Ti and TiN thin-films fabricated on <100> P-type Si wafers via magnetron sputtering are characterized by transmission Kikuchi Diffraction (TKD) in an FEI Sirion scanning electron microscope (SEM). A simple and cost-effective sample preparation combined with the TKD experiment at working distance of ~ 3 mm and a tilt of 35° produced indexable Kikuchi patterns. Pole figures generated by automatic and manual indexing revealed <0001>|| normal direction (ND) fiber texture and <111>||ND fiber texture in monolithic Ti and TiN thin-films, respectively. Glancing angle x-ray diffraction results corroborated the texture in the samples. Spatial resolution of ~ 60 nm is achieved that is close to the limit of resolution in conventional electron backscatter diffraction (EBSD). It is demonstrated that a simple sample preparation method in TKD technique combined with selected area electron diffraction (SAED) in transmission electron microscope (TEM) may reveal fundamental knowledge about different mechanisms occurring during fabrication and processing of the thin-film systems.

The fact that electron diffraction techniques can be successfully utilized in the determination of crystal structure, lattice strain and the grain orientations is now well established and has matured quite significantly after its first experimental observation in 1927 [1]. These diffraction experiments are now routinely performed in scanning electron microscope (SEM) as electron backscattered diffraction (EBSD) (or with electron channeling pattern (ECP)) and transmission Kikuchi diffraction (TKD); or in transmission electron microscope (TEM) as selected area electron diffraction (SAED) and convergent beam electron diffraction (CBED). EBSD utilizes the backscattered electrons for generating backscattered Kikuchi pattern (BKP) whereas TKD uses transmission Kikuchi patterns in the SEM. The SAED technique uses a parallel electron beam whereas a more detailed analysis is performed using CBED that uses a conical electron beam allowing to reveal full 3D symmetry of a crystal. Although TEM offers a good tool for working in the high resolution regime (few nm or less), it has its limitations such as reduced specimen area observation, specimen thickness limits, poor cost-effectiveness and is preferred mostly for the study of crystal defects. While TKD in the SEM has some of these same limitations, TKD offers a greater field of view and thus provides larger specimen area observation, which often makes TKD a precursor before going further to TEM operations. Furthermore, the crystal orientation measured in TEM is less accurate than in EBSD, latter of which has a stable accuracy of ~ 0.5° [2]. Therefore, recently, there has been a tremendous effort to increase the spatial resolution achievable in SEM for EBSD and TKD techniques [3]. A resolution of 20 nm has been achieved in EBSD [4] whereas in the case of TKD, a much better resolution of ~ 2-5 nm has been reported [5, 6]. However, in general, a spatial resolution of ~20 nm is difficult to achieve with ease in most modern SEMs and EBSD systems. Additionally, the reported resolution values of ~ 2-5 nm in TKD are for the best-case scenario and not an expected practical range for TKD. Nonetheless, due to a specimen tilt angle of ~ 70° in EBSD, the resolution down the tilt direction is about three times lower than the lateral resolution. Therefore, success of EBSD in characterizing nano-crystalline (NC) and nanostructured materials becomes limited.

The fundamental difference between TKD and EBSD is that the former exploits the very small thickness (tens of nm) of the specimen and a tilt of ~20-40° away from the EBSD detector to transmit and emit the inelastically scattered electrons from the back of the specimen. Consequently, Kikuchi pattern is generated from a small diffraction volume that leads to an improved spatial resolution.

High resolution becomes important in the case of characterization of thin-films as they mostly consist of small grain sizes. Metal-ceramic multilayer (MCM) thin-film systems are well-known for their promising mechanical and chemical properties as a result of which they are used widely for extreme environment applications. Exceptional properties are usually achieved in MCMs at nano-scales having domain sizes (i.e. layer thickness or grain sizes) < 10 nm [7, 8]. These remarkable properties are much dependent on their crystal/interface structure. In perspective of the abovementioned limitations of the use of EBSD at NC sizes, we apply TKD technique to successfully analyse NC grain structure and crystallographic texture in thin-films consisting of a metal (Ti) and ceramic (TiN) that leads to a spatial resolution of ~ 60 nm.

Thin-films consisting of monolithic Ti and TiN as well as an MCM consisting of alternating layers of Ti and TiN were fabricated via physical vapour deposition (PVD) by magnetron sputtering on <100> P-type Si wafers. TiN was deposited non-reactively by radio-frequency (RF) sputtering using a TiN compound target (2 in. diameter, Goodfellow, 99.99 % purity) at 300 W and an Ar pressure of 1.7×10^{-3} Torr. Ti was deposited via direct current (DC) sputtering using a Ti target (2 in. diameter, Goodfellow, 99.99 % purity) at 100 W and an Ar pressure of 1.6×10^{-3} Torr. A high modulation ratio (η) (i.e. h_{TiN}/h_{Ti}) of ~ 17.5 was ensured for the MCM sample whereas thickness of both monolithic Ti and TiN films was ~ 90 nm. Thickness of individual Ti and TiN layers in the MCM was ~ 1.7 nm and 33 nm, respectively. A high modulation ratio between ceramic/metal was chosen because of the fact that it facilitates high strain hardening rate [7] leading to remarkable mechanical properties in MCMs, in general.

Structure of the samples were characterized via glancing-angle x-ray diffraction (GXRD) at an incidence angle of 3°, and by diffraction and imaging techniques in SEM and TEM, details of which will be mentioned in the following sections. TEM sample preparation involved scratching of the thin-films by a diamond scribe and placing the powders on carbon coated Formvar® TEM grids. Mechanical characterization is conducted to determine the hardness by Nanoindentation technique using a Berkovich tip as instructed by the Oliver-Pharr method [9]. Monolithic and the multilayer films were re-deposited in order to have the total thickness close to ~ 600 nm to minimize the effect of the substrate during Nanoindentation.

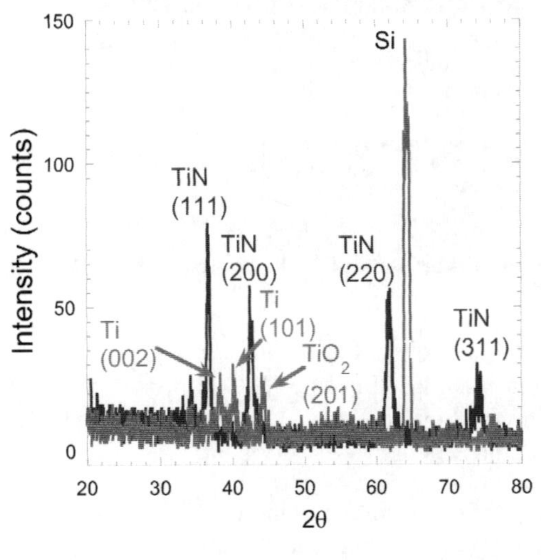

Figure 1. GXRD scan from monolithic TiN and Ti films deposited on <100> P-type Si. Crystalline nature of films is validated. Bragg reflections are clearly indexed showing a weak (111) orientation in TiN film.

Figure 1 shows the glancing angle x-ray diffraction (GXRD) 2θ-scans on monolithic Ti and TiN. Clearly indexed Ti (002), (101) and TiN (111), (200), (220) and (311) Bragg reflections are observed and it is realized that the films are crystalline. Rutile-TiO_2 phase is also detected which is due to high affinity of Ti towards oxygen. A weakly preferred (111) orientation in the monolithic TiN film is observed. Ti film shows a weakly preferred (0001) orientation. Approximate crystallite size was roughly estimated by using Scherrer equation ($D = K\lambda/\beta\cos\theta$), where K is a dimensionless shape factor, λ is the x-ray wavelength, β is full width at half maximum intensity and θ is the Bragg angle. Estimated crystallite domain size is ~ 15 nm for TiN and ~ 40 nm for Ti. It is concluded from the x-ray diffraction experiments that the monolithic films consist of NC structure and follow growth by surface energy minimization.

Owing to the presence of small grain sizes in the thin-films, TKD was conducted on the monolithic Ti and TiN thin-films in an FEI Sirion SEM to evaluate the crystallographic texture and the microstructure. Recently, TKD has been successfully performed on numerous thin-films including that of Ni [10, 11], Al [12], Cu [13], Si [14] and oxide thin-films [15, 16, 17]. A spatial resolution of less than 10 nm has been achieved in several of these experiments. Common sample preparation technique for TKD consists of producing thin foils of the films using a focused ion-beam (FIB). However, in the current investigation, TKD experiments were performed on samples produced by breaking the Ti and TiN thin-films deposited on Si wafer followed by mounting the small wafer pieces on a sample holder consisting of a sweep cut 45° wedge (figure 2a) away from the EBSD detector with a silver paste. Thin-film side of the wafer always pointed away from the incident electron beam. Thorough search for locations with electron transparent sections near to the Si wafer edges consisting of protruding thin-films was conducted afterwards. Although this technique relies substantially on statistical distribution of the electron transparent sections over the edges, given a systematic routine of random observations, it is demonstrated in the current paper that this technique can provide reliable crystallographic data. Similar sample preparation techniques have been implemented in analysing nanoparticles [10, 16]. Due to the presence of 45° wedge on the sample holder, an SEM tilt of 10° was required to achieve a specimen tilt of ~ 35° as shown in figure 2b. A working distance of ~3 mm and beam accelerating voltage ranging from 20-30 kV, depending on the possible different sample thicknesses at different locations, allowed sufficient scattering to produce detectable transmission Kikuchi patterns.

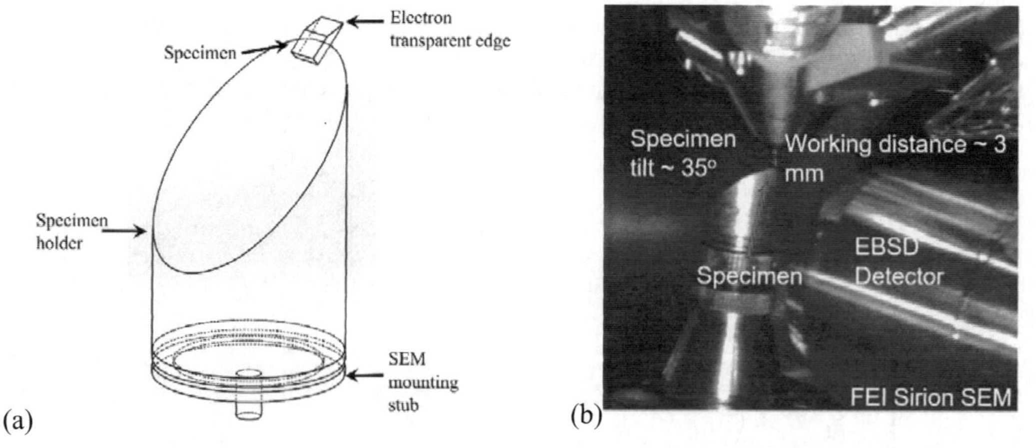

Figure 2. Setup to perform transmission Kikuchi diffraction in an FEI Sirion SEM. (a) broken thin-film sample mounted on a specimen holder and (b) tilted to 35° away from the detector at a working distance of ~ 3 mm. 45° wedge on the specimen holder allows 10° tilt in the SEM.

Figure 3 displays the SEM image of electron transparent sections from the monolithic Ti and TiN thin-films, respective TKD patterns and an image quality map from the Ti thin-film. figure 3a and figure 3b show the region of interest on Ti thin-film (circled) and the produced TKD pattern, respectively. It is realized that the Kikuchi pattern is clearly detectable with (0001) pole pointing towards the normal direction (ND) of the film. figure 3c displays the image quality (IQ) map from the region of interest. The TKD patterns were generally of poor quality in this sample. This is attributed to the sample preparation technique due to which the sample surface normals, where the scattered electrons exit the backside of the sample, don't point coherently towards the detector. Thus, the orientation maps reconstructed from the TKD measurements are of insufficient quality to provide a representative estimate of the microstructure or texture. Nevertheless, the spatial resolution in Ti film can be successfully assessed from the IQ map. IQ map represents the sharpness/detectability of Kikuchi patterns, which is representative of crystalline defects in the sample. Although IQ differences from grain orientations are usually smaller than those due to the presence of grain boundaries, lattice strains and different phases, it is generally understood that grain boundaries produce diffused scattering and hence a low IQ. Therefore, from figure 3c, it can be deduced that relatively brighter regions are separated by dark regions signifying that they may be separated by grain boundaries. It can be noticed that the lateral grain size of ~ 60 nm is detected in Ti thin-film, which is not very different from the value (~ 40 nm) obtained from estimation of crystallite size from x-ray diffraction described earlier.

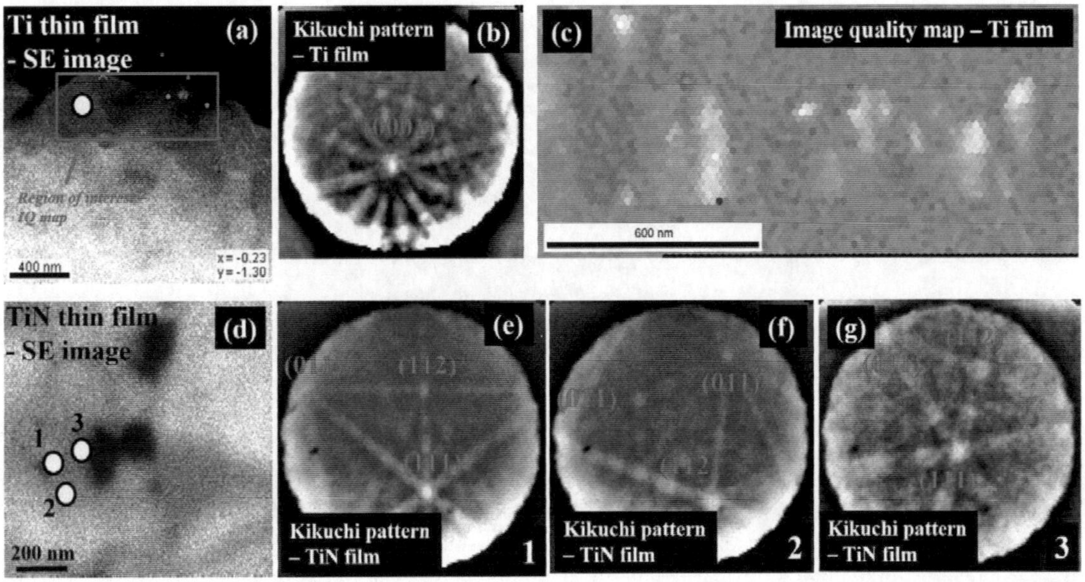

Figure 3. Results from transmission Kikuchi diffraction (TKD) on monolithic Ti and TiN thin-films showing region of electron transparent section in (a) Ti and (d) TiN, produced TKD patterns from (b) Ti and (e,f,g) TiN and (c) image quality map from the Ti thin-film. Region 1 and 2 produced faint TKD patterns (e,f) while from a thinner section at region 3 produced brighter TKD pattern (g). Spatial resolution of ~60 nm in Ti and ~80 nm in TiN thin-film can be estimated.

Figure 3d shows similarly an SEM image from the electron transparent region from the TiN thin-film. figure 3e and 3f show the indexed TKD patterns that are obtained from region 1 and 2 shown in figure 3d, respectively. These regions are relatively thicker as they are far from the edge of the electron transparent region of the TiN thin-film. It is noticed that the TKD patterns are less sharp and smeared out compared to that obtained in figure 3b. Region 3 is closer to edge which results in diffraction from a thinner section of the sample producing a sharper TKD pattern as shown in figure 3g. A lateral resolution of ~ 80 nm is obtained from the TiN thin-film. Additionally, detectable patterns were produced only from a limited region as shown in figure 3a and 3d and not from relatively thicker sections of the sample due to a higher as-deposited film thicknesses of ~ 90 nm for both Ti and TiN films. Furthermore, as the sample preparation technique involved breaking of coated Si wafers, thus facilitating an irregular path of crack propagation, that resulted in a majority of sample surface NDs facing away from the EBSD detector. These limitations combined with smeared TKD patterns inhibits automatic indexing of the patterns. This limits the diffraction data obtained from the TKD experiment and prevents a precise calculation of spatial mapping of the orientations. Relatively low spatial resolution (~ 60 nm for Ti film and ~ 80 nm for TiN film) obtained in the current study is a result of such an artefact. The resolution values have been calculated respectively via the smallest grain in the IQ map and the average distance between points 1, 2 and 3 in figure 3d. These points are separated by the least distance where Kikuchi patterns start to deviate significantly from each other. Nevertheless it can be concluded that in the case of thin-films, a relatively simple sample preparation process combined with TKD experiments can be used to estimate microstructural information from NC grains in a thin-film system. Operation of conventional EBSD in these samples failed to produce any indexable Kikuchi patterns, thus implying an important role of TKD in the characterization of the thin-films.

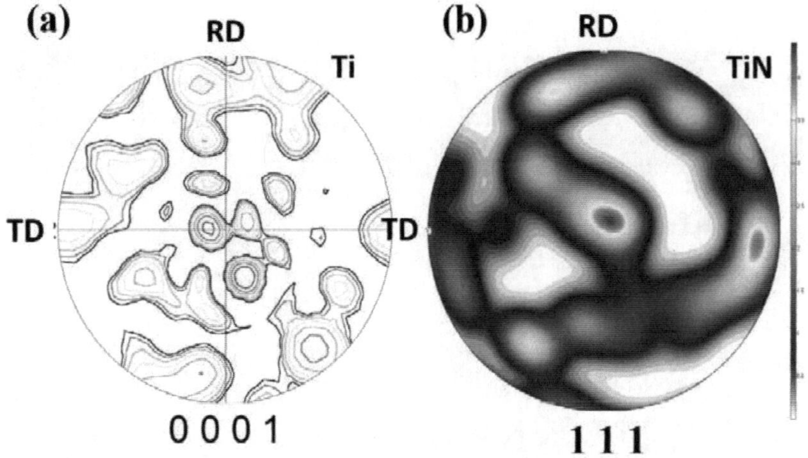

(c)

	φ_1	Φ	φ_2
1	38	48	27
2	34	51	216
3	24	113	189
4	17	19	0
5	17	19	0
6	309	74	51
7	285	108	45
8	329	84	72
9	120	125	144
10	120	125	144
11	120	125	144
12	120	125	144
13	195	125	144
14	216	132	168

Figure 4. Pole figures generated from automatic pattern indexing from (a) monolithic Ti and (b) from manual indexing from monolithic TiN thin-films via transmission Kikuchi diffraction (TKD). Approximate <0001>||ND and <111>||ND fiber texture are observed in Ti and TiN, respectively. (c) shows the raw Euler angles obtained from manual indexing in TiN thin-film to create the pole figure in (b).

Figure 4 shows pole figures obtained from analysis of the TKD patterns generated from monolithic Ti and TiN thin-films. Analyses were done using TSL OIM© v 7 and MTEX (in MATLAB©) software packages. As mentioned earlier, only a limited region on the sample produced detectable TKD patterns out of which only a few were automatically indexable. Therefore, a statistically significant data for pole figure calculation is contingent on the amount of data collected which is restricted by the relatively rigorous procedure of manual indexing. Nevertheless, as shown in figure 4a, automatic indexing of the TKD patterns produced crystallographic texture data representing a <0001>||ND fiber texture in monolithic Ti thin-film. This is in coherence with a weak basal texture obtained from x-ray diffraction data in figure 1. Pole figure for the monolithic TiN thin-film shown in figure 4b was generated after manually indexing of the TKD patterns. Crystallographic texture representing a <111>||ND fiber texture is observed. This is also in coherence with the x-ray diffraction data representing a weak (111) texture in TiN film. Figure 4c shows the Euler angles obtained after manual indexing of the patterns from TiN film, used to create pole figure in figure 4b. It should be noted that as the number of TKD patterns are not statistically significant, the pole figure data is not representative of the true crystallographic texture of the thin-films. This insufficiency is observed in approximate fiber texture represented in figure 4a and 4b. Nonetheless, texture calculations from the aforementioned procedure of sample preparation combined with TKD also enables an estimation of true crystallographic texture in thin-film system consisting of NC grains and nanostructures.

Figure 5 represents results from TEM and Nanoindentation analysis on the samples. It is clear from the cross-sectional bright field TEM image (figure 5a) of the multilayer sample that the microstructure consists of NC grains in TiN layer with approximate grain size of ~ 5-10 nm. The grain size is however very different than that obtained from the TKD analysis of monolithic TiN film. This can either be attributed to the effect of multilayering of the Ti and TiN layers inhibiting grain growth during the deposition process or to a relatively low resolution (~ 80 nm) obtained in the TKD analysis of the TiN sample as a consequence of limitations due to the sample preparation technique. The inset in figure 5a shows the selected area diffraction pattern (SADP) obtained from diffracting a significant volume of the multilayer sample in TEM. Clearly indexed diffraction rings as (111), (200), (220) and (311) pertaining to TiN layer are distinguished.

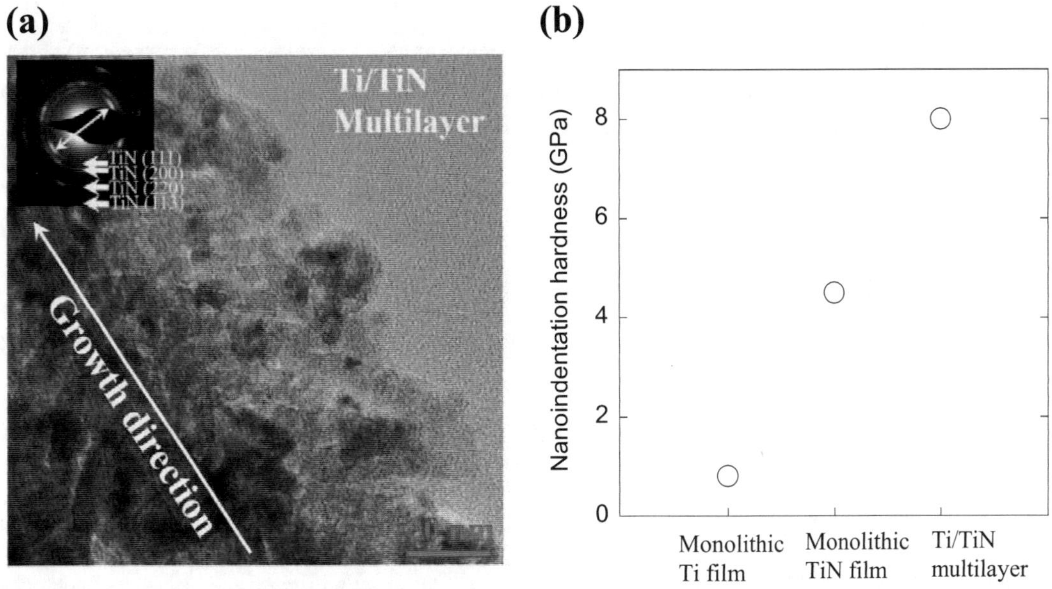

Figure 5. Characterization of the thin-film samples via (a) TEM showing in the inset the SADP image and (b) Nanoindentation showing hardness of the samples. It is realized that TiN layers in the multilayer sample has crystallite size ~ 5 nm with an in-plane orientation of (111) crystallographic planes leading to expected texture softening. Growth occurs by interface energy minimization in the multilayer sample. Multilayer hardness shows opposite trend depicting interface strengthening mechanisms.

It is however noted that the TiN layer does no longer show a fiber texture. Instead, (111) planes lie parallel to the ND of the film indicating an in-plane orientation of the closest packed planes. This transition of texture in TiN layer depicts a possible transition of growth mechanism during deposition from surface energy minimization in monolithic thin-films and a interface energy minization during multilayering of the two films. Also, the orientation of (111) planes lying parallel to the ND in TiN layer would predict texture softening in the multilayer sample as opposed to a hard orientation in the monolithic TiN thin-film pertaining to a low Schmid factor (i.e. <111>||ND) in the case compressive loading. figure 5b shows the Nanoindentation hardness of the samples. It is readily observed that the hardness of the samples increases significantly in the multilayer sample compared to its monolithic counterparts. However, the softening effect due by crystallographic texture predicts a lower hardness for the multilayer sample than predicted by the rule-of-mixtures. This indicates that there are other strengthening mechanisms such as interface acting as barriers to dislocations transmission during the deformation of the multilayer than in the monolithic Ti and TiN samples. It has been shown that interfaces act as significant obstacles to dislocation propagation for nano-scale multilayered structures, owing to the application of classical Hall-Petch relation extended to the realm of multilayered structures [8, 18].

In summary, it is shown that thin-films consisting of NC grains and nanostructures can be characterized to obtain approximate information about the microstructure and crystallographic texture via a simple and cost-effective sample preparation technique combined with TKD technique in an SEM. In spite of several shortcomings of this procedure, a spatial resolution of ~ 60 nm is achieved, making TKD an essential tool for investigating detailed crystallography of NC thin-films. Moreover, combination of TKD technique with other diffraction routines such a SAED in TEM in conjunction with mechanical characterization may reveal fundamental knowledge about different mechanisms occuring during fabrication and processing of the thin-film systems.

Acknowledgements

The authors acknowledge the support provided by Qatar National Research Fund (QNRF) under NPRP grant # 7-1470-2-528 for this work. QNRF is a constituent member of the Qatar Foundation.

References

[1] Thomson G P and Reid A 1927 *Nature* **119** 890
[2] Humphreys F J 2001 *J. Mater. Sci.* **36** 3833–54
[3] Suzuki S 2013 *JOM* **65** 1254–63
[4] Troost K and Kamminga J –D 1994 *Proc. Microscopy and Microanalysis, The Microscopy Society of America, USA* 606
[5] Trimby P W 2012 *Ultramicroscopy* **120** 16–24
[6] Brodusch N, Demers H and Gauvin R 2013 *J. Microsc.* **250** 1–14
[7] Wang J and Misra A 2014 *Curr. Opin. Sol. St. Mater. Sci.* **18** 19–28
[8] Misra A, Hirth J P and Hoagland R G 2005 *Acta Mater.* **53** 4817–24
[9] Oliver W C and Pharr G M 2011 *J. Mater. Res.* **7** 1564–83
[10] Keller R R and Geiss R H 2012 *J. Microsc.* **245** 245–51
[11] Kacher J, Hattar K and Robertson I M 2016 *Mater. Sci. Eng. A* **675** 110–9
[12] Zhang S Y, Zhang Y J, Kwek W M, Goi L S, Trigg A D and Tang L J 2014 *Proceedings of the 16th Electronics Packaging Technology Conference, EPTC*
[13] Lee S Y, Guim H U, Kim D I, Joo Y C, Shim C H, Ahn J P, Choi I S and Abbasi M 2017 *Scr. Mater.* **138** 52–6
[14] Fundenberger J J, Bouzy E, Goran D, Guyon J, Yuan H and Morawiec A 2016 *Ultramicroscopy* **161** 17–22
[15] Mortazavi N, Esmaily M and Halvarsson M 2015 *Mater. Lett.* **147** 42–5
[16] Geiss R H, Rice K P and Keller R R 2013 *Microsc. Microanal.* **19** 696–7
[17] Garner A, Gholinia A, Frankel P, Gass M, MacLaren I and Preuss M 2014 *Acta Mater.* **80** 159–171
[18] Misra A, Verdier M, Lu Y C, Kung H, Mitchell T E, Nastasi M and Embury J D 1998 *Scr. Mater.* **39** 555–60

IOP Conf. Series: Materials Science and Engineering **375** (2018) 012010 doi:10.1088/1757-899X/375/1/012010

Pseudo-texture memory in AISI 321 austenitic stainless steel

A A Tiamiyu*, J A Szpunar, A G Odeshi
Department of Mechanical Engineering, University of Saskatchewan, Saskatoon, SK, Canada
E-mail: ahmed.tiamiyu@usask.ca (A.A. Tiamiyu).

Abstract. In this study, a 45 μm-sized AISI 321 austenitic stainless steel was cryo-rolled at -196 °C to 50 % thickness reduction to completely transform the γ-austenite phase to α′-martensite. The cryo-rolled sample was subsequently annealed in the temperature range of 923 – 1073 K for 30-28800s (0.5 – 480 mins) to reverse the α′-martensite back to γ-austenite and develop ultrafine grain (UFG) structure at an optimum annealing condition. Average austenite grain sizes of 0.22 and 0.31 μm were obtained at an optimum annealing conditions of 923 K for less than 480 minutes and 1023 K for 10 minutes, respectively. High resolution EBSD and XRD texture results shows that ζ-fibre ({110}<uvw>) is the major texture component of austenite grains in the UFG structure and it is stronger in specimen annealed at 923 K than those of higher temperatures (973, 1023 and 1073 K) used in this study. The strong intensity of ζ-fibre in UFG is attributed to pseudo-texture memory in AISI 321 i.e. the ability of the steel to memorize the crystallographic orientation of the deformed austenite, rather than the conventional as-received austenite phase that is random.

1 Introduction

Texture memory effect (TME) is a phenomenon which occurs when the texture of material strengthens or remains unchanged after the phase transformation cycle from one phase to another and back to the initial phase [1,2]. TME has mainly been observed after the α′ (bcc) → γ (fcc) → α′ (bcc) transformation in BCC metals [2,3] and α (hcp) → β (bcc) → α (hcp) transformation in HCP metals [4,5] during a heating and cooling cycle. On the completion of transformation cycle in 0.1% C-1% Mn steel [2], a BCC metal, {100}<011>, {311}<011> and {332}<113> texture components reappeared with little decrease in intensity. Other alloys in which TME has been reported includes interstitial free (IF) steel [3], Ti–6Al–4V [4] and zirconium [5]. The orientation relationship (OR) between γ and α′ in steels is the Kurdjumov–Sachs (K–S) relation, $\{111\}_\gamma \| \{110\}_{\alpha'}$ and $<110>_\gamma \| <111>_{\alpha'}$, or the Nishiyama–Wasserman (N–W) relation, $\{111\}_\gamma \| \{110\}_{\alpha'}$ and $<112>_\gamma \| <110>_{\alpha'}$. There are about 24 and 12 crystallographic variants in K–S and N–W relations, respectively. In the aforementioned metals that exhibit TME, crystallographic OR is obeyed between parent and product phases. Thus, implying that out of the dozens of the equivalent crystallographic orientations allowed under the OR, only specific ones are selected after the completion of the phase transformation cycle. This phenomenon is termed variant selection [2] and its description can be found elsewhere [6,7]. Although the mechanism behind TME is still not very clear, mechanisms related to transformation strain, role of precipitates, and development of special boundaries during heating have been proposed [1,2].

In the present study, UFG structure was developed in AISI 321 austenitic stainless steel (ASS) to improve its mechanical properties. AISI 321 is a titanium (Ti)-stabilized austenitic stainless steel that is widely used in chemical process plants [8] and fabrication of pressure vessels [9]. To the best of the authors' knowledge, the development of UFG structure and the accompanying crystallographic texture in AISI 321 austenitic stainless steel has not been well addressed. The objective of this study is to investigate texture evolution during the development of UFG structure in AISI 321 steel. For the first time, the mechanism of texture memory in AISI 321 is addressed.

2 Material and methods

The nominal composition of the hot-rolled AISI 321 austenitic stainless steel plate used in this study is 17.61 Cr, 9.17 Ni, 1.56 Mn, 0.42 Mo, 0.40 Si, 0.36 Ti, 0.30 Cu, 0.15 Co and 0.044 C in weight percent. The average grain diameter of the as-received steel is 45±5μm. As-received sample was solution treated by soaking at 1373 K for 30 minutes and subsequently quenched in water at room temperature. The solution-treated sample was soaked in liquid nitrogen (77 K) for 25 minutes, followed by multi-pass unidirectional rolling. The steel specimen was intermittently re-soaked to avoid temperature rise during the multiple pass rolling. To obtain complete transformation and crushing of α′-martensite, cryo-rolling up to 50% thickness reduction (r = 50%) was conducted on 105 mm x 33 mm x 11 mm plates. The cryo-rolled samples were reversed to austenite (γ_r) by annealing in the temperature range of 923 – 1073 K for 30-28800s (0.5 – 480 mins). UFG structure developed in specimens annealed at 923 K and 1023 K. For brevity, only the results of specimens annealed at 923 K and 1023 K for 60-28800s (1-480 mins) and 30-1200s (0.5-20 mins), respectively, are presented in section 3. The specimens were electrolytically polished and subsequently characterized by X-ray diffraction (XRD) and electron back-scatter diffraction (EBSD) techniques. EBSD measurements were conducted using an SU 6600 Hitachi field emission scanning electron microscope coupled with an Oxford Instruments Nordlys Nano EBSD detector. The volume fraction of BCC α′ was obtained using a Feritscope MP30E. The actual martensite content was determined by multiplying the Feritscope reading by a correction factor of 1.7 [10]. Hardness measurement was conducted using Mitutoyo Micro Vickers hardness testing machine (MVK-H1). All microstructural analysis and measurements in this study were conducted on the rolling plane (RP).

3 Results and discussion

Figure 1a shows a random texture in the γ-austenite phase of the as-received AISI 321 ASS. The as-received AISI 321 sample was slightly deformed (early stage of rolling) to avoid the start of martensitic transformation and the resulting deformed structure is hereafter referred to as the deformed austenite (γ_d). The EBSD IPF map of the γ_d grains (Fig. 1b) showed a significant texturing near the ζ-fibre ({110}<uvw>). The strong development of {110}<uvw> could be explained using grain B in Fig. 1b. This grain shows a possible orientation interplay between {110}||ND and {111}||ND i.e. the gradual rotation of {111}||ND grains during deformation to {110}||ND, which is the most stable end-orientation. Schmid factor (SF) map using {111}[1-10] as the active slip system was plotted to establish the correlation between plastic deformation mechanism (slip) and the orientation of γ_d (Fig. 1c). The SF varies with orientations in the following order: {100}||ND (highest) > {110}||ND > {111}||ND. This indicates that the ease at which slip will occur on {111}[1-10] slip system for the grains with {111}||ND orientation is lowest and as a matter of fact, these set of grains are near hard orientation. The {111}||ND oriented-grains are then "forced" to rotate during deformation to a more stable orientation ({110}||ND orientation with higher SF) [11]. Comparing the spread of orientation in the inverse pole figure (IPF) of undeformed

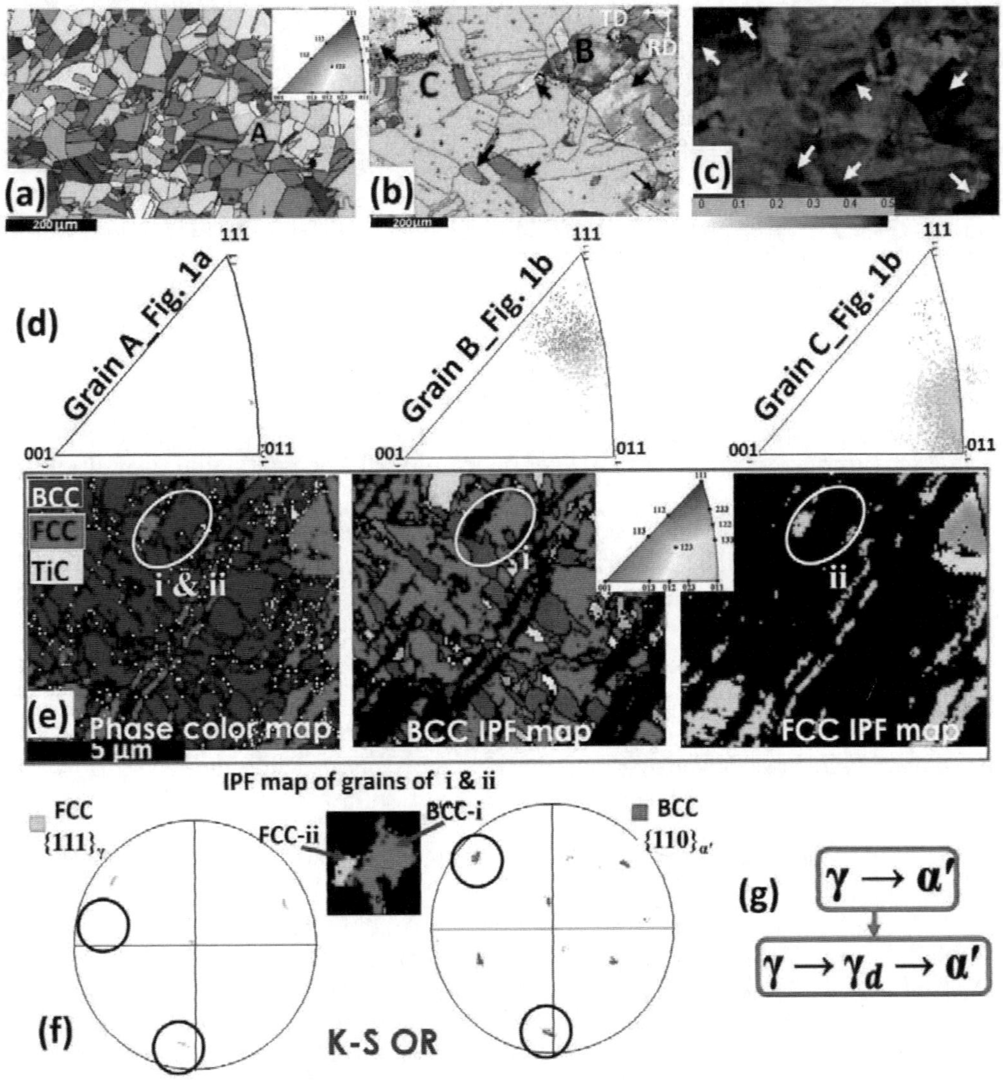

Fig 1. (a) EBSD-IPF map of the as-received sample, (b) IPF and (c) Schmid factor maps of deformed austenite (γ_d), (d) inverse pole figures of selected undeformed and deformed austenite grains. (e) EBSD-phase color, BCC and FCC IPF color maps of γ_d and α', (f) $\{111\}_\gamma$ and $\{110\}_{\alpha'}$ pole figures of selected grains in (e), (g) revised sequence of strain-induced martensitic transformation (Color figure online).

γ (grain A) and deformed γ_d (grains B and C) in Fig. 1d shows that the γ_d has undergone plastic deformation by slip. The orientation of grain A converges to a point in the IPF triangle; suggesting a strain free grain. The grains B and C show the widest orientation spread, which is an indication of large plastic deformation by slip. The IPF triangle could also be a useful tool in showing the gradual rotation of {111}||ND to {110}||ND as in grain B (a spread of blue color towards green) and C (rotation nearly complete as confirmed by the predominant green color). Upon 15% thickness reduction during cryo-rolling, 80% of the γ phase is transformed to α'-martensite (Fig. 1e). While the BCC IPF map in Fig. 1e shows a significant texturing of α'-phase around {100}||ND, the remaining γ_d is textured around {110}||ND. To validate the existing OR between the γ_d and α' phases, grains i and ii in Fig. 1e are selected and analyzed in Fig. 1f. The comparison of $\{110\}_{\alpha'}$ and $\{111\}_{\gamma}$ pole figures for grains i and ii, respectively, in Fig. 1f shows the coincidence of poles (circled). These confirms that K-S orientation relationship exists between the γ_d and α' and not γ and α'. Hence, the $\gamma \rightarrow \alpha'$ sequence is revised to $\gamma \rightarrow \gamma_d \rightarrow \alpha'$ as also expressed in Fig. 1g.

The results of the X-ray measurement that confirms the complete martensitic transformation of the austenite phase in the as-received sample after cryo-rolling at 77 K are presented in Fig. 2a. Although, γ-austenite of the as-received steel was transformed completely to α'-martensite at 20% thickness reduction, it was increased to 50% to generate finer martensite grains (insets in Fig. 2a) that accumulate enough stored energy. This promotes the development of ultra-fine strain-free austenite grains during subsequent annealing. The sigmoidal isothermal reversion curves and associated hardness profiles at two annealing temperatures are presented in Fig. 2b. Feritscope reading showed that austenitic reversion had barely occurred for samples annealed at 923 K and 1023 K for 60 s. As annealing time increases, specimens showed an increase in austenite fraction and a corresponding decrease in hardness values with annealing time. A notable observation in Fig. 2b is the sharp drop in hardness value of specimens annealed beyond 600s at 1023K. This is suspected to signal the onset of grain growth. Therefore, annealing at 1023 K for 600s may be the optimum reversion condition to develop UFG structure via the martensitic shear reversion mechanism. Similarly, annealing at 923 K for 28800s may be the optimum reversion condition via the diffusional reversion mechanism since the volume fraction of the unreversed martensite is approximately 6%.

The results of macrotexture analyses show that the UFG structure is significantly textured with strong Brass {110}<112>, Goss {110}<100> and Copper {112}<111> components (Fig. 3a). Specimen annealed at 923 K for 28800s (8 h) developed stronger texture than the specimen annealed at 1023 K for 600s (10 min). The EBSD maps of a specimen annealed for 180s at 923 K are presented in Fig. 3b-d, which indicate an early stage of reversion. Fairly equiaxed austenite grains with an average size of 0.16 μm randomly nucleated in the α'-martensite lath boundaries (Fig. 3b). At the annealing time of 300s, the area fraction, morphology and texture of the phases are significantly changed (Fig. 3e-g). The shape of the α'-martensite grains had changed from lath to near equiaxed after annealing for 300s. It is apparent in Fig. 3e-g that the morphology of the growing austenite grains is influenced by the morphology of their neighboring martensite grains.

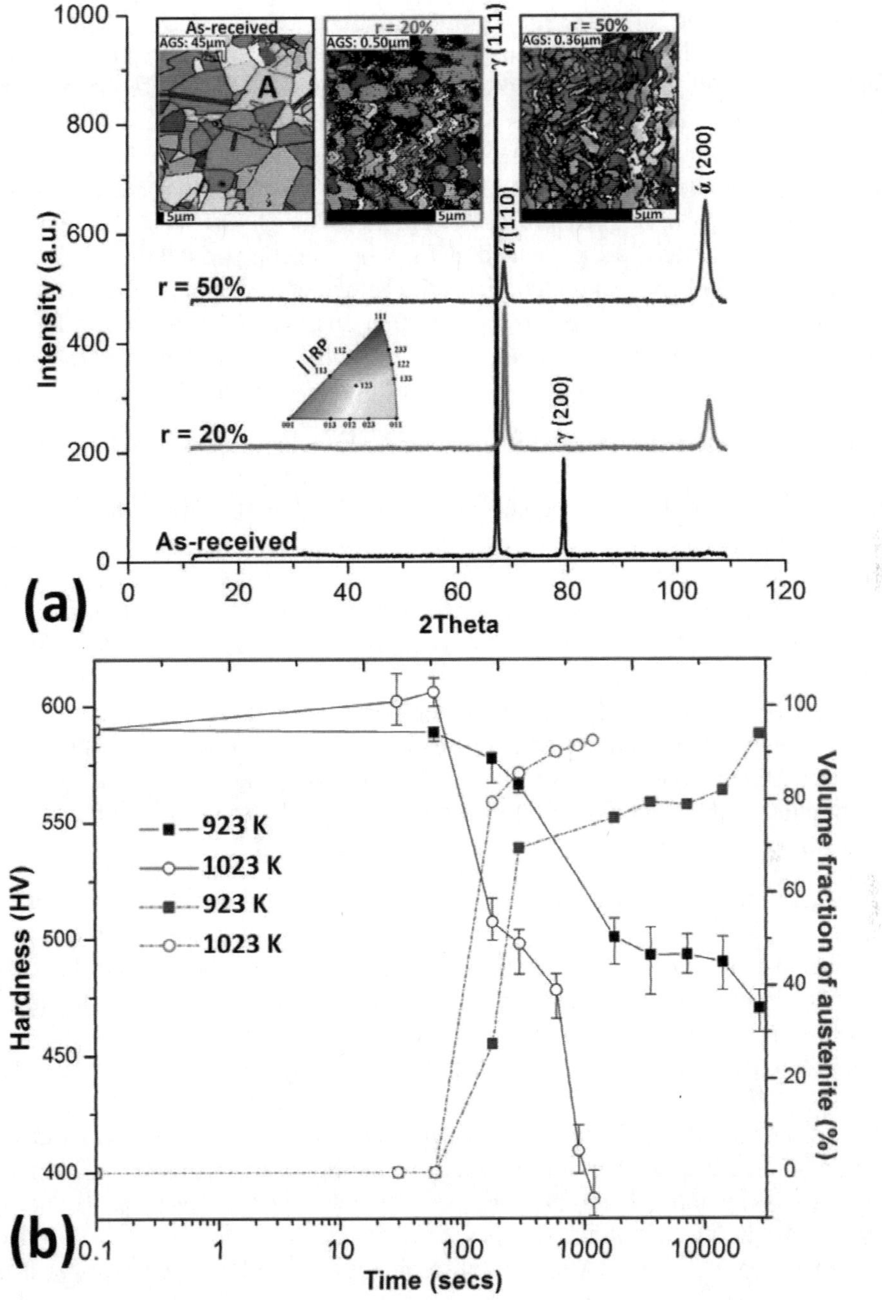

Fig 2. (a) X-ray diffraction patterns of the as-received sample and cryo-rolled specimens. (b) Volume fraction of reversed austenite and corresponding hardness profile during reversion. Inset in (a) is the associated EBSD inverse pole figure (IPF-Z) maps of both the FCC-austenitic phase of the as-received and the BCC-martensitic phase of the cryo-rolled specimens.

To expatiate on this, four grains (1, 2, 3 & 4) in Fig. 3e-g are considered. The equiaxed martensite grain 1 has a corresponding equiaxed austenite grain 3 as its neighbour. Similarly, grains 2 and 4 are neighbours and both elongated. This shows that a growing austenite grain boundary may be hindered by the boundaries of their neighbour grains; forcing the austenite grain to take the morphology of the martensite. Elsewhere, the morphology of transformed α′ have been reported to influence the shape of the reversed austenite (γ_r) [12]. On the crystallographic orientation of grains, few variants are present in both BCC and FCC phases (Fig. 3h and i). All the variants developed during reversion in this study are presented in Table 1. The comparison of $\{111\}_\gamma$ and $\{110\}_{\alpha'}$ pole figures for all grains in Fig. 3j and k confirms the coincidence of poles. Grains 5 ({100} oriented-α′) and 6 ({110} oriented-γ) are selected and further analyzed as presented in Fig. 3l. One of the poles of $\{111\}_\gamma$ is parallel to one of the poles of $\{110\}_{\alpha'}$. This confirms that K-S orientation relationship exists between the reversed γ-austenite and α′-martensite.

The EBSD maps of specimens annealed for 28800s (Fig. 4a-c) and 600s (Fig. 4d-f) at 923 K and 1023 K, respectively, shows the development of UFG structure with very low and acceptable fraction of α′-martensite. The average austenite grain sizes are 0.22 μm and 0.31 μm in 923 K and 1023 K-annealed specimens, respectively. Although, the orientation of γ_r grains are close to {110}<uvw> (ζ-fibre), texture is stronger in specimen annealed at 923 K (Fig. 4c) than specimen annealed at 1023 K (Fig. 4f). This agrees with the result of the macrotexture measurement in Fig. 3a. It is worth noting that most of the unreversed α′-martensite and TiC particles are located at the austenite triple junctions. These contribute to the stability of the UFG structure via Zener pinning of grain boundaries [13]. At higher annealing temperature and times, the larger fraction of the TiC particles dissolve and the volume fraction of α′-martensite was near-zero. This leads to abnormal grain growth by the breaking away of high angle boundaries and loss of texture [14]. To develop an UFG structure without abnormal grain growth, it may therefore be impractical to completely reverse the α′ to γ without retaining some negligible fraction of the α′.

While the texture of the γ-phase (parent) in the as-received AISI 321 ASS is random (Fig. 1a), it is apparent that the ζ-fibre texture of the γ_r (product) in the UFG structure memorizes the texture of the deformed austenite (γ_d) on the completion of γ (parent) → α′ → γ_r (product) transformation cycle. Since γ_d is intermediated between γ (parent) → α′ and conventional texture memory like those observed in BCC and HCP metals require that the texture of the product phase 'remembers' that of the initial parent phase, the phenomenon in AISI 321 steel is hereby referred to as Pseudo-texture memory. Hence, the transformation cycle is revised to $\gamma \rightarrow \gamma_d \rightarrow \alpha' \rightarrow \gamma_r$ as shown in Fig. 4g. The mechanism behind the pseudo-texture memory involves microstructural sites such as special grain boundaries. The variant that acted on the first transformation ($\gamma_d \rightarrow \alpha'$) in the special grain boundary is memorized and the same path becomes more accessible than the others for the $\alpha' \rightarrow \gamma_r$ reversion during annealing. This is schematically shown in Fig. 4h-l. At the early stage of cryogenic rolling of AISI 321 ASS, the austenite grains will first be deformed (Fig. 4h). As cryogenic rolling proceeds, two α′ grains for instance nucleate along the same γ_d boundary (Fig. 4i).

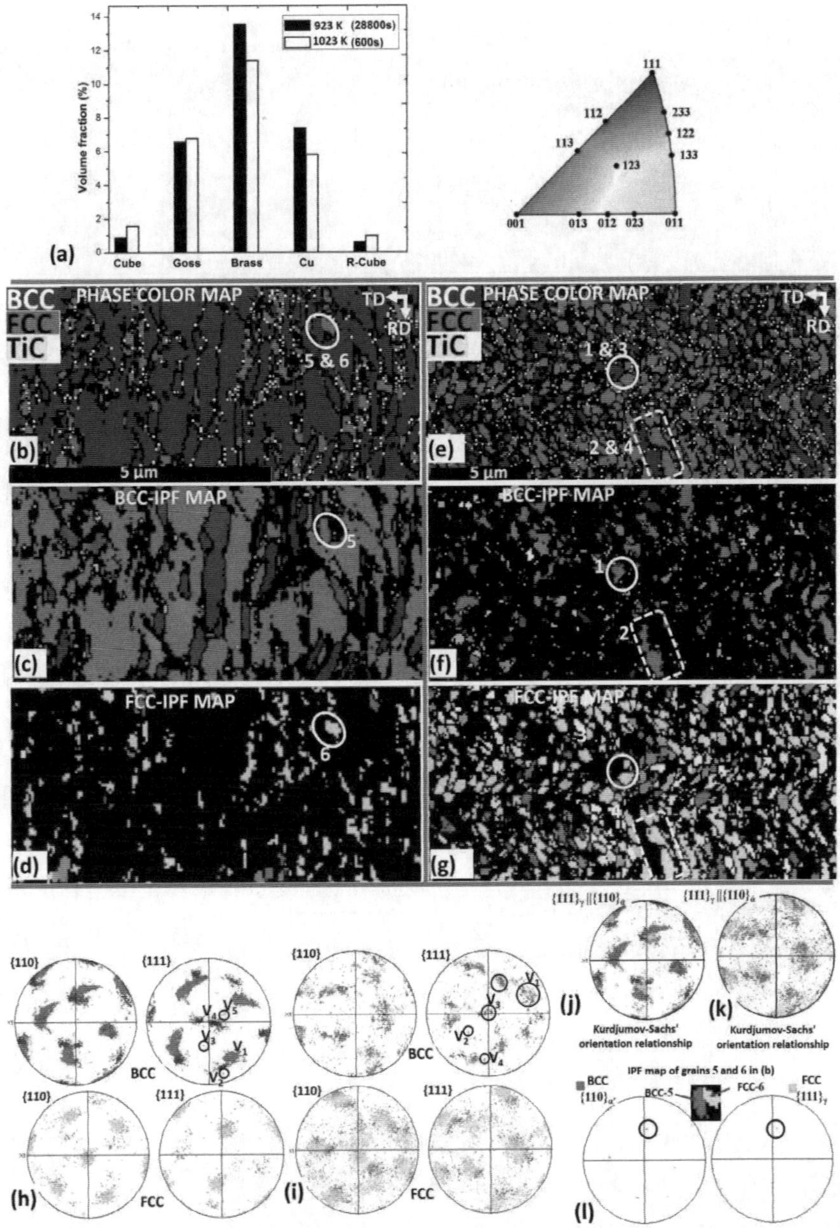

Fig 3. (a) Volume fraction of selected texture components. (b-g) EBSD maps of specimen annealed at 923 K (650 °C) for (b-d) 180s and (e-g) 300s: (b and e) phase color maps, (c, d, f, g) IPF color maps. {110} and {111} pole figures of specimens annealed at (h) 180s and (i) 300s. Superimposed $\{111\}_{\gamma}$ and $\{110\}_{\alpha'}$ pole figures of all grains in specimen annealed for (j) 180s and (k) 300s, (l) $\{111\}_{\gamma}$ and $\{110\}_{\alpha'}$ pole figures of selected grains in specimen annealed for 180s (Color figure online).

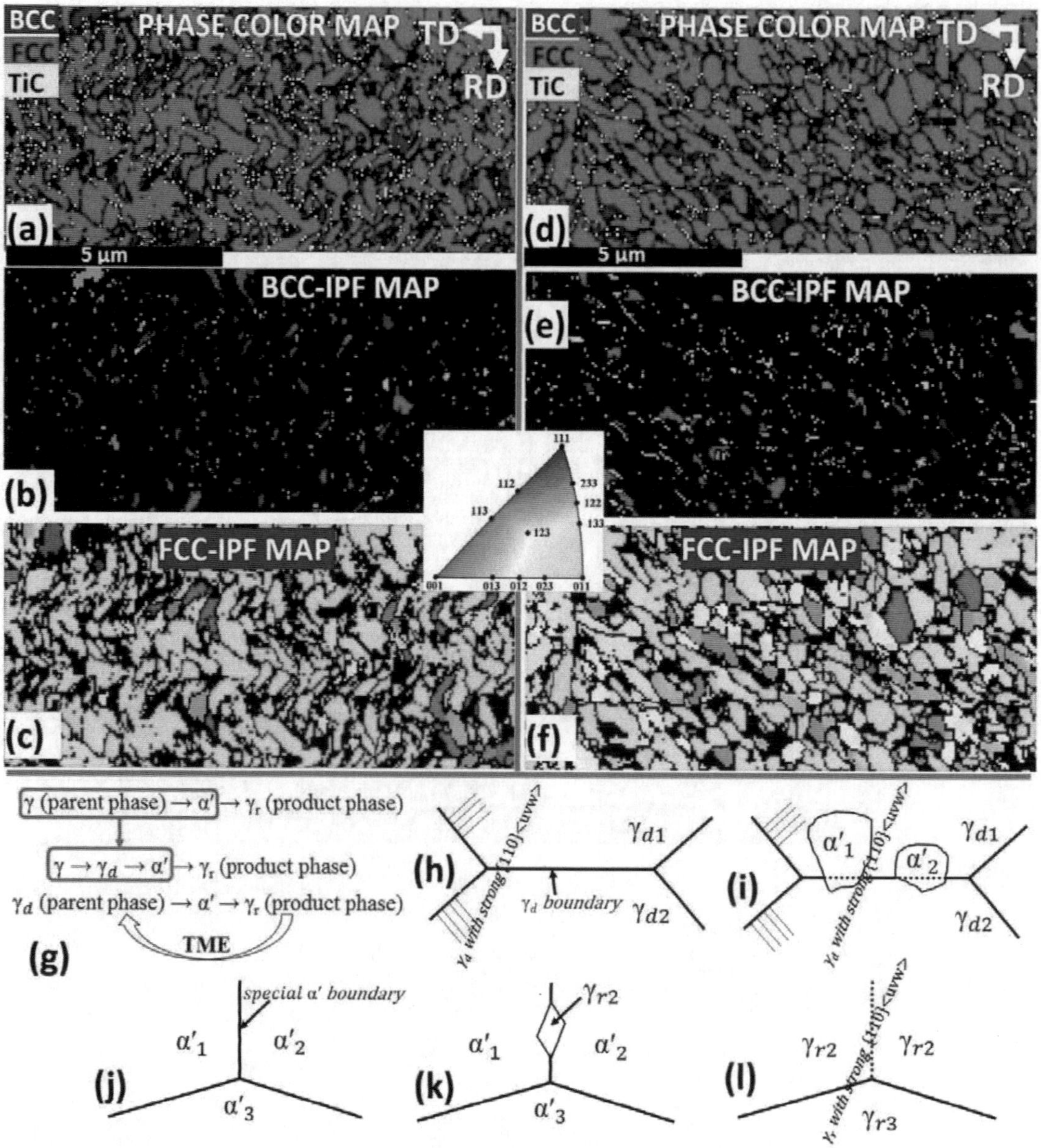

Fig 4. EBSD maps of specimen cryo-rolled and subsequently annealed at (a-c) 923 K (650 °C) for 28800s and (d-f) 1023 K (750 °C) for 600s. (g) sequence of transformation/reversion, (h-l) schematic of the pseudo-texture memory process adapted from Ref. [1] (Color figure online).

IOP Conf. Series: Materials Science and Engineering **375** (2018) 012010 doi:10.1088/1757-899X/375/1/012010

Table 1 The {hkl}<uvw> of all identified variants in the EBSD pole figures and ODFs.

Variants	α'-martensite	γ_r-austenite	α'-martensite	γ_r-austenite	α'-martensite	γ_r-austenite
	923 K / 180s		923 K / 300s		923 K / 28800s	
V_1	{001}<100>	{001}<201>	{001}<201>	{110}<3-11> {001}<201>	{001}<100>	{110}<100>
V_2	{111}<2-31>	{110}<110>	{110}<2-11>	{113}<2-51>	{110}<1-10>	{001}<101>
V_3	{110}<-331>	{113}<3-16>	{111}<-1-21>	{112}<0-21>	{332}<1-22>	{111}<-1-34>
V_4	{112}<-2-33>	{212}<1-51>	{113}<1-41>	{111}<-1-23>	-	{211}<0-32>
V_5	{123}<-2-33>	-	{213}<1-30>	{213}<-1-32>	-	-
	1023 K / 180s		1023 K / 600s		1023 K / 900s	
V_1	{023}<1-32>	{110}<-331> {001}<102>	{011}<0-11> {001}<100>	{011}<0-11> {031}<0-13>	{011}<1-11>	{011}<0-11>
V_2	{112}<111> {113}<-1-21>	{113}<1-10>	{111}<3-41>	{114}<1-10>	{001}<102>	{001}<3-10>
V_3	{012}<2-41>	{112}<110> {012}<1-33>	{213}<0-41>	{213}<0-41>	{111}<1-21>	{223}<3-41>
V_4	{014}<-1-71>	{015}<-4-31>	-	-	{210}<2-41>	{210}<1-34>
V_5	-	-	-	-	{213}<0-21>	{213}<0-21>

These two α′ grains obey the K-S orientation relationship with γ_{d2}, but exhibiting different variants of the 24 possible K-S conditions. As the percentage thickness reduction increases, both grains should grow into the neighbor γ_{d1} grain and then merge to establish a 'special grain boundary' in the new α′ grain structure (Fig. 4j). When the new α′ grain structure is annealed, the special grain boundary will be highly favored for the nucleation of γ_r grains that possess the same orientation as the γ_{d2} since it is the orientation that fulfills the K-S condition (Fig. 4k), as indicated in the experimental results presented in Fig. 3b-d, h-l. The available special grain boundaries in the specimen therefore act as crystallographic orientation memory sites during austenitic reversion. This process is more pronounced at lower annealing temperature (923 K) where TiC precipitates are more stable than at higher annealing temperature (1023 K). This explains why the intensity of ζ-fibre is stronger at lower temperature (Figs. 3a and 4c) and weaker at higher annealing temperature (Figs. 3a and 4f). TME in Mn-IF steel has been attributed to the formation of TiC precipitates [1]. It can therefore be inferred that TiC precipitates played a significant role in the TME exhibited by AISI 321 steel. The schematic of pseudo-texture memory and stages of martensitic transformation and austenitic reversion during the development of UFG structure is shown in Fig. 5.

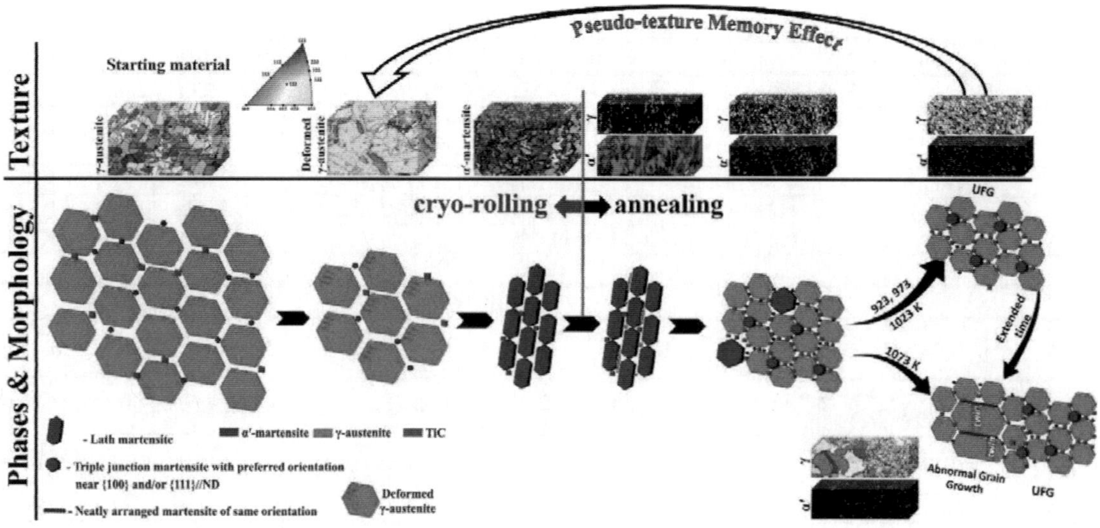

Fig. 5 Schematic of texture memory and stages of martensitic transformation and austenitic reversion. Shape and size of grains and particles are exaggerated to show the sequence of UFG development (Color figure online).

4 Conclusion

The strong ζ-fibre ({110}<uvw>) observed in the austenite phase of the UFG structure can be traced to AISI 321 ASS's ability to memorize the orientation of the deformed austenite.

Acknowledgment

The authors wish to acknowledge the Natural Sciences and Engineering Research Council of Canada (NSERC) and Vanier Canada Graduate Scholarship for the financial support of this study. The support of ACUREN Group Inc. for the use of Fischer Feritscope MP30E is well appreciated.

References

[1] Hutchinson B and Kestens L A I 2008 *Appl. Texture Anal.*, Rollett A D (New Jersey: John Wiley 81 Sons Inc) p 281.

[2] Tomida T, Wakita M, Yasuyama M, Sugaya S, Tomota Y and Vogel S C 2013 *Acta Mater.* **61** 2828.

[3] Yoshinaga N, Inoue H, Kawasaki K, Kestens L and De Cooman B C 2007 *Mater. Trans.* **48** 2036.

[4] Beladi H, Chao Q and Rohrer G S 2014 *Acta Mater.* **80** 478.

[5] Romero J, Preuss M and Fonseca J Q 2009 *Acta Mater.* **57** 5501.

[6] Nakada N, Tsuchiyama T, Takaki S and Hashizume S, 2007 *ISIJ Int.* **47** 1527.

[7] Humbert M, Petit B, Bolle B and Gey N 2007 *Mater. Sci. Eng. A* **454–455** 508.

[8] Wang J, Lin Y, Yan J, Zen D, Zhang Q, Huang R and Fan H 2012 *Surf. Coatings Technol.* **206** 3399.

[9] Llewellyn D T and Hudd R C 1998 *STEELS-Metallurgy and Applications* (Oxford, UK: Reed Educational and Professional Publishing Ltd) p 337.

[10] Talonen J, Aspegren P and Hänninen H 2004 *Mater. Sci. Technol.* **20** 1506.

[11] Nesterenko V F, Meyers M A, LaSalvia J C, Bondar M P, Chen Y J and Lukyanov Y L 1997 *Mater. Sci. Eng. A* **229** 23.

[12] Tomimura K, Takaki S and Tokunaga Y 1991 *ISIJ Int.* **31** 1431.

[13] Zhang X F, Terasaki H and Komizo Y 2001 *Philos. Mag. Lett.* **91** 491.

[14] Rajasekhara S and Ferreira P J 2011 *Acta Mater.* **59** 738.

IOP Conf. Series: Materials Science and Engineering **375** (2018) 012011 doi:10.1088/1757-899X/375/1/012011

Analyses of texture memory, surface-effect-induced transformation texture and variant selection in low graded electrical steels

P YANG, C GU and WM MAO

School of Materials Science and Engineering, University of Science and Technology Beijing, Beijing, China, 100083
Email: yangp@mater.ustb.edu.cn

Abstract. This study investigates the retaining of both {100} texture and topographic structure of columnar grains in cast slabs of industrial low graded electrical steels in terms of texture memory and variant selection. In addition, phase transformation treatment is conducted on industrial low graded electrical steels to explore the possibility of producing {100} texture by surface effect. The results show that the texture memory in columnar grains of cast slab can be explained by the directional cooling and internal thermal stress. The abundant Σ3 grain boundaries of small equiaxed grains within coarse columnar grains are detected, which can reduce transformation strain effectively. After phase transformation annealing, {111} and {100} texture memories with equiaxed grains in recrystallization type are resulted in commercial sheets due to the presence of P and Al elements which restrict surface-effect induced transformation texture. In comparison, strong {100} texture and columnar grains are formed in a laboratory-melt sheet as surface effect induced transformation texture by preferred nucleation at surface and preferred growth into sheet centre layer.

1. Introduction

The textures in commercial non-oriented electrical steels are generally optimized by the control of deformation and recrystallization parameters irrespective of their grades or silicon contents. The favorite {100} texture in such condition takes at maximum only about 20% in volume fraction. In contrast nearly 80% volume fraction of beneficial {100} texture can be obtained by phase transformation in pure Fe or slightly alloyed Fe with columnar-grained structure in final sheets, which is confirmed by many researchers[1-4]. The texture memory effect is often detected in steels during recrystallization and $\alpha \rightarrow \gamma \rightarrow \alpha$ transformation, and the crystallographic orientation relationships are usually maintained between the parent and new phase which is often subjected to variant selection. Without such variant selection, texture will be normally weak during transformation. Yet texture memory or variant selection is not equal to the presence of strong phase transformation texture, it is often seen that phase transformation texture is quite different to the rolling texture of ferrite before transformation. If phase transformation proceeds uniformly in steel sheets, {111} texture or nearly random orientations can also be resulted. If nucleation preferentially appears at sheet surface in pure hydrogen after rolling and successive growth into sheet center takes place, columnar structure with strong {100} new texture can be obtained (the so-called surface effect induced transformation texture), so that magnetic properties can be improved.

Based on the fact that strong {100} texture can be produced by rolling and subsequent phase transformation annealing in low graded electrical steels, this work aims to investigate the role of texture memory and variant selection during $\delta \rightarrow \gamma \rightarrow \alpha$ transformation in cast slabs of low graded electrical steels containing dominant columnar grains. The phase transformation treatment is performed on the industrially low graded electrical steels to explore the possibility of producing {100} texture by surface effect. The purpose of this study is to provide fundamental principles for the potential industrial application of low graded non-oriented steels by special transformation treatment.

2. Experimental

The starting material is a columnar-grained industrial low graded electrical steel cast slab containing 0.35%Si, 0.002%C, 0.13%Mn and 0.14%P. A dilatometer was used to measure the critical phase transformation temperature: A_{c1}=980°C, A_{c3}=1090°C; A_{r1}=938°C, A_{r3}=973°C. The deformation process for four samples is listed in table 1. The cast slab was heated at 1100°C for 30min and hot rolled from 10mm to 2mm thick in three rolling passes. The finish rolling temperature is about 800°C (finish hot rolling in ferrite area). After different deformation processes, the phase transformation annealing of samples is conducted in a tube furnace under H_2 atmosphere with a flow rate of 8L/min. Samples were pushed into the furnace at 1100°C and held for 5min, subsequently the samples were cooled to 900°C at a cooling rate of 300°C/h. After cooled to 900°C, the samples were pulled out from the furnace and cooled in a chamber at room temperature. Textures and microstructures in the cross sections of different samples were measured by the electron back-scatter diffraction (EBSD) system of HKL Oxford-Instruments, mounted on a ZEISS ULTRA55 field-emission SEM. The orientation of maximum deviation angle is set to 15° in this experiment.

Table 1. Deformation process before phase transformation annealing.

Samples	deformation process
Sample 1	No deformation(original cast slab)
Sample 2	Hot rolling(2.0mm)
Sample 3	Hot rolling(2.0mm)-cold rolling(0.2mm)
Sample 4	Hot rolling(2.0mm)-cold rolling(0.6mm)

3. Result and discussion

Two types of texture memories are observed in this steel. The first type occurs in cast slab containing coarse columnar grains and is free from external strain, whereas the second one appears in the strongly rolled and recrystallized sheet subjected to a final special phase transformation. Figure 1 shows the grain orientations and microstructure of cast slab which was subjected to twice transformation of $\delta \rightarrow \gamma \rightarrow \alpha$ during cooling. The EBSD data reveal that the area fraction of columnar grains with nearly {100} orientation (the maximum deviation is 15°) is about 70%. As the twice transformations are free from external strain, the reason for such texture memory is attributed to the directional cooling from slab plate surface to plate centre and the internal strain due to cooling. It is well known that there are 24 variants in the case of a K-S orientation relationship ($\{111\}_{fcc}//\{110\}_{bcc}$ and $\langle 110\rangle_{fcc}//\langle 111\rangle_{bcc}$). In order to reduce the transformation strain as much as possible, plenty of $\Sigma 3$ variants (red lines in Figure.2b) which generate lower strain are detected between coarse columnar grains and small equiaxed grains. It indicates that variant selection is strong when columnar {100} δ-grains transform into columnar {100} α-grains.

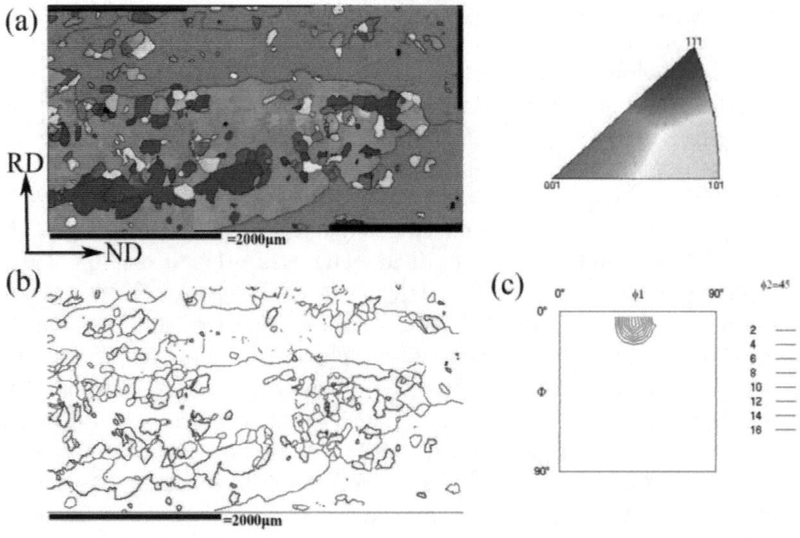

Figure 1.EBSD data for the cast slab: (a)IPF-Z map; (b) GB distribution map (red lines represent Σ3GB); (c) ODF at $\varphi_2 = 45°$ section.

Figure 2 presents the EBSD data for the further treated cast slab. A further circle of phase transformation annealing ($\alpha_1 \rightarrow \gamma \rightarrow \alpha_2$) at 1100°C is performed without external straining on the cast slab, thus four times of transformation occur in the steel. It is seen that after this treatment, still 65% near {100} (the maximum deviation is 15°) columnar grains together with plenty of Σ3 variants are resulted. Thus the texture memory and variant selection in coarse columnar grains are strong.

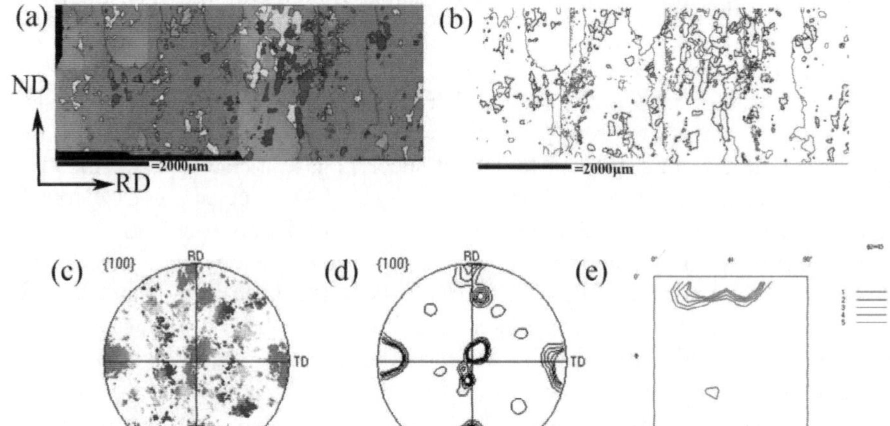

Figure 2. EBSD data for sample1 after phase transformation annealing: (a)IPF-Z map; (b) GB distribution map (red lines represent Σ3GB); (c, d){100} pole figures; (e)ODF at $\varphi_2 = 45°$ section.

Figure 3 displays the EBSD data for the cross sections of hot rolled sheet (sample2), cold rolled sheets (sample3 and sample4), respectively. It shows that {110}<001>/<112> and {111}<112>/<110> orientations are resulted at surface region and centre layer in the hot rolling sheet respectively (Figure 3a,b). The cold rolled sheets with different thicknesses (0.6mm for sample3 and 0.2mm for sample4) present similar rolling textures comprising the α-fibre and the {111}<110> textures.

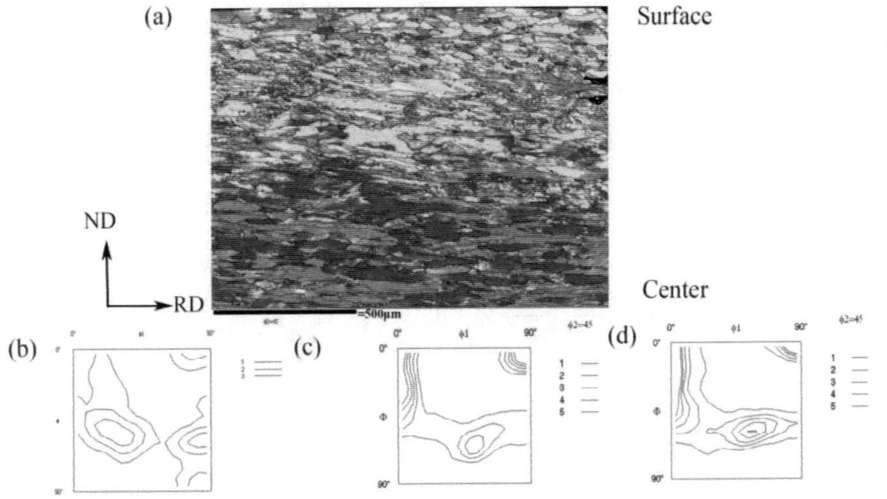

Figure 3. EBSD data for the cross sections through thickness of the deformed sheets: (a) IPF-Z map for Sample2 (half of the thickness); (b) ODF at φ_2 = 45° section for Sample2; (c) ODF at φ_2 = 45° section for sample3; (b)ODF at φ_2 = 45° section for Sample4.

Figure 4 and Figure 5 display the EBSD data for sample3 and sample4 after transformation annealing ($\alpha_1 \rightarrow \gamma \rightarrow \alpha_2$) at 1000°C in pure H_2, respectively. Instead of columnar grains with near {100} texture as reported in the study of Zhang[4], only equiaxed grains with mainly {111}<112> and {100}<021> orientations are obtained in the sheets after phase transformation annealing, as seen in Figure 4(e) and Figure 5(e), namely the transformation texture here belongs to the recrystallization type, which is also a kind of texture memory effect. The reason for this kind of texture memory lies in the uniform transformation throughout the sheet thickness, rather than only in the surface for the columnar grain structure. In addition, less Σ3 variants are produced, see Figure 4(b) and Figure 5(b).

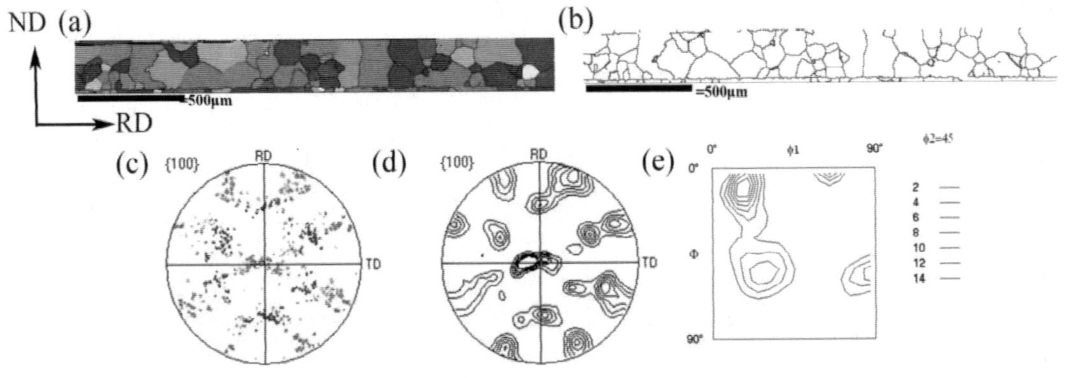

Figure 4. EBSD data for sample3 after phase transformation annealing: (a)IPF-Z map; (b) GB distribution map (red lines represent Σ3GB); (c, d){100} pole figures; (e)ODF at φ_2 = 45° section.

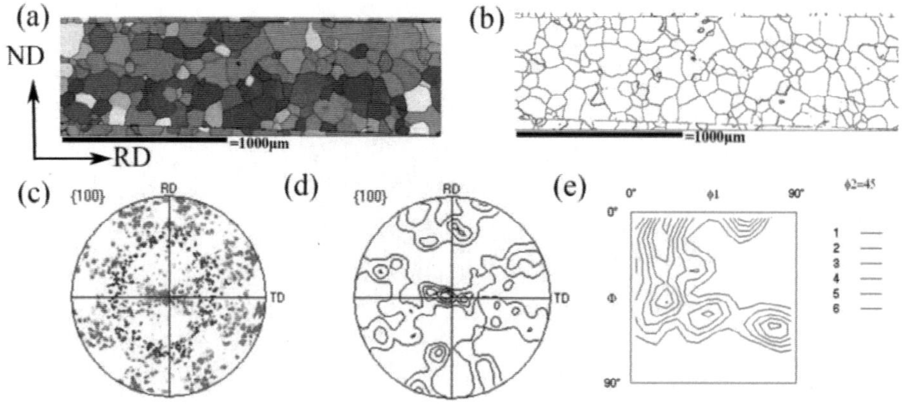

Figure 5. EBSD data for sample4 after phase transformation annealing: (a)IPF-Z map; (b) GB distribution map (red lines represent Σ3GB); (c, d){100} pole figures; (e)ODF at $\varphi_2 = 45°$ section.

For comparison, Figure 6 shows the EBSD data for Fe-0.5Si-0.5Mn steel sheet (a laboratory-melt steel, 0.5mm in thickness) after phase transformation annealing. The near {100} texture and columnar grains were formed in this laboratory-melt steel during phase transformation annealing in H_2 atmosphere, where surface effect induced transformation texture is prevailing and the near {100} texture is a new texture component formed by preferred nucleation at surface and preferred growth into sheet centre layer. The variant selection is also strong because only one new variant is formed during the growth of {100} columnar grain.

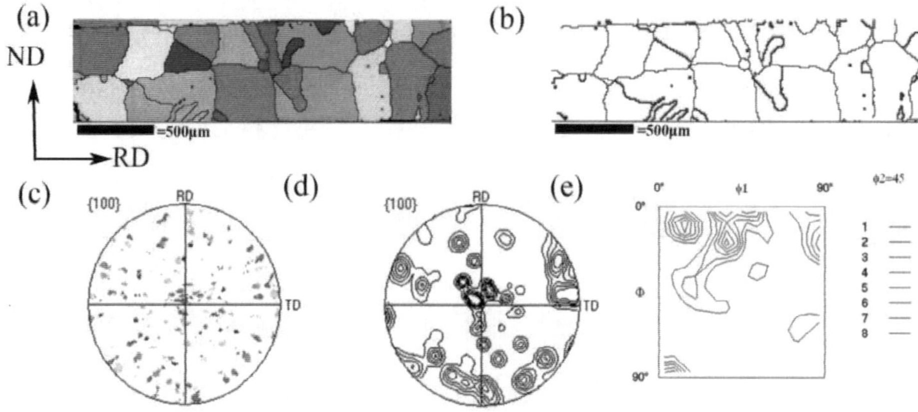

Figure 6. EBSD data for Fe-0.5Si-0.5Mn steel sheet (laboratory-melt steel, 0.5mm in thickness) after phase transformation annealing: (a)IPF-Z map; (b) GB distribution map (red lines represent Σ3GB); (c, d){100} pole figures; (e)ODF at $\varphi_2 = 45°$ section.

By comparing the tèxtures and microstructures of commercial steels with the laboratory-melt steels after phase transformation annealing in pure H_2, it is seen that surface-effect induced phase transformation and columnar grain microstructure are less significant in the commercial steel sheet containing 0.12%Al and 0.14%P, which may influence the surface effect during phase transformation annealing. In order to analyze the effect of element distribution on microstructure development, a glow discharge spectrometer (GDS) is used to determine the depth profiles of different elements from surface to inside, as shown in Figure 7. The results indicate that SiO_2 film forms in the surface and Al and P are segregated in the surface. The surface effect may weaken due to the SiO_2 film and the segregation of Al and P in the surface. The presence of P and Al element segregation at surface leads to the uniform nucleation in whole sheets, thus the columnar grains are hard to nucleate preferentially at sheet surface during $\alpha \rightarrow \gamma \rightarrow \alpha$ transformation.

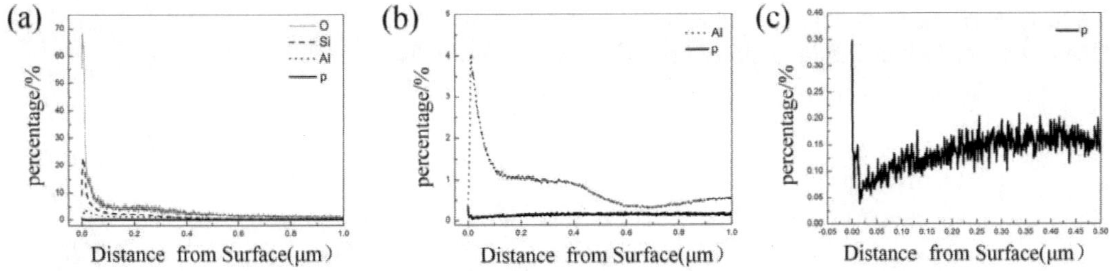

Figure 7. Glow discharge date profiles for the sample3 after phase transformation annealing.

4. Conclusion

1) {100} texture memory in coarse columnar grains during $\delta \rightarrow \gamma \rightarrow \alpha$ and $\alpha_1 \rightarrow \gamma \rightarrow \alpha_2$ transformation in the cast slab is observed. The texture memory of strong {111} and weak {100} in recrystallization type is also detected after phase transformation treatment of rolled sheets in pure H_2. The difference between them is attributed to the fact that the directional transformation occurred in the first case and throughout in whole sheet at same time in the latter case. The variant selection is strong in both cases.
2) Surface effect induced transformation texture of mainly {100} and columnar-grained structure are formed by preferred nucleation at surface and preferred growth into sheet center layer. Variant selection is also strong in this case.
3) The presence of P and Al elements in commercial electrical steels restricts surface-effect induced transformation texture by probably segregation at sheet surface, which leads to the uniform nucleation in whole sheets.

Acknowledgements
This work is financially supported by the National Nature Science Foundation of China (No. 51771024).

References

[1] Tomida T 1996 *J.Mater.Eng.Perform.***5** 316–322
[2] Sung J K, Lee D N, Wang D H and Koo Y M 2011 *Isij. Int.* **51** 284-290
[3] Xie L, Yang P, Zhang N, Zong C, Xiang D S and Mao W M 2014 *J. Magn. Magn. Mater.* **356** 1-4
[4] Zhang L W, Yang P, Wang J H and Mao W M 2016 *J. Mater. Sci.* **51** 8087-8097

IOP Conf. Series: Materials Science and Engineering **375** (2018) 012012 doi:10.1088/1757-899X/375/1/012012

Crystallographic Reconstruction of Parent Austenite Twin Boundaries in a Lath Martensitic Steel

S Cluff[1], E Homer[1], T Nelson[1] and R Song[2], D Fullwood[1]
[1]*Department of Mechanical Engineering, Brigham Young University, Provo UT, USA.*
[2]*ArcelorMittal Global R&D—East Chicago Automotive Products, East Chicago, IN, 46312, USA*

Abstract
The study of post-transformation microstructures and their properties can be greatly enhanced by studying their dependence on the grain boundary content of parent microstructures. Recent work has extended the crystallographic reconstruction of parent austenite in steels to include the reconstruction of special boundaries, such as annealing twins. These reconstructions present unique challenges, as twinned austenite grains share a subset of possible daughter variant orientations. This gives rise to regions of ambiguity in a reconstruction. A technique for the reconstruction of twin boundaries is presented here that is capable of reconstructing 60° <1 1 1> twins, even in the case where twin regions are comprised entirely of variants that are common between the twin and the parent. This technique is demonstrated in the reconstruction of lath martensitic steels. The reconstruction method utilizes a delayed decision-making approach, where a chosen orientation relationship is used to define all possible groupings of daughter grains into possible parents before divisive decisions are made. These overlapping, inclusive groupings (called clusters) are compared to each other individually using their calculated parent austenite orientations and the topographical nature of the overlapping region. These comparisons are used to uncover possible locations of twin boundaries present in the parent austenite. This technique can be applied to future studies on the dependence of post-transformation microstructures on the special grain boundary content of parent microstructures.

1. Introduction

Many crystalline materials undergo displacive or diffusionless transformations in which the product phase is dependent on the orientation of the parent phase. Various properties of the daughter microstructure are inherited from the parent phase due to this dependence, however the daughter structure typically consumes the parent structure making it difficult to observe how the processing of the parent phase affects the final post-transformation material. The martensite transformation in steel is a common example of this. When the orientation dependence of the daughter phase on the parent phase is known, it may be possible to "reconstruct" the parent phase computationally, identifying parent grains that may have generated the observed daughter grain structure. If this can be done accurately, the pre- and post-transformation structures can be compared directly. This can be of value in improving the understanding of microstructural inheritance in materials such as steel, where it is difficult to observe the history of the pre-transformed phase—including the application of complex loading conditions at austenitic temperatures. Ultimately the crystallographic reconstruction of a parent phase can facilitate optimization of the thermomechanical processing of the pre-transformed structure. [1-3]

Reconstruction of austenite in steel presents unique challenges. Crystallographic reconstruction is based on a known Orientation Relationship (OR) between the two phases, and the quality of the daughter orientation map obtained from electron backscatter diffraction (EBSD) measurements. This information is essential to a parent phase reconstruction, which is often complicated by various uncertainties. These uncertainties arise in the reconstruction of austenite in steel due to noisy datasets, deviations from a given OR, and other ambiguities [3, 4]. Methods must be developed to deal with these uncertainties. While prior austenite reconstructions in steel can be made difficult by these ambiguities, they can potentially be resolved through a semi-automated delayed-decision making approach detailed here.

Reconstructions depicted in subsequent sections of this work are performed on two martensitic steels. The first EBSD dataset was provided by US Steel and Carnegie Mellon University. The martensitic microstructure is from a seamless tubular steel intended for use as oil country tubular goods. Most of the production parameters are proprietary and the information provided is as specific as possible. The steel composition is similar to AISI 4130, and the product was heat treated as follows: heated to above the austenitizing temperature (>850 °C), then quenched by water sprays, and finally tempered at a temperature between 500 and 700 °C [5]. The second EBSD dataset was provided by ArcelorMittal. It is M1700 steel, a commercially available tempered martensitic sheet steel. Its composition is 0.3 wt% C, 0.45 wt% Mn, 0.01 wt% P, and 0.015 wt% S.

2. Background

Parent phase reconstruction has been performed on various materials and has been applied in the literature by various authors on materials for which an OR can be defined. Germain, Humbert, et al. have performed prior-phase reconstruction on austenite in steel, as well as the beta phase of titanium alloys [3, 6, 7]. Cayron, et al. have developed a method of parent austenite reconstruction in steels that utilizes a groupoid approach, defining mathematical operators that link different daughter variants to each other [1, 8, 9]. Miyamoto, Abbasi, and other authors have proposed methods, and addressed the challenges of prior phase reconstructions in different steels [2, 4, 10]. Central to each of these methods is the definition of the orientation relationship between the parent and daughter phases. Various ORs have been experimentally determined in steels and are well documented in the literature [3, 11, 12]. The determination of which OR is followed has been shown to depend on factors such as material composition, processing of the austenite, and resulting martensite morphology [11, 12]. While ORs are typically expressed as sets of crystallographic planes and directions that are coincident between the two phases, measured ORs are almost always irrational [13]. Different material chemistries will tend towards specific ORs, however; low carbon steels have been shown to exhibit the KS OR, while Fe-Ni systems often exhibit the NW OR [12]. Table 1 lists commonly observed ORs in steel. For martensite, the KS and NW ORs can be seen as “bounding” ORs, where the OR of the formation of lath martensite has been observed to range between the two [12, 13]. Figure 1 demonstrates the resulting set of orientational variants predicted by the KS OR.

Table 1

OR	Parallel Plane	Parallel Direction
Kurdjumov-Sachs (KS)	$\{1\ 1\ 1\}_{\gamma} \| \{1\ 1\ 0\}_{\alpha}$	$\langle 1\ 1\ 0\rangle_{\gamma} \| \langle 1\ 1\ 1\rangle_{\alpha}$
Nishiyama-Wassermann (NW)	$\{1\ 1\ 1\}_{\gamma} \| \{1\ 1\ 0\}_{\alpha}$	$\langle 1\ 1\ 2\rangle_{\gamma} \| \langle 1\ 1\ 0\rangle_{\alpha}$
Greninger-Troiano (GT)	$\{1\ 1\ 1\}_{\gamma}$ 1° from $\{1\ 1\ 0\}_{\alpha}$	$\langle 2\ 1\ 1\rangle_{\gamma}$ 2.5° from $\langle 1\ 1\ 0\rangle_{\alpha}$

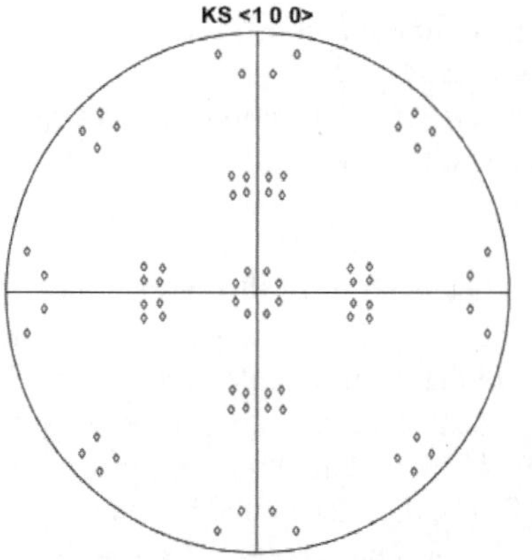

Figure 1: Pole figure centered on [001] demonstrating the <100> poles of each of the 24 variants of the KS orientation relationship, assuming a parent austenite orientation unrotated from the pole figure's axes.

A common method of parent phase reconstruction is a neighbor-to-neighbor comparison of daughter variants, in search of groupings of daughter grains that have the same parent. Orientation relationships are used in this comparison. For a given parent orientation, a set $\{i\}$ of possible daughter α orientations is defined by an OR according to equation 1. Likewise, a set of possible γ orientations for a single daughter grain can be defined using the inverse of the orientation relationship (represented as a single rotation T_{OR}). The symmetries of each cubic phase are represented by the set of symmetry operators S_i.

$$g_i^\alpha = T_{OR} \cdot S_i^\gamma \cdot g^\gamma \qquad\qquad g_i^\gamma = T_{OR}^{-1} \cdot S_i^\alpha \cdot g^\alpha \tag{1}$$

Where two daughter grains share a possible parent, they can be grouped together as having possibly originated from a parent austenite grain of that orientation. The two possible parents "overlap" when their misorientation angle ε as defined by equation 2 is within a pre-set tolerance. Groups can be defined for which a single orientation is within tolerance of possible parents for many daughter grains, defining a region of prior austenite. This process is depicted in figure 2. This type of neighbor-to-neighbor comparison using an OR is well outlined in papers by Germain and Humbert [3, 7]. The final reconstruction of a parent austenite grain consists of a "cluster" of daughter grains from the EBSD map, and a collection of possible parent orientations in orientation space - one possible parent for each daughter grain. The orientation of the possible parent is back-calculated by computing the average of the collection of possible parents.

$$Tr(g_2 g_1^T) = 2\cos(\varepsilon_{12}) + 1 \tag{2}$$

Equation 2: For two orientations g_1 and g_2, their misorientation is ε_{12}

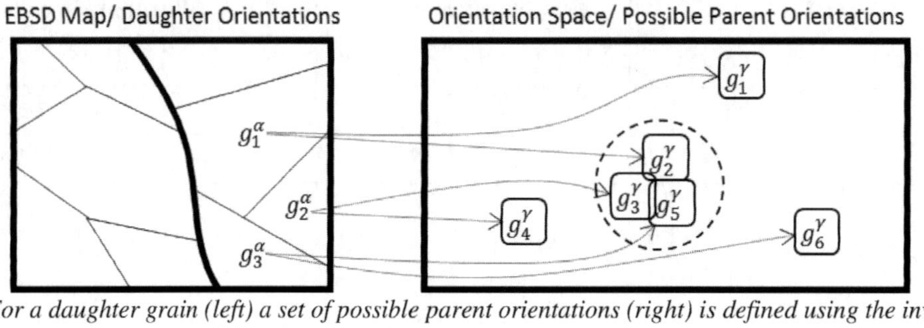

Figure 2: For a daughter grain (left) a set of possible parent orientations (right) is defined using the inverse of an orientation relationship. A viable solution for the grouping of daughter grains is found where each daughter grain has one possible parent within a dense local grouping of possible parents in orientation space. The local grouping of orientations in orientation space is defined by their misorientation as defined in equation 2.

IOP Conf. Series: Materials Science and Engineering **375** (2018) 012012 doi:10.1088/1757-899X/375/1/012012

A neighbor-to-neighbor approach to prior austenite reconstruction often leads to ambiguities. The sets of daughter grains belonging to two clusters may not be mutually exclusive; there can be subregions of daughter grains that satisfy the OR for both clusters simultaneously (i.e. each daughter grain in the ambiguous subregion has two possible parents, each fitting with one of the back-calculated parent orientations for the two clusters). When performing a reconstruction, this ambiguous region must be either assigned to one cluster or divided between the two clusters. This process is difficult to automate, and often requires manual decision making [4].

A common cause of ambiguities in the reconstruction of prior austenite in steel is the presence of twin boundaries in the parent austenite. The twin relationship of the two parent crystals results in the theoretical sets of possible daughter orientations having a subset in common. According to the KS OR, two twinned austenite crystals will share 6 of their 24 possible daughter orientations. When the transformation takes place along the twin boundary, the daughter variants that are "selected" to nucleate are often from this common set of 6 variants due to the energetic favorability of having a daughter variant satisfy the OR with both parent austenite orientations simultaneously. These daughter grains are said to satisfy a "double-KS" or "KS-KS" orientation relationship. These ambiguous regions are often very difficult to resolve, as the parent austenite twin boundary is often partially or completely consumed by daughter grains that traverse the twin boundary. Figure 3 demonstrates the often complex morphology of the ambiguous region, as well as the uncertainty as to where the twin boundary actually lies. The orientations of daughter variants for two twinned austenite grains, as predicted by the KS OR, are shown in figure 4. Variants that are common between two twinned austenite orientations are shown to overlap in the pole figure.

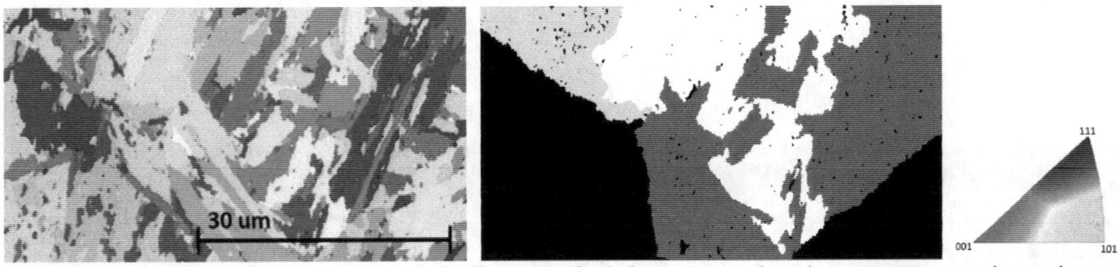

Figure 3: For the martensite grains in the IPF map to the left, two groupings into parent austenite grains can be found and are demonstrated in the IPF map to the right. The region highlighted in white belongs to both clusters. Each martensite grain within the highlighted region satisfies the KS OR with both parent austenite orientations. This EBSD map is a subsection of the data provided by Carnegie Mellon University.

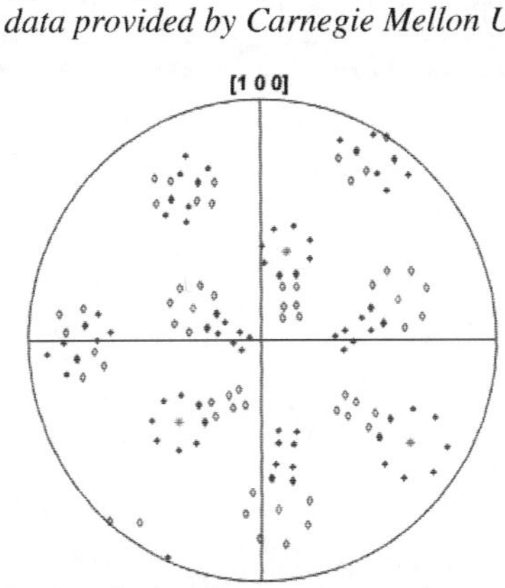

Figure 4: The orientations of the two back-calculated parent austenite grains from figure 3 (red), and their theoretical daughter variants as predicted by the KS OR (blue). Sets of poles belonging to either cluster are differentiated by marker (diamond vs. asterisk). Six variants overlap, as denoted by asterisks and diamond markers overlapping. This suggests that the parent orientations are twinned.

3. Method

The reconstruction method employed here utilizes a neighbor-to-neighbor approach within a delayed-decision making framework. During a reconstruction effort, when an ambiguous region is found, the resolution of that ambiguous region is delayed until after all possible groupings of daughter grains into possible parents have been found. This approach is in contrast to methods that attempt to resolve the ambiguity as they are found (i.e. making a choice of where to assign an ambiguous daughter grain in the moment based on some additional metric). Instead, reconstructed regions are allowed to overlap each other, some groupings completely enveloping others. This provides a list of many possible groupings or "clusters", with some of the groupings ultimately providing a more insightful view of the pre-transformed structure. This method has significant advantages when reconstructing regions that satisfy a KS-KS relationship.

The core of this reconstruction method lies with the independent "growth" of clusters of daughter grains. Each cluster is started from a seed comprised of 3 daughter grains that share a possible parent orientation (see figure 2). Prior to a reconstruction, all viable seeds for a given region are found by comparing the lists of possible parents for all sets of 3 neighboring daughter grains. Once a seed has been found, the 3 possible parent orientations that are within tolerance of each other are averaged to define a starting PA orientation for the cluster to be grown. Neighboring daughter grains are iteratively added to the cluster if they have a possible parent within a tolerance misorientation of this averaged austenite orientation, which is then re-averaged. This process is repeated until no new neighbors can be added to the cluster. Once this has been achieved, the final parent austenite orientation assigned to the cluster is taken as the average of the collected possible parent orientations- one from each daughter grain that is a member of the cluster. This average is computed as the vector average of the set of collected parent orientations represented as quaternions [14, 15]. This process is described in figure 6, a flowchart outlining the algorithm's basic logic. Figure 5 depicts the growth of a cluster visually, as neighbor martensite grains are added.

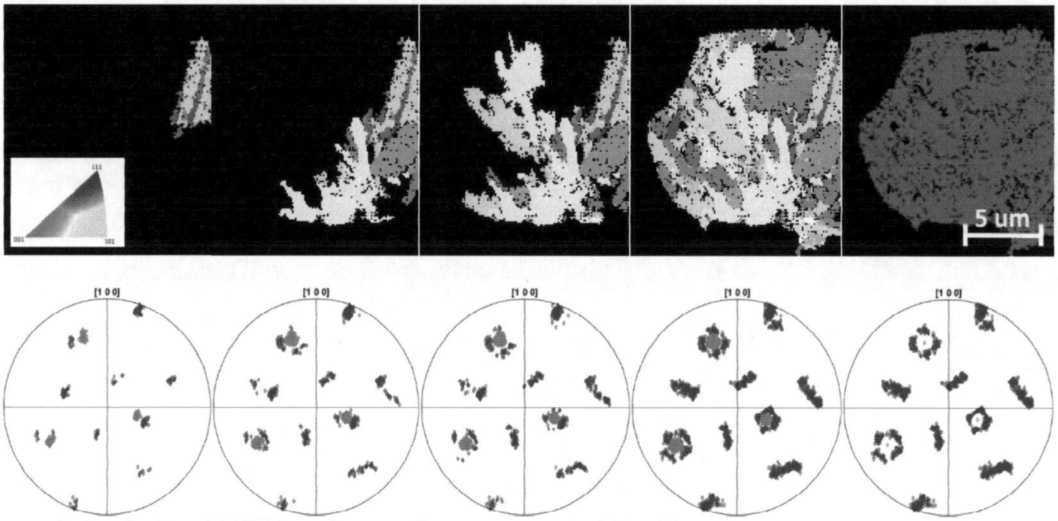

Figure 5: ArcelorMittal M1700 steel, original scan spanning a 14 by 16 μm area at a step size of 0.1 μm. Progressing from left to right- individual stages of the cluster growth algorithm, showing growth of a cluster in the IPF map "space". The pole figures demonstrate the collection of possible parent orientations (red) in orientation space. As neighboring martensite grains are added to the cluster, their orientations (blue), and their selected possible parents are added to the pole figure. Once no more grains can be added, the group of possible parents are averaged into a single orientation.

Each cluster is grown independently from all other clusters. Any daughter grain that was added to a previous cluster may also be added to a subsequent cluster. By creating an exhaustive list of seeds and

IOP Conf. Series: Materials Science and Engineering **375** (2018) 012012 doi:10.1088/1757-899X/375/1/012012

allowing each to grow independently, an extended list of both clusters and ambiguities is created. While each of these ambiguities must be considered, the full list of possible groupings of daughter grains into clusters provides a large dataset that can be explored to find the best reconstruction.

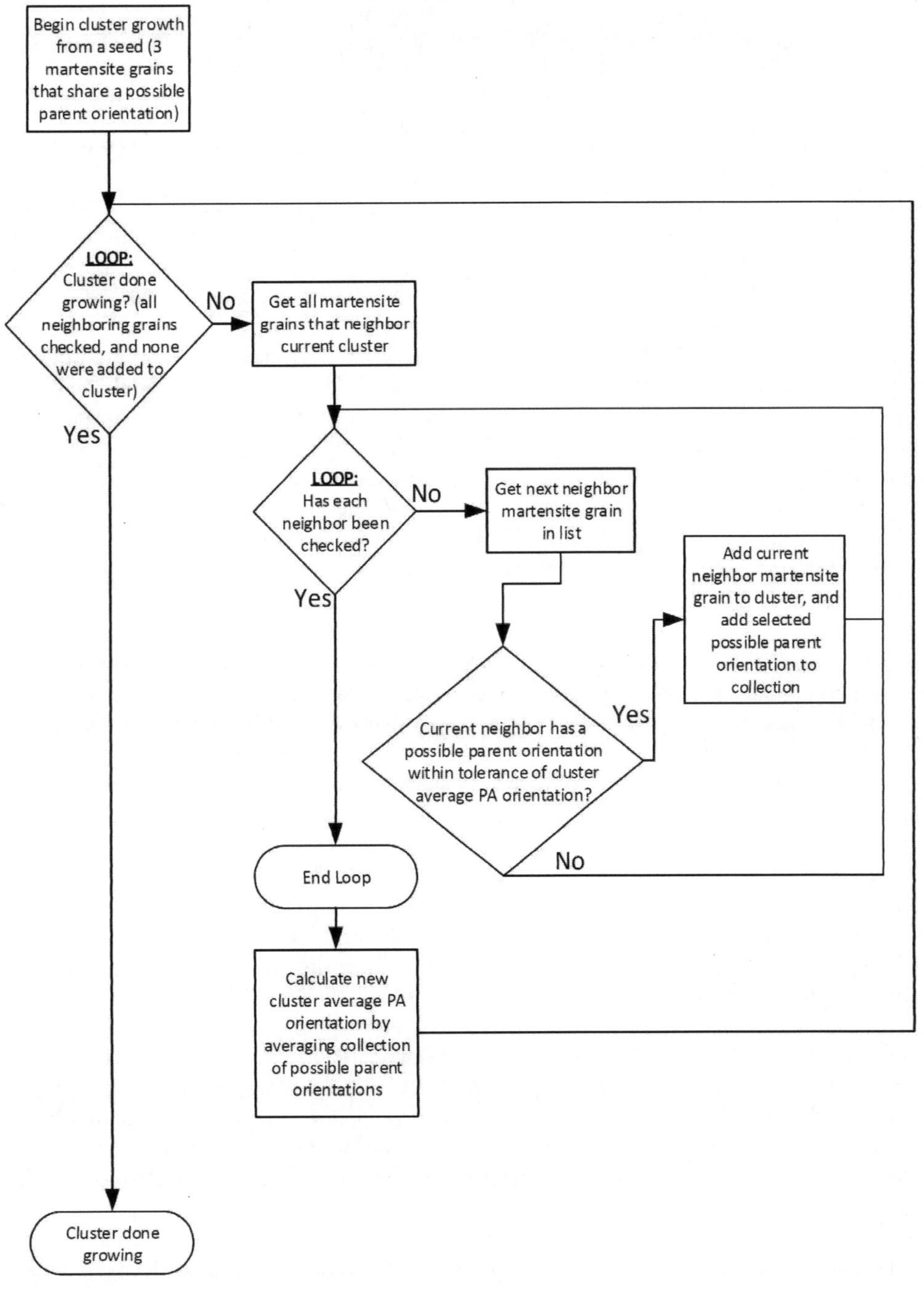

Figure 6: A flowchart describing how a cluster is grown from a "seed"

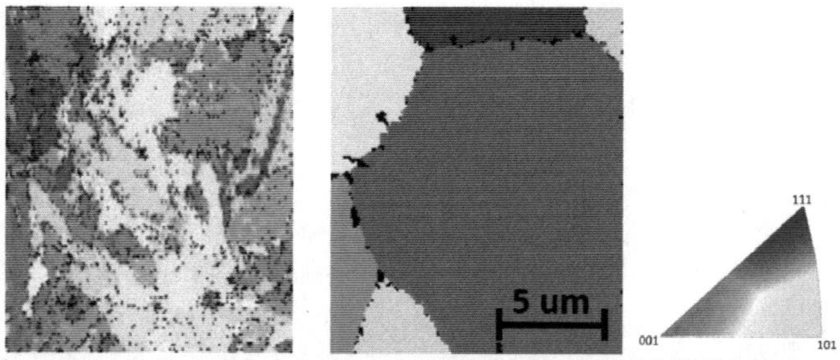

Figure 7: A completed reconstruction of the same region of ArcelorMittal M1700 depicted in figure 5. Each cluster is mutually exclusive and no ambiguities exist.

Various automated and semi-automated methods are used to resolve ambiguities. Ambiguities are classified by the topological relationship between the two overlapping clusters. Some ambiguities are in the form of clusters that are complete subsets of other clusters, where an additional possible austenite orientation can be assigned to that region. Some ambiguities are caused by two clusters with the same set of daughter grains but different parent orientations. These are cases where the set of variants is insufficient for a single unique parent orientation to be calculated. Other ambiguities are caused when daughter grains are erroneously assigned to a neighboring cluster by simply being within tolerance of that clusters calculated parent orientation- which may happen in a noisy dataset. Each of these ambiguities are unique in origin and must be handled in different ways. The determination of the nature of each ambiguity can only be automated to an extent, and often manual intervention is required for their resolution. A delayed decision making process allows the full extent of each ambiguity to be explored. The following sections detail the results of reconstructions containing ambiguities of different types. The approach to resolving them is described.

4. Results

Several reconstructions of martensitic steels demonstrate the different topological natures of ambiguous regions caused by twinning in the parent austenite. The common set of daughter grains belonging to two overlapping clusters as well as the spatial distribution of the clusters within the space of the EBSD map are important factors to consider when assessing the nature of an ambiguity. One cluster may be a subset of another cluster, or they may share a subset. Furthermore, the external boundaries of each cluster (as seen by the human eye) may suggest a relationship not reflected by their topologies. Various reconstructed datasets demonstrate this here.

One such dataset, demonstrated in figure 8b, shows several large clusters with ambiguous regions obscuring the boundaries between the two parents that contain them. In both cases, the calculated parent orientations of the clusters containing the ambiguity can be shown to satisfy the misorientation relationship of a pair of twinned FCC crystals—a 60° rotation about a <111> axis. From the back-calculated orientations for either overlapping cluster the twinning plane can be defined. In figure 8c, the twinning plane for both ambiguities is plotted where it best fits the existing boundary between the two overlapping clusters. In one case (the leftmost white region), the twinning plane is reasonably well oriented with existing boundary between the two overlapping clusters. In the case of the ambiguity to the right, however, the boundary between both clusters seems to be completely consumed by the ambiguous region. To resolve the ambiguities, the best division of daughter grains was chosen that resulted in a boundary that was both straight and reasonably well oriented with respect to the twinning plane. This example demonstrates the variation in degree of overlap an ambiguity may have with the clusters that define it—an ambiguous region may be evenly distributed between both overlapping clusters, or it may exist mostly to one side of the real parent boundary.

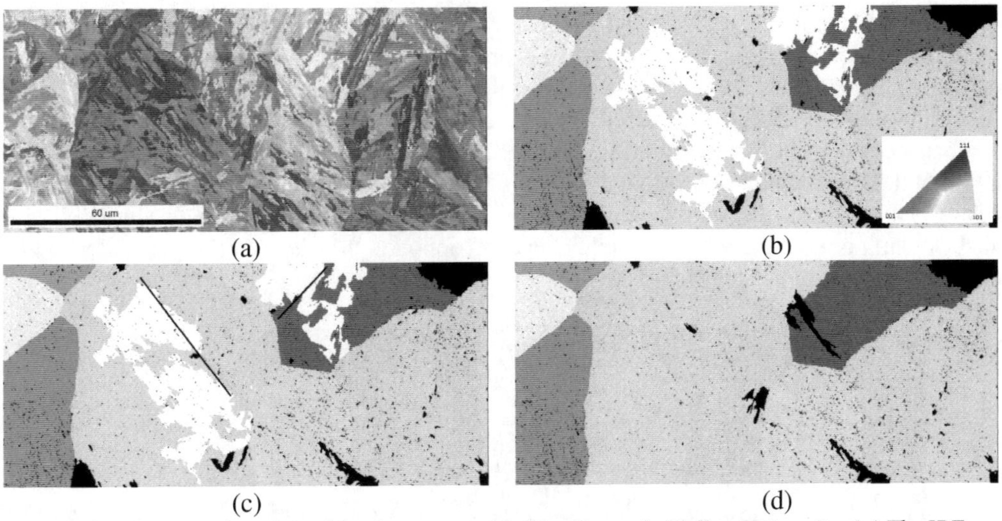

Figure 8:Another, larger subsection of the dataset provided by Carnegie Mellon University. (a) The IPF map of a martensitic steel, as produced by OIM analysis. (b) A reconstruction of the parent structure (IPF map generated in MATLAB), each cluster colored according to its back-calculated parent orientation. Two large ambiguous regions are found, highlighted in white. In both cases, the parent clusters that overlap satisfy the misorientation relationship of two twinned FCC grains to within 2 degrees. (c) The (111) plane that would be the twinning plane (if the two parent grains were indeed twinned) is plotted for each ambiguity. In the case of the left ambiguity, the proposed twinning plane is in good agreement with existing boundary between the two clusters. (d) The ambiguities are split between their clusters according to the best fit with the proposed twinning planes. In this way, other such twins may be found in this dataset.

Another topological case for an ambiguity is when a cluster is found as a complete subset of another cluster. This is demonstrated in figure 9. Here, all the members of the smaller cluster are also members of the larger surrounding cluster. The cluster representing the subset again satisfies the misorientation relationship of a twin with the cluster that contains it. Furthermore, the major axis of an ellipse fitted to the sub-cluster is in close agreement with the orientation of the twinning plane between both clusters' orientations. This suggests the strong possibility that an annealing twin existed within this parent grain, as demonstrated by the sub-cluster stretching across the larger cluster. In this case, each member of the smaller cluster is one of the common variants between the two twinned austenite orientations.

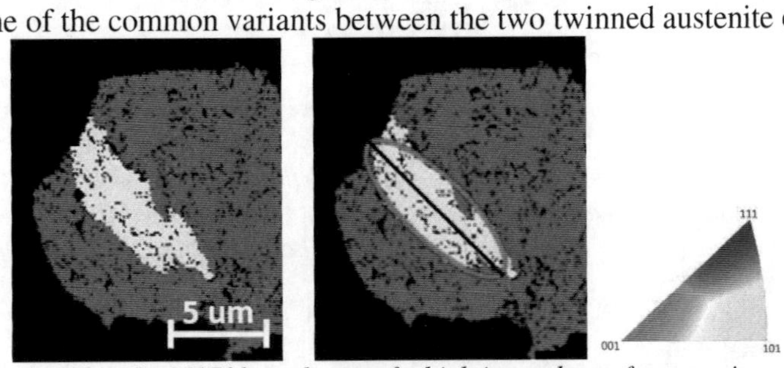

Figure 9: Two clusters found in the M1700 steel, one of which is a subset of martensite grains belonging to the larger cluster. In the case of the smaller cluster, the back-calculated parent orientation satisfies a twin misorientation relationship with the other cluster's orientation to within 3 degrees. The long axis of the cluster (taken as the major axis of a fitted ellipse) is also within 3 degrees of the (111) twinning planes of either austenite orientation.

A final example demonstrates the relationship of two clusters that share a subset, but where one cluster appears to exist only "inside" the physical boundaries of a larger cluster. Figure 10 Shows the larger cluster, and then the superposition of the smaller cluster. There are small sets of daughter grains that belong to only the smaller cluster, but that are also within the boundary that would form a convex hull around the larger cluster. Again, these two clusters are within a close misorientation of a twin relationship. The smaller cluster appears to suggest the presence of a twin band—this time large enough that some small regions of material within the twin are comprised of variants that were not common between both clusters' orientations, but belonging only to the twinned orientation. The orientation of the twinning plane as plotted is also in agreement with sections of boundary between the two clusters. Of interest is the region not covered by the twin cluster. This region is composed of variants belonging only to the *larger* cluster. This casts doubt on the idea that the twin extended fully across the larger grain.

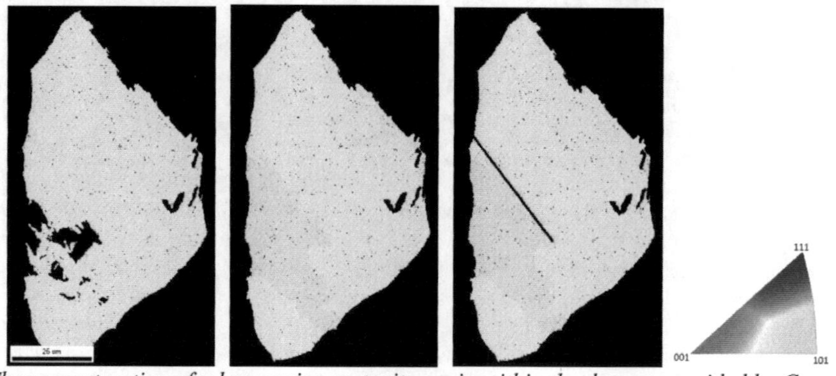

Figure 10: The reconstruction of a large prior austenite grain within the dataset provided by Carnegie Mellon University (IPF created in OIM). Another example of an interior cluster that satisfies a twinned misorientation relationship with the cluster that contains it. The twinning plane is plotted and is shown to be in good agreement with the existing boundary.

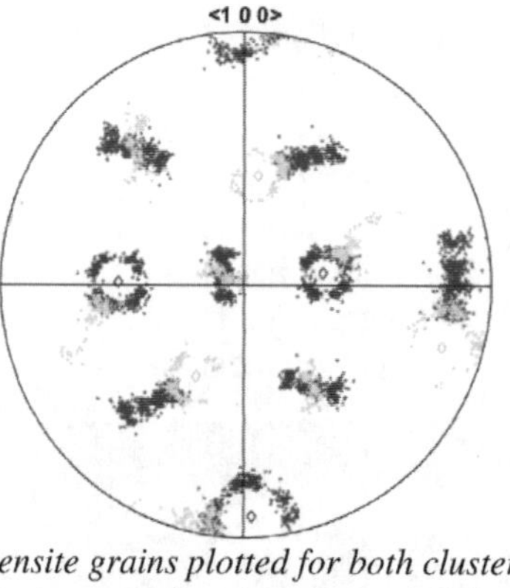

Figure 11: The orientations of the martensite grains plotted for both clusters shown in figure 9. Martensite orientations belonging to the interior clusters are plotted in green, and the orientations belonging to the larger cluster are plotted in blue. The interior cluster contains variants that are not held in common between both twinned cluster orientations.

5. Discussion and Conclusions

The reconstruction of parent austenite was performed here using a cluster growth algorithm within a delayed-decision framework. Many possible groupings of daughter grains into parent clusters are considered, and the nature of the ambiguous overlapping regions are explored. The ambiguous regions caused by the presence of twin boundaries in the parent austenite are of particular interest, and their various topologies and morphologies are demonstrated. While this method of reconstruction is automated, post-processing and interpretation of results is left to semi-automated or manual intervention. The complex nature of each ambiguous region makes it difficult to resolve each one via automated methods, though many ambiguities can potentially be dealt with via automated techniques.

The goal of this method is to explore all possible groupings of daughter grains into possible parent grains. This allows the user to take advantage of the information made available by a full collection of possible groupings, and a full list of ambiguities contained in the data set.

The delayed decision making approach used here was found to have some advantages over a reconstruction that makes assignments prematurely. A cluster growth algorithm that employs a "first come first serve" type approach would be very unlikely to correctly assign members of an ambiguous region to the appropriate cluster. This is demonstrated by the various topologies of the ambiguous regions created by twins, and the different extent to which an ambiguous region may extend into one cluster or the other. Twin bands composed entirely of common variants between both twinned austenite orientations would very likely be missed by a reconstruction method that does not perform a more extensive search of possible groupings. More complex still are the overlapping clusters that have both inclusive and exclusive subsets, but where the nature of their physical boundaries suggest that one cluster is spatially "contained" by the other. Each of these cases demonstrate the advantage a delayed-decision making approach has to exposing the full extent of the ambiguity.

While the reconstruction of twin boundaries is not precise in the sense that it results in a straight boundary, the knowledge of the presence and approximate orientation of the boundary is of great value in a reconstruction attempt. In figure 9, where a large interior twin is demonstrated, there are still questions that must be answered regarding the nature of the boundary between the two reconstructed clusters. The region of the larger cluster that seems as though it should belong to the twin orientation is composed of variants that are *not* common between the two twinned orientations, but only to the larger cluster. Such information must be considered when deciding whether or not an interior cluster truly represents a twin.

Future work will continue to automate specific tasks, as well as study different aspects of the martensite transformation. Reconstructions using this method may shed light into variant selection behaviors. Other nucleation and growth behaviors may be shown to result in similar ambiguities to twins. Variants can be grouped according to the close-packed habit plane on which they form, nucleating and growing in packets. This type of delayed-decision making approach may aid in a study of the formation and size of these packets.

The advantages of this method are demonstrated in the type of information made available to a person performing a reconstruction of parent austenite. The large number of clusters found provide a valuable set of data. Though searching this dataset is only semi-automated and the generation of many clusters can be computationally intensive, the resulting information that is made available is of high value.

6. Acknowledgements

This work was performed in collaboration with and supported financially by ArcelorMittal. This work was also supported by the National Science Foundation under contract no. DMR-1507095. The authors also thank Anthony Rollett for providing EBSD data used in this work.

7. References

1. Cayron, C., B. Artaud, and L. Briottet, *Reconstruction of parent grains from EBSD data.* Materials Characterization, 2006. **57**(4-5): p. 386-401.
2. Abbasi, M., et al., *An approach to prior austenite reconstruction.* Materials Characterization, 2012. **66**: p. 1-8.
3. Germain, L., et al., *An advanced approach to reconstructing parent orientation maps in the case of approximate orientation relations: Application to steels.* Acta Materialia, 2012. **60**(11): p. 4551-4562.
4. Abbasi, M., et al., *EBSD and reconstruction of pre-transformation microstructures, examples and complexities in steels.* Materials Characterization, 2014. **95**: p. 219-231.
5. Ranger, C.T., V.; Farjami, S.; Merwin, M.; Germain, L.; Rollett, A.; , *Austenite Reconstruction Elucidates Prior Grain Size Dependence of Toughness in a Low Alloy Steel.* Submitted (2018).
6. Germain, L., et al., *Determination of parent orientation maps in advanced titanium-based alloys.* Journal of Microscopy-Oxford, 2007. **227**(3): p. 284-291.
7. Humbert, M. and N. Gey, *The calculation of a parent grain orientation from inherited variants for approximate (b.c.c.-h.c.p.) orientation relations.* Journal of Applied Crystallography, 2002. **35**: p. 401-405.
8. Cayron, C., *Groupoid of orientational variants.* Acta Crystallographica Section A, 2006. **62**: p. 21-40.
9. Cayron, C., *Crystallographic Reconstruction Methods to Study Phase Transformations by EBSD.* Microscopy and Microanalysis, 2009. **15**: p. 396-397.
10. Miyamoto, G., et al., *Reconstruction of Parent Austenite Grain Structure Based on Crystal Orientation Map of Bainite with and without Ausforming.* Isij International, 2011. **51**(7): p. 1174-1178.
11. Verbeken, K., L. Barbe, and D. Raabe, *Evaluation of the Crystallographic Orientation Relationships between FCC and BCC Phases in TRIP Steels.* Isij International, 2009. **49**(10): p. 1601-1609.
12. Pereloma, E. and D.V. Edmonds, *Phase transformations in steels.* Woodhead Publishing in materials. 2012, Oxford ; Philadelphia: Woodhead Publishing. 2 volumes (xxii, 634 xxiv, 648 pages).
13. Miyamoto, G., N. Takayama, and T. Furuhara, *Accurate measurement of the orientation relationship of lath martensite and bainite by electron backscatter diffraction analysis.* Scripta Materialia, 2009. **60**(12): p. 1113-1116.
14. Cho, J.H., A.D. Rollett, and K.H. Oh, *Determination of a mean orientation in electron backscatter diffraction measurements.* Metallurgical and Materials Transactions a-Physical Metallurgy and Materials Science, 2005. **36a**(12): p. 3427-3438.
15. Tari, V., A.D. Rollett, and H. Beladi, *Back calculation of parent austenite orientation using a clustering approach.* Journal of Applied Crystallography, 2013. **46**: p. 210-215.

IOP Conf. Series: Materials Science and Engineering **375** (2018) 012013 doi:10.1088/1757-899X/375/1/012013

Effect of Annealing Temperature on the Texture and Magnetic Barkhausen Noise of a Non-oriented Electrical Steel (0.88 wt% Si) after Inclined Cold Rolling

Youliang He[1*], Mehdi Mehdi[1, 2], Erik J. Hilinski[3], Afsaneh Edrisy[2]

[1] CanmetMATERIALS, Natural Resources Canada, Hamilton, Ontario, Canada
[2] Department of Mechanical, Automotive, and Materials Engineering, University of Windsor, Windsor, ON, Canada
[3] Tempel Steel Co., Chicago, IL, USA

[*]Corresponding author: youliang.he@canada.ca

Abstract. *Inclined cold rolling* was employed in this study to process a 0.88 wt% Si non-oriented electrical steel. After conventional hot rolling and annealing, the steel was cold rolled at various angles (i.e. 0°, 45°, and 90°) to the hot rolling direction (HRD), and the cold-rolled steel sheets were then annealed at different temperatures from 600°C to 750°C for 30 seconds to investigate the effect of annealing temperature on the texture and magnetic response of the material. The texture was measured by electron backscatter diffraction (EBSD), and the magnetic response of the steel was evaluated by magnetic Barkhausen noise (MBN) analysis. It was found that all the cold-rolled steels partially recrystallize at temperatures below 750°C, but the progress of recrystallization differs in steels cold rolled at different angles to the HRD, i.e. samples rolled at 45° to the HRD recrystallize faster than those after conventional rolling (0° to HRD) or cross rolling (90° to the HRD). The initial cold rolling texture (mainly the α-fibre and γ-fibre) may change to cube, rotated cube, rotated Goss or {111}<112> depending on the rolling scheme and the annealing temperature. The MBN root mean square values of the samples cold rolled at different angles to the HRD show substantial differences during the annealing process. At low annealing temperatures (600°C and 650°C), the anisotropy of MBN in the conventionally rolled steel is much higher than those after inclined rolling or cross rolling.

1. Introduction

Non-oriented electrical steels are widely used for the manufacturing of lamination cores for electric motors. The crystallographic texture of the final steel sheets has a considerable effect on the magnetic properties of the steel core, thus affecting the performance and energy efficiency of the motors. To improve energy efficiency and maximize the output torque of the motor, the steel core is required to possess low core loss and high permeability, which are affected by the chemistry, thickness, microstructure, texture and stress state of the sheets. The desired crystallographic orientations for electric motors are the <001>//ND (normal direction) fiber texture, which maximizes the easy <001> axes in the sheet plane. To achieve the desired magnetic properties, a good control of the steel making/alloying as well as the thermomechanical processing is necessary.

Annealing is usually the last step of the thermomechanical processing, and the microstructure and texture obtained from the final annealing process have a direct effect on the magnetic properties of the lamination core. It has been shown [1] that it was quite challenging to produce the desired <001>//ND texture through conventional rolling-annealing procedures. Thus in this research, an unconventional rolling process, i.e. *inclined cold rolling* [2], was adopted to alter the conventional cold-rolling texture by changing the starting texture before cold rolling. Inclined cold rolling differs from conventional rolling in that, the cold rolling is conducted at an *angle* to the *hot rolling direction* (instead of *along* the HRD as in conventional rolling), i.e. after conventional hot rolling, smaller plates were cut from the hot band at different angles to the HRD and cold rolled in a conventional manner [3]. In this way, the initial texture was conveniently altered (rotated around the ND) while the chemistry, microstructure and processing history of the material were kept the same. This enabled one to exclusively investigate the effect of initial texture on the final texture without changing other parameters.

Previous studies [2-4] have shown that using this method, it was able to produce a strong cube texture at some inclination angles. It was also noticed that, not only did the initial texture before cold rolling affect the final annealing texture, but the annealing conditions (e.g. temperature and holding time) also influenced the texture. Thus, in this study, a series of annealing temperatures (600°C, 650°C, 700°C, and 750°C) were selected to conduct the annealing experiments on the incline-rolled steel to investigate the effect of annealing temperature on the microstructure/texture and the magnetic properties of the steel. Moreover, different from conventional magnetic property measurements by standard Epstein frame technique [5] or single sheet method [6], the magnetic response of the steel in this study was measured by a relatively new technology, i.e. magnetic Barkhausen noise (MBN) [7-9], which enabled one to measure angular MBN response. The effects of the cold rolling scheme and annealing temperature on the magnetic Barkhausen noise were examined.

2. Material and Experimental Procedures

The material investigated in this study was a low silicon non-oriented electrical steel. The chemical composition of the steel was (wt%): 0.88 Si, 0.002 C, 0.31 Mn, 0.46 Al, 0.001 S and 0.01 P. Detailed procedures to melt and process the steel were shown elsewhere [2]. After conventional hot rolling and annealing, samples (180 mm × 50 mm × 2.5 mm) were cut from the hot band at 0°, 45° and 90° to the HRD and cold rolled to a thickness of ~0.5 mm (~78% reduction), which corresponded to conventional rolling (0°), inclined rolling (45°) and cross rolling (90°), respectively. The cold-rolled steel sheets were then annealed at various temperatures (600°C, 650°C, 700°C and 750°C) in an argon-protected environment for 30 seconds (with a heating rate of 500°C/h), and furnace cooled. EBSD scans were performed on the ND-RD (normal direction - rolling direction in cold rolling) sections using an EDAX OIM 6.2 EBSD system under a field emission gun scanning electron microscope. The textures (ODF's) were then calculated from the orientation data using a harmonic series expansion method. The grain sizes of the recrystallized crystals were also calculated based on the EBSD data.

Magnetic Barkhausen noise technique was utilized in this research to evaluate the magnetic response of the material. MBN is a noise-like signal generated from a ferromagnetic material when an external magnetic field with varying intensity and/or direction is applied to the material [10]. MBN is closely related to the interaction between the motion of magnetic domain walls and the stress configuration and microstructural features in ferromagnetic materials [10]. The Barkhausen noise is usually detected as a voltage signal (in millivolts), and the MBN may be affected by a number of material characteristics, such as the presence of elastic, plastic or residual stresses, the microstructure of the material (carbon content, phases, precipitates, dislocations, grain boundaries, etc.), the hardness and texture, etc.

The MBN measurements were conducted on the surfaces of the cold-rolled and annealed samples using a commercial MBN analyzer [9, 11]. A general purpose sensor was employed for all the measurements. To eliminate the "lift-off" problem generally encountered in MBN measurements, a testing platform with three-dimensional translation, two-axis tilting and two-directional rotation was built, which made the measurements repeatable and reliable. A picture of the MBN system used for this study is shown in Figure 1. For each sample, the MBN signals were collected along all the directions (0°-360°) in the sheet plane, with an interval of 15°/30° to the cold rolling direction. The average value of the measurements in all the directions on each sample was then taken as the overall MBN. At each direction, the measurement consists of 20 bursts of MBN and the average root mean square value of the 20 bursts was reported. The frequency and voltage used for the magnetic excitation were 125 Hz and 5 V, respectively. The raw MBN signals collected from the material were filtered to a frequency range of 0.1 - 1000 kHz. During data processing, these were further filtered to 10-1000 kHz to smooth the data [11].

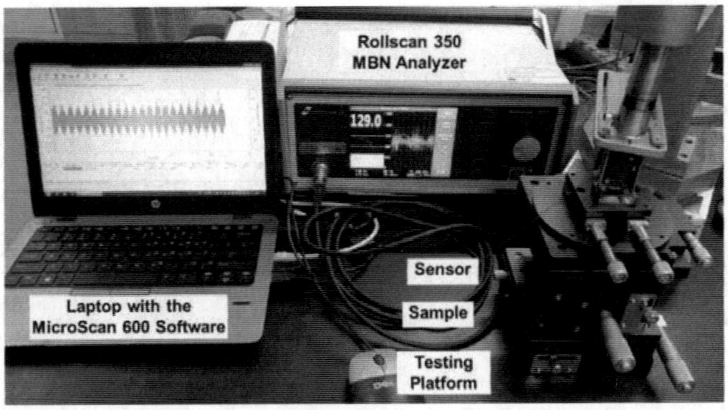

Figure 1. The magnetic Barkhausen noise analysis system used in this study.

3. Results and Discussion

3.1. Crystallographic Texture

3.1.1. Cold deformation texture. When the steel plate was cold rolled at 0°, 45° and 90° to the hot rolling direction, the material was actually subjected to conventional, inclined and cross rolling, respectively. Although the starting microstructure was identical, due to the rotation (45° and 90° around ND) of the initial texture for inclined and cross rolling, the resulted deformation textures are noticeably different from conventional rolling (Figure 2). The maximum texture intensity after cross rolling is much higher than that after conventional rolling, while the texture intensity after inclined rolling is considerably lower than that after conventional rolling. Although the textures after the three rolling processes have common components such as rotated cube ({001}<110>), α-fibre (<011>//RD) and γ-fibre (<111>//ND), the relative intensities among these components are quite different. Conventional rolling produces a strong rotated cube texture (maximum intensity: 13.4), a relatively strong and continuous α-fibre, and a weak γ-fibre. Inclined rolling reduces the intensity of the rotated cube while enhancing the γ-fibre. Thus the texture is featured by a considerably lower maximum intensity but a more uniform distribution of components such as rotated cube, α-fibre and γ-fibre. The strongest texture is observed near the {112}<110> orientation on the α-fibre. A strong {111}<110> component on the γ-fibre is also produced, which is not obvious in either conventional rolling or cross rolling. Cross rolling created a very strong rotated cube texture (intensity: 21.3), but the γ-fibre and α-fibre are weakened.

3.1.2. Annealing microstructure and texture. The variations of the fraction of recrystallization and grain size with respect to the annealing temperature and cold rolling direction are illustrated in Figure 3. Below 750°C, the fractions of recrystallization under the same annealing temperature are quite different among the three cold rolling schemes (Figure 3a), especially after annealing at 650°C (varying from 45% to 77%). The relationship between the fraction of recrystallization and the annealing temperature for cross-rolled, conventionally-rolled and incline-rolled samples can be fitted to a linear, a 2nd order polynomial and a 3rd order polynomial function, respectively. Compared to conventional rolling, cross rolling retards the recrystallization process since the fractions of recrystallization at all the temperatures are smaller than those after conventional rolling. By contrast, inclined rolling accelerates the recrystallization process, since the fraction of recrystallization under the same annealing temperature is higher than those after conventional rolling. Apparently, cold rolling at various directions to the HRD created different microstructures, textures and stored energies or different distributions of those in the steel, which, in turn, lead to different recrystallization behaviours during the annealing process. The fast recrystallization of the incline-rolled steel may be attributed to the strong {111}<110> texture in the deformed structure, which was associated with a higher stored energy and numerous nucleation sites for the start of recrystallization.

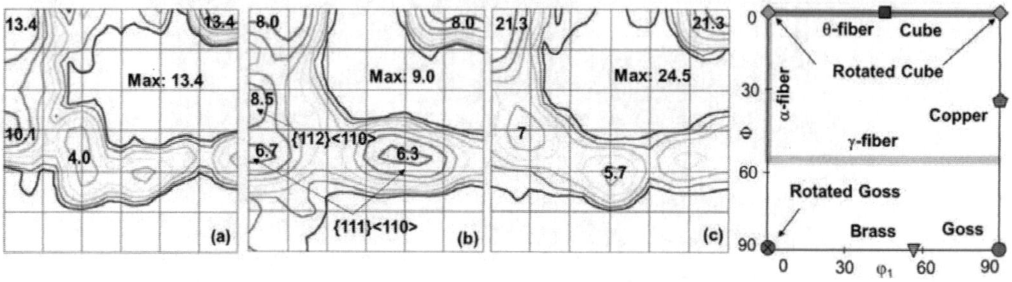

Figure 2. The textures after cold rolling: (a) conventional rolling (0° to HRD), (b) inclined rolling (45° to HRD), (c) cross rolling (90° to HRD). $\varphi_2 = 45°$ sections of the Euler space (Bunge notation).

The grain size of the recrystallized crystals also shows considerable differences among the three rolling schemes (Figure 3b). At a low annealing temperature (600°C), the grain size is almost the same for the three schemes, but with the increase of the temperature, the grain sizes of conventionally and cross rolled samples increase gradually with the temperature, while that of the incline-rolled sample increases quickly at 650°C, then stops at 700°C, and finally increases again at 750°C. The final grain sizes after complete recrystallization (750°C) are also different: the cross-rolled steel has a considerably larger grain size (~34 μm) than that (~21 μm) after conventional rolling or inclined rolling. Thus, the cold rolling with different initial textures also induced different grain growth behaviours during the annealing process.

The variation of recrystallization texture at various annealing temperatures is shown in Figure 4. For the steel after conventional rolling, annealing at 600°C produces the cube and rotated cube textures on the θ-fiber, which are the desired textures for non-oriented electrical steels. There is also a {110}<665> component on the <110>//ND fiber together with a weak rotated Goss texture. When the annealing temperature is increased to 650°C, the overall texture intensity is considerably increased and the rotated Goss texture is strengthened. In the meantime, a relatively strong cube texture is created and the rotated cube texture disappears. If the annealing temperature is 700°C, a very strong {111}<112> texture (on the γ-fiber) forms and, the cube texture becomes very weak. Apparently, annealing the conventionally rolled steel at this temperature should be avoided since it will generate the undesired <111>//ND texture. If the annealing temperature is further increased to 750°C, a {001}<120> texture forms on the θ-fiber, and a rotated Goss is also produced.

The steel after inclined rolling (45° to the HRD) shows considerably different textures during annealing: when the annealing temperature increases from 600°C to 700°C, there is always a strong cube texture together with a relatively strong rotated Goss texture. When the annealing temperature is further increased to 750°C, the cube texture disappears, and a strong {110}<111> texture is produced on the <110>//ND fiber. Essentially no γ-fibre is produced at any of the annealing temperatures. Thus inclined cold rolling can prevent the formation of the γ-fibre texture during the annealing process, although in the deformed state, it has the strongest γ-fiber. Apparently, for incline-rolled steel, the optimum annealing temperature is 700°C since a strong cube texture can be produced.

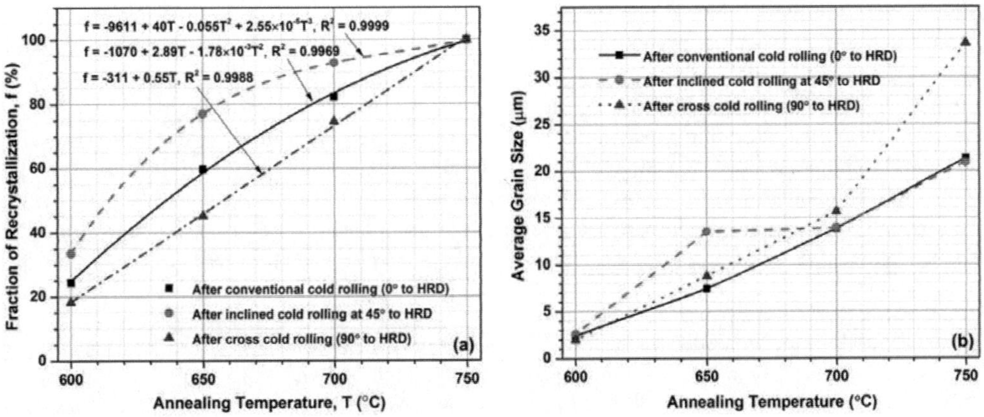

Figure 3. Variation of recrystallization with respect to the annealing temperature: (a) fraction of recrystallization, (b) grain size of the recrystallized crystals.

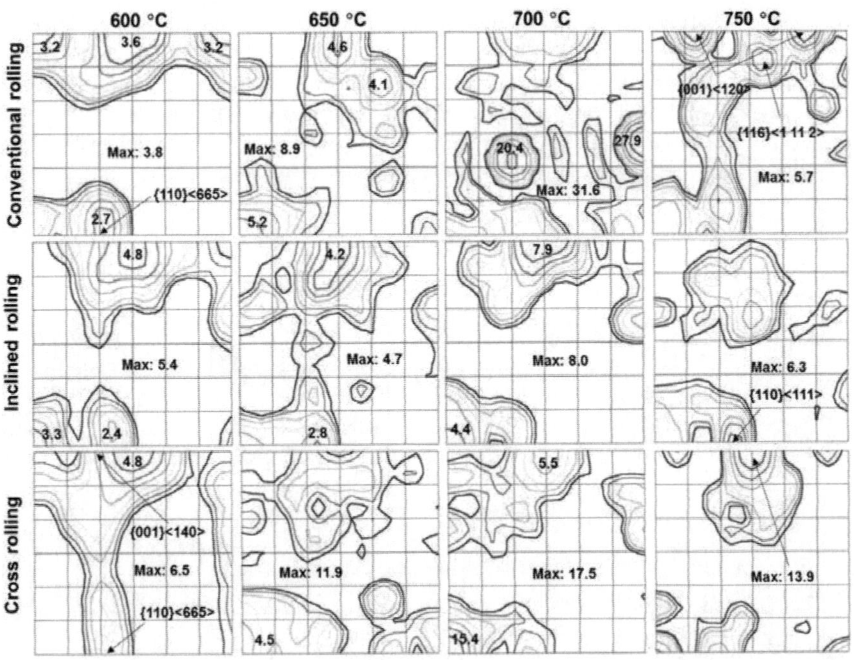

Figure 4. Comparison of the recrystallization textures after annealing at various temperatures for the samples after conventional, inclined and cross rolling.

For the steel after cross rolling, annealing at 600°C produces a cube texture plus a fibre extending from {001}<140> on the θ-fiber to {110}<665> on the <110>//ND fiber. At 650°C, the texture on the

φ_2 =45° section is quite weak, although the maximum intensity (on other sections) is relatively high. Only a relatively strong rotated Goss texture is noticed. When the temperature is increased to 700°C, the rotated Goss texture is significantly strengthened, and a relatively strong cube component is also created. Further increasing the annealing temperature to 750°C, a very strong cube texture forms, and the rotated Goss texture is weakened. Again, no <111>//ND component is formed if the annealing temperature is 650°C or higher. Apparently, for cross-rolled steel the optimum annealing temperature is 750°C because a strong cube texture forms at this temperature.

It is thus seen that, after cold rolling at different angles to the HRD, the deformed steels behave differently during the annealing process. To produce the required <001>//ND texture, the steel after conventional should not be annealed at 700°C, because the texture produced is a very strong {111}<112> component on the γ-fibre. On the other hand, for the incline-rolled steel, it is better annealed at 700°C since at this temperature, a strong cube component is produced. For the cross-rolled steel, annealing at 750°C is suggested since a strong cube texture can be produced.

3.2. Magnetic Barkhausen Noise

The original MBN signal recorded by the testing system is the distribution of the noise (a voltage signal) with respect to time (Figure 5). The raw data is usually smoothed and statistically analyzed, and a number of MBN parameters can be derived from the smoothed MBN envelope [11]. One of the most commonly used parameters is the root mean square (rms) of the MBN signal, which is the square root of the sum of all the voltages squared divided by the total number of events recorded. The MBN_{rms} represents the overall intensity of the MBN signal, and the rms value of a complete MBN burst is fairly stable from one burst to another. In this research, the MBN_{rms} is utilized to evaluate the magnetic response of the material. The MBN_{rms} value is highly dependent on the stress state and the microstructure of the material. For example, the two MBN bursts (recorded from cold-rolled and annealed samples) shown in Figure 5 have significantly different shapes and peak voltages. The plastic deformation induced a huge amount of dislocations and residual stresses in the material, thus the MBN voltages are much higher than those from a completely recrystallized steel (with much less dislocations and stresses).

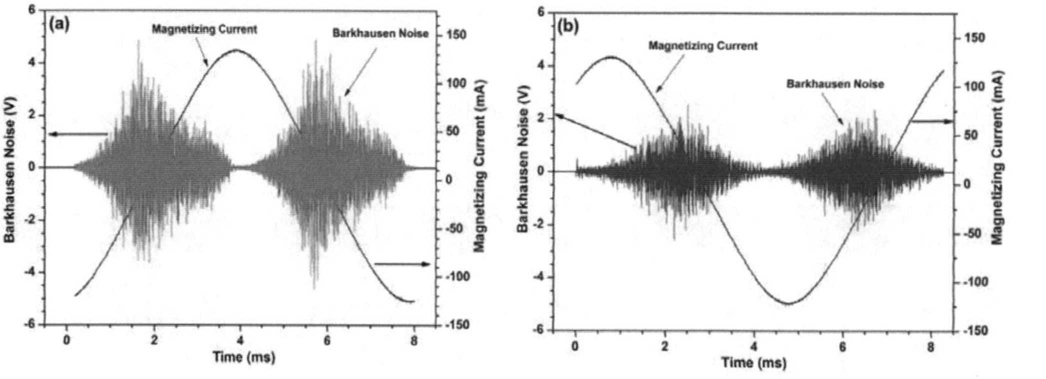

Figure 5. Typical MBN bursts recorded from steel samples: (a) after cold rolling at 45° to the HRD, (b) after annealing at 750°C for 30 seconds.

The angular MBN_{rms} distributions with respect to the direction of magnetization for the steels after cold rolling at different angles to the HRD are shown in Figure 6. Although the overall MBN_{rms} values after cold rolling with all the three schemes are quite similar (around 700), the angular MBN_{rms} distributions after annealing at various temperatures are quite different. For the conventionally-rolled steel (Figure 6a), annealing at 600°C causes a considerable drop of the MBN_{rms} at all the directions, but the anisotropy of the MBN_{rms} is similar to that after cold rolling, i.e. with a slightly larger MBN_{rms} in the RD than in the TD. Increasing the annealing temperature to 650°C, the overall MBN_{rms} values go back to the level of the cold-rolled steel. This is unusual as it is normally expected that the recrystallization of the steel would result in the decrease of the MBN because of the decrease of dislocation density and the release of stress. The reason for this abnormal increase of the MBN_{rms} at 650°C is not clear. Further increasing the annealing temperature to 700°C and 750°C, the MBN_{rms} values gradually decrease, and the degree of anisotropy is also reduced (the distributions are close to perfect circles). After complete recrystallization at 750°C, apparent MBN_{rms} anisotropy is still noticed, but the polarity has changed, i.e. the minimum MBN_{rms} is observed at 30°/210° from the RD, while the maximum MBN_{rms} is seen at 120°/300° from the RD. The anisotropy observed in the completely recrystallized steel is mainly due to the magnetocrystalline anisotropy induced by the crystallographic texture [12].

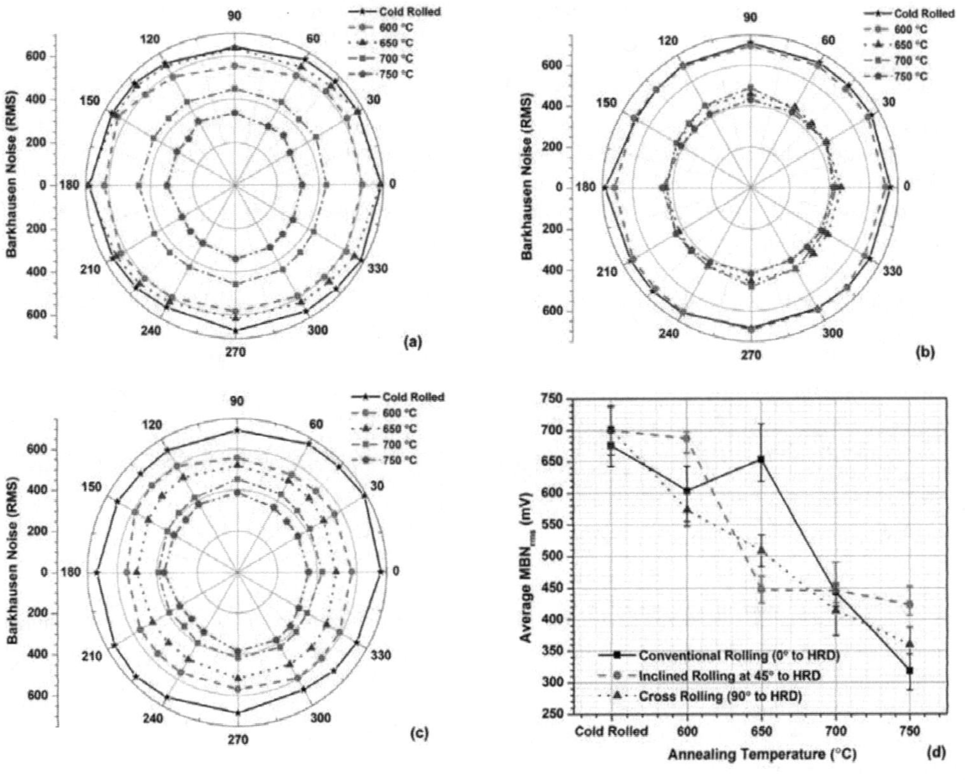

Figure 6. Variation of the MBN_{rms} with respect to the annealing temperature and the cold rolling scheme: (a), (b), and (c) angular MBN_{rms} for the conventionally (0° to HRD), incline (45° to HRD), and cross rolled (90° to HRD) steels, respectively, (d) average MBN_{rms} of all the directions.

The incline-rolled steel shows considerably different MBN when subjected to annealing (Figure 6b). At 600°C, although with the largest fraction of recrystallization (35%) among the three steels (Figure 3a), the measured angular MBN_{rms} has essentially the same values as the cold-rolled steel, i.e. annealing at this temperature essentially does not change the MBN of the material, although the steel has already partially recrystallized. However, when the annealing temperature is increased to 650°C (the fraction of recrystallization reaches ~80%), the MBN_{rms} suddenly drops to a very low level (i.e. from ~700 to ~420), and further increasing the annealing temperature essentially does not change the MBN_{rms} anymore. It is also noted that the anisotropy of the MBN_{rms} is reduced. The MBN_{rms} values (~400) after complete recrystallization (750°C) are considerably higher than those (~320) of the conventionally-rolled steel annealed at the same temperature.

The MBN_{rms} of the steel after cross rolling is different from both conventionally and incline rolled samples (Figure 6c). Instead of having a maximum MBN_{rms} in the RD, this is observed at 30°/210° to the RD, and the minimum MBN_{rms} is seen at 120°/300° to RD. When the annealing temperature increases, the overall MBN gradually decreases (Figure 6d), i.e. there exists a correspondence relation between the MBN_{rms} and the annealing temperature. With the increase of the fraction of recrystallization, the anisotropy of the MBN_{rms} also changes, i.e. after complete recrystallization (750°C), the maximum MBN_{rms} is at 120°/300° to RD, while the minimum MBN_{rms} is at 30°/210° to RD. The change of the average MBN_{rms} with respect to the annealing temperature is shown in Figure 6d, which clearly shows the trends of the MBN after cold rolling using the three different schemes.

4. Conclusions

Cold rolling at different angles to the hot rolling direction produces noticeably different deformation textures that lead to different recrystallization behaviors during the annealing process.

Upon annealing, the incline rolled steel recrystallizes considerably faster than the conventionally or cross rolled steels. However, the cross-rolled steel has the largest final grain size when the recrystallization is complete. The annealing textures are also quite different at different temperatures.

After complete recrystallization at 750°C, the cross-rolled steel forms a strong cube texture and the conventionally rolled sample shows a relatively strong {001}<120> texture, both being the desired textures. The steel after inclined rolling, however, does not produce such textures.

While the MBN_{rms} of the cross-rolled steel decreases gradually with the increase of the annealing temperature, the MBN_{rms} values of the conventionally rolled or incline rolled steels show some abnormal behaviors, i.e. the MBN_{rms} of conventionally-rolled steel after annealing at 650°C goes back to the cold-rolled values, while the MBN_{rms} of incline-rolled steel is essentially constant with the increase of the temperature when the temperature is 650°C or higher.

Acknowledgements

The Program of Energy Research and Development, Natural Resources Canada, and the Natural Sciences and Engineering Research Council of Canada are gratefully acknowledged for financial support. Renata Zavadil and Jian Li are thanked for their help on the EBSD measurements.

References

[1] L. Kestens, S. Jacobs, Texture, Stress, and Microstructure (2008), Article ID 173083, 1-9.
[2] Y. He, E. Hilinski, J. Li, Metall. Mater. Trans. A. 46 (2015) 5350-5365.
[3] M. Mehdi, Y. He, E.J. Hilinski, A. Edrisy, J. Magn. Magn. Mater. 429 (2017) 148-160.
[4] Y. He, E.J. Hilinski, J. Magn. Magn. Mater. 405 (2016) 337-352.
[5] ASTM Standard A343/A343M, ASTM International, West Conshohocken, PA (2014).
[6] S. Harada, Y. Ishihara, T. Todaka, Electrical Engineering in Japan. 107 (1987) 1-8.
[7] H. Barkhausen, Physikalishe Zeitschr. 20 (1919) 401-403.
[8] S. Tiitto, IEEE Transactions on Magnetics. MAG-12 (1976) 855-857.
[9] Stresstech Group, Rollscan 350 Operating Instructions, V1.0b (2013) 1-39.
[10] C.G. Stefanita, The Many Sides of Magnetism, Springer, Berlin, 2008, pp. 19-40.
[11] Stresstech Group, MicroScan 600 Operating Instructions, V.5.4b (2015) 1-37.
[12] T.W. Krause, L. Clapham, D.L. Atherton, J. Appl. Phys. 75 (1994) 7983-7988.

IOP Conf. Series: Materials Science and Engineering **375** (2018) 012014 doi:10.1088/1757-899X/375/1/012014

Texture Evolution of a 2.8 wt% Si Non-Oriented Electrical Steel during Hot Band Annealing

Mehdi Mehdi[1, 2], Youliang He[1*], Erik J. Hilinski[3], Afsaneh Edrisy[2]

[1]CanmetMATERIALS, Natural Resources Canada, Hamilton, ON, Canada L8P 0A5
[2]Department of Mechanical, Automotive, and Materials Engineering, University of Windsor, Windsor, ON, Canada N9B 3P4
[3]Tempel Steel Co., Chicago, IL, USA 60640-1020

[*]Corresponding author: youliang.he@canada.ca

Abstract. To optimize the magnetic properties of non-oriented electrical steels, it is necessary to carefully control all the stages of thermomechanical processing during the production of the steel sheets, since the microstructure and crystallographic texture at an early step will usually affect those at the subsequent steps. Many studies have shown that hot band annealing may have a positive effect on the texture of the final sheet, but it is not clear what are the optimal annealing conditions to obtain the desired final textures. In this study, the evolution of microtexture and microstructure of a 2.8 wt% Si non-oriented electrical steel during hot band annealing is studied by using electron backscatter diffraction (EBSD) techniques. A region of the hot-rolled sample was marked by micro-indents so that the same area could be examined at various annealing times (under the same temperature) to investigate the evolution of the microstructure and microtexture during recrystallization. In this way, the mechanisms of texture evolution was elucidated, and optimal annealing conditions can be determined.

1. Introduction

Electrical steel is a soft magnetic material that is commonly used in transformers, motors, generators, alternators, etc. [1-6]. Superior magnetic properties are thus needed for such applications which involve decreasing hysteresis and eddy current losses in addition to increasing permeability [2]. The magnetic properties of electrical steel are dependent on many factors including silicon content, grain size, sheet thickness, residual stress as well as crystallographic texture. It is well known that the magnetic properties of electrical steel are orientation dependent due to the fact that the easy magnetization directions are along the <001> axes and the hard magnetization directions are along the <111> axes [2-6]. Therefore, to improve the magnetic properties of electrical steel, the <001>//ND (θ-fibre) texture (ND is the normal direction) should be enhanced which includes both the cube {001}<010> and rotated cube {001}<110> orientations, while the <111>//ND (γ-fibre) texture must be suppressed in the final sheet [3, 4].

The final texture of electrical steel is highly dependent on the manufacturing process. The texture evolves during solidification, plastic deformation, dynamic recrystallization, phase transformation, static recrystallization and grain growth [3]. During cold rolling of non-oriented electrical steel, a strong γ-fibre <111>//ND and an α-fibre <110>//RD (RD is the rolling direction) usually develop especially around the {001}<110> and {112} <110> orientations. After annealing there is a significant decrease in the α-fibre and an increase in the {111}<112> component (on the γ-fibre). The observed textures after annealing are generally related to the deformation texture with an orientation of 20-30° rotation about the <110> and <100> axes [7]. Several unconventional methods have been examined in the literature to optimize the texture of electrical steels which included columnar grain growth, cross rolling, two stage cold rolling, inclined and skew rolling [3, 5-6].

In the literature, several studies have examined the effect of the hot rolling texture on the final texture developed after cold rolling and final annealing [8-12]. Some of them found that the increase in grain size of the hot band played a significant role in increasing the magnetic permeability of electrical steel while simultaneously decreasing the core losses [9, 10]. Others observed a correlation between the formation of shear band and micro-band and the recrystallization of magnetically favorable crystal orientations (e.g. Goss and Cube), which decreases the volume fraction of the magnetically detrimental <111>//ND orientations. On the other hand, some studies [8] have shown that annealing after hot rolling has no effect or has a negative effect on the magnetic properties of the final sheet after cold rolling and annealing. Thus, it is important to understand the texture formation during hot rolling and the subsequent annealing to be able to make understand the texture evolution after cold rolling and final annealing. In this paper the texture evolution was tracked after hot rolling and annealing at the same area up to 160 minutes at 800 °C. Correlations were observed between the initial texture formed during hot rolling and the annealing texture at several different regions of the hot-rolled electrical steel plate.

2. Experimental

The chemical composition of (wt%) of the electrical steel used in this study is: 2.77 Si, 0.52 Al, 0.3 Mn, 0.01 P, and 0.003 C. It was melted in a vacuum furnace and cast into ingots with a cross-sectional area of 200 × 200 mm^2. It was then heated up to a nominal temperature of 1038 °C and hot rolled to a thickness of 25 mm in six passes with ~87.5% reduction. The surface oxides were then machined before the second hot rolling process. The steel plates were heated up to 1038 °C again and hot rolled in six passes down to a final thickness of ~3.5 mm (~85% reduction). The plates were finally pickled in a HCl solution to remove surface oxides generated during the second hot rolling.

One hot-rolled plate was cut into a small rectangular piece which was then prepared using conventional metallographic procedures for EBSD characterization. To track the same region throughout the entire study, a rectangular area in the ND-RD plane was marked with several indentations that could be readily located during the SEM analysis. EBSD characterizations were performed in a field emission gun SEM (Nova NanoSEM, FEI) which is equipped with an EDAX Orientation Imaging Microscopy system (OIM 6.2). The scans covered the entire thickness of the plate. Orientation distribution functions (ODFs) were calculated using harmonic series expansion method with a Gaussian half-width of 5° and a series rank of 22. Texture components were plotted on the $\varphi_2 = 45°$ section of the Euler space (Bunge notation).

The sample was then annealed at 800 °C for various times in argon protected atmosphere and subsequently water quenched to freeze the microstructures. The furnace was first raised to the desired temperature and the sample was inserted into a region inside the furnace that had a uniform temperature distribution. Following each annealing process, an EBSD scan was conducted at the exact same location that was marked with indentations. A very slight polishing step was applied using a colloidal silica solution after each annealing experiment.

3. Results and Discussion

The microstructure and texture of the hot-rolled plate is shown in Figure 1, which showed apparent heterogeneity across the thickness of the sample. The microstructure can be divided into three regions, i.e. A, B and C. Region A starts from the surface and extends about 800 μm towards the center. This region is composed of both deformed coarse grains at the surface and dynamically recrystallized fine grains beneath them. The dominant textures (Figure 2a) in region A include Copper {112} <111>, Goss {110} <001> and Brass {110} <112>. These textures are often observed in cold rolled fcc metals such as Al, Ni, Cu and Ag [7]. Region B is immediately below Region A and extends ~250 μm towards the center. This region is composed of several elongated grains with <011>//ND orientations (green) and some other smaller grains embedded between them. The texture (Figure 2b) is essentially a single orientation at {110}<001> (the Goss texture). Region C is at the middle of the rolled plate and spreads about 1/3 of the thickness. The microstructure is composed of several elongated grains with different orientations (different colors). Some of these grains have very large ingrain misorientations (>10°), e.g. left side of Region C. The texture (Figure 2c) of this region mainly consists of an α-fibre (<110>//RD) and a γ-fibre (<111>//ND), and the strongest orientation is the rotated cube {001}<110>. This texture is very similar to the typical cold-rolling texture of bcc metals [1,7].

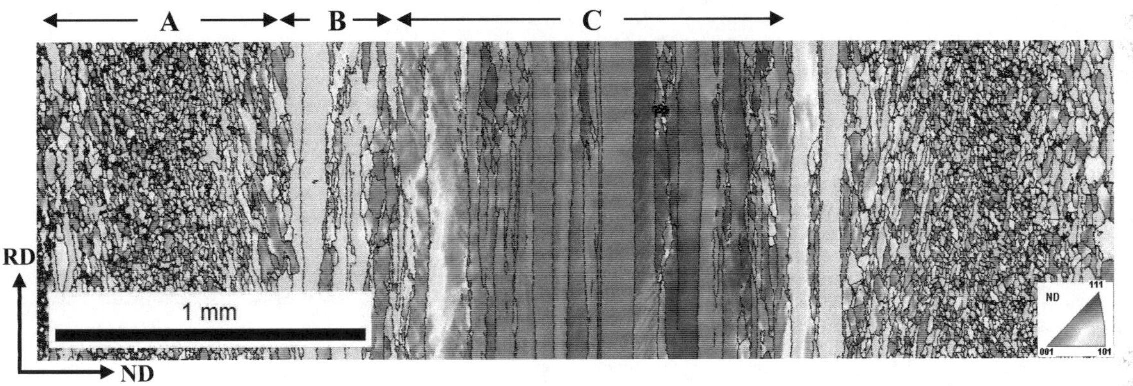

Figure 1. EBSD inverse pole figure (IPF) map of a hot-rolled 2.8% Si steel on the ND-RD plane showing the three different regions A, B and C.

The near surface textures (regions A and B) are developed due to the strong shear stress between the rolls and the surfaces of the material. This shear stress is the result of friction and is parallel to the rolling direction, which will rotate the crystal around the transverse direction (TD). In fact, a simple 90°-rotation of the texture at the center region (Region C) around the TD will give rise to a texture very similar to what was observed in Region A. Holscher et al. [13] explained the relationship between the rolling textures and the shear textures in bcc metals as a result of slip symmetry, which arises from the slip system with a 0 Schmidt factor turning into a maximum Schmidt factor system when a rotation of the lattice occurs by 45°. The overall texture of the material in the ND-RD plane (including all regions) is shown in Figure 2d, which shows a strong rotated cube, an α-fibre, a Goss component and some other orientations.

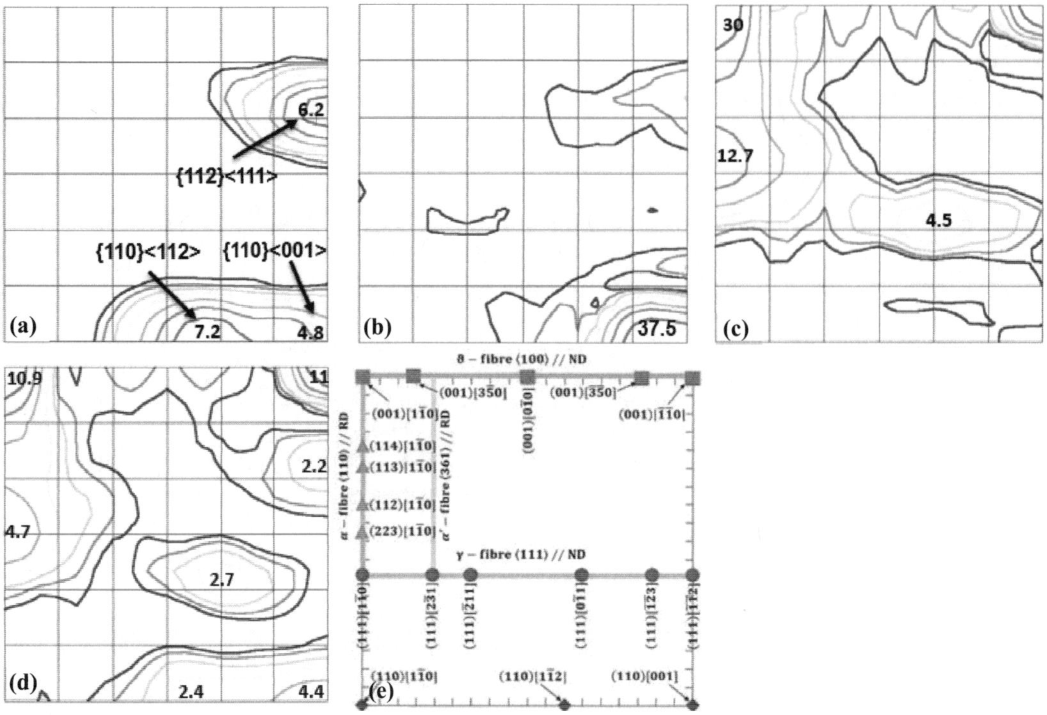

Figure 2. $\varphi_2 = 45°$ sections of the ODF for different regions of the hot-rolled plate: (a) Region A, (b) Region B, (c) Region C, (d) overall including all regions, and (e) texture key.

The evolution of texture in the surface regions (Regions A and B) during annealing is shown in Figure 3. As the annealing time increases, the overall texture intensity generally decreases. The Goss texture apparently decreases with the annealing time, while the copper ({112}<111>) and brass ({110}<112>) components essentially do not change until the annealing time increases to 110 minutes. After 160 minutes, all the components are significantly weakened. The copper and brass orientations have a higher Taylor factor than the Goss orientation, thus having more stored energy than Goss, which caused them to recrystallize first. During recrystallization, new copper and brass crystals nucleate from the originally deformed copper and brass crystals, and they grow into the deformed Goss matrix and consume the Goss grains. This is due to the fact that the misorientation relationship between the deformed Goss and the copper and brass orientations is all within 25-35°<110>, which is very close to high mobility CSL boundaries, i.e. Σ19a (26.5°<110>) and Σ9 (38.9°<110>).

As the annealing time increases to more than 30 minutes, new orientations, e.g. {111}<148>, start to appear (Figure 3b). This orientation is close to the stable bcc recrystallization {111}<011> orientation. It becomes more prominent and stable with the increase of annealing time (Figure 3c). A {443}<338> orientation also starts to appear after 110 minutes, which is very close to the two stable deformation orientations {111}<112> and {441}<118>. These latter orientations usually form from plane-strain deformation of the {110}<001> orientation [11]. After 60 minutes (Figure 3c), a {555}<135> orientation starts to appear and it becomes stronger with the increase of annealing time (Figures 3d and 3e). After 160 minutes, the original shearing textures change to final components such as {110}<112>, {555}<135>, {443 <338> and {110}<118>.

IOP Conf. Series: Materials Science and Engineering **375** (2018) 012014 doi:10.1088/1757-899X/375/1/012014

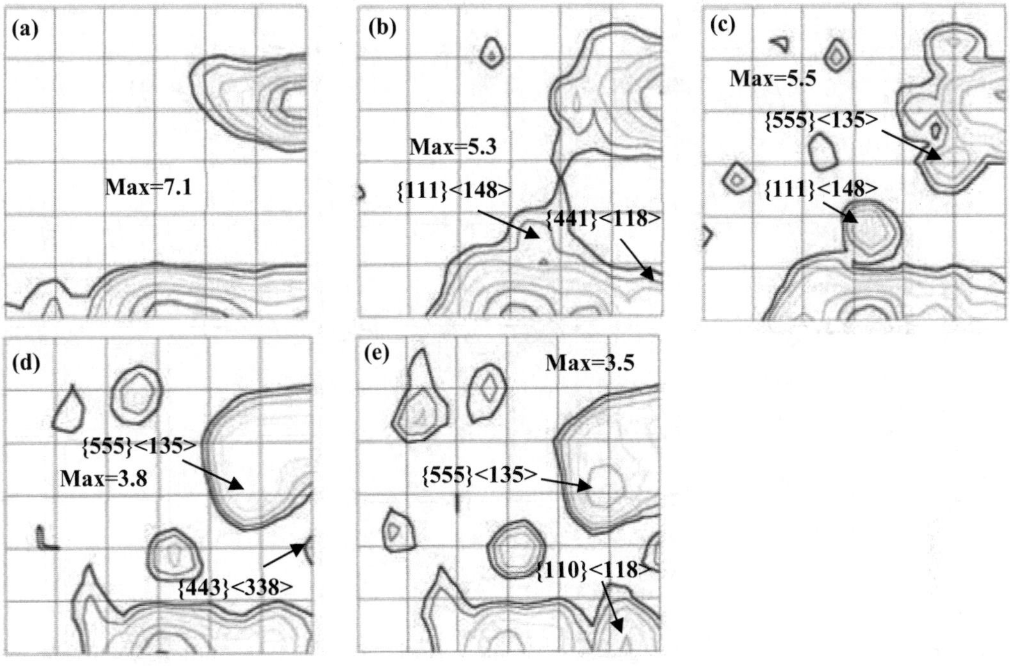

Figure 3. $\varphi_2 = 45°$ sections of the ODF after annealing at 800 °C for various times (Regions A and B): (a) as hot rolled (0 minutes), (b) 30 minutes, (c) 60 minutes, (d) 110 minutes, and (e) 160 minutes.

The evolution of texture in the middle region of the hot band (Region C) is shown in Figure 4. Recrystallization starts at the deformed γ-fibre grains (either at shear bands, microbands or grain boundaries) and they are the first to be consumed as seen from the decrease of their intensity from Figure 4a to Figure 4b. Again it is due to the high Taylor factor and high stored energy of these grains. As the annealing time increases to 60 minutes (Figure 4c), a {113}<361> orientation starts to appear which is on the α*-fibre ($\{11h\}\langle 12\frac{1}{h}\rangle$). This fibre becomes strong and prominent when the annealing time is increased to 110 minutes (Figure 4d). The α-fibre generally weakens with the increase of annealing time, and the only exception is that the {111}<110> component (which belongs to both the α- and γ- fibres) first weakens with time and then strengthens again after 110 minutes. The rotated cube orientation remains stable and strong until 110 minutes. Beyond that, it weakens significantly with the formation a {001}<350> component. A {118}<081> component also starts to form when the annealing time is 30 minutes, and it strengthens with the increase of the annealing time.

An interesting feature to be noticed in this region (Region C) is the strengthening of the Goss orientation with annealing time. It should be noted that this orientation was not present in the central region in the deformed state as it was not stable under plane-strain deformation. Several studies [9, 14] have found that the Goss is only present in the shear bands and microbands of the {111}<112>, {111}<110> or {112}<110> deformed grains. During hot rolling, no shear bands were noticed in these grains in Region C. Thus, the nucleation and growth of the {110}<001> orientations cannot be attributed to the same mechanism that gives rise to this orientation after cold rolling and primary recrystallization.

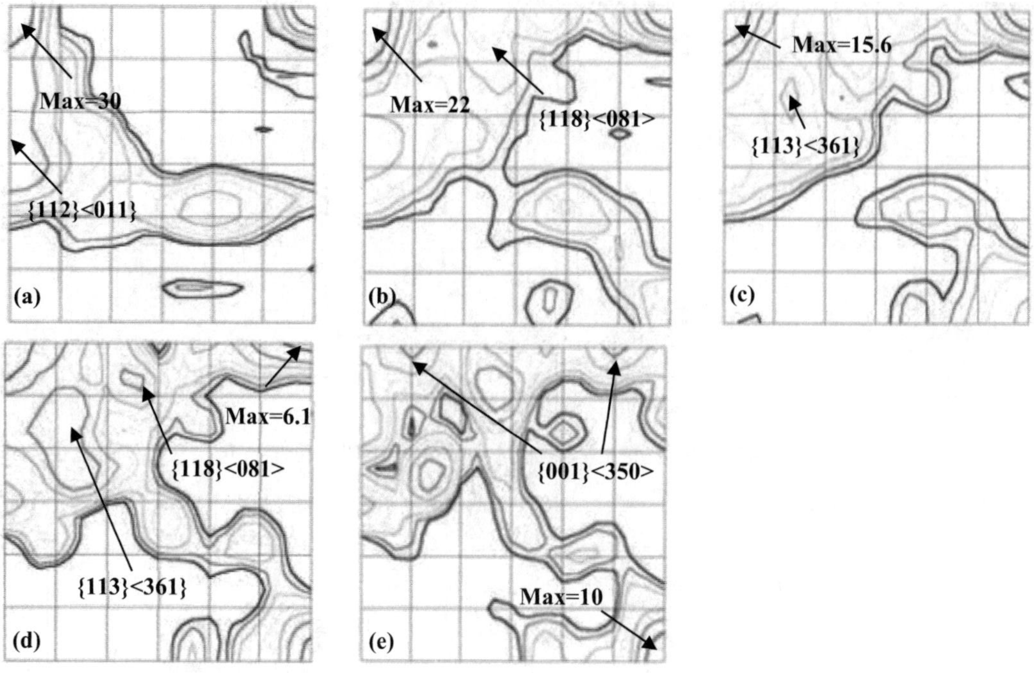

Figure 4. $\varphi_2 = 45°$ sections of the ODF showing the evolution of texture (Region C) during annealing at 800 °C for various times: (a) as hot rolled (0 minutes), (b) 30 minutes, (c) 60 minutes, (d) 110 minutes, and (e) 160 minutes.

The microstructure during and after hot rolling is very complicated due to the fact that several processes occur simultaneously, e.g. plastic deformation, dynamic recovery and recrystallization, static recovery and recrystallization. In an attempt to understand the reason for the formation of a large shear-textured zone in the microstructure, the following mechanism is proposed. The shear stress generated during hot rolling cannot be large enough to influence the grains ~1.2 mm below the surface. Thus, a thinner shear deformation zone is believed to be generated during the first hot rolling process (when the thickness was reduced from 200 mm to 25 mm). The consequent annealing at a temperature of 1038 °C allowed the surface textures to grow well below the surface of the hot-rolled plate. This growth of the shearing textures can be attributed to two factors: (a) the high strain and elevated temperature that are present near the surface allow the region to recrystallize and grow before the central region; the recrystallized grains grow and consume the inner deformed matrix and thus extending the shear-textured region; (b) the Goss orientation which forms at the bottom of the shear-textured zone (corresponding to Region B in Figure 1) is well known to have a faster mobility than other orientations, which is due to the fact that the Goss {110}<001> possess a high angle grain boundary with respect to the major texture components (copper and brass) in the deformed region above it and below it (α-fibre and γ-fibre). The brass and copper orientations both have a misorientation angle of ~35° about the <110> axis to the Goss orientation. This misorientation has been quoted in several studies [1, 9] to have a high angle mobility in comparison with other misorientations. A schematic summary of the suggested mechanism is shown in Figure 5a. A similar mechanism was observed in the evolution of the texture in Region C. The Goss intensity increased with annealing time despite the absence of any observable shear bands in the central region. This once again could be explained as the Goss orientation growing into the deformed regions rather than nucleating therein.

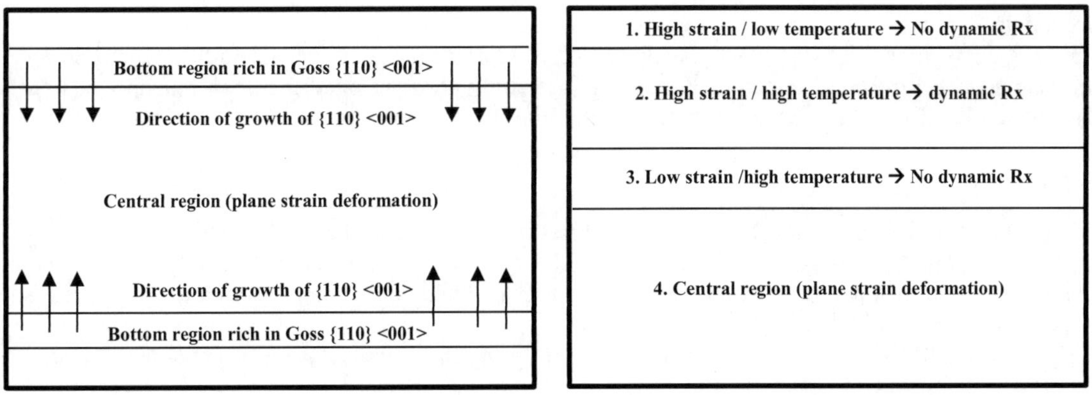

(a) (b)

Figure 5. The proposed mechanism for the formation of texture and microstructure during hot rolling of electrical steel: (a) the growth of the Goss grains into the deformed central region after the first hot rolling process, (b) the difference in microstructure and texture in the upper 1/3 of the plate due to variation in strain and temperature.

Another interesting feature in the hot-rolled electrical steel is the existence of multiple dissimilar regions in the upper 1/3 of the hot-rolled plate. A schematic representation of the several inhomogeneous layers is shown in Figure 5b. The differences in microstructure and texture among these regions can be explained as follows: the upper surface region (Region 1 in Figure 5b) is composed of long elongated grains without fine grains embedded in the microstructure. This is due to the fact that despite the strain being extremely high in this region, the direct contact of the surfaces with the rolls decreases the surface temperature and thus does not allow dynamic or static recrystallization to occur. The central region of the top 1/3 of the plate (Region 2 in Figure 5b) is very close to the surface, thus the shear strain is high. On the other hand, the temperature in this inner region is high enough (not in contact with the rolls), which combines with the high strain to result in dynamic and static recrystallization (the fine grains shown in Region A in Figure 1). The bottom region of the upper 1/3 of the hot-rolled plate (Region 3 in Figure 5b) is far away from the surface and thus has a low strain but a high temperature. These combined conditions prevent the occurrence of recrystallization. Finally the central region of the hot-rolled plate is far away from the rolls, thus the shear strain is negligible, which results in the typical bcc rolling texture and microstructure.

4. Summary

The microstructure and texture across the thickness of hot-rolled electrical steel plate are significantly different due to the differences in deformation mode (i.e. shear or plane strain) and temperature.

For dynamic or static recrystallization to occur at the surfaces, both high shear-strain and high temperature should be present. These two conditions are only present in regions immediately below the surfaces, thus recrystallized microstructure and texture exist in these regions.

The center region of the plate is not subjected to shear deformation, therefore it does not recrystallize during hot rolling, which results in the typical bcc rolling microstructure and texture.

During annealing, the intensity of the {110}<001> decreases in the surface layer (Region A) and increases in central region (Region C) with annealing time.

Acknowledgements
The Program of Energy Research and Development, Natural Resources Canada, and the Natural Sciences and Engineering Research Council of Canada are gratefully acknowledged for financial support. Renata Zavadil and Jian Li are thanked for their help on the EBSD measurements.

References

1. Humphreys, F. J., & Hatherly, M. (2012). Recrystallization and related annealing phenomena. Elsevier.
2. Mehdi, M., He, Y., Hilinski, E. J., & Edrisy, A. (2017). Effect of skin pass rolling reduction rate on the texture evolution of a non-oriented electrical steel after inclined cold rolling. Journal of Magnetism and Magnetic Materials, 429, 148-160.
3. Kestens, L., & Jacobs, S. (2008). Texture control during the manufacturing of nonoriented electrical steels. Texture, Stress, and Microstructure, 2008.
4. Rollett, A. D., Storch, M. L., Hilinski, E. J., & Goodman, S. R. (2001). Approach to saturation in textured soft magnetic materials. Metallurgical and Materials Transactions A, 32(10), 2595-2603.
5. He, Y., & Hilinski, E. J. (2017). Skew rolling and its effect on the deformation textures of non-oriented electrical steels. Journal of Materials Processing Technology, 242, 182-195.
6. He, Y., Hilinski, E., & Li, J. (2015). Texture evolution of a non-oriented electrical steel cold rolled at directions different from the hot rolling direction. Metallurgical and Materials Transactions A, 46(11), 5350-5365.
7. Hu, H. (1974). Texture of metals. Texture, Stress, and Microstructure, 1(4), 233-258.
8. Pedrosa, J. S. M., da Costa Paolinelli, S., & Cota, A. B. (2015). Influence of initial annealing on structure evolution and magnetic properties of 3.4% Si non-oriented steel during final annealing. Journal of Magnetism and Magnetic Materials, 393, 146-150.
9. Park, J. T., & Szpunar, J. A. (2009). Effect of initial grain size on texture evolution and magnetic properties in nonoriented electrical steels. Journal of Magnetism and Magnetic Materials, 321(13), 1928-1932.
10. R. Takanohashi, F.J.G. Landgraf, Effect of hot-band grain size and intermediate annealing on magnetic properties and texture of non-oriented silicon steels, J. Magn. Magn. Mater. 304 (2006) 608–610.
11. T. Haratani, W.B. Hutchinson, I.L. Dillamore, P. Bate, Contribution of shear banding to origin of Goss texture in silicon steel, Met. Sci. 18 (1984) 57–65.
12. S.C. Paolinelli, M.A. Cunha, A.B. Cota, The influence of shear bands on final structure and magnetic properties of 3% Si non-oriented silicon steel, J. Magn. Magn. Mater. 320 (2008) 641–644.
13. Hölscher, M., Raabe, D., & Lücke, K. (1994). Relationship between rolling textures and shear textures in fcc and bcc metals. Acta metallurgica et materialia, 42(3), 879-886.
14. Dorner, D., Zaefferer, S., & Raabe, D. (2007). Retention of the Goss orientation between microbands during cold rolling of an Fe3% Si single crystal. Acta materialia, 55(7), 2519-2530.

IOP Conf. Series: Materials Science and Engineering **375** (2018) 012015 doi:10.1088/1757-899X/375/1/012015

Recrystallization Texture Evolution of Cold Rolled and Asymmetrically Warm Rolled Austenitic Stainless Steel Sheets

S. Umehara[1], H. Inoue[1] and J. Hamada[2]

[1]Osaka Prefecture University, Sakai, Japan.
[2]Nippon Steel & Sumikin Stainless Steel Corporation, Hikari, Japan.
Corresponding author's e-mail:inoue@mtr.osakafu-u.ac.jp

Abstract. A {111} texture leads to good deep drawability but does not generally develop in face-centered cubic metals. One of the authors previously succeeded in {111} recrystallization texture evolution by cold rolling, asymmetric warm rolling and subsequent solution treatment for Al-Mg-Si alloy sheets. In this study, rolling and recrystallization textures of austenitic stainless steel with low stacking fault energy have been investigated to reveal whether the {111} texture can be formed by similar processing. Rolling texture changed from the α-fiber texture in 70% cold rolled sheets to an asymmetric texture on the TD axis consisting of {331}<116>-{111}<112> by additional 40% asymmetric warm rolling at 873 K. Correspondingly, pole density at the center of {111} pole figure increased from 2.16 to 3.18. In addition, microstructural observation showed that there were two kinds of shear bands inclined at about 30° and 150° to RD on the longitudinal section. One was micro shear bands within grains and the other was shear bands passing through a number of grains. Recrystallization texture after annealing also showed an asymmetric texture on the TD axis, but consisted of {431}<257>-{331}<116> near to {110}<112>-{110}<001>. The 1173 K-1800 s annealing decreased pole density at the center of {111} pole figure to 0.78. In conclusion, the rolling texture with near-{111}<112> orientation was obtained in cold rolled and asymmetrically warm rolled austenitic stainless steel sheets. The recrystallization texture, however, changed to near-{110}<112> orientation as a main component.

1. Introduction

Control of recrystallization texture is necessary for improving sheet metal formability. {111} textures with {111} planes parallel to the sheet surface lead to good deep drawability[1)] but does not generally develop in f.c.c. metals. One of the authors[2)] has recently succeeded in {111} recrystallization texture evolution in mid-thickness by cold rolling, asymmetric warm rolling and subsequent solution treatment for Al-Mg-Si alloy sheets with high SFE (stacking fault energy).

Recrystallization texture depends on rolling texture. There are two kinds of rolling texture in f.c.c metals, namely pure-metal and alloy types. The SFE of aluminum and copper showing pure-metal type rolling texture are 166 mJm^{-2} and 78 mJm^{-2}, respectively, and the SFE of stainless steel and brass showing alloy type rolling texture are 21 mJm^{-2} and 20 mJm^{-2}, respectively[3)]. 95% cold rolled pure copper has the β-fiber texture ({112}<111>-{123}<634>-{011}<211>)[3)]. In contrast, 95% cold rolled brass has the α-fiber texture ({011}<211>-{011}<100>)[3)]. The recrystallization texture of pure copper consists mainly of cube orientation {001}<100>, and that of brass has {236}<385> or {113}<121> as a preferred orientation[3)]. Morii et al.[4)] explained the formation process of rolling texture in brass. Rolling texture at low reductions in thickness was formed by slip and deformation twining. Crystallographic rotation caused the {111} twin planes to be parallel to the sheet surface with shear banding in twin-deformed grains at middle rolling reductions ranging from 40% to 80%. Two sets of large shear bands crossing each other decrease the twins at high rolling reductions exceeds 80%. As the area of the shear bands increases, {011}-oriented grains increase. Nezakat et al.[5)] also showed that $(111)[1\bar{1}0]$ and $(111)[\bar{1}\bar{1}2]$ orientations as weak components formed at cold rolling reductions of 70% or more in SUS 316L austenitic stainless steel. Williams[6)] mentioned that simple shear deformation in brass formed a shear texture consisting of $(111)[01\bar{1}]$, $(111)[11\bar{2}]$ and (100)[011] orientations similar to simple shear deformation in aluminum.

Recently, hydrogen energy is expected as clean energy. Stable austenitic stainless steel for which hydrogen embrittlement hardly occurs is useful as parts in hydrogen[7)]. However, the improvement in deformability by the TRIP (transformation induced plasticity) effect observed in metastable austenitic stainless steel[8)] cannot be used in stable austenitic stainless steel. This is unfavorable for deep drawing. In this study, rolling and recrystallization textures of SUS316L stable austenitic stainless steel with low stacking fault energy were investigated in order to reveal whether the {111} texture favorable for deep drawing can be developed through cold rolling and additional asymmetric warm rolling or not.

2. Experimental

Table 1 shows the chemical composition of SUS316L used in this study. The starting materials were hot-rolled sheets with a thickness of 5 mm. The thickness of the hot-rolled sheets was reduced to 70% by cold rolling at room temperature. These sheets were further rolled by asymmetric warm rolling using the laboratory two-high rolling mill with different roll diameters of 96 and 144 mm (in upper and lower rolls, respectively) at heating temperature of 873 K in a furnace. The asymmetric warm rolling was performed without changing the front and back of sheet surface and with using the exact rolling direction of not 180° but 0°. Embrittlement due to the sigma phase[9)] would not occur at this rolling temperature. The rolling reduction of asymmetric warm rolling was set to a low reduction of 40%. In fact this reduction required many passes of 30 to 32 on the above rolling mill.

Annealing for recrystallization was conducted after the above combined rolling. Annealing temperature was determined from a softening curve drawn by the results of micro Vickers hardness tests of isochronally annealed samples for 1800 s. Annealing time was determined from a softening curve obtained by isothermal annealing at 1173 K. The annealing time of 0 s means the time when samples were put in a furnace. Microstructures of rolled and annealed samples were observed by an optical microscope. The grain size was measured by the intercept method. {111}, {100} and {110} incomplete pole figures viewed from the side of a roll with larger diameter were measured by the Schulz reflection method[10)] using MoKα radiation in order to examine rolling and recrystallization textures.

Table 1. Chemical composition of SUS316L stainless steel used in this study (mass %).

C	Si	Mn	P	S	Ni	Cr	Mo	N	Fe
0.01	0.50	0.84	0.03	0.00	12.0	17.4	2.0	0.01	Bal.

3. Results

3.1. Microstructures

Figure 1 shows Vickers hardness of samples annealed for 1800 s at various temperatures after cold rolling and asymmetric warm rolling (combined rolling). Since a slope of the softening curve varied at 1123 K, the finished temperature of recrystallization is considered to be more than 1123 K. Figure 2 shows the relationship between annealing time and Vickers hardness of the annealed samples at 1173 K. It is considered that recrystallization was finished after annealing of 300 or 1800 s when the Vickers

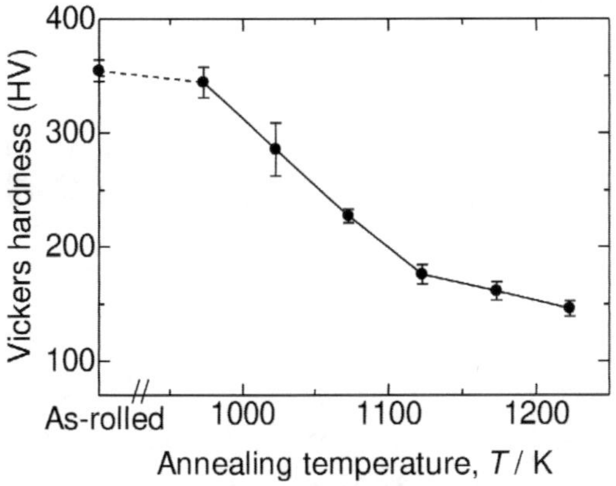

Fig. 1 Variation of hardness during isochronal annealing for 1800 s of cold rolled and asymmetrically warm rolled samples.

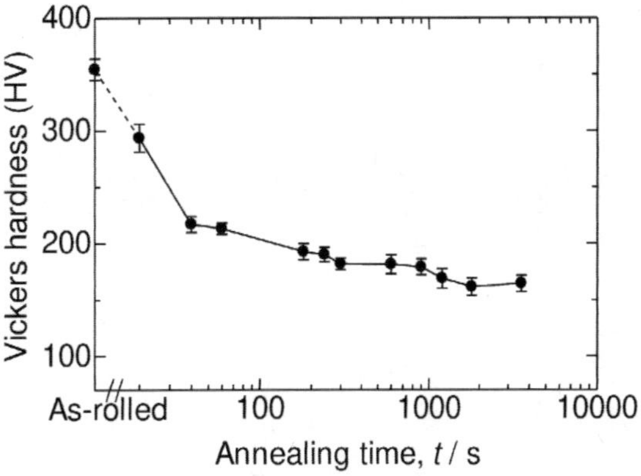

Fig. 2 Variation of hardness during isothermal annealing at 1173 K of cold rolled and asymmetrically warm rolled samples.

hardness became constant. As a result of microstructural observation, recrystallization was finished after annealing of 1800 s. Thus an annealing condition for recrystallization in the combined rolled sample was determined to 1173 K-1800 s. Figure 3 shows microstructures of (a) a cold rolled sample, (b) a cold rolled and asymmetrically warm rolled sample and (c) a cold rolled, asymmetrically warm rolled and annealed sample. Figure 3a exhibited elongated grains and no definite shear bands. Figure 3b showed large shear bands inclined by about 30° counterclockwise from RD indicated by white arrows and micro shear bands within grains inclined by about 30° clockwise from RD indicated by black arrows. Figure 3c showed equiaxed recrystallized grains with an average grain size of 4.2 μm and an aspect ratio of 1.1, including annealing twins.

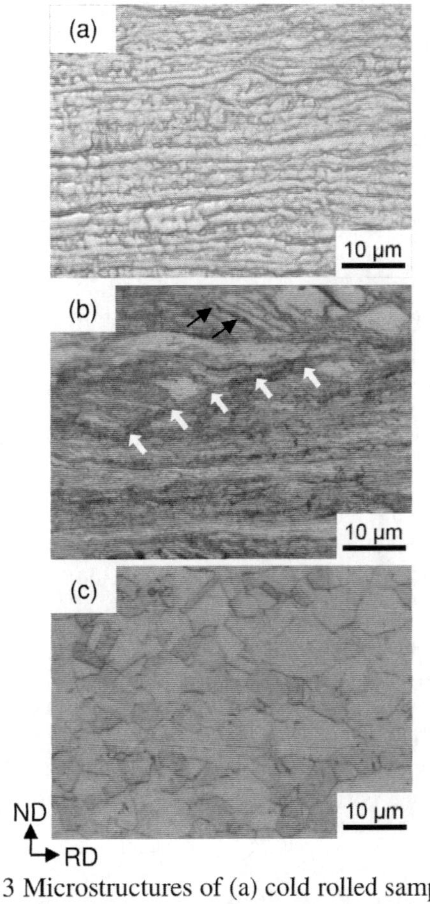

Fig. 3 Microstructures of (a) cold rolled sample, (b) cold rolled and asymmetrically warm rolled sample, and (c) annealed sample at 1173 K for 1800 s after the combined rolling. White arrows indicate shear band and black arrows indicate micro-shear bands within grains in (b).

3.2. Rolling texture

Figure 4 shows {111} and {110} pole figures measured in mid-thickness of samples. These pole figures were viewed from the side of a roll with larger diameter for a 70% cold rolled sample, and for 20% or 40% asymmetrically warm rolled samples after the cold rolling. As Nezakat et al.[5] reported, the cold rolled sample had an alloy-type rolling texture composed of α-fiber (<011>// ND) with $(111)[1\bar{1}0]$ and $(111)[\bar{1}\bar{1}2]$ orientations as weak components. In the cold rolled and 40% asymmetrically warm rolled sample, rolling

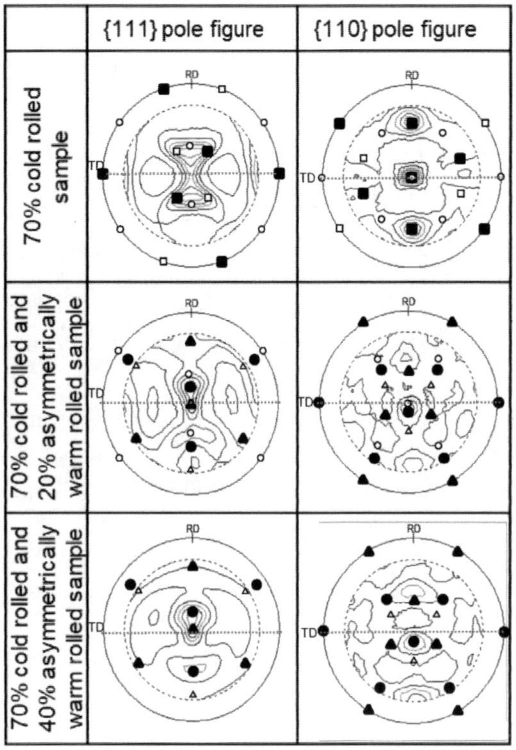

□ (110)[$\bar{1}12$] ■ (110)[$1\bar{1}2$] ○ (110)[001]
▵ (111)[$\bar{1}\bar{1}2$] ▲ (111)[$11\bar{2}$] ● (331)[$\bar{1}\bar{1}6$]

Fig. 4 {111} and {110} pole figures of 70% cold rolled sample (top), and 70% cold rolled and 20% (middle) or 40% (bottom) asymmetrically warm rolled samples. Contour levels are 0.5 to 5.0 at a step of 0.5.

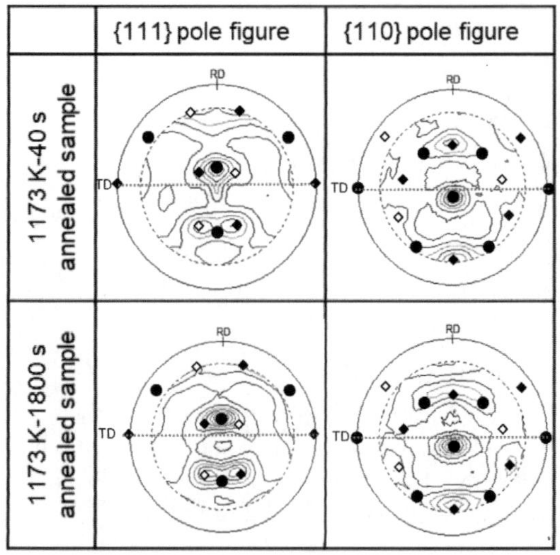

◇ (341)[$\bar{5}27$] ◆ (431)[$2\bar{5}7$] ● (331)[$\bar{1}\bar{1}6$]

Fig. 5 {111} and {110} pole figures of annealed samples at 1173 K after 70% cold rolling and 40% asymmetrically warm rolling. Contour levels are 0.5 to 4.0 at a step of 0.5.

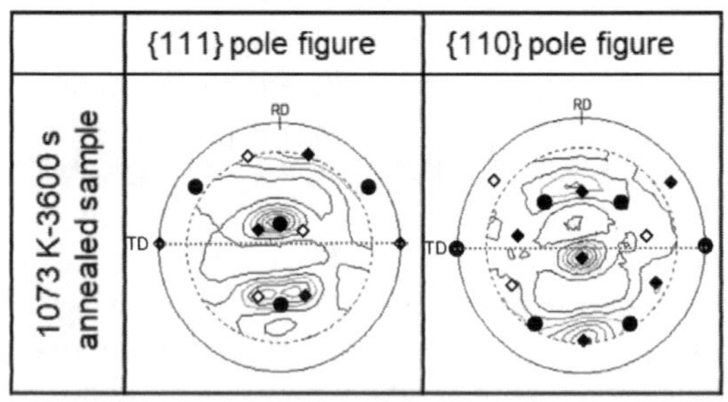

◇ (341)[$\bar{5}27$] ◆ (431)[$2\bar{5}7$] ● (331)[$\bar{1}\bar{1}6$]

Fig. 6 {111} and {110} pole figures of 70% cold rolled, 40% asymmetrically warm rolled and 1073 K-3600 s annealed sample. Contour levels are 0.5 to 4.0 at a step of 0.5.

texture of (331)[$\bar{1}\bar{1}6$]-(111)[$\bar{1}\bar{1}2$] orientation group asymmetrical on the TD axis was recognized. The (331)[$\bar{1}\bar{1}6$] orientation corresponds to the orientation that {110}<001> was rotated by about 13° around the TD axis. In the cold rolled and 20% asymmetrically warm rolled sample, rolling texture had an intermediate texture of the cold rolled sample and the additionally 40% asymmetrically warm rolled sample. The pole density at the center of {111} pole figure indicating the existence of {111}<*uvw*> components increased from 2.2 to 3.2 due to additional 40% asymmetric warm rolling. In addition, the pole density at the center of {110} pole figure decreased considerably from 5.1 to 1.3 by adding 40% asymmetric warm rolling.

3.3. Recrystallization texture

Figure 5 shows {111} and {110} pole figures of samples annealed at 1173 K after cold rolling and 40% asymmetric warm rolling. Formation of the (341)[$\bar{5}27$]-(331)[$\bar{1}\bar{1}6$]-(431)[$2\bar{5}7$] orientation group is recognized in the pole figures of a sample annealed under a condition of 1173 K-1800 s. The pole density at the center of the {111} pole figure was 0.8. The pole density at the center of the {110} pole figure was 2.0. The {111}<112> orientation decreased considerably and the {431}<257> orientation was formed by annealing for 40 s. After annealing of 40 s, the orientation group from the {331}<116> to the {431}<257> orientation exhibited an annular spread around the <110> axis inclined by about 10° from ND to the opposite direction of RD.

The rolling texture of the cold rolled and 40% asymmetrically warm rolled sample had a preferred orientation close to {111}<112>. In order to retain the {111}<112> orientation in recrystallization texture as well as in rolling texture, in-situ recrystallization can be expected. The annealing condition for recrystallization completion was decided as 1073 K-3600 s under which the embrittlement due to sigma phase[9] hardly occurred. Figure 6 shows {111} and {100} pole figures of the 1073 K-3600 s annealed sample. The center of {111} pole figure had pole density of 1.0, which slightly increased compared to 1173 K-1800 s. However, the main orientation of this recrystallization texture was basically the same as that of the 1173 K-1800 s annealed sample. Figure 7 shows {111} and {110} pole figures of an annealed sample at a lower heating rate. The pole density at the center of {111} pole figure was 0.5. The orientation group from the {331}<116> to the {431}<257> orientation developed more strongly than that of the 1173 K-1800 s annealed sample.

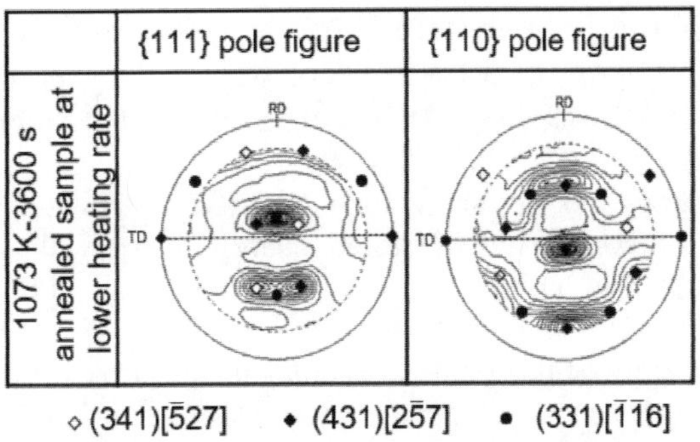

Fig. 7 {111} and {110} pole figures of 70% cold rolled, 40% asymmetrically warm rolled and 1073 K-3600 s annealed sample at a lower heating rate. Contour levels are 0.5 to 7.0 at a step of 0.5.

4. Discussion

4.1. Rolling texture evolution

In general, shear bands formed by cold rolling appear not in one direction but in symmetric two directions with respect to RD. Additional asymmetric warm rolling formed large-scale shear bands in one direction, although the shear bands were not seen after cold rolling. It is thought that strain was stored, although it was recovered in a furnace during the warm rolling process. Morii et al.[4], in the case of brass, described crystallographic rotation that causes the {111} twin planes to be parallel to the sheet surface with shear banding in twin-deformed grains. The large-scale shear bands generated by the twin-deformed grains reduce twin regions and oriented grains parallel to the sheet surface rapidly increases on the {011} face. The unidirectional large-scale shear bands had little influence on the decrease of the {111} twin plane, and it is considered that the {111} twin plane increased with the activity of shear bands generated in twin-deformed grains. In addition, it is thought that shear texture[6] such as {111}<112> and {111}<011> orientations was formed by additional shear deformation during asymmetric warm rolling.

Shear bands within the grains were formed in a direction crossing the large-scale shear bands in one direction in the microstructure. Noda[11] reported that {110}<001> orientation was formed by shear bands in cold rolling. Formation of asymmetric texture on TD was also observed in cold rolled and asymmetrically warm rolled Al-Mg-Si alloy sheets[2]. Deformation by asymmetrically warm rolling consists of plane strain compression and additional shear deformation. It is thought that (331)[$\bar{1}\bar{1}$6] orientation was formed due to rotation of the {110}<001> orientation around the TD axis by plane strain compression and additional shear deformation.

4.2. Recrystallization texture evolution

For alloy type metals, {236}<385> or {113}<121> orientation is formed[3]. Chowdhury et al.[12] reported that there was a twin relationship between the {236} <385> orientation and the {110} <001> orientation, and that the {236} <385> orientation had a rotation relation of 40° around the common <111> axis with the {110} <112> (40° <111> relationship). In addition, the 60° <111> twinning can be described as 70.5° <110>[13]. The (331)[$\bar{1}\bar{1}$6] orientation and (341)[$\bar{5}$27] or (431)[2$\bar{5}$7] orientation had ± 35° <110> relationship. However, since an approximately 70° <110> relationship was satisfied between (341)[$\bar{5}$27] and (431)[2$\bar{5}$7] orientations, the formation of these two variants would result from annealing twins.

Sumitomo[14] reported about SUS304 austenitic stainless steel that recrystallized nuclei of {211} and {311} having higher stored energy than {110}, which were generated at a lower temperature, encroached on the recrystallized nuclei of {110}. In this study, grains having a {331}<116>-{431}<257> orientation group relatively close to {110} developed as a main component of recrystallization texture. It is thought that stored energy was changed due to additional shear deformation and recovery in asymmetric warm rolling. It is possible that release of stored energy was performed earlier at the {331}<116>-{431}<257> orientation group because the {331}<116>-{431}<257> orientation group was strongly formed by low-speed annealing. In order to clarify the mechanism of evolution of recrystallization texture consisting of the {331}<116>-{431}<257> orientation group close to {110}, research on stored energy may be required in the future.

Amane et al.[15] reported that for cold rolled, asymmetrically warm rolled and subsequently annealed Al-Mg-Si alloy sheets, {111} oriented grains formed and grew at the early stage of annealing. Combined rolling consisting of cold rolling and asymmetric warm rolling was effective for metals with high SFE. Goodman and Hu[16] reported that the rolling texture of austenitic stainless steel changed depending on rolling temperature, and a pure metal type rolling texture similar to that of high SFE metals developed when rolled at high temperature. From these facts, asymmetric rolling at a temperature higher than 873 K might have a possibility of {111} recrystallization texture evolution like Al-Mg-Si alloy having a pure metal type rolling texture.

5. Conclusions

SUS 316L austenitic stainless steel was 40% asymmetric rolled at 873 K after 70% cold rolling, and subsequently annealed for recrystallization. The main results are as follows:

(1) Rolling texture with near-{111}<112> orientation was obtained in cold rolled and asymmetrically warm rolled austenitic stainless steel sheets.

(2) Recrystallization texture changed to near-{110}<112> orientation as a main component. The preferred orientation group exhibited an annular spread around the <110> axis inclined by about 10° from ND to the opposite direction of RD.

References

[1] H. Inoue and T. Takasugi: Mater. Trans., **48** (2007), 2014-2020.
[2] H. Inoue, M. Hori, T. Komatsubara, H. Tanaka and T. Takasugi: Mater. Sci. Forum, **558-559** (2007), 207-212.
[3] F.J. Humphreys, M. Hatherly: Recrystallization and Related Annealing Phenomena, Second Edition, Elsevier, Oxford, (2004).
[4] K. Morii, M. Mera and Y. Nakayama: J. Japan Inst. Metals, **44** (1980), 555-561.
[5] M. Nezakat, H. Akhiani, M. Hoseini and J. Szpunar: Mater. Charact., **98** (2014), 10-17.
[6] R.O. Williams: Trans. Metall. Soc. AIME, **224** (1962), 129-140.
[7] M. Kikuchi: DENKI-SEIKO ELECTRIC FURNACE STEEL, **86** (2015), 37-44
[8] N. Tsuchida, Y. Morimoto, S. Okamoto, K. Fukaura, Y. Harada and R. Ueji: J. Japan Inst. Metals, **72** (2008), 769-775.
[9] H. Suzuki: Tetsu-to-Hagané, **46** (1960), 498-520.
[10] L.G. Schulz: J.Appl. Phys., **20** (1949), 1030-1033.
[11] T. Noda: Bulletin of the Japan Institute of Metals, **19** (1980), 638-644.
[12] S.G. Chowdhury, S. Das, B. Ravikumar and P.K. De: Metall. Mater. Trans. A, **37** (2006), 2349-2359.
[13] H. Kaneko: Tetsu-to-Hagané, **56** (1970), 622-635.
[14] H. Sumitomo: Tetsu-to-Hagané, **77** (1991), 558-565.
[15] R. Amane and H. Inoue: J. Japan Inst. Light Metals, **66** (2016), 627-620.
[16] S.R. Goodman and H. Hu: Trans. Metall. Soc. AIME **230** (1964), 1413-1419.

Segregation affecting the evolution of primary recrystallization textures in a ternary Fe-Si-Sn alloy

N Mavrikakis[1,2,*], P R Calvillo[3], W Saikaly[3], M Descoins[1], D Mangelinck[1] and M Dumont[1]

[1] Aix Marseille Université, CNRS, IM2NP UMR 7334, 13397, Marseille, France
[2] ArcelorMittal Maizières Research SA, Voie Romaine, BP 30320, 57283, Maizières les Metz, France
[3] ArcelorMittal Global R&D, Technologiepark 935, BE-9052, Zwijnaarde, Belgium
[*]email address: nikolas.mavrikakis@im2np.fr

Abstract

The effect of Sn addition on the primary recrystallization of cold rolled Fe-3 wt. % Si alloys is investigated. Texture evolution and misorientation distributions are analyzed on partially recrystallized samples using the electron backscatter diffraction technique (EBSD). Sn was found to affect the microstructure, throughout the thermal treatment of the materials, by refining the grains and altering the texture. In the presence of Sn, the intensity of {111}<*uvw*> grains is reduced through all stages of recrystallization, while that of {100}<*uvw*> and {*hkl*}<100> grains is increased. The favored growth of these latter grains is most likely due to a combination of mechanisms that involve nucleation site preferences and probably the presence of high mobility CSL boundaries. Σ5 boundaries were observed to increase in frequency with Sn through all stages of recrystallization. Additionally {100} grains were found to be most frequently correlated with Σ5 interfaces, which might be due to geometrical considerations. Site-specific characterization of grain boundaries was conducted with three-dimensional atom probe tomography (3D-APT) to measure the grain boundary segregation of Sn.

1. Introduction

The Fe-Si system is well known for studying annealing phenomena, as it is single-phase ferritic for Si content above 1.9 wt. % [1]. The overall recrystallization process, during annealing of the as cold rolled sheet, is characterized by a strong γ-fiber texture with the peak of intensity depending on the rolling reduction [2,3]. Nevertheless, adding trace elements such as P, Sb and Sn has a great impact on the texture development thanks to the promotion of selective grain growth, due to surface and grain boundary segregation phenomena [4].

More specifically, Sn segregation has been found to strongly affect the recrystallization texture of iron, by selectively promoting the grain growth of grains with specific texture. {111}<uvw> texture components were often weakened whilst the growth of {110} grains was mainly favored [4–6]. In some other studies, Sn favored the {100}<uvw>, {hkl}<100> oriented grains, in most cases strengthening the Goss component [7]. Nakashima et. al observed that Sn facilitates secondary recrystallization and increases the migration rate of special (CSL) boundaries, like Σ9 [7]. In the literature, special properties are often attributed to CSL boundaries such as low energy and high mobility. The low energy aspect of CSL boundaries is believed to be the most important by Hutchinson and Hasson [8,9], while Lücke and others highlight the enhanced mobility of low Σ boundaries, as the most important aspect of these interfaces [10]. Harase [11] reports that CSL boundaries become more mobile in the presence of solutes or precipitates, as the latter inhibit less mobile grain boundaries from growing. He also suggests that Σ5 boundaries become more mobile than Σ9 boundaries at lower temperatures.

Literature on Sn interaction with the microstructure during annealing is scarce, mainly the effect of Sn after the recrystallization and grain growth are completed, is addressed. Such studies are mainly focused on the contribution of Sn on secondary recrystallization, which assists the Goss texture formation. While, the interfacial segregation of Sn was usually analyzed with Auger Electron Spectroscopy (AES), in randomly opened grain boundaries, after brittle intergranular fracture [12]. Therefore, the present study evaluates the texture evolution during the phenomenon of recrystallization, from the nucleation stages and through the growth of the recrystallized grains, considering the effect of Sn grain boundary segregation and its potential impact on the as developed crystallographic texture. For the means of this research, the site-specific technique that combines EBSD, focused ion beam (FIB) and 3D-APT, will be employed to analyze the grain boundary segregation of Sn, at the near-atomic scale and in combination with orientation data.

2. Experimental

A binary Fe – 3Si and a ternary Fe – 3Si – 0.1Sn (wt. %) were laboratory vacuum cast with maximum impurity level of 200 ppm. The binary alloy provided a reference material to isolate the silicon contribution and be able to characterize the Sn effect in the microstructure evolution of iron. Both materials were similarly processed, with a final cold rolling step at a Von Mises true strain of about $\varepsilon = 2$.

Subsequently, the as cold rolled materials were heat treated using a BAHR dilatometer, at a constant heating rate, following interrupted annealing schemes to monitor the primary recrystallization kinetics at 700°C. After the isothermal soaking, samples were quenched to room temperature to ensure that no segregation occurs during cooling and in order to freeze the microstructure. Following the processing, materials were metallography prepared and optically examined in a Keyence VHX-5000 digital microscope. The recrystallization volume fractions were first quantified using the point counting method, considering from 400 to 500 points [13]. Samples at fraction of volume recrystallized, X_v = 0.2, 0.5 and 0.8 were selected to be further analyzed in EBSD.

The EBSD scans were acquired covering a minimal area of 900 x 300 μm^2 with a step size of 0.5μm. The number of recrystallized grains included in the mappings, and consequent analysis, range from 2400 to 4000 grains, which are statistically sufficient to derive reliable conclusions. Recrystallized (RX) grains were defined as new, almost dislocation free grains, bounded by high-angle grain boundaries (HAGBs), $\theta \geq 15^o$, with a grain internal misorientation not exceeding 2°. The grain shape should be close to equiaxed, having a maximum aspect ratio of 3:1. The post-processing was performed in both HKL Channel 5 and OIM Analysis platforms. From the EBSD scanned areas, grain boundaries of the recrystallization front, migrating boundaries between RX and the surrounding not-recrystallized (not-RX) structure, were selected and prepared for 3D-APT, in a FEI Helios NanoLab 600 FIB, following the lift-out technique [14]. Atom probe experiments were conducted in a Cameca LEAP 3000XHR instrument, with a detector efficiency of 37%. Measurements were performed in voltage mode, while the sample temperature and fraction of electric pulses were selected at 70K and 20% respectively. 3D-APT data were reconstructed after proper calibration of the reconstruction parameters, as described by Gault et. al [14]. Finally, the quantification of segregation has been performed with the, well-established, Gibbsian excess method [15].

3. Results

3.1. Recrystallization kinetics

Before analyzing the kinetics, the conversion from non-isothermal (during continuous heating) to isothermal time was made, according to the Whittaker and Sellars approach [16]. The Johnson-Mehl-Avrami-Kolgomorov (JMAK) equation was used to characterize the recrystallization kinetics:

$$f_{REX} = 1 - exp(-Kt^n) \quad (1)$$

where f_{REX} is the recrystallized volume fraction at a given time, *t*, *n* is the Avrami exponent and *K* is a kinetic coefficient. The JMAK description of isothermal kinetics was fitted, using the least-squares analysis, to obtain a best-fit for the Avrami plot, ln(ln(1/(1-X))) vs. ln(t) plot. The Avrami exponent, *n*, was found experimentally to be $n \approx 1$ for both alloys, and according to Christian [17] the recrystallization is then a phenomenon of interface controlled growth in a condition of grain boundary nucleation with site saturation. Sn addition was found to retard the recrystallization kinetics, figure 1(a).

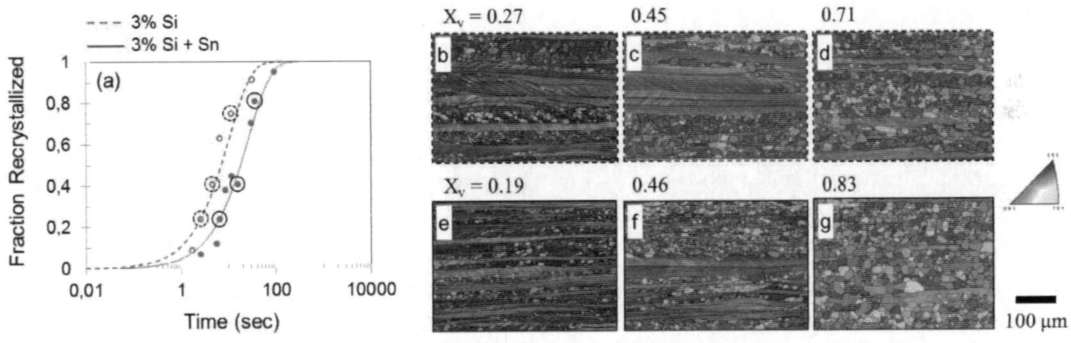

Figure 1. (a) Recrystallization kinetics of the materials, quantitative metallography data fitted with the JMAK equation. (b-d) Partial recrystallization EBSD ND-IPF mappings for the binary alloy and (e-g) for the ternary alloy. The recrystallization volume fractions measured by EBSD are shown for comparison and marked with circles on the kinetics plot.

The EBSD ND-IPF maps are illustrated for the two materials in figure 1(b-g), where all restored grains (recovery and recrystallization products) are highlighted, while the not recrystallized structure is shown in grey scale. The recrystallized volume fractions as determined by the EBSD were in very good agreement with the quantitative metallography values and are indicated in figure 1 as black circles. It is observed that large colonies of primary recrystallized grains are present within certain deformation bands while other regions remain in an entirely deformed state. Such behavior is well known in many cases of recrystallization in ultra-low carbon and other bcc steels and is considered to arise due to the variation in stored energy between grains of different orientations in the deformed structure [18,19].

3.2. Recrystallization textures

Analyzing the recrystallized texture from the EBSD, figure 2; the binary alloy, figure 2(a-c), is found to have strong recrystallization texture [3]. From the early stages of recrystallization, the binary alloy texture, has a strong {111}<112> component that remains during growth. The texture does not change significantly as recrystallization advances. On the other hand, in the ternary alloy, figure 2(d-f), the texture of recrystallized grains is significantly weaker. At the early stages, intensities of the γ-fiber are dispersed quite evenly, with the {111}<112> component clearly weaker compared to the binary. At the same time, the strongest texture component is the Goss {110}<100>. As recrystallization progresses, the more random character of texture is maintained, the γ-fiber is weakened at the {111}<110> component while the Goss is still present.

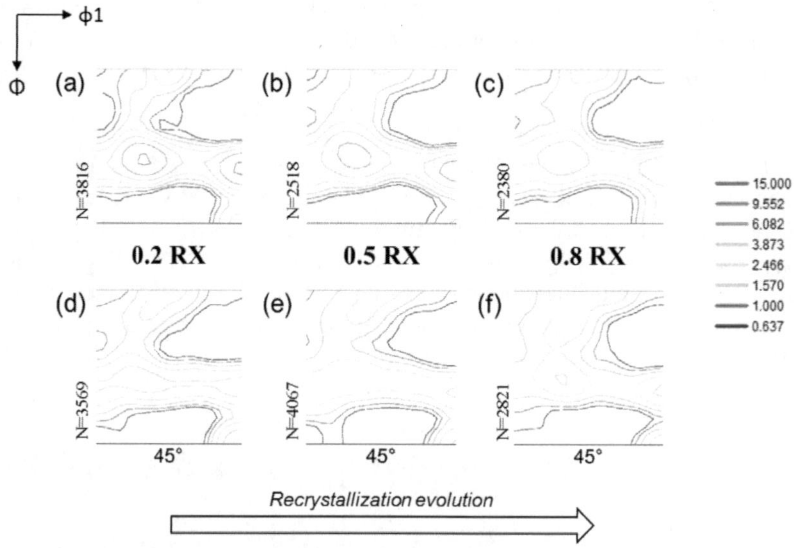

Figure 2. Micro-texture ODFs at the φ_2=45° section of the Euler space of all restored grains during the progress of recrystallization. (a-c) The texture of the binary and (d-f) the ternary alloy, along with the number of grains that were included in the analysis.

Figure 3 shows the texture densities along the main fibers, η, α and γ-fibers, and can provide a more quantitative insight on the stages intensities of the various texture components. It can be seen that the effect of Sn is stronger at the early of recrystallization, where the Goss component with Sn is at about x6 random intensity, compared to about x2 in the case of the binary alloy. Also, the most pronounced decrease of the {111}<112> component is found at this stage; the ternary alloy was found to have a x4 random intensity compared to the binary which has a relative intensity of about x14 random.

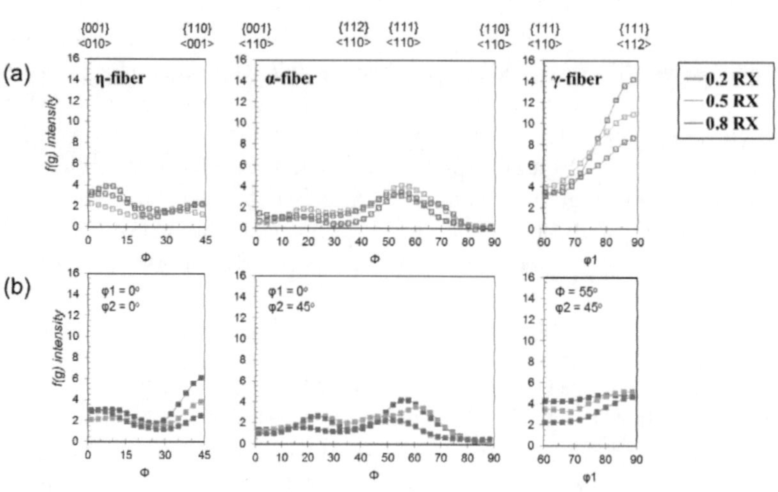

Figure 3. Texture densities along the main fibers observed in bcc steels (η, α, γ-fibers) for the binary (a) and the ternary (b) alloys.

As recrystallization advances these differences in texture remain, however weaken. It is thus believed that Sn addition might have a stronger effect during the nucleation stage of recrystallization.

3.3. CSL boundary distribution

Analyzing the frequency of CSL boundaries of the entire scanned areas, it is seen in figure 4(a) that the CSL fraction is increasing with the progress of recrystallization in both binary and ternary alloy. Σ3 boundaries were not displayed here as there is no twinning observed in these alloys, and Σ3 are generally known to be less mobile. The Σ3 boundaries in this case represent a 60° <111> rotation that is frequently found between two γ-fiber grains and between the {111}<112> and the {001}<110> texture components.

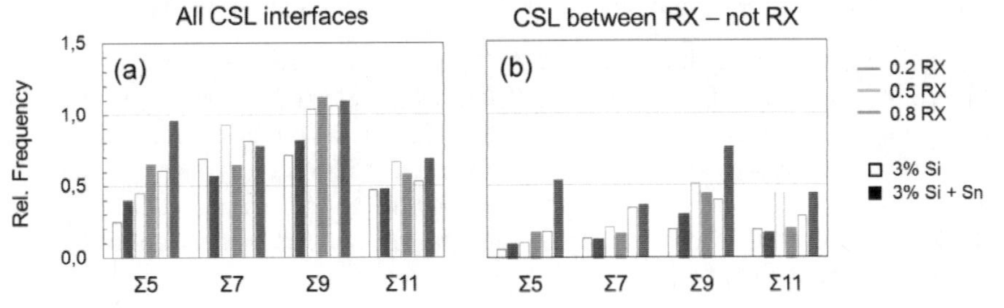

Figure 4. The relative frequencies of low Σ, CSL boundaries as measured in the binary (empty bars) and the ternary (solid bars) alloys, shown as a function of recrystallized fraction (increasing fraction from left to right). (a) is the CSL boundary distributions in the entire scan, (b) is the CSL distributions only between RX and not RX grains (for interpretation of the references to color in this figure, the reader is referred to the web version of this article).

To better quantify the CSL fraction, an analysis was made within the data to evaluate the CSL between RX grains and between RX and not-RX grains. The former classification illustrated a change in CSL fraction as RX grains grow during recrystallization however, it was not very different between the binary and ternary alloy. On the contrary, by analyzing the relative frequency of CSL boundaries of the recrystallization front, figure 4(b), it can be seen that as recrystallization progresses, and especially at 0.8 RX, the total fraction of CSL boundaries is increased with Sn. In particular, it is observed that Σ5 36.87° <100> boundaries have a steady increase in the ternary alloy. A systematic evaluation of the CSL fraction at various temperatures and stages of restoration between the two alloys (unpublished work), showed that the ternary alloy has a consistently higher fraction of Σ5 boundaries.

4. Discussion

Textures were measured in a binary (Fe-Si) and a ternary (Fe-Si-Sn) alloy, to evaluate the effect of Sn on texture evolution during static recrystallization. On one hand, the binary alloy illustrated typical recrystallization textures of low carbon steels [3], with strong rolling texture due to prior cold rolling. On the other hand, the texture of the ternary alloy was found to have less features of the rolling texture, the γ-fiber was remarkably reduced in the recrystallized texture, and at the same time, the peak of intensity was at the Goss component. Previous authors have reported such behavior with the addition of Sn in the fully annealed texture, and also when alloying with other surface active elements as P or Sb [5–7].

The texture observations suggest that Sn has a more profound effect during the early stages of recrystallization. In rolled bcc steels, the sub-grain growth process often takes place preferentially in the vicinity of prior grain boundaries and at the shear bands of {111} deformed grains [19]. It may

IOP Conf. Series: Materials Science and Engineering **375** (2018) 012016 doi:10.1088/1757-899X/375/1/012016

be suggested that Sn segregation at prior grain boundaries may decrease the probability of recrystallization nucleation from the prior boundaries. Prior grain boundaries, are known to be the main nucleation site of γ-fiber grains [18], while at shear bands of {111}<112> grains, nucleation of mainly Goss {110}<001> grains is taking place [20]. This possible effect of Sn on nucleation of recrystallization has been previously mentioned in literature [5].

Moreover, Sn segregation on the recrystallization front might have an added effect on the recrystallized microstructure, through affecting the grain boundary mobility. Special grain boundaries are considered to have higher mobility than general ones [21]. Aust and Rutter [22] have studied grain boundary migration velocity under the effect of Sn in pure Pb. For general and special (CSL) HAGBs, they observed that Sn, even at trace levels (a few ppms), decreases the boundary migration velocity considerably. The effect is more remarkable in the case of general grain boundaries. In the CSL boundaries, although there is an observed solute drag, the effect is weaker.

The segregation of Sn was quantified with 3D-APT in a site-specific manner. The grain boundary character is taken into account before the region of interest is selected. Initially, general HAGBs of the recrystallization front (schematic shown in figure 5a), were preferred for analysis, as these boundaries represent the majority of the migrating interfaces. Figure 5(c) shows a typical volume from 3D-APT, view from two different perspectives, for the Sn atoms and their enrichment at a general HAGB. Results show that Sn segregates at general HAGBs, at concentrations between 1-1.4 at. % (0.4-0.6 at. nm^{-2} excess number of atoms).

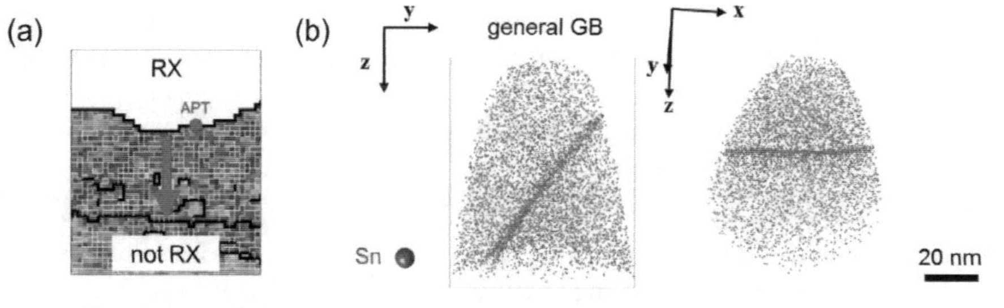

Figure 5. (a) Schematic of the recrystallization front where segregation analysis was made with 3D-APT and (b) a 3D volume for Sn atoms showing segregation at a general GB.

In this study, the segregation of Sn at general HAGBs was measured and assuming that Sn decreased the migration velocity of grain boundaries in bcc α-Fe in a similar trend as Aust and Rutter observed, it is suspected that its segregation at general grain boundaries will increase the contrast in grain boundary mobility between general and CSL boundaries. Therefore, orientations associated with high mobility CSL boundaries might exhibit a preferential growth, which might be the case of the Σ5 36.87° <100> that were found to consistently increase with Sn.

In general, it was observed that in the ternary alloy there is a significant increase in all CSL boundaries at the recrystallization front (figure 4b). In order to evaluate the potential contribution of CSL on the texture development, the case of Σ5 was explored. The deformed state texture was taken as the matrix texture, figure 6(a), and the product orientations were calculated applying a Σ5 36.87° <100> rotation to the main bcc texture components, present in the matrix texture. Figure 6(b) shows the product orientations superimposed on the recrystallization (0.8) texture of the ternary alloy. Careful examination of this figure shows that some of these components are found to be Σ5 related orientations that are present in the texture after recrystallization, however, the recrystallization texture cannot be explained only by the Σ5. Additionally, it can be observed that most of these Σ5 rotations, result along the {100}<uvw> (θ-fiber) and the {hkl}<100> (η-fiber) orientations, figure 6(c). Thus, it is observed that Σ5 have an orientation relationship with {100}<uvw> and {hkl}<100>

grains, as statistically have a higher possibility to bound these fibers due to their common <100> axis of rotation.

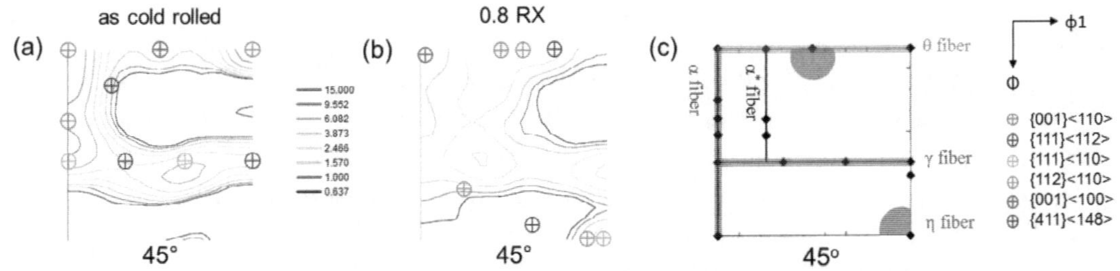

Figure 6. The calculated peak positions for Σ5 relationships, from the matrix (a) as cold rolled to the recrystallization texture (b) in the ternary alloy, recrystallization texture at $X_v=0.8$. The symbols indicate the calculated Σ5 rotations from the deformed to the recrystallization texture for main texture components often observed in bcc steels and (c) is an illustration of the main texture fibers.

Further investigation needs to be made on the potential Σ5 role during recrystallization, and whether their orientation relationship with {100} grains can give to these latter grains, the sufficient size advantage to survive the grain growth process. This could occur as Sn is prone to segregate in HAGBs and will therefore decrease their migration rate, as Aust and Rutter observed. Additional grain boundary segregation analyses is needed to be performed along with an evaluation of the potential solute drag effect of Sn on migrating grain boundaries.

5. Conclusions

Research study on Fe – 3Si binary and Fe – 3Si – 0.1Sn (wt. %) ternary alloy was conducted to evaluate the texture evolution in the presence of Sn, during static recrystallization after cold deformation. It was observed that Sn addition resulted in a recrystallization texture with weaker rolling components. {111}<112> component was decreased meanwhile a clear promotion of Goss texture component was suggested. Sn segregation could affect recrystallization texture at two stages of the process:

- at the stage of nucleation

By modifying the nucleation site preferences. If Sn segregates to prior grain boundaries it would reduce the grain boundary energy, thus it could decrease the nucleation of recrystallized grains from such sites. As an aftermath, nucleation of γ-fiber grains which tend to nucleate at prior grain boundaries will be reduced. In the meantime, nucleation at the shear bands is not affected, Goss component, which nucleates preferably at shear bands, will be favored with Sn addition.

- at the recrystallization front

By modifying the boundary migration preferences. Σ5 boundaries might be assisting the texture development of {100} grains, as there is an orientation relationship of these grains with the Σ5. Such boundaries are known to have higher mobility than general ones. These boundaries were found to have consistently higher relative frequency with Sn, at all stages of primary recrystallization but also during early stages of subsequent grain growth. More work is needed however, to validate such a hypothesis.

Acknowledgements

This project (ANRT – Project n° 1073-2015) was possible with the support of National Association of Research and Technology.

IOP Conf. Series: Materials Science and Engineering **375** (2018) 012016 doi:10.1088/1757-899X/375/1/012016

References

[1] Kubaschewski O 1982 *Iron-Binary phase diagrams*

[2] L. Kestens J J J 2005 Transformation And Recrystallization Textures Associated With Steel Processing *ASM International Volume 14A* vol 14A (ASM International) pp 685–700

[3] Raabe D and Lücke K 1992 Annealing textures of BCC metals *Scr. Metall. Mater.* **27** 1533–8

[4] Suzuki S, Kuroki K, Kobayashi H and Takahashi N 1992 Sn Segregation at Grain Boundary and Interface between MnS and Matrix in Fe-3 mass%Si Alloys Doped with Tin *Mater. Trans. JIM* **33** 1068–76

[5] Kubota T. K K Matsuo Y .Takahashi N 1996 Effect of Sn on Primary Recrystallization Texture of Fe-Si Alloy *Mater. Sci. Forum*

[6] Godec M, Jenko M, Grabke H J and Mast R 1999 Sn Segregation and Its Influence on Electrical Steel Texture Development *ISIJ Int.* **39** 742–6

[7] S. Nakashima K T JHarase, TKamijo 1996 Effect of Tin Addition on Primary and Secondary Recrystallizations of Silicon Steel *Mater. Trans.*

[8] Homma H and Hutchinson B 2003 Orientation dependence of secondary recrystallisation in silicon–iron *Acta Mater.* **51** 3795–805

[9] Hasson G C and Goux C 1971 Interfacial energies of tilt boundaries in aluminium. Experimental and theoretical determination *Scr. Metall.* **5** 889–894

[10] Shimizu R 1989 Coincidence grain boundary and texture evolution in Fe-3%Si *Acta Metall.* **37** 1241–9

[11] Harase J 1995 Coincidence boundary and secondary recrystallization in fcc and bcc metals *Can. Metall. Q.* **34** 185–93

[12] Seah M P and Hondros E D 1973 Grain Boundary Segregation *Proc. R. Soc. Math. Phys. Eng. Sci.* **335** 191–212

[13] Rossi P O and Sellars C M 1997 Quantitative metallography of recrystallization *Acta Mater.* **45** 137–148

[14] Gault B, Moody M P, Cairney J M and Ringer S P 2012 *Atom Probe Microscopy* vol 160 (New York, NY: Springer New York)

[15] Krakauer B W and Seidman D N 11 Subnanometer scale study of segregation at grain boundaries in an Fe(Si) alloy *Acta Mater.* **46** 6145–61

[16] Sellars C M and Whiteman J A 1978 Recrystallization And Grain Growth In Hot Rolling **13** 187–94

[17] Christian J W 1965 *The Theory of Transformations in Metals and Alloys* (Pergamon)

[18] Hutchinson W B 1984 Development and control of annealing textures in low-carbon steels *Int. Met. Rev.* **29** 25–42

[19] Dillamore I L, Smith C J E and Watson T W 1967 Oriented Nucleation in the Formation of Annealing Textures in Iron *Met. Sci. J.* **1** 49–54

[20] Ushioda K and Hutchinson W B 1989 Role of shear bands in annealing texture formation in 3%Si-Fe(111)[112] single crystals. *ISIJ Int.* **29** 862–7

[21] Lejcek P 2010 *Grain Boundary Segregation in Metals* (Elsevier)

[22] Aust K T and Rutter J W Grain Boundary Migration in High Purity Lead and Dilute Lead-Tin Alloys *Trans. Met. Soc. AIME* **215** 119–27

IOP Conf. Series: Materials Science and Engineering **375** (2018) 012017 doi:10.1088/1757-899X/375/1/012017

Preferred orientation formation in Al-3mass%Mg subjected to shear deformation and subsequent annealing

Y. Takayama, T. Kanemaru, K. Wachi and H. Watanabe

Department of Mechanical and Intelligent Engineering,
Graduate School of Engineering, Utsunomiya University,
7-1-2 Yoto, Utsunomiya, Tochigi 321-8585, Japan

takayama@cc.utsunomiya-u.ac.jp

Abstract. Shear deformation is imposed on materials by several advanced techniques for microstructural and textural control. Shear texture evolved during deformation tends to remain after subsequent annealing. Preferred orientation formation with shear deformation and annealing in Al-3mass%Mg alloy rolled sheet has been investigated by SEM/EBSD technique. S-shaped specimen was prepared from the alloy sheet to impose shear strain in compression. The central objective part of the S-shaped specimen was sheared with rotation about the normal axis so as to develop preferred orientations of {111} parallel to the original shear plane and <110> parallel to the original shear direction before deformation. In-situ EBSD analysis revealed that the preferred orientations formed by shear deformation remain and evolve as recrystallization texture during subsequent annealing. The evolution of the shear recrystallization texture depended obviously upon sample direction parallel to shear direction.

1. Introduction

Severe plastic deformation (SPD) has attracted great interests owing to both formation of extremely fine microstructure and improvement of related properties. Numerous studies have been performed to produce the ultrafine grains and make clear their formation mechanism in various SPD processes such as equal channel angular pressing (ECAP)[1], accumulative roll bonding (ARB)[2] and high pressure torsion (HPT)[3]. Crystallographic textures often appeared in the SPDed materials, because extremely large strain should be accompanied with necessary directional priority in slip deformation. The texture evolution during ECAP by only the two shear systems {111}<110> and {001}<110> ({shear plane}<shear direction>) in an Al-Zn-Mg-Cu alloy having a strong initial texture was reported [4]. To an aluminum single crystal ECAP was applied with consistence of theoretical shear plane and direction to crystal slip system [5], and moreover, a crystal plasticity finite element method model showed the textural change with the lattice rotation around the transverse axis of ECAP angle[6]. These imply correspondence of microscopic shear system with macroscopic shear deformation.

Sakai *et al.*[7] pointed out that the shear texture developed by the shear rolling had a broad one spreading around the typical shear components after annealing. Shi *et al.* [8, 9] reported that the shear texture formed in the commercial purity titanium sheet subjected to a friction roll surface processing (FRSP), which was proposed as a SPD process for the surface of the sheet materials in order to control microstructure and texture, remained preferential after annealing.

IOP Conf. Series: Materials Science and Engineering **375** (2018) 012017 doi:10.1088/1757-899X/375/1/012017

In the present study, preferred orientation formation with shear deformation and annealing in Al-3mass%Mg alloy rolled sheet has been investigated by SEM/EBSD technique.

2. Experimental procedure

Al-3mass%Mg alloy was prepared as a 90% rolled sheet of 1mm thick. Chemical composition of the alloy is listed in Table 1. S-shaped specimen as shown in Fig. 1 was cut from the sheet with its longitudinal direction parallel to rolling or transverse direction (RD or TD) for the compressive shear test. A central neck part was sheared by compressing the specimen in vertical or longitudinal direction during the test after annealing at 653K for 3.6ks in Ar and scanning electron microscope/ back scatter diffraction (SEM/EBSD) analysis. Shear strain was calculated from change in relative positions of Vickers hardness indents, which had been sited on the neck part as 750μm square lattice of four fiducial markers before the test.

Table 1 Chemical composition of Al-3mass%Mg alloy used (mass%).

Si	Fe	Cu	Mn	Mg	Cr	Zn	Ti	Al
0.0104	0.0057	0.0028	0.0043	2.9312	0.0004	0.0021	0.0005	Bal

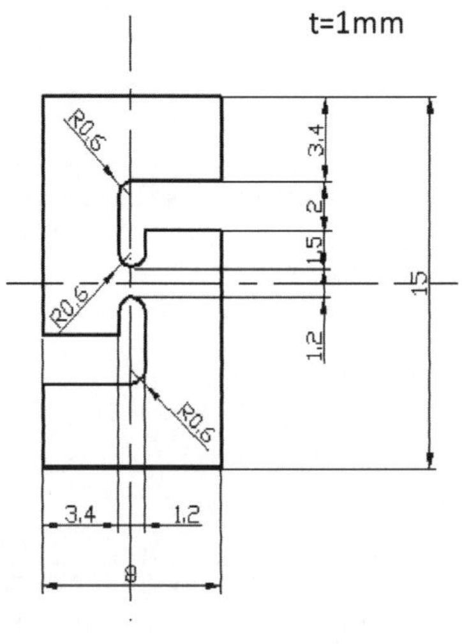

Fig. 1 S-shaped specimen.

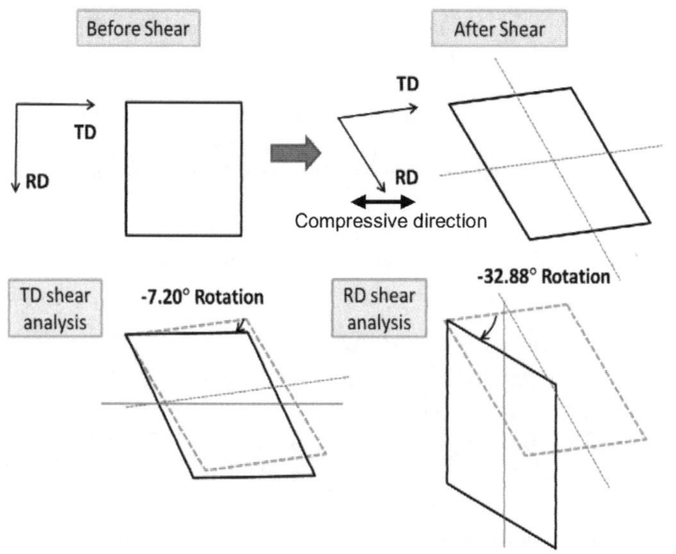

Fig. 2 Objective squares before and after shear deformation (in TD-compressive shear test). Grain orientation was analyzed on shear plane and direction for the measured data rotated by each angle about ND axis.

To examine changes in microstructure and texture the S-shaped specimen was annealed for 2.7ks and in-situ analyzed by SEM/EBSD repeatedly during step heating controlled with heater temperature.

IOP Conf. Series: Materials Science and Engineering **375** (2018) 012017 doi:10.1088/1757-899X/375/1/012017

3. Experimental results and discussion

3.1. Macroscopic change in the central part of the S-shaped specimen

The objective square in the central part of the S-shaped specimen (which was composed of the four fiducial markers with an interval of 750μm) deformed into the parallelogram with rotation about normal direction (ND) axis as illustrated in Fig. 2 (TD compressive shear). In the figure, TD and RD were rotated by 7.20° and 32.88° in counter clock-wise to original directions, respectively. Because shear deformation takes place by conjugate shear stresses along TD and RD in the square of the figure, change in grain orientation with shear deformation should be analyzed for the measured data reversely rotated by each angle about ND axis as TD or RD corresponds to the original one. The former is called as *TD shear analysis* here while the latter is called as *RD shear analysis*.

3.2. Change in microstructure through TD-compressive shear and step heating

3.2.1. TD-compressive shear test. Figure 3 displays the inverse pole figure (IPF) maps before and after TD-compressive shear test with a shear strain γ=0.47, which was calculated from change in relative positions of the indent markers. The grain orientation in TD is expressed as color. Grain size before the test was measured as 63.2μm by the planimetric method [10]. The map after the deformation (Fig. 3(b))

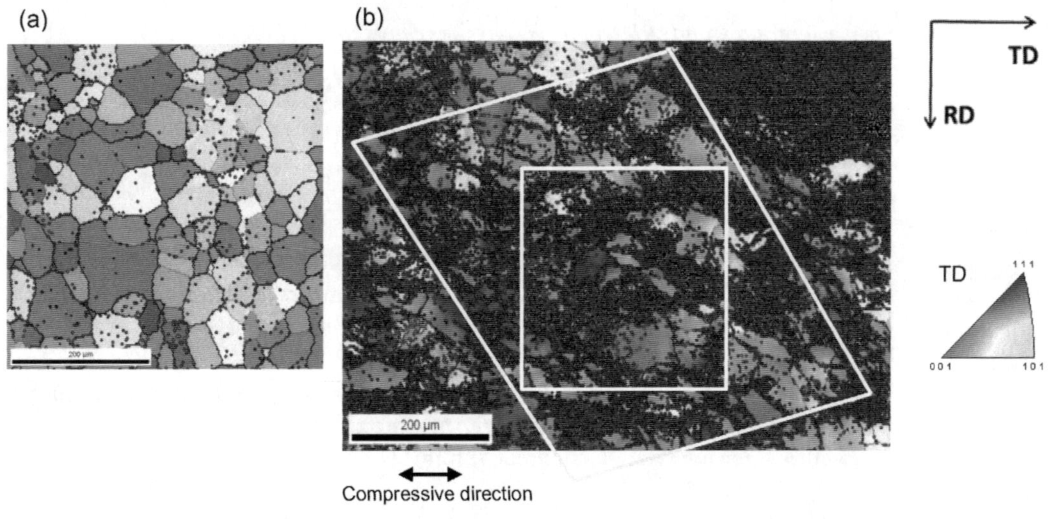

Fig. 3 IPF maps representing microstructures (a) before and (b) after TD-compressive shear deformation. Yellow parallelogram refers to the square region before the deformation. White square indicates objective area for the in-situ SEM/EBSD analysis during step heating.

reveals local grain rotation as change in color and stored shear strain in grain interior as gradation. Larger strain appears as unanalysed points in the figure.

3.2.2. In-situ analysis during heating. Figure 4 shows IPF and kernel average misorientation (KAM) maps in restoration process during step heating in the square region of Fig. 3(b). It is known that intragranular misorientation or KAM increases with increment of imposed strain at ambient temperature and decreases with decrement of dislocation density in recrystallization process [11]. Recrystallized grains with low KAM or strain relief appear at 559K and cover larger area gradually to 607K. Mean KAM lowered from 1.95 degrees at 543 to 0.83 degrees at 607K for the whole area.

Preferred orientation evolution was investigated from the EBSD data, which was rotated by -7.2° or -32.88° about ND axis to return TD or RD to the original direction as shown in Fig. 2. Changes in fractions of <111>$\perp TD_O$ (TD_O: original TD before deformation) and <110>//TD_O are displayed in Figs. 5(a) and 5(b). The fractions were calculated for the whole analyzed area before step heating and for the recrystallized area (KAM<1°) during step heating. As shown in the figure, the shear textures <111>$\perp TD_O$ and <110>//TD_O, which mean that slip plane and direction are consistent with shear plane and direction, respectively, evolved considerably to fractions of 19.3% and 23.0% after shear deformation in the specimen subjected to TD-compressive shear test. The fractions of the shear texture increase markedly during step heating and reach 27.4% and 44.2% at 607K. The recrystallization textures include more fractions of components close to ideal orientations compared with the shear deformation textures. Moreover, the fraction of <110>//TD_O, amounts to three forth of the whole recrystallization area at 607K. These results support the fact that shear deformation texture remains preferential after annealing.

On the other hand, Figs. 5(c) and 5(d) show changes in fractions of <111>$\perp RD_O$ (RD_O: original RD before deformation) and <110>//RD_O in the same way. The shear textures <111>$\perp RD_O$ and <110>//RD_O, also evolve by shear deformation and subsequent step heating. However, their fractions are obviously smaller than those of <111>$\perp TD_O$ and <110>//TD_O. In particular, the fractions of the recrystallization texture are only less than a half of those on TD. Besides, there appear less fractions of components close to ideal orientations.

3.3. Change in microstructure through RD-compressive shear and step heating

3.3.1. RD-compressive shear test. A shear strain γ=0.50 was imposed in RD-compressive shear test. Grain size before the test was measured as 53.7μm. The square in the central area of the S-shaped specimen deformed into a parallelogram with rotated RD and TD by 6.9° and 33.5°.

3.3.2. In-situ analysis during heating. Figure 6 displays changes in fractions of <111>$\perp RD_O$, <110>//RD_O, <111>$\perp TD_O$ and <110>//TD_O. Even In RD-compressive shear specimen the shear textures on RD, <111>$\perp RD_O$ and <110>//RD_O, do not evolve remarkably. The fractions after shear deformation are only 10.2% and 15.7%, and moreover, less than 10% after the heating at 607K. In contrast to this, the fractions of <111>$\perp TD_O$ and <110>//TD_O rise up to 19.5% and 23.8%, respectively after shear deformation in the RD-compressive shear specimen also. The fractions of the shear recrystallization texture increase acceleratingly during step heating and reach 35.9% and 35.3% at 607K. Recrystallization proceeds somewhat lately compared with in TD-compressive shear specimen.

3.4. Evolution of E_{TD}{111}<110> through compressive shear and step heating

Evolution of E_{TD}{111}<110>{shear plane}<shear direction> on TD (or TD shear analysis) is supposed from the results in clauses 3.2 and 3.3. Orientation E {111}<110> [12] was examined as one of shear texture components in asymmetric rolling process. Figure 7 shows orientation maps of E_{TD} with its fraction during step heating in both of TD- and RD- compressive shear specimens. Orientation E_{TD}{111}<110> grew up to more than 25% in annealing after attaining fraction of 10% by shear deformation.

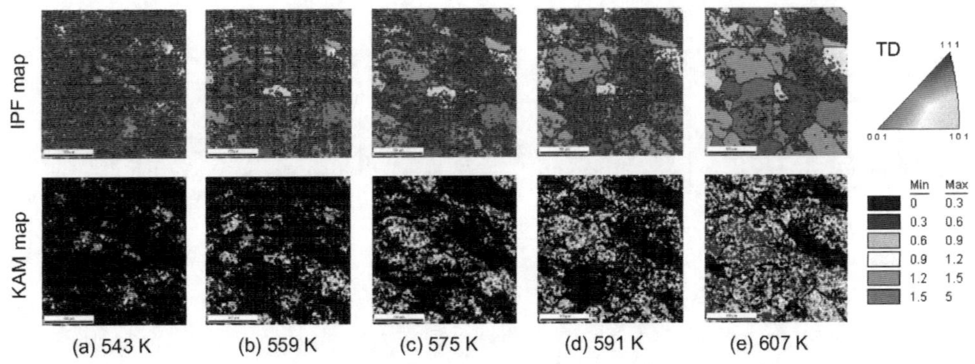

Fig. 4 IPF and KAM maps representing microstructural changes during step heating after TD-compressive shear deformation. KAM lowers from 1.95, 1.38, 1.11, 0.99 and 0.83 degrees.

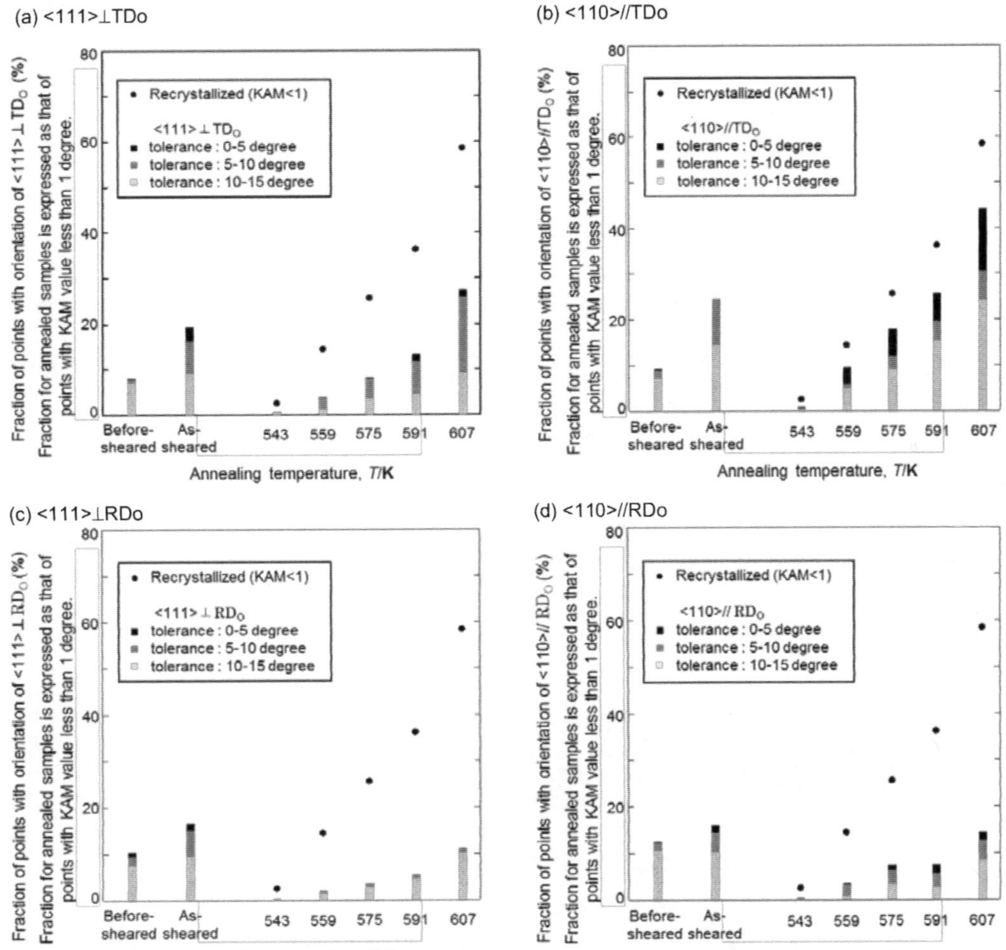

Fig. 5 Evolutions of shear preferred orientations before and after shear deformation and subsequent step heating: fractions of (a) $<111>\perp TD_O$, (b) $<110>//TD_O$, (c) $<111>\perp RD_O$ and (d) $<110>//RD_O$ in TD-compressive shear specimen.

IOP Conf. Series: Materials Science and Engineering **375** (2018) 012017 doi:10.1088/1757-899X/375/1/012017

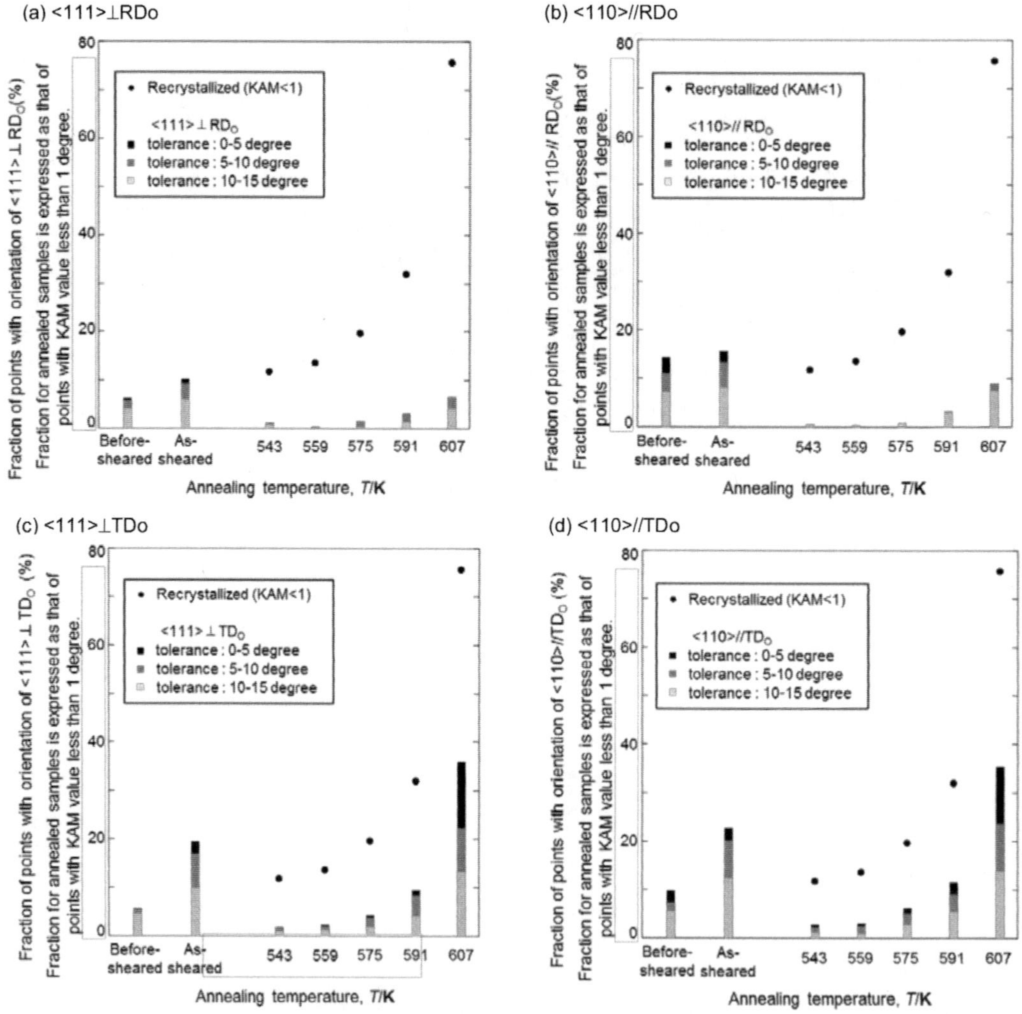

Fig. 6 Evolutions of shear preferred orientations before and after shear deformation and subsequent step heating: fractions of (a) <111>$\perp$RD$_O$, (b) <110>//RD$_O$, (c) <111>$\perp$TD$_O$ and (d) <110>//TD$_O$ in RD-compressive shear specimen.

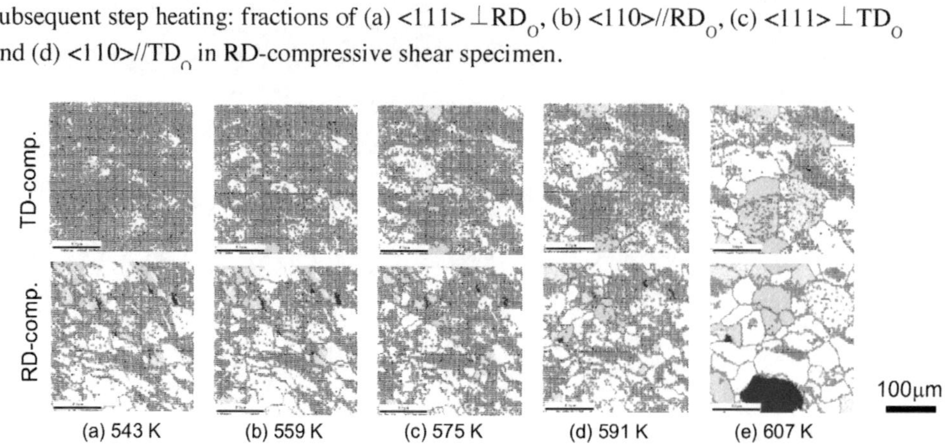

Fig. 7 Evolution of E_{TD}\{111\}<110> during step heating after TD- and RD- compressive shear deformation. Fractions of E_{TD} increase from 6.2% to 26.2% and from 13.0% to 28.0%.

IOP Conf. Series: Materials Science and Engineering **375** (2018) 012017 doi:10.1088/1757-899X/375/1/012017

4. Conclusions

In the present study, preferred orientation formation with shear deformation and annealing in Al-3mass%Mg alloy rolled sheet has been investigated by SEM/EBSD technique. The obtained results are summarized as follows:

1. Grain orientation with its slip plane parallel to shear one or its slip direction parallel to shear one tends to evolve through shear deformation and subsequent annealing.

2. For both of TD- and RD- compressive shear specimens, remarkable increases in shear texture components on TD plane are found, which indicates that both shear systems are active equivalently for preferred orientation formation and the components on TD have predominance.

3. Orientation E_{TD}\{111\}<110> grows up to more than 25% in annealing after attaining fraction of 10% by shear deformation.

References

[1] Y. Iwahashi, J.Wang, Z. Horita, M. Nemoto and T.G.. Langdon: Scr. Mater. 35 (1996) 143-146.
[2] Y. Saito, N.Tsuji, H. Utsunomiya, T. Sakai and R.G. Hong: Scr. Mater. 39 (1998) 1221-1227.
[3] A.P. Zhilyaev, G.V. Nurislamova, B.-K. Kim, M.D. Baro, J.A. Szpunar and T.G. Langdon: Acta Mater. 51 (2003) 735-765.
[4] S.C. Wang, M.J. Starink, N.Gao, X.G. Qiao, C. Xu, T.G. Langdon: Acta Mater. 56 (2008) 3800-3809.
[5] Y. Fukuda, K. Oh-ishi, M. Furukawa, Z. Horita, T.G. Langdon: Acta Mater. 52 (2004) 1387-1395.
[6] C. Lu, G.Y. Deng, A.K. Tieu, L.H. Su, H.T. Zhu, X.H. Liu: Acta Mater. 59 (2011) 3581-3592.
[7] T. Sakai, S. Hamada and Y. Saito: Scripta Mater. 44 (2001) 2569-2573.
[8] M. Shi, Y. Takayama, T. Umetsu, H. Kato, H. Watanabe, H. Inoue: Mater. Trans. 50 (2009) 210-214.
[9] M. Shi, Y. Takayama, C. Ma, H. Watanabe, H. Inoue: Trans. Nonferrous Met. Soc. China 22 (2012) 2616–2627.
[10] Z.Jeffries, A.H.Kline and E.B.Zimmer: Trans.AIME, 57(1916), 594-607.
[11] Y. Takayama and J.A. Szpunar: Mater. Trans., 45(2004), 2316-2325.
[12] J. Sidor, A. Miroux, R. Petrov, L. Kestens: Acta Mater. 56 (2008) 2495-2507.

IOP Conf. Series: Materials Science and Engineering **375** (2018) 012018 doi:10.1088/1757-899X/375/1/012018

Effect of strain path change on the texture evolution after cold rolling and recrystallization of Nickel-40 wt. %Cobalt alloy

Gyan Shankar, Satyam Suwas

Department of Materials Engineering, Indian Institute of Science, Bangalore-560 012, India

Email: gyanshankar@iisc.ac.in

Abstract

Effect of change in strain path during rolling and its role in the development of crystallographic texture after deformation and then after recrystallization has been studied for Ni-40wt%Co which is a medium stacking fault energy (SFE) material. Results indicate that textures developed after unidirectional rolling and multi-step cross-rolling are quite different which also leads to different texture evolution after recrystallization. Microstructure of both UDR and cross rolled sample shows band type of feature within the grain having lesser band density for cross rolled samples. After recrystallization unidirectionally rolled sample shows bimodal whereas cross rolled sample shows equiaxed grain size distribution. But all the samples show large fraction of annealing twins. It has been shown that in all cases formation of annealing twins have large influence on the texture transition after recrystallization.

Keywords: Deformation texture; Cold rolling, Unidirectional rolling; Cross-rolling; Texture transition; Recrystallization texture; annealing twin

1. Introduction

The mechanisms of evolution of deformation texture in face centred cubic (FCC) metals and alloys have been investigated by many researchers [1–8]. One of the most important material property that controls the micro-mechanisms of deformation in FCC materials is the stacking fault energy (SFE). It is well known that rolling texture in a closed packed face centred cubic materials show texture transition from Cu (or pure metal) type to Bs (or alloy type) with a decrease in stacking fault energy. In general, high and medium SFE material shows Cu type texture which consist of prominent Cu {112} <111> component, a continuous spread of orientation from Bs {110} <112> to S {123} <634> and a weak Goss {110} <001> components. Low SFE material shows Bs type texture which consists of strong Bs {110} <112> and Goss {110} <001> components. This transition in deformation texture with a change in SFE is also accompanied by deformation heterogeneities such as deformation twins, micro-bands and shear bands. In a low SFE material, it was found that twinning occurs in the initial stage of deformation process, at the intermediate level twinning activity ceases and later at high strain formation of shear bands take place. Experiments conducted on single crystals revealed that Cu orientation is very prone to shear banding, whereas Goss orientation is not susceptible to heterogeneities during deformation [9, 10].

Apart from the material property, the processing variables are also equally important in the formation and stability of a particular texture [5, 11, 12]. It was found that change in strain path during rolling has pronounced effect on the evolution of texture [13–15]. Rotation of the

IOP Conf. Series: Materials Science and Engineering **375** (2018) 012018 doi:10.1088/1757-899X/375/1/012018

sample after every pass during rolling rotates the substructure along the new rolling direction. It leads to destabilizations of the substructures formed during the previous pass of rolling and thereby alters the evolution of texture.

Formation of recrystallization texture strongly depends on the initial deformation texture which is dependent upon the stacking fault energy of material and the processing parameters of deformation such as, mode of deformation, strain, strain path etc. Previous investigations on the development of recrystallization texture show that Cu type rolling texture typically results in cube recrystallization texture and Bs type rolling texture usually recrystallize in brass orientation [16]. Annealing twin is also an important factor which contributes to the formation of nuclei during early stages of recrystallization and later development of texture which was not present in the initial deformed material [17].

The aim of the present work is to understand the effect of strain path on the development of texture during deformation and subsequent recrystallization in Ni-40wt. %Co alloy. Ni-Co alloys form completely substitutional FCC solid solution up to 65 wt. %Cobalt. Stacking fault energy of the system decreases with cobalt addition from high SFE of (pure Ni, 150 mJ m^{-2}) to very low SFE of (Ni-60Co, 30 mJ m^{-2}) [18]. Ni-40Co alloy, used in the present work is a medium SFE material with SFE value is about 60 mJ m^{-2}. The micro-mechanisms of texture evolution in Ni-40Co are elaborated in this investigation.

2. Experimental Procedure

An ingot of nickel-cobalt alloy containing 40 wt. %Co was melted in vacuum arc melting furnace. During melting magnetic stirring was used to ensure composition uniformity. Quantitative composition analysis was carried out by energy dispersive spectroscopy (EDS). To break the cast dendritic structure and to weaken the initial solidification texture, the as-cast ingot was cross-rolled to 50% thickness reduction. After that, the rolled material was further annealed at 700 °C for 6 hours, to get completely homogenised strain free equiaxed microstructure having random texture. It was taken as starting material for our study. The initial material was then subjected to 70% unidirectional rolling (UDR) and multistep cross rolling (CR), equivalent to a true strain of ~1.2 at room temperature. A schematic representation of rolling schedules followed in the present study is shown in Fig.1. Both the Unidirectional and the cross rolled material was then subjected to isothermal annealing at 550 °C for 30 minutes for recrystallization.

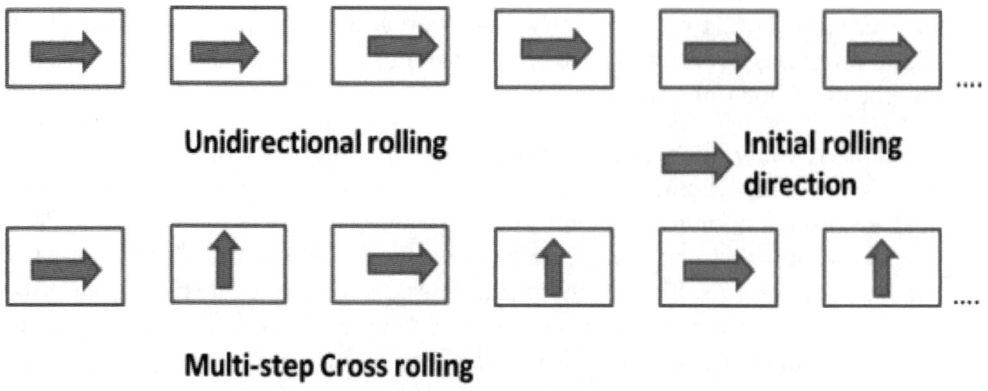

Fig. 1. A schematic representation of rolling pass followed in the present study. Rolling direction in each pass is shown by arrow. 0.1 true strain has been given in each pass of rolling.

Bulk texture measurements for all the rolled and recrystallized samples were carried out by the Schulz reflection method, using an X-ray texture goniometer with Co K_α radiation (D8 Discover, Bruker AXS, Germany). Four incomplete pole figures (α = 0-75°) were measured from the 111, 200, 220 and 311 peaks on the rolling plane of the sample. Because of coarse grain for initial sample texture measurements were done at multiple locations to improve the statistics of the measurements. The three-dimensional orientation distribution function (ODFs) was calculated using Labotex® software [19].

The microstructure of all rolled and recrystallized samples was taken in a field emission scanning electron microscope, equipped with EBSD system. All microstructure were taken on the transverse plane of rolled samples. EBSD data has been rotated to ND-plane to see the important texture component. Samples for EBSD scans were mechanically polished using silicon carbide emery sheets and further electropolished.

3. Results and Discussion

Fig. 2 shows (111) pole figures for all the starting, deformed and recrystallized Ni-40wt. %Co alloy. A key pole figure depicting the important texture components is also presented along with. The Miller indices of ideal FCC rolling and recrystallized texture component are given in Table 1.

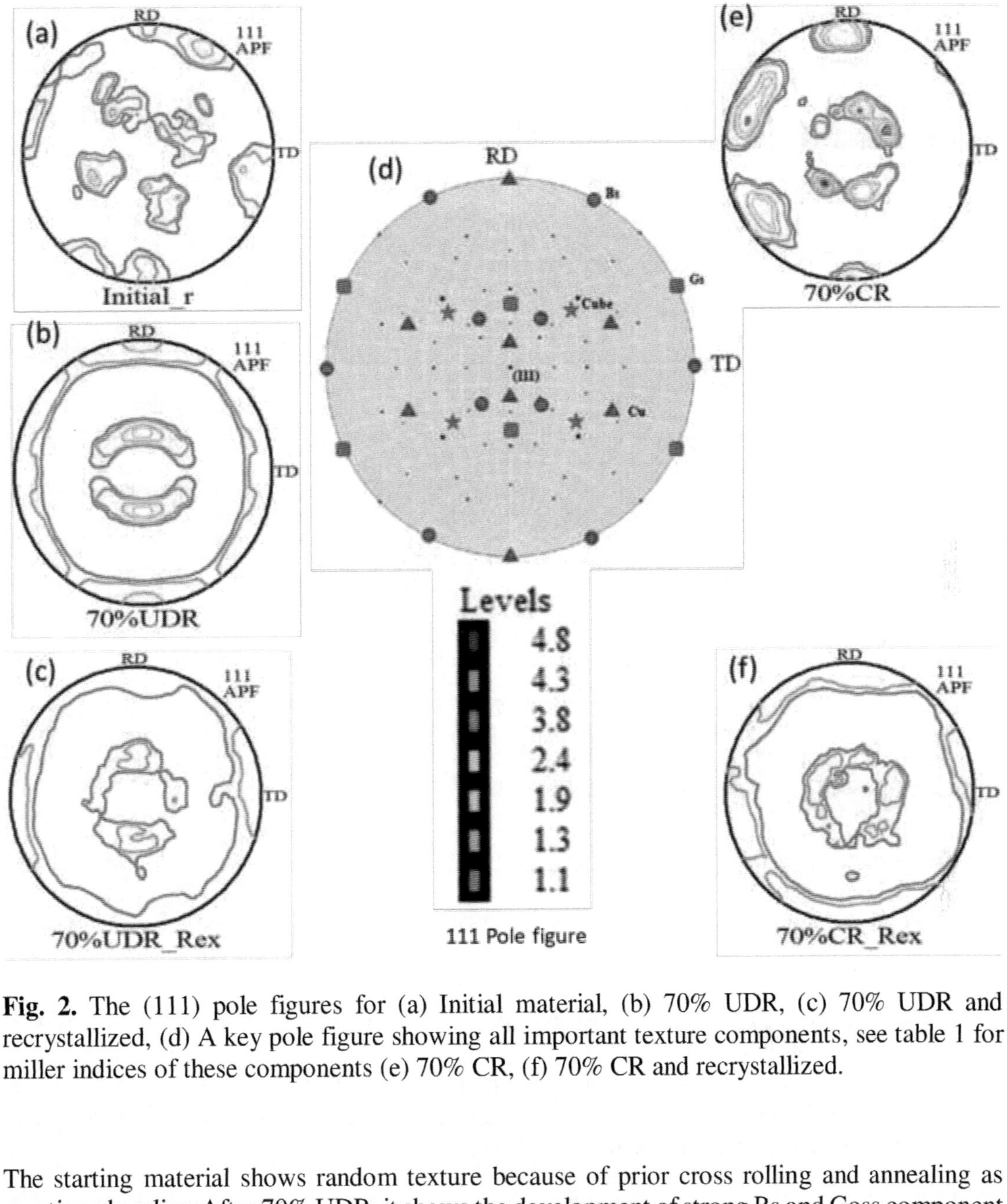

Fig. 2. The (111) pole figures for (a) Initial material, (b) 70% UDR, (c) 70% UDR and recrystallized, (d) A key pole figure showing all important texture components, see table 1 for miller indices of these components (e) 70% CR, (f) 70% CR and recrystallized.

The starting material shows random texture because of prior cross rolling and annealing as mentioned earlier. After 70% UDR, it shows the development of strong Bs and Goss component with weak Cu and S component. It also shows some amount of Goss twin orientation which developed as a result of unidirectional rolling. On the other hand, the 70% Cross rolled sample shows weaker texture as compared to UDR which is also reported by many authors [20]. It shows remarkable Goss and {110} <111> (A) component. The development of 'A' component after cross rolling has also been reported for copper [20]. The intensity of Cu and S component was found very weak as compared to UDR sample. Cross rolled samples also show retention of Bs, G/B and rotated Goss component which was present in the initial material. The volume fraction of important deformation and recrystallization texture component was calculated after

IOP Conf. Series: Materials Science and Engineering **375** (2018) 012018 doi:10.1088/1757-899X/375/1/012018

computing the orientation distribution functions. The volume fractions of different texture components are shown in Fig. 3.

Table 1. Ideal FCC rolling and recrystallization texture components.

Texture component	Miller indices
Cu	{1 1 2} <1 1 1>
Bs	{1 1 0} <1 1 2>
S	{1 2 3} <6 3 4>
Goss	{1 1 0} <0 0 1>
Cube	{0 0 1} <1 0 0>
Taylor	{4 4 11} <11 11 8>
R	{1 2 3} <4 1 2>
G/B	{1 1 0} <1 1 5>
Rotated Goss (Rt-G)	{1 1 0} <0 1 1>
Rotated Cu (Rt-Cu)	{1 1 2} <1 1 0>
Rotated Cube (Rt-C)	{0 0 1} <1 1 0>
Goss Twin (GT)	{1 1 3} <3 3 2>
Cu Twin (CuT)	{5 5 2} <1 1 5>
A	{1 1 0} <1 1 1>
F	{1 1 1} <1 1 2>
E	{1 1 1} <0 1 1>
α - fiber	{1 1 0} <0 0 1> - {1 1 0} <1 1 2>
β - fibre	{1 1 2} <1 1 1> - {1 2 3} <6 3 4> - {1 1 0} <1 1 2>
τ- fibre	{1 1 2} <1 1 1> - {5 5 2} <1 1 5> - {1 1 0} <0 0 1>
ϒ - fibre	[1 1 1] II ND ({1 1 1} <0 1 1> - {1 1 1} <1 1 2>)

After recrystallization of 70% UDR sample shows strong Brass, cube, Goss and Rt-cube component in almost equal proportion. It can also be seen that the intensity of Cu, Bs, S and Goss component went down as compared to deformed samples whereas the intensity of cube and Rt- cube component got increased. The 70% Cross rolled and recrystallized sample shows completely different texture as compared to UDR recrystallized samples. The development of strong rotated Goss component is clearly noticed, which was not present in the UDR recrystallized sample. It also shows little increases in Cu and S component as compared to cross rolled sample which was decreasing when UDR sample was recrystallized. One similarity were found in texture transition during recrystallization of both UDR and cross rolled sample is that Goss component always decreases after recrystallization.

To get the better understanding of texture transition upon rolling and after recrystallization, important fibres were plotted from the ODFs. The fibre plots are shown in Fig. 4. The α and τ- fibres show orientations <110> || ND and <110> || TD respectively. The α and τ- fibres are quite inhomogeneous. The β- fibre also does not show much variation in differently processed samples except for 70% cross rolled sample which exhibits a maximum between S/Bs and Bs orientation. The 70% UDR sample shows very high intensity near Bs and Goss orientation in α- fibre plot but after recrystallization, their intensity decreases. However, it maintains a broad peak between G/Bs and Bs orientation. The retention of some Brass orientation after recrystallization can be explained by nucleation with orientation close to those of deformed matrix. It can take place by twinning into crystallographic identical variants influenced by annealing twin since lots of annealing twins can be observed in the microstructure after recrystallization (Fig. 6c) [21]. The 70% UDR recrystallized sample also shows a remarkable peak between CuT and Goss orientation with some intensity at Rt- Cube orientation in τ- fibre which cannot be ignored. The presence of CuT orientation could be attributed to the annealing twin of Cu orientation which was present in the deformed matrix. The development of rotated cube (Rt-C) component may take place by transformation of CuT orientation since it possesses $\Sigma 3$ twin relation with it.

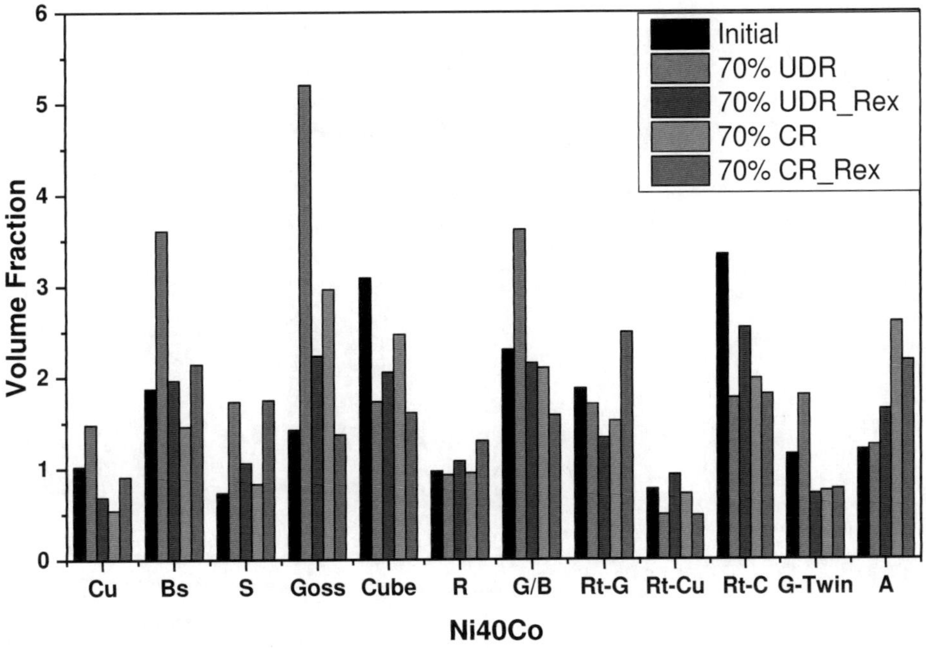

Fig. 3. The volume fraction of important texture component of initial rolled and recrystallized material.

On the other hand, 70% cross rolled sample shows a strong peak near Goss and {110} <111> (A) orientation in α- fibre and very high intensity between S/Bs and Brass orientation in β- fibre. The reason behind the occurrence of high intensity between S/Bs and Brass orientation could be destabilization of substructure because of rotation of sample after subsequent passes of rolling. Since Ni40Co is a medium SFE material, initial unidirectional passes leads to develop some Brass orientation but because of rotation of sample previous orientation destabilize and try to rotate towards other very preferential rolling texture component S orientation. As a result finally, high intensity develops at a location between S/Bs and Brass orientations which can be seen in β- fibre (Fig 4c). After recrystallization of 70% cross rolled sample it shows high intensity near brass orientation and a maximum at rotated Goss (Rt-G) orientation in α- fibre. It can also be seen that there is a sharp decrease in Goss orientation after recrystallization. However, near Bs orientated grains recrystallized in the same manner as explained earlier for the unidirectional rolled samples through the formation of annealing twin, since recrystallized cross rolled sample also shows a large fraction of annealing twins. Overall it can be observed that the formation of annealing twins has a large influence on the texture transition during recrystallization.

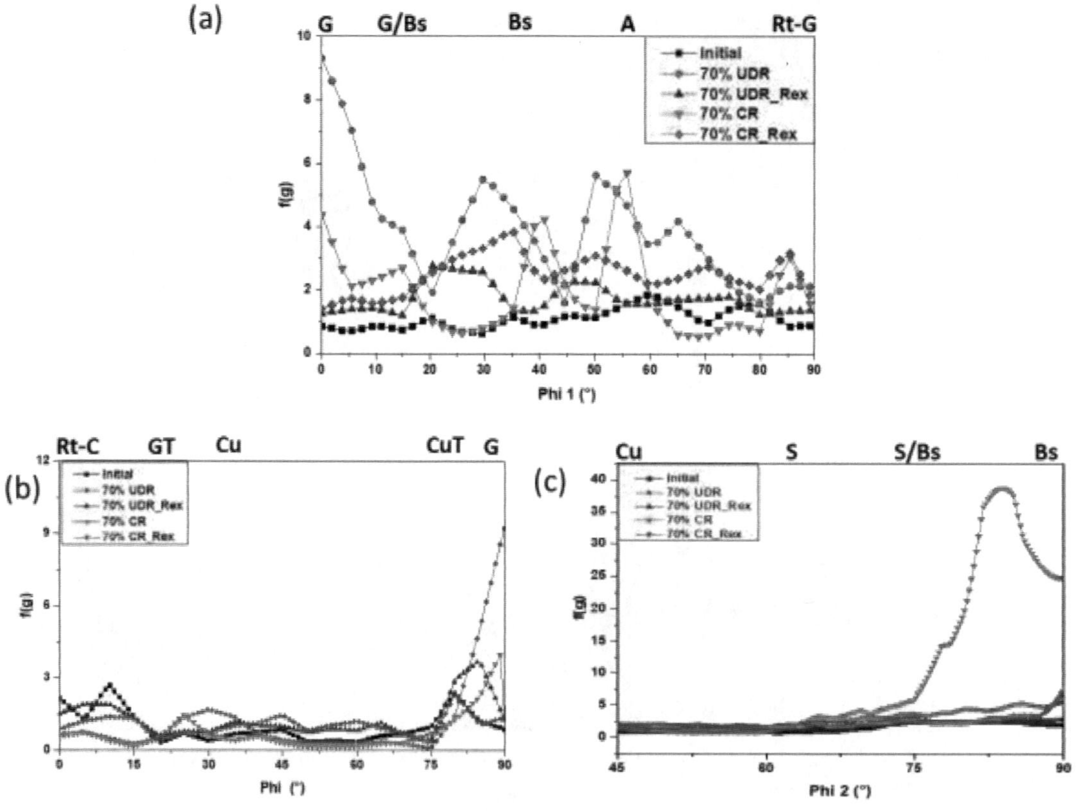

Fig. 4. Texture fibre plot of the initial Ni-40Co alloy, after 70% UDR, CR and after recrystallization (a) α – fibre, (b) τ – fibre and (c) β – fibre.

The inverse pole figure (IPF) map for initial and all differently processed samples are shown in Fig.5. The initial microstructure shows a grain size of about 350 microns and consists of some annealing twins which was formed during homogenisation (Fig. 5a). After 70% unidirectional rolling microstructure shows lots of band type of feature inside the grain (Fig. 5b). These banded features are similar to the ones observed earlier in Al-Mg alloy by Hurley and Humphreys [22, 23], and later also in Ni40Co alloy reported by Madhvan [18]. They called these bands as "micro-shear bands", because of its microscopic nature and confined within a single grain.

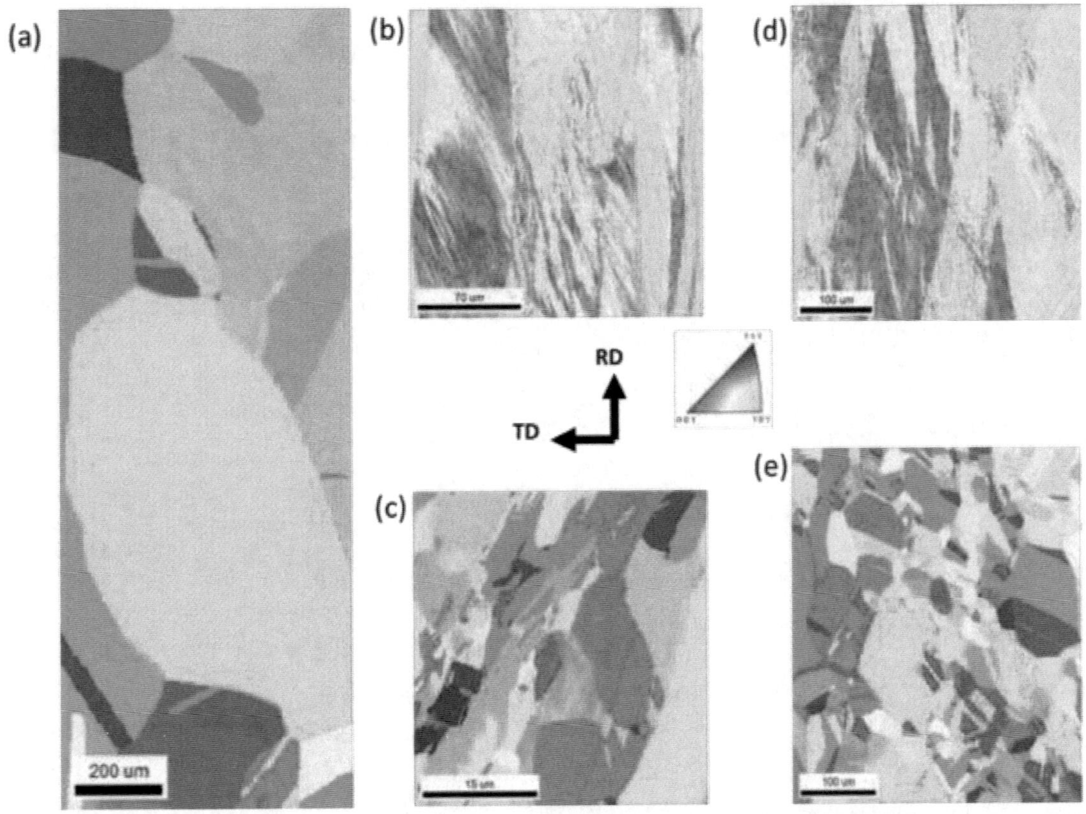

Fig. 5. Orientation map presented on ND- plane for (a) Initial material, (b) 70% UDR, (c) 70% UDR and recrystallized, (d) 70% CR, (e) 70% CR and recrystallized.

4. Conclusions

The effect of strain path on the evolution of deformation and recrystallization texture, as well as microstructure in a medium-SFE Ni–40Co alloy, has been investigated for 70% rolling reduction equivalent to a true strain of 1.2. The alloy showed quite different bulk texture and microstructure of UDR and cross rolled sample. Even after recrystallization, the effect of strain path during deformation can be seen in the formation of recrystallization texture. Marked differences from other medium-SFE materials, e.g. Cu, is clearly noticed. The important conclusions of this investigation are:

1. Unidrectionally rolled sample shows extended α- fibre having strong Bs and Goss component with some Goss twin orientation whereas cross rolled sample shows weaker texture having remarkable Goss and {110} <111> (A) component. Cross rolled sample also shows retention of Bs and rotated Goss component.

2. After recrystallization strong Bs, Goss, cube and Rt-cube orientation develop in almost equal proportion for UDR sample whereas strong rotated Goss component with little increase in Cu and S orientation formed for cross rolled samples.

3. For differently processed sample α and τ- fibres are quite inhomogeneous but β- fibre shows does not show much variation except for 70% cross rolled sample showing a maximum between S/Bs and Bs orientation.

Acknowledgements

The authors acknowledge the extensive use of microscopy facilities at the Advanced Facility for Microscopy and Microanalysis (AFMM), Department of materials engineering, Indian Institute of Science.

References

1. Duggan BJ, Hatherly M, Hutchinson WB, Wakefield PT (1978) Met Sci 12:343–351.
2. Hutchinson WB, Duggan BJ, Hatherly M (1979) Met Technol 6:398–403.
3. Suwas S, Singh AK, Rao KN, et al (2011) Acta Metall 36:4827–4840.
4. LEE CS, DUGGAN BJ, SMALLMAN RE (1993) J Phys Iv 3:2027–2032.
5. Hirsch J, Lücke K (1988) Acta Metall 36:2863–2882.
6. C. D, Valle R, Dervin P, Penelle R (1989) Acta Met 37:1547–1571.
7. Engler O, Kong XW, Lücke K (1999) Scr Mater 41:493–503.
8. Engler O (2000) Acta Metall 48:4827–4840.
9. Nakayama Y, Morii K (1987) Acta Metall 35:1747–1755.
10. Nakayama Y (1985) Acta Metall 33:379–386.
11. J. HIRSCH KL and MH (1988) Acta Metall 36:2905–2927.
12. Engler O, Hirsch J, Cke KLI (1995) Acta Metall 43:121–138.
13. Suwas S, Singh AK (2003 Mater Sci Eng A 368–371.
14. Gurao NP, Sethuraman S, Suwas S (2011) Mater Sci Eng A 528:7739–7750.
18. Madhavan R, Ray RK, Suwas S (2014) Acta Mater 74:151–164.
19. K. Pawlik (1986) phys. stat. 801. (b) 134, 477.
20. Gurao NP, Kapoor R, Suwas S (2010) Metall Mater Trans A 41:2794–2804.
21. Saleh AA, Pereloma E V, Gazder AA (2011) Mater Sci Eng A 528:4537–4549.
22. Hurley PJ, Humphreys FJ (2003) Acta Mater 51:1087–1102.

23. Hurley PJ, Humphreys FJ (2003) Acta Mater 51:3779–3793.

24. Madhavan R, Ray RK, Suwas S (2014) Acta Mater 78:222–235.

25. Hansen N, Jensen DJ (1999) Philos Trans R Soc A Math Phys Eng Sci 357:1447–1469.

26. Garg R, Ranganathan S, Suwas S (2010 Mater Sci Eng A 527:4582–4592.

Quantifying Crystallographic Texture Variation in a Titanium Billet

P S Davies[1,2], B P Wynne[1*], M J Thomas[3], W M Rainforth[1]

1. Department of Materials Science and Engineering, The University of Sheffield, Mappin Street, Sheffield S1 3JD, UK.
2. Sheffield Forgemasters RD26 Ltd, 286 Brightside Lane, Sheffield S9 2RW, UK.
3. Timet UK Ltd., P.O.Box 704, Witton, Birmingham, West Midlands B6 7UR, UK.

* b.wynne@sheffield.ac.uk

Abstract. Ti-834 and other titanium alloy forgings can contain regions of similarly orientated primary alpha grains called macrozones that can have a deleterious effect on performance in-service. Understanding the origin of macrozones and how they change during processing is an important step in optimising process routes for these alloys. In this work a detailed characterisation of crystallographic texture variation was carried out on the cross-section of a 250 mm diameter Ti-834 billet using optical metallography, EBSD and neutron texture analysis. Results showed there was a variation in texture and macrozone size and shape through the billet with each macrozone composed of a single hcp component, with varying degrees of orientation spread. The texture of the majority of macrozones can be classed as either axial (c-axis parallel to the billet axis) or transverse (c-axis perpendicular to the billet axis). In the centre of the billet cross-section, the global texture was dominated by a fibre-like $\{10\bar{1}0\}$ texture due to a dominance of transverse macrozones. At the edge, there was a greater proportion of axial macrozones which gave rise to an additional {0002} component.

1. Introduction

This paper reports on work that was carried out to investigate and characterise microstructure, microtexture and global textures in a billet of the titanium alloy Ti-834 (Ti–5.8Al–4Sn–3.5Zr–0.7Nb–0.5Mo–0.35Si–0.06C in wt%). Several titanium alloys, including Ti-834, Ti6Al4V and Ti6242, etc., in multiple product forms have been reported to contain regions with strong local textures referred to as macrozones (e.g., [1, 2, 3]). The main focus of this work was to study the effect of location on the macrozones in terms of their size, shape, texture intensity and crystallographic orientation. This was accomplished using a combination of optical microscopy, EBSD and neutron texture analysis to study a cross-sectional slice from the billet.

2. Experimental Procedure

An 8 mm thick transverse slice was obtained from a 250 mm diameter billet of Ti-834. The billet was sectioned for analysis using optical microscopy, neutron texture analysis and EBSD utilising a range of step sizes. The locations analysed are indicated in Figure 1. Specimens labelled 1-7, with dimensions 15 x 10 x 3 mm, were examined using EBSD. Specimens labelled A-Q, with dimensions 15 x 15 x 8 mm, were used for the neutron texture analysis. The macroscopic coordinate system used for this analysis is

IOP Conf. Series: Materials Science and Engineering **375** (2018) 012019 doi:10.1088/1757-899X/375/1/012019

also shown in Figure 1. It should be noted that the two perpendicular radial directions (R1 and R2) were arbitrarily chosen and are not related to specific forging directions, since these were not known.

A detailed description of the experimental setup for the EBSD and neutron texture analysis can be found in Davies et al. [4].

3. Results

3.1. Characterisation at the Scale of the Macrozones

Optical microscopy (Figure 2) at lower magnification showed that the billet contained regions of developed yet aligned primary alpha αp grains.

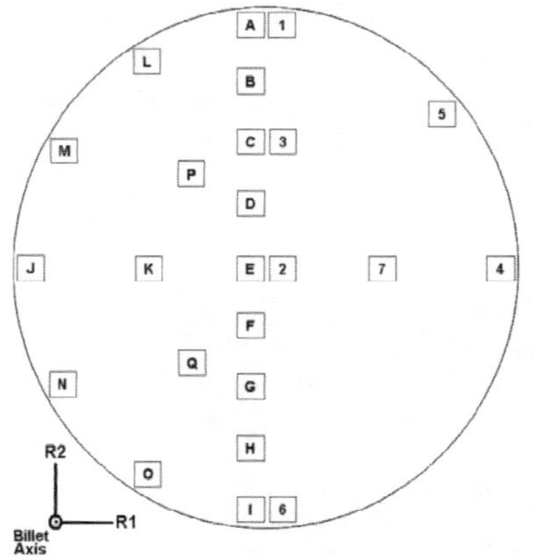

Figure 1. Specimen positions and macroscopic coordinates for the billet texture characterisation.

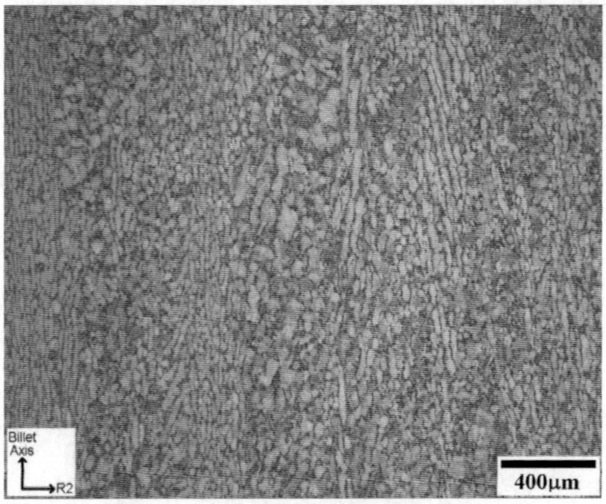

Figure 2. Longitudinal section showing columnar shaped groupings of aligned αp grains. Both the groupings and the αp grains themselves are predominantly aligned with the billet axis.

To provide information on the crystallographic orientation of the aligned regions, an EBSD map was acquired from the centre of the billet in the longitudinal plane using a step size of 3µm. The map is displayed in Figure 3a using Euler colouring and Figure 3b using inverse pole figure (IPF) colouring (note that the IPF colouring is with respect to the billet axis instead of the map normal, which means that the colours show which crystallographic planes of the α phase lie in the transverse plane of the billet). It can be seen that the regions of aligned grains also share common crystallographic orientations. This demonstrates that regions of aligned αp grains are in fact macrozones, regions with strong local textures. In the longitudinal section, the macrozones have thicknesses ranging from approximately 100 to 1000 µm and are up to several thousand microns in length.

To provide more detailed information on the crystallographic orientations of the α phase in the aligned regions, two EBSD maps were acquired from specimen 2 (centre) and 1 (edge) using a step size of 3 µm. These maps were acquired in the transverse plane in order to maximise the number of aligned regions contained in the maps. The map from the centre of the billet (specimen 2) is shown in Figure 3c using a combination of band contrast and IPF colouring to reveal microstructure as well as orientation information. It can be seen that viewed in the transverse plane, the macrozones are much more equiaxed. The macrozones in the map from the centre of the billet are mainly blue or red in colour. In the blue macrozones, the prismatic $\{10\bar{1}0\}$ planes lie in the transverse plane of the billet, which means that the c-axis must also lie in the transverse plane in a direction perpendicular to the billet axis. We can therefore describe these macrozones in terms of the c-axis direction, as transverse type. In the red macrozones, the {0002} planes lie in the transverse plane, which means the c-axis is aligned with the billet axis. In terms of the c-axis direction, these macrozones can be described as axial type. More diffuse groups of green grains are also present. These are also transverse type but in this case, the $\{11\bar{2}0\}$ planes lies in the transverse plane. The basal transverse macrozones tend to be larger and occupy a greater portion of the scanned region compared to the basal axial macrozones. Figure 4 shows a typical example of a basal transverse macrozone in greater detail, together with a misorientation map and pole figure of the measured orientations in the selected macrozone. It can be seen from the figure that almost every αp grain in the macrozone has a similar crystallographic orientation, which results in an extremely strong local texture. For the basal transverse macrozones, there was a very strong tendency for the c-axis of the crystal to be in the same direction as the direction of grain alignment in the transverse plane. The misorientation map shows misorientations between adjacent α grains because of the thin layers of β phase between the α grains was not indexed. The clear lack of boundaries within the basal transverse macrozone shows that most of the grains within the macrozone are misorientated from their neighbours by less than 15°. Figure 5 shows a typical example of a basal axial macrozone in greater detail. It can be seen from the pole figure that there is a larger spread in the grain orientations compared to the basal transverse macrozone, which results in a weaker local texture. The misorientation map shows that most of the grains within a basal axial macrozone are misorientated from their neighbours by more than 15°.

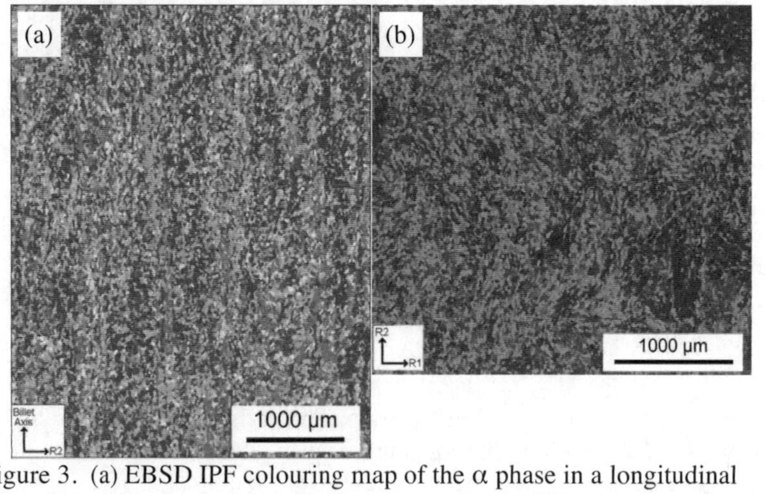

Figure 3. (a) EBSD IPF colouring map of the α phase in a longitudinal section from the centre of the billet (specimen 2). (b) IPF colouring + band contrast map of the α phase in a transverse section from the centre of the billet (specimen 2).

The map from the edge of the billet (specimen 1) is shown in Figure 6. Unlike the centre of the billet, the grains are more or less equiaxed in the transverse plane, so there is no clear alignment of the grain structure in the map. However, macrozones can once again be identified as areas with a dominant IPF colour. Compared to the centre of the billet, macrozones at the edge appear to be more diffuse, which is at least partly due to a higher volume fraction of secondary alpha αs at the edge. This αs will be expected to have a spread of orientations due to the β→α transformation even if there is a strong variant selection. The largest macrozones are of the basal axial type. Smaller groupings of grains with basal transverse type orientations with both $\{10\bar{1}0\}$ or $\{11\bar{2}0\}$ parallel are also present. Detailed examination of both the blue and red macrozones showed that their strengths are similar but neither are as strong as the blue transverse macrozones that featured prominently in the centre of the billet. The boundary maps for the macrozones at the edge also showed that most of the α grains in the macrozones are misorientated from their neighbour grains by more than 15°.

IOP Conf. Series: Materials Science and Engineering **375** (2018) 012019 doi:10.1088/1757-899X/375/1/012019

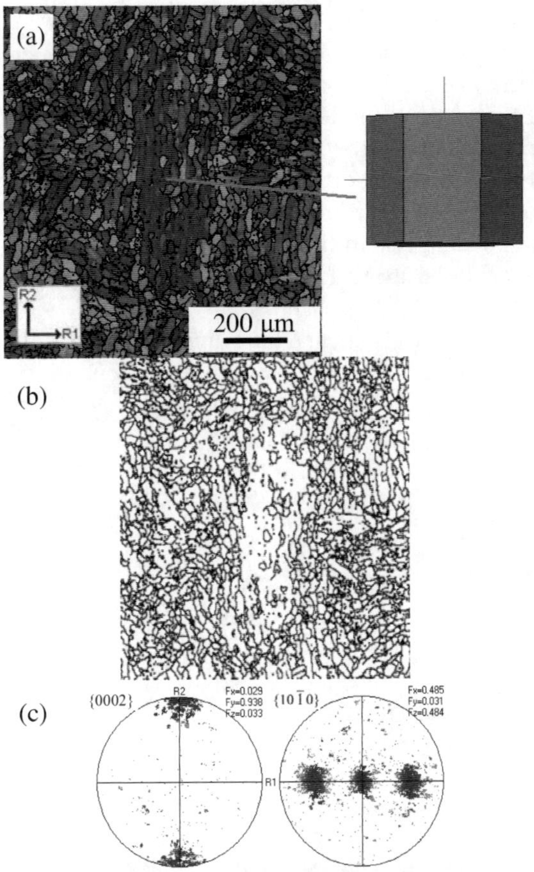

Figure 4. Typical example of a blue transverse macrozone from the centre of the billet. (a) IPF of selected macrozone, (b) 15° misorientations, (c) pole figures

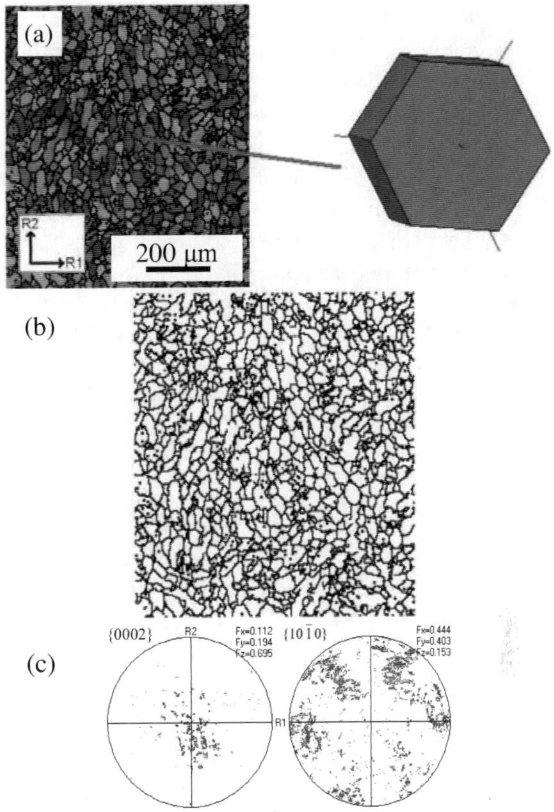

Figure 5. Typical example of a red axial macrozone from the centre of the billet. (a) IPF of selected macrozone, (b) 15° misorientations, (c) pole figures

Figure 6. EBSD map of the α phase in a transverse section (perpendicular to billet axis) from the edge of the billet (specimen 1). IPF colouring + band contrast.

3.2. Characterisation of Global Textures and Macrostructural Alignment

The results of the neutron and EBSD texture analysis are presented together in Figure 7, in which each texture measurement is superimposed on the billet slice according to its location. Each texture measurement is represented by an {0002} and a $\{10\bar{1}0\}$ pole figure contour plot in equal area projection. Orientation image maps for the EBSD analysis on specimens 1 to 7 are displayed in Figure 8. IPF colouring is used with respect to the billet axis, which means that the colours show which crystallographic planes of the α phase lie in the transverse plane of the billet.

A direct comparison can be made between the EBSD and neutron texture analysis techniques by comparing the textures of specimens taken from adjacent regions in the billet cross-section i.e. 1 with A, 3 with C, 2 with E, and 6 with I. It can be seen that there is reasonably good agreement between the EBSD and neutron texture measurements, apart from relatively small differences in the intensity of some of the poles. The close agreement suggests that the neutron texture analysis has successfully reproduced the global textures, and also indicates that the areas analysed using EBSD were large enough for representative global textures to be acquired.

In general, the global billet textures were found to be quite weak, with maximum multiples of random distribution (m.r.d.) values of between 2 and 3. However, there are clearly global textures present, which indicates that the macrozones tend to have preferred orientations. Furthermore, these textures change depending on the location in the billet.

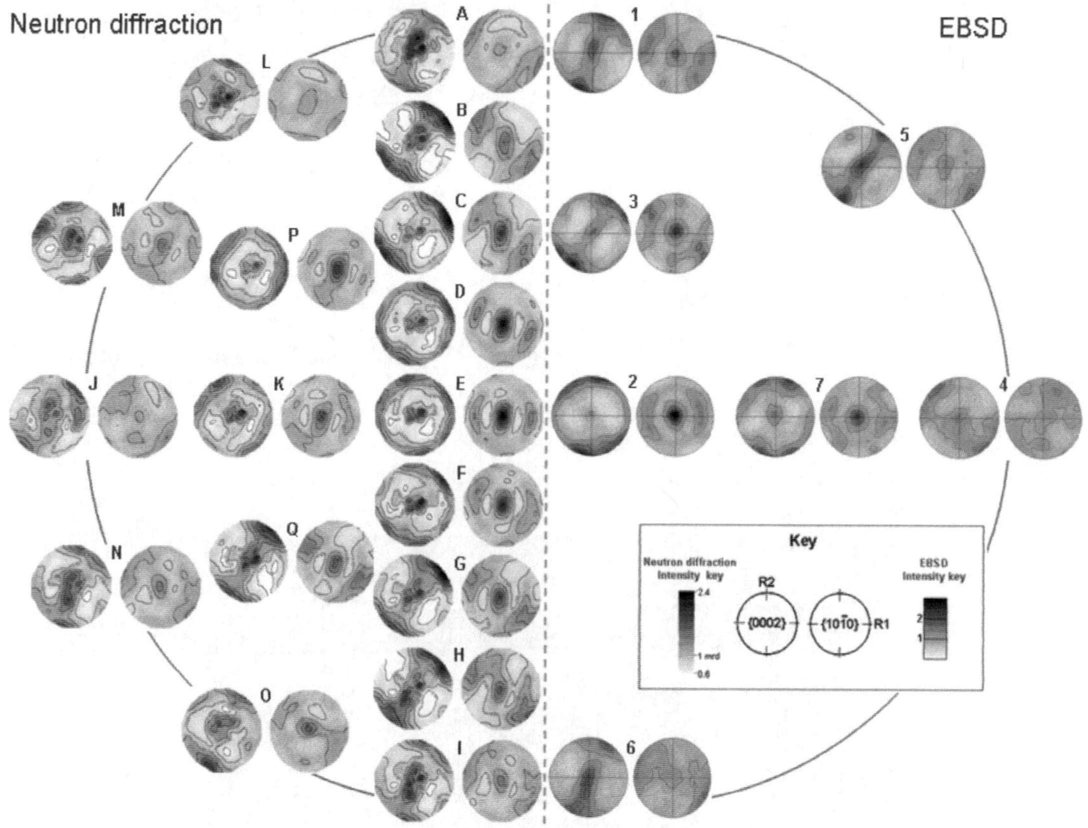

Figure 7. Billet texture results. Specimens A-Q analysed by neutron diffraction, specimens 1-7 analysed by EBSD.

When the texture results (Figure 7) are viewed in conjunction with the EBSD maps (Figure 8), the most obvious trend is a centre to edge variation in the relative amounts of the transverse and axial macrozones. In the centre of the billet, the radial blue type gives rise to a dominant pole at the centre of the $\{10\bar{1}0\}$ pole figure and blue macrozones occupy the greatest portion of the map. A smaller amount of the axial red type is also present in the centre of the billet but the weaker local textures of the red macrozones means there is only a weak pole at the centre of the {0002} pole figure. As one moves from the centre of the billet to the edge, the relative amounts of the two types of macrozone change, so that at the edge of the billet the dominant macrozones are axial. This means that the pole at the centre of the $\{10\bar{1}0\}$ pole weakens, while the pole at the centre of the {0002} pole figure strengthens and red macrozones occupy the greatest portion of the maps.

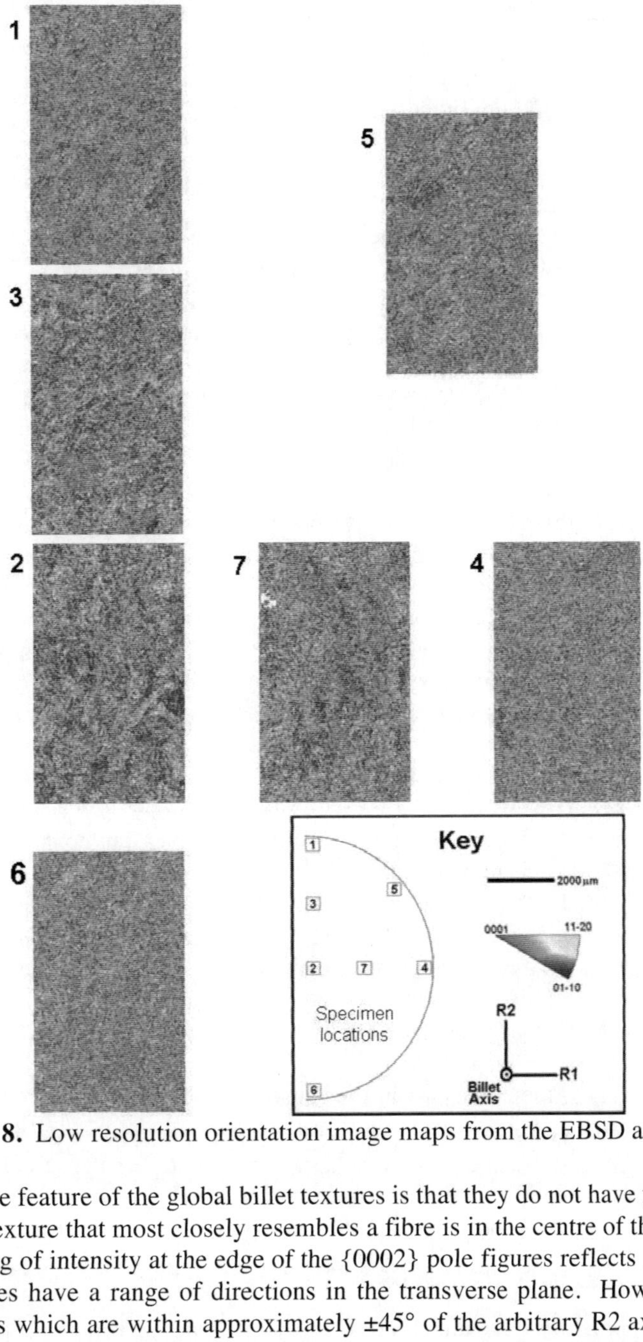

Figure 8. Low resolution orientation image maps from the EBSD analysis.

Another noticeable feature of the global billet textures is that they do not have fibre symmetry about the billet axis. The texture that most closely resembles a fibre is in the centre of the billet (specimens 2 and E). Here, the ring of intensity at the edge of the {0002} pole figures reflects how the c-axes in the transverse macrozones have a range of directions in the transverse plane. However, there is higher intensity in directions which are within approximately ±45° of the arbitrary R2 axis, which shows that there is preferential alignment of the transverse macrozones in these directions. Further away from the centre of the billet, there is a stronger preferential alignment of the c-axes in the transverse plane. In many locations in the cross-section, there is a similar direction for this preferred alignment, with the c-axes lying in a radial direction approximately 30° clockwise from the arbitrary R2 axis. Where the preferential alignment is strongest, a band of intensity connects the {0002} poles at the edge of the pole figure, with the {0002} pole in the centre.

As well as the alignment in the pole figures, there is also a subtle alignment of the macrostructure in some of the low resolution EBSD maps (Figure 8), most notably in specimens 1, 3, 5 and 6. The direction of this macrostructural alignment is similar and corresponds to the direction of the bands of intensity in the {0002} pole figures i.e. a radial direction approximately 30° clockwise from the arbitrary R2 axis. These results indicate that in addition to the very strong primary alignment of the macrostructure along the billet axis, in some locations in the billet, there is also a secondary alignment of the macrostructure in a specific direction in the transverse plane.

4. Discussion

The results of the low resolution EBSD analysis (Figure 8) have revealed that the billet contains macrozones throughout the cross section. The macrozones are related to prior colonies of aligned α lamellae, which have the same crystallographic orientation and as a result tend to deform and globularise in a similar manner, thus maintaining similar crystallographic orientations in the globularised αp grains. The macrozones are large columnar shaped regions elongated along the billet axis. In certain locations of the billet, some macrozones also show a secondary elongation in the transverse plane.

Regardless of the location in the cross section, the α phase texture of a macrozone is always composed of a single hcp component, with varying degrees of spread up to around 30°. This is in agreement with macrozone textures reported by Thomas [5] and Pilchak [3] but differs from the macrozones analysed by Germain et al. [1], which consisted of a dominant component and in addition, several weaker ones. The absence of the weaker components in this study and Thomas [5] can be explained by the different proportions of αp and αs in the billet that was analysed by Germain et al. [1]. The billet in the work of Germain et al. [1] contained 70% αs, and by analysing the contribution of the αpand αs separately, it was shown that the αs gave rise to the weaker components, while the dominant component was due to overlapping of a single αp component and the main αs component. In the present study, the highest volume fraction of αs was only 30% at the edge of the billet, which is too low to produce significant secondary components in the macrozone textures. Consequently, the high volume fraction of αp in the billet means that the overall α texture only shows the orientation of the single αp component and thus a separation of the contribution of the αp and αs was not considered to be necessary for the macrozone textures in this work.

Analysis of the orientation image maps in conjunction with measurements of global texture, have shown that there is a tendency for the macrozones to have certain preferred orientations. The majority of macrozones can be divided into two types: axial and transverse. In axial macrozones, the c-axes are aligned with the billet axis, which means that the basal {0002} planes lie parallel to the transverse plane. In the IPF maps obtained in the transverse plane, the axial macrozones appear as clusters of red grains. The individual αp grains in the axial macrozones tend to be plate shaped with the plane of the plates parallel to the billet axis. This means that the c-axis direction tends to lie in the plane of the plate shaped grains.

In transverse macrozones, the c-axes lie in the transverse plane i.e. at 90° to the billet axis. The largest and sharpest macrozones in the billet were of this transverse type and also had a $\{10\bar{1}0\}$ plane lying in the transverse plane, thus appearing in the transverse plane IPF maps as clusters of blue grains. The individual αp grains in the blue transverse macrozones tend to be plate shaped, with the plane of the plates parallel to the billet axis and aligned in the same transverse direction as other αp grains within the same macrozone. The $<10\bar{1}0>$ direction and the c-axis direction tend to lie in the plane of the plate shaped grains.

The macrozones reported by Germain et al. [1] in Ti-834 billet and Bantounas et al. [4] in Ti-6Al-4V bar were of the transverse type. Thomas [3], in Ti-834 billet reported mainly the transverse type but also one instance of the axial type. In this study, it was found that the relative amount of axial and transverse macrozones varies depending on the location in the cross-section. The most obvious trend is the centre to edge variation in the relative amounts of the two types of macrozone. The centre of the billet consists mainly of the transverse type, which have $\{10\bar{1}0\}$ planes lying in the transverse plane, and a lesser amount of the axial red type. As one moves from the centre to the edge of the billet, the relative amounts change so that the edge contains more of the axial macrozones and a lesser amount of the blue transverse type. This work suggests that axial macrozones are much more significant than suggested by previous studies, particularly at the edges of a billet.

5. Conclusions

The crystallographic texture of a cross-section of Ti-834 billet has been investigated and the following conclusions can be drawn:

1. The billet contained microtexture heterogeneities (macrozones) throughout the billet cross-section.
2. The α phase texture of a typical macrozone is composed of a single hcp component, with varying degrees of spread up to around 30°.
3. Macrozones are typically columnar shaped regions elongated along the billet axis. In certain locations of the billet, some macrozones also show a secondary elongation in the transverse plane.
4. Macrozones are typically 100-1000μm in diameter in the transverse plane and up to several thousand microns in length.
5. The majority of macrozones can be classed as either axial (c-axis parallel to the billet axis) or transverse (c-axis perpendicular to the billet axis).
6. In the centre of the billet cross-section, the global texture was dominated by a fibre-like $\{10\bar{1}0\}$ texture due to a dominance of transverse macrozones. At the edge, there was a greater proportion of axial macrozones which gave rise to an additional {0002} component.
7. In this product form, areas analysed using EBSD were large enough for representative global textures to be acquired.

References

[1] Germain, L, Gey, N, Humbert, M, Bocher, P, and Jahazi, M 2005, *Acta Materialia*, **53**, 3535

[2] Woodfield AP, Gorman MD, Corderman RR, Sutliff JA, Yamrom B (1996) *Titanium'95: Science and Technology,* UK:1116.

[3] Pilchak AL, Szczepanski CJ, Shaffer JA, Salem AA, Semiatin SL (2013) *Metall Mater Trans A* **44**, 4881.

[4] Davies P S et al 2008 *Meas. Sci. Technol.*, **19** 034002

[5] Thomas M J 2007 *The Effect of Thermomechanical Process Parameters on the Microstructure and Crystallographic Texture Evolution of Near-α Aerospace Alloy Timetal 834* PhD Thesis, The University of Sheffield, UK.

[6] Bantounas I, Dye D, Lindley T C *Crack growth, microstructure and texture in Ti 6Al 4V*. In Proceedings of the 11th World Conference on Titanium, Kyoto, Japan 4–7 June 2007.

IOP Conf. Series: Materials Science and Engineering **375** (2018) 012020 doi:10.1088/1757-899X/375/1/012020

Texture evolution in medium Mn containing TWIP steel: Experiments and Simulation

Rajib Kalsar, Satyam Suwas

Department of Materials Engineering, Indian Institute of Science, Bangalore-560 012, India

Email: satyamsuwas@iisc.ac.in

Abstract

In the present work, deformation mechanisms have been studied by evaluating texture and microstructure under cold rolling for medium Mn containing Fe-14Mn-0.5C-3.5Al twinning induced plasticity (TWIP) steel. Rolling has been performed up to true strain ~2.3. Microstructural features reveal that deformation takes place by twinning. Texture evolution shows α-fiber type texture with a continuous α-fiber up to very large strains. Experimentally observed deformation micro-mechanisms have been verified with the visco-plastic self-consistent (VPSC) simulation. Effects of twin volume fraction in texture evolution also been studied by incorporating volume transfer scheme in the VPSC simulation. Simulation results show very good agreement with the experimental results.

Keywords: Texture, microstructure, deformation mechanisms, visco-plastic self-consistent (VPSC) simulation

1. Introduction

The interest in twinning induced plasticity (TWIP) steels has increased over the last few years due to the excellent combination of strength and ductility that renders these materials suitable for many applications. TWIP steels exhibit very high strain hardening rate due to the formation of very high fraction of deformation twins which can lead to dynamic Hall-Petch effect [1-3]. TWIP steels are high manganese containing steels with face centred cubic (FCC) crystal structure. In conventional TWIP steels, the manganese content generally varies between 20 – 30 weight percent. Although such a high Mn content leads to some very desirable properties, it also poses a challenge in materials production. The addition of Mn to such a high extent leads to delayed cracking of the materials after processing to shapes [4]. This could be detrimental for industrial applications. Therefore, it has been proposed to reduce the Mn content in TWIP steels to lower levels. The addition of Al also helps in reducing the tendency for delayed cracking. Since the Mn content in TWIP steels influences the stacking fault energy (SFE) and so does Al-addition turn dictates deformation micromechanisms, it is important to examine the effect of SFE in the case of lower Mn Al-containing TWIP steels [5]. SFE is reported to play a major role in twin mediated deformation mechanisms [6, 7]. It has been reported that for steels with SFE in the range of 20-50 mJm^{-2}, deformation take place by twinning whereas for materials with SFE below 15 mJm^{-2}, deformation is governed by martensitic transformation [8, 9]. The critical value of SFE for the transition from strain induced martensite transformation to mechanical twinning has been reported as > 9 mJm^{-2}, and the maximum value of SFE for the formation of ε-martensite is 12 mJm^{-2} [7].

It is well known that texture development in FCC materials strongly depends on the SFE. Materials with higher SFE (i.e. Al, Ni and Cu) do not undergo by twinning during deformation and develop a strong Cu-type texture (comprises of Brass, Cu and S components), while the materials with low SFE (i.e. Ni-60Co, Cu-30Zn and Ag) generally deform by twinning and develop Brass-type texture (comprises of Goss and Brass components) [10-13]. Recently, Madhavan et al. [14-16] reported Cu-type to Brass-type texture transition in some Ni-Co alloys, which have SFE in the same range as TWIP steels, at large rolling strains. TWIP steels are categorized as low SFE materials.

In the present investigation, the evolution of deformation texture and microstructure has been carried out for a TWIP steel with the composition Fe-14Mn-0.5C-3.5Al to understand the deformation mechanisms during rolling to large strain. The SFE of this steel is approximately ~50 mJm^{-2}. In order to understand the individual contribution of slip-twin systems in texture evolution visco-plastic self-consistent (VPSC) simulation was performed.

2. Experiments

2.1 Material and processing

TWIP steel with a composition Fe-14Mn-0.5C-3.5Al (wt%) was melted in a arc melting furnace under an argon atmosphere. The charge was melted several times to get adequately homogeneous composition. As-melted button shaped sample was homogenized in a furnace at 1100°C for 5 hrs under high vacuum (<10^{-5} mbar). The homogenised button was cross rolled in many steps to 50 % reduction in thickness and then annealed at 830°C for 30 minutes. An equiaxed microstructure with average grain size ~50 μm was obtained after this treatment. The so-obtained plate was treated the starting material. The starting plate was rolled to thickness reductions ~ 40%, 70% and 90% and subsequently subjected to investigations pertaining to texture and microstructural evolution.

2.2 Characterizations

Structural characterization of the material at different stages was carried out by X-ray diffraction. Textures were measured for the starting material as well as for the cold rolled samples using a Bruker D8 discover diffractometer based on the Schultz reflection geometry, equipped with a four circle goniometer employing Co k_α (λ=0.1791 nm) target. Four incomplete pole figures (111), (200), (220) and (311) were measured. Measurement was performed at the mid section parallel to the rolling plane. All the measurements were carried out in the range phi=0-360° and chi=0-75° with a step size 5° and exposure time 5 sec. The orientation distribution functions (ODFs) were calculated from the four measured pole figures using JTex software [17]. Microstructural characterization was performed in a FEI ESEM-Quanta scanning electron microscope (SEM), equipped with electron backscatter diffraction (EBSD) system. EBSD scans were recorded on the RD-ND plane (the plane normal to TD).

The rolled samples were metallegraphically polished followed by electropolishing. Struers ElectroPol-5 electropolishing system was used for EBSD sample preparation. The analysis of the EBSD scans was done using TSL-OIM™ software. The mechanical properties were evaluated by carrying out tensile tests on the annealed strips. For tensile tests, flat dog-bone shaped samples with 6 mm gauge length, 2 mm width and ~0.7 mm thickness were used. Tensile tests were carried out in a Servo-hydraulic UTS machine at room temperature with strain rate of 10^{-3} s^{-1}. Tests were carried out till failure.

3. Visco-plastic self-consistent simulation

Visco-plastic self consistent (VPSC) model was used to investigate the contribution of different microscopic deformation mechanisms and consequent texture evolution. The details VPSC model is given in Ref. [18]. In the present simulation, perfect slip and twinning systems are have been activated to account for the deformation according to the critical resolved shear stress (CRSS) associated with each of the deformation process. Bronkhorst type hardening law was used to account for microscopic hardening [19]. Bronkhorst hardening law can be written as:

$$H^{ij} = q^{ij} h_0 \left(1 - \frac{\tau_0^i}{\tau_{sat}}\right)^a \quad (1)$$

where h_0 and a are hardening exponents, τ_{sat} is saturation stress. The q^{ij} is symmetric hardening matrix which expresses interaction between different slip systems. Geometric configuration of different slip systems are co-planar slip (q_1), collinear slip (q_2), slip with perpendicular Burgers vector (q_3) and for all other cases (q_4).

Volume transfer scheme was incorporated in simulation to adopt the statistical approach in calculation to account the twin volume effects in texture evolution [20]. According to this scheme, twin volume transfer from a mother grain to the given twin during at a time increment Δt, can be expressed as:

$$\Delta V^s = V_{mother} \times \frac{\dot{\gamma}^s_{twin} \Delta t}{\gamma_{twin}} \quad (2)$$

where, V_{mother} is the volume of mother grain, $\dot{\gamma}^s_{twin}$ is the rate of twinning, γ_{twin} is twin shear. Single twin variant was activated in the present simulation. The advantage of this scheme over the predominant twin re-orientation (PTR) is that it accounts for the thickening of twins as well as it considers the twin morphology [21]. Twin boundary rotation during strain also can be accounted for in volume transfer method, which helps in approaching towards more realistic situation.

4. Results and discussion

4.1 Microstructural evolution during rolling

Fig. 1 displays the microstructures of the starting material as well as the deformed samples. Starting microstructure shows typical recrystallized equiaxed grain structure with high fraction of annealing twins (Fig. 1a). The 40% cold deformed microstructure, however,

IOP Conf. Series: Materials Science and Engineering **375** (2018) 012020 doi:10.1088/1757-899X/375/1/012020

reveals the presence of deformation twins and a large strain gradient in some grain. Some region in the microstructure could not get indexed properly due to very high strain localization (Fig. 1b). Localized strain distribution is depicted using grain reference orientation deviation (GROD) plot, which represents the misorientation development inside the grain due to simultaneous generation of geometrically necessary dislocations (GNDs), that are required to accommodate the imposed plastic strain. GROD plot indicates the presence of maximum strain at near to grain boundary (Fig. 1c). Large strain gradient inside the grains reveals that plastic deformation does not solely takes place by twinning, rather dislocation slip also play a crucial role. Another important observation is that the grains, which does not deformed by twinning, show higher GROD value, whereas an opposite trend is observed for the twinned grains. The characterization of the deformed microstructure has been carried out up to 40% rolling reduction, beyond which the microstructure could not be indexed properly in EBSD. The presence of twins in microstructure has been characterized by plotting misorientation distribution across the grains. Misorientations of ~60° with matrix orientation indicates the presence of twins in microstructure (Fig. 1d).

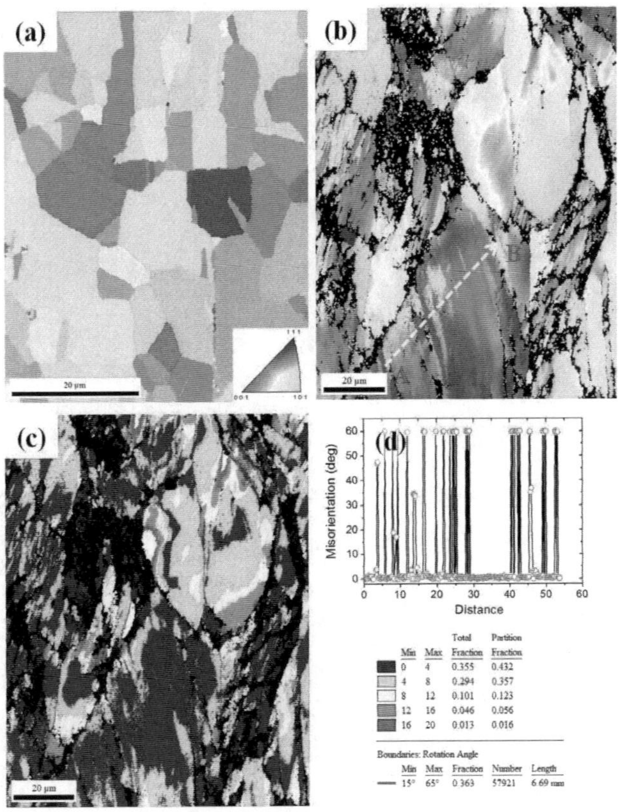

Fig. 1: (a) EBSD microstructure of starting material, (b) 40% cold rolled microstructure, (c) GROD map of 40% rolled sample and (d) misorientation distribution shows presence of deformation twins in microstructure. Misorientation distribution is plotted from position A to B shown in (b). Misorientation at 60° indicates the presence of twins.

4.2 Evolution of deformation texture

The Φ_2 = 0°, 45° and 65° sections of the orientation distribution function (ODF) plots for the deformed samples are shown in Fig. 2. The ODF sections show α-fiber type texture in all the cold rolled samples at early stages of rolling deformation. For intermediate to large rolling strain, maximum intensity is located at Goss ({110}<001>) and Goss/Brass position along α-fiber, which suggests texture evolution as Brass-type. Negligible intensities observed at Cu ({112}<111>} and S ({123}<634>) positions. Many authors have reported Bs-type texture evolution during rolling in high Mn TWIP steels [22-24]. In the ODFs, in addition to α-fibre, a weak γ-fiber ({111}||ND) is also observed in the Φ_2 = 45° section. The γ-fiber develops at medium deformation level (~50-70% rolling reduction). The γ-fiber intensity remains very weak up to large deformation levels. The development of a weak γ-fiber is reported by many authors earlier for high Mn TWIP steels [22, 25] and also for some other non-ferrous low SFE materials [14, 15, 26, 27].

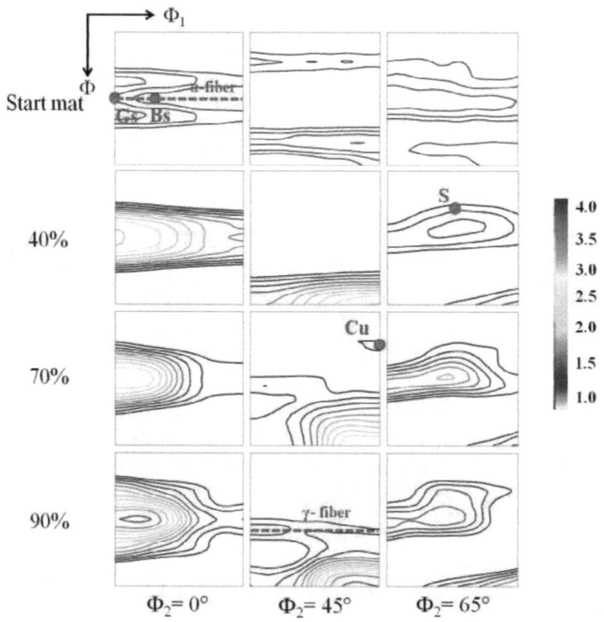

Fig. 2: Φ_2 = 0°, 45° and 65° ODF sections of starting material and 40% to 90% cold rolled TWIP steel. Some important rolling texture components and fibres are marked (in red colour) in ODF sections.

4.3 Visco-plastic self-consistent (VPSC) simulation

Texture simulation was carried out to understand the individual contributions of slip and twinning in texture evolution. A random texture with spherical grain shape was taken as starting texture. Self-consistent mode with strain rate sensitivity 0.16 was used for texture simulation. As mentioned earlier, Bronkhorst type hardening and symmetric latent hardening parameters are listed in Table 1. Fig. 3 shows the simulated texture for 40%, 70% and 90%

rolling reductions with CRSS ratio for twin to slip as 1.25. In the present case, texture simulation has been tuned with experimentally determined tensile stress-strain curve (Fig. 4a). A very good fit between tensile data and simulated stress-strain curve is observed. Twin/slip activity ratio reveals the maximum twin activity at ~30% reduction level, then decreases and above 50% reduction levels twinning gets saturated (Fig. 4b). Fig. 4c shows the imposed twin volume fraction in texture evolution. In the present case, approximately 45% twin volume has been incorporated for 90% rolling reduction. Twin deviation from the ideal position was (60° misorientation with respect to matrix) has also been calculated, and the results show a continuous deviation from the ideal position with rolling strain (Fig. 4d).

Table 1: Bronkhorst hardening, symmetric hardening parameters and velocity gradient for rolling

τ_{sat}	τ_0	h_0	a	q_1	q_2	q_3	q_4	Velocity gradient
2000	3.0	1500	5.25	1	1.2	2.0	1.5	$\begin{bmatrix} 1 & 0 & 0 \\ 0 & 0 & 0 \\ 0 & 0 & -1 \end{bmatrix}$

The simulated textures are in very good agreement with the experimentally observed texture. In the simulated texture, maximum intensity is located at Goss and Brass positions throughout the deformation level. Weak Cu and S components are also observed similar to experimentally evaluated texture. Simulated results indicate significant role of twinning for Brass-type texture evolution in TWIP steels. Earlier researchers have performed texture simulation using VPSC modelling with contributions of slip and twinning on texture evolution and have reported that twinning plays a predominant role in the formation of Brass-type texture [22, 23, 28].

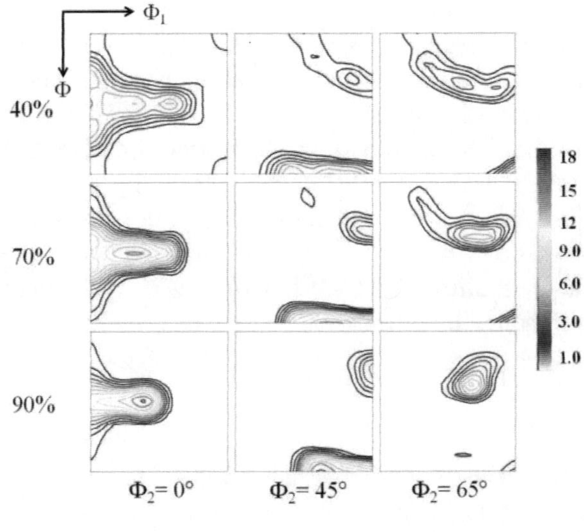

Fig. 3: Simulated Φ_2 = 0°, 45° and 65° sections of ODF for 40% to 90% cold rolled TWIP steel.

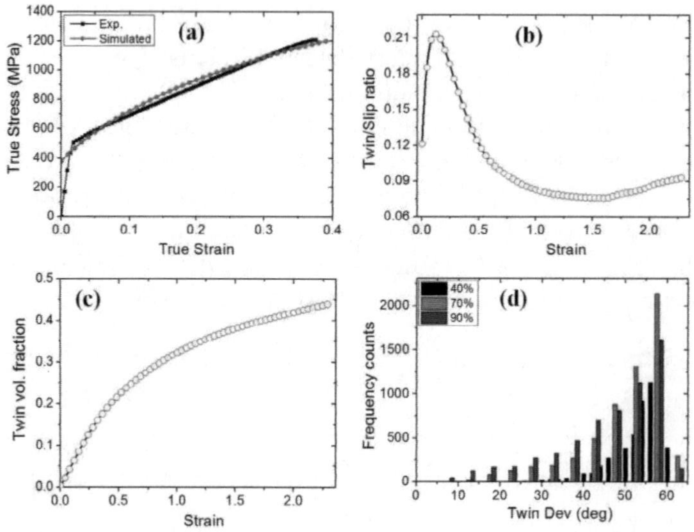

Fig. 4: (a) Experimental vs. simulated stress-strain curve, (b) activity ratio of twin/slip, (c) imposed twin volume fraction in simulation and (d) twin boundary deviation from ideal position (60°) with rolling reduction.

5. Conclusions

The evolution of deformation texture and microstructure were studied for a medium Mn TWIP steel with Al. A combination of experimental and crystal plasticity simulation was employed to understand the deformation behaviour of this alloy. Following are the important conclusions can be drawn from the analyses experimental and simulation results:

1. Texture evolution in medium Mn Al-containing TWIP steels is brass type with continuous α-fiber throughout the deformation regime. The intensity maxima are located at Goss and Brass positions.
2. Microstructure of the deformed material exhibit deformation twinning as well as large strain gradient inside grains which indicates both slip and twins take part in plastic deformation, with their relative contributions varying with strain.
3. Simulated texture shows a very good agreement with the experimentally observed texture. Continuous shortening of α-fiber observed in simulated texture which also observed in experimental texture.

Acknowledgements

This work was inspired by a grant from R & D Division, Tata Steel India Ltd. The authors are grateful to Drs. S. Chandra and S. Kundu for many useful discussions. Authors acknowledge Prof. Laszlo S. Toth for providing VPSC code. The extensive use of microscopes at the Advanced Facility of Microscopy and Microanalysis (AFMM) facility and XRD at the Institute X-ray facility, Indian Institute of Science, Bangalore is also acknowledged.

IOP Conf. Series: Materials Science and Engineering **375** (2018) 012020 doi:10.1088/1757-899X/375/1/012020

References

1. O. Bouaziz, S. Allain and C. Scott 2008 Scr. Mater. **58** 484-487.
2. L. Remy 1978 Acta Metall. **26** 443-451.
3. I. Karaman, H. Sehitoglu, A. J. Beaudoin, Y. I. Chumlyakov, H. J. Maier and C. N. Tomé 2000 Acta Mater. **48** 2031-2047.
4. K.-G. Chin, C.-Y. Kang, S. Y. Shin, S. Hong, S. Lee, H. S. Kim, K.-H. Kim and N. J. Kim 2011 Mater. Sci. Eng. A **528** 2922-2928.
5. J.-E. Jin and Y.-K. Lee 2012 Acta Mater. **60** 1680-1688.
6. L. Bracke, K. Verbeken, L. Kestens and J. Penning 2009 Acta Mater. **57** 1512-1524.
7. L. Remy and A. Pineau 1977 Mater. Sci. Eng. **28** 99-107.
8. G. Olson and M. Cohen 1972 J. Less-Common Met. **28** 107-118.
9. S. Allain, J-P. Chateau, O. Bouaziz, S. Migot and N. Guelton 2004 Mater. Sci. Eng. A **387** 158-162.
10. J. Hirsch and K. Lücke 1988 Acta Metall. **36** 2883-2904.
11. T. Leffers and R. Ray 2009 Prog. Mater. Sci. **54** 351-396.
12. R. Smallman and D. Green 1964 Acta Metall. **12** 145-154.
13. R. Garg, S. Ranganathan and S. Suwas 2010 Mater. Sci. Eng. A **527** 4582-4592.
14. R. Madhavan, R. Ray and S. Suwas 2014 Acta Mater. **78** 222-235.
15. R. Madhavan, R. Ray and S. Suwas 2014 Acta Mater. **74** 151-164.
16. R. Madhavan, R. Ray and S. Suwas 2016 Philos. Mag. **96** 3177-3199.
17. J. Fundenberger and B. Beausir 2015 *JTEX–Software for Texture Analysis.* Universite'de lorraine-Metz.
18. R. A. Lebensohn and C. Tomé 1993 Acta Metall. Mater. **41** 2611-2624.
19. S. R. Kalidindi, C.A. Bronkhorst and L. Anand 1992 J. Mech. Phys. Solids **40** 537-569.
20. C. Gu, L.S. Toth and M. Hoffman 2014 Acta mater. **62** 212-224.
21. I. J. Beyerlein, L. S. Tóth, C. N. Tomé and S. Suwas. 2007 Philos. Mag. **87** 885-906.
22. A. A. Saleh, C. Haase, E. V. Pereloma, D. A. Molodov and A. A. Gazder 2014 Acta Mater. **70** 259-271.
23. N. P. Gurao, P. Kumar, B. Bhattacharya, A. Haldar and S. Suwas 2012 Metall. Mater. Trans. A **43** 5193-5201.
24. C. Haase, S. Ghosh Chowdhury, L. A. Barrales-Mora, D. A. Molodov and G. Gottstein 2013 Metall. Mater. Trans. A **44** 911-922.
25. S. Vercammen, B. Blanpain, B. C. De Cooman and P. Wollants 2004 Acta Mater. **52** 2005-2012.
26. J. Hirsch and K. Lücke 1988 Acta Metall. **36** 2863-2882.
27. R. Madhavan and S. Suwas 2014 Philos. Mag. Lett. **94** 548-555.
28. R. Kalsar and S. Suwas 2017 Mater. Sci. Eng. A **700** 209-219.

Crystallographic analysis on the primary and secondary extension twin variants in Mg alloys

Q L Huang[1], **G Chen**[2] **and R L Xin**[2,3]

[1] School of Computer and Information Science, Southwest University, Chongqing, 400715, China
[2] College of Materials Science and Engineering, Chongqing University, Chongqing, 400044, China
[3] Corresponding author.

Email: rlxin@cqu.edu.cn (R. Xin)

Abstract. {10-12}-{10-12} double twins (DTs) are observed in AZ31 Mg alloys after one path uniaxial tension. Secondary twins (STs) are produced mainly from the intersecting points of two primary twins (PTs), forming the PT^{ing}-PT^{ed}-ST grouping. The relative orientations among these variants are analyzed. Statistical analysis reveals that the two primary variants have a specific orientation relationship with the geometric compatibility factor (m') of 0.562. The detwinning of PT^{ed} and ST also have such type of orientation. Moreover, the PT^{ing} and the selected ST have almost opposite shear directions with the angle of 151.1°. Tensor analysis conforms that these geometric orientation relationships are beneficial for minimizing the local strain incompatibility.

1. Introduction

Twinning is an important strain accommodation mechanism for hexagonal close packed (hcp) metals such as Mg, Ti and Zr [1-12]. For Mg alloys, {10-12}<-1011> and {10-11}-<10-1-2> are two most popular twinning modes, which hereafter are referred to as extension twin (ET) and contraction twin (CT), respectively. Under appropriate deformation conditions, secondary twinning can further be activated in primary twins, forming double twins (DTs). For example, {10-11}-{10-12} is the CT-ET type of DTs, which is very popular in AZ31 Mg alloy sheets after rolling at room temperature [1,3-5]. {10-12}-{10-12} is the ET-ET type of DTs, which likely forms in Mg alloys after strain-path change compression. The variant selection of the ET-ET type of DTs have been examined in a previous study [7]. However, it should be addressed that the stress state and hence Schmid factors (SFs) for twinning and dislocation slip are changed when altering the strain path. Moreover, detwinning may become one important deformation mode [13]. These factors largely complicate the analysis of the variant selection rule.

A few studies reported that the ET-ET type of DTs can also be formed in Mg alloys after one path strain deformation [14-19]. In some cases, the applied stress is complicated, which hinders understanding the mechanism of variant selection. Note that knowledge on the selection rule of ET-ET variants in one path deformation is still limited compared to that of CT-ET variants. In recent work [19], we found the formation of {10-12}-{10-12} DTs in AZ31 Mg alloys after simple uniaxial tension at room temperature. The crystallographic aspects of the secondary variants were analyzed and linked to the crystallographic features of the intersecting primary twins. Here, we aim to extend

IOP Conf. Series: Materials Science and Engineering **375** (2018) 012021 doi:10.1088/1757-899X/375/1/012021

understanding the selection rule for these variants. Being different from the previous work, here the relative geometric relationships between the intersecting primary variants and that between one primary variant and the secondary twin are analyzed. Then the selection of secondary variants is explained from the point view of local strain accommodation. Moreover, the displacement gradient tensors associated with the primary and secondary variants are calculated and analyzed. This provides some insights into understanding the activation of these DTs.

2. Material and method

A hot rolled AZ31 Mg alloy plate (~70 mm in thickness) is used as the starting material. Fig. 1a is the typical EBSD map of this plate. The average grain size is determined to be ~32 μm by a linear intercept method. Some grains are larger than 100 μm. A cylindrical tensile sample is machined from the AZ31 plate with the axis direction parallel to the normal direction (ND) of the plate. Then the sample is subjected to uniaxial tension at room temperature and at a strain rate of 10^{-3} s^{-1}. The tensile test is interrupted at ~7%. The microstructure of the deformed sample is shown in Fig. 1b. It is clear that many twins are formed in the deformed sample. At the intersecting points of some primary twins (PTs), secondary twins (STs) are produced in the thicker PTs. All the primary and secondary twins are confirmed to be of {10-12}<-1011> type.

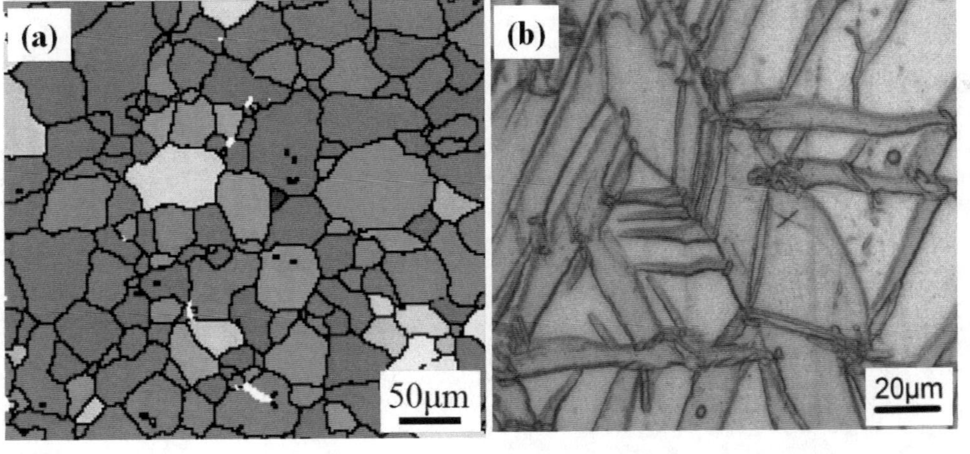

Figure 1. (a) EBSD map showing the original microstructure, (b) Metallurgical graph showing the deformed microstructure.

Due to the hexagonal crystal symmetry, each PT has six crystallographically equivalent variants. There are three types of relative orientations among the six variants [20]. As shown in Fig. 2, these relative orientations are referred to as ortho-position (OP), meta-position (MP) and para-position (PP), respectively [21]. For OP, the angles between the two habit plane normals (ψ) and between the two shear directions (κ) are 40.0° and 42.8°, respectively. Thus, the geometric compatibility factor (defined as $m' = \cos\psi \cos\kappa$) is ~0.562. In previous studies, m' has been used to evaluate the local strain accommodation between slip and twins [22-26]. The m' value ranges from -1 to 1. The larger the m' means the better the local strain compatibility. The ψ, κ and m' for the other types of relative orientations are also calculated, as shown in Fig. 2b. It reveals that the m' for MP and PP is very small, being 0.060 and -0.004, respectively. Previous studies indicated that the variants with OP orientation likely form in the same grain to reduce the local plastic anisotropy. As revealed in Fig. 1b and some previous work [14-16], STs likely initiate from the intersecting points of two PTs. Hereafter, the one containing ST is referred to as PT^{ed} (meaning this PT is impinged by another one); another is referred to as PT^{ing} (meaning this PT is impinging another one). These two PTs together with the ST form a PT^{ing}-PT^{ed}-ST grouping. In the following analysis, the ET variants 1-6 correspond to the indices of (10-12)-[-1011], (01-12)-[0-111], (-1102)-[1-101], (-1012)-[10-11], (0-112)-[01-11], (1-102)[-1101], respectively.

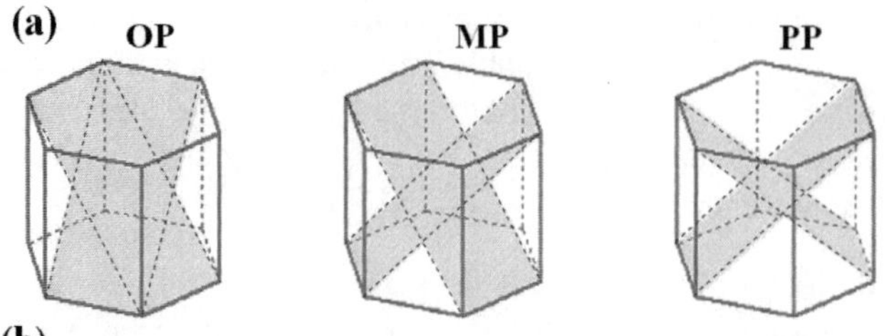

(b)

Geometric Orientations	Ortho-Position (OP)	Meta-Position (MP),	Para-Position (PP)
ψ	40.0°	72.6°	86.3°
κ	42.8°	78.4°	93.7°
m'	0.562	0.060	-0.004

Figure 2. (a) Illustrating the relative orientations of any two variant habit planes, (b) Geometric orientation data between the variants illustrated in (a).

3. Analysis and discussion

The EBSD map in Fig. 3a shows a typical {10-12}-{10-12} DT, where a is the matrix, b, c and d are PTs, and e is a ST formed in b. The ST e is connected with the PT c (or e) at twin b boundary. The unit cells (hexagons) shown in Fig. 3b are constructed based on the Euler angles of a-e. Fig. 3c illustrates the relationships of the different generations of variants. Note that twins c and d have similar orientations and thus correspond to the same variant.

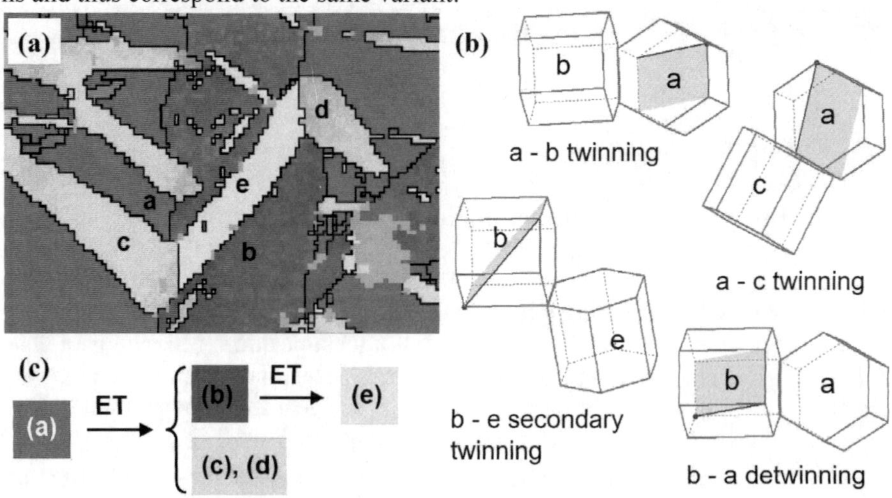

Figure 3. (a) EBSD map showing one typical PT^{ing}-PT^{ed}-ST grouping formed after tension deformation, (b) illustrating the relative orientations between some variants involved the grouping, (c) listing the different generations of twin variants.

The primary and secondary variants are determined based on misorientation analysis. Fig. 4a shows the (0002) stereographic projections for a-d and the six potential PTs. Those for b, e and the six potential STs are shown in Fig. 4b. Here, the potential variant nearest to the observed twin is considered as the active one. In this way, the active twin systems are determined, and their habit planes and shear directions are highlighted in the unit cells as shown in Fig. 3b. It indicates that twins b and c have OP orientation, meaning that their co-activations can minimize the local strain anisotropy. The detwinning of b can be considered as the "twin" a formed in the "matrix" b. Crystallographic calculations indicate that ψ and κ between detwin b and twin e is 80° and 151.1°, respectively. Therefore, the detwin b also has OP orientation with e. Moreover, as indicated in Fig. 3b, twins c and e have close shear directions. Their angle is calculated to be ~151°. It is worth to mention that the orientation relationship between e and c (or d) is of 49.7°<-909-4>. According to the previous work, this geometric relationship favors the activation of STs in terms of basal slip glide [19].

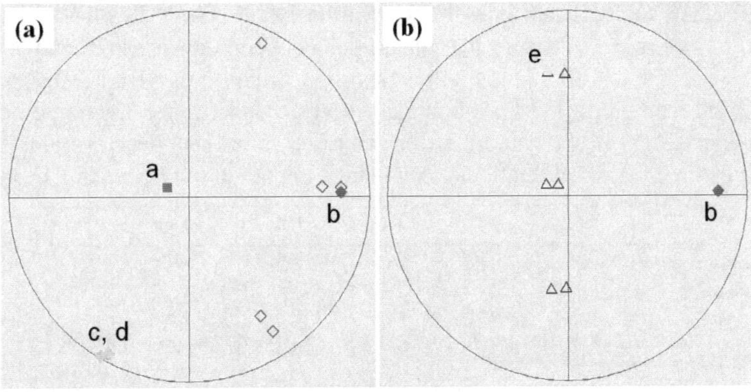

Figure 4. (a) Orientations of twin variants present in Figure 3a: (a) the primary twins, (b) the secondary twins.

Fig. 5a shows the EBSD map of another {10-12}-{10-12} DT. Similarly, b, c and d are PTs, and e is a ST formed in b. The relative orientations of these twin systems are schematic in Fig. 5b. It is seen that twin b has OP orientation with c (or d). Detwin b and twin e also have such kind of orientation. The relationships between the first and second generations of variants are summarized in Fig. 5c. Similar crystallographic analyses are performed on more than 20 PT^{ing}-PT^{ed}-ST groupings. It consistently confirms that the PT^{ing} and PT^{ed} have OP orientation with m'=0.562. The detwinning of PT^{ed} and ST also have such relative orientation. These analyses imply that the relative orientation between PT^{ing} and PT^{ed} and that between PT^{ed} detwin and ST determine the selection of secondary variants. Such kind of grouping is beneficial for minimizing the local strain incompatibility.

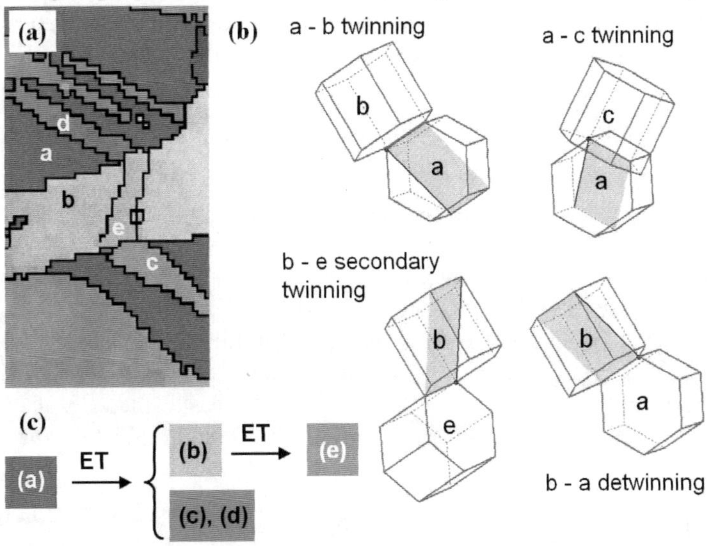

Figure 5. (a) EBSD map showing another PT^{ing}-PT^{e}d-ST grouping formed after tension deformation, (b) illustrating the relative orientations between some variants involved the grouping, (c) listing the different generations of twin variants.

As illustrated in Table 1, six potential STs can be formed in PT_1 (the PT^{ed}). There are 36 relative orientations between the potential ST and PT_1 detwin or the other PTs (i.e. PT_2-PT_6). The ψ, κ and m' for the 36 combinations are calculated and listed in Table 1. In our experiment, more than 90% of the observed PT^{ing}-PT^{ed}-ST groupings have ψ, κ and m' of 80.0°, 151.1° and -0.15, respectively. This corresponds to the combinations of PT_2-ST_{12} or PT_6-ST_{16} in Table 1. According to the indices of twin variants, PT_2 (or PT_6) has OP orientation with PT_1. Similarly, ST_{12} (or ST_{16}) has OP relationship with PT_1 detwin. In addition, Table 1 indicates that the angles between the shear directions of ST_{12} and PT_2 (or of ST_{16} and PT_6) is close to 180°. This implies that the formation of ST and PT^{ing} pairs is helpful to minimize the strain incompatibility at PT^{ed} boundary. On the other hand, the angle between the habit plane normals of ST and PT^{ing} is less important for determining the secondary variants.

Table 1. The theoretical orientation relationships (ψ, κ and m') between PTi and PT1j.

	PT1	PT2	PT3	PT4	PT5	PT6
ST11	0.0 180.0 -1.00	40.0 137.2 -0.56	72.7 101.6 -0.06	86.3 86.3 0.00	72.7 101.6 -0.06	40.0 137.2 -0.56
ST12	40.0 137.2 -0.56	**80.0** **151.1** **-0.15**	108.0 111.1 0.11	101.5 72.8 -0.06	66.9 64.0 0.17	28.9 94.4 -0.07
ST13	72.7 101.6 -0.06	108.0 111.1 0.11	145.3 83.1 -0.10	136.7 40.6 -0.55	96.9 23.3 -0.11	66.9 64.0 0.17
ST14	86.3 86.3 0.00	101.5 72.8 -0.06	136.7 40.6 -0.55	172.6 7.4 -0.98	136.7 40.6 -0.55	101.5 72.8 -0.06
ST15	72.7 101.6 -0.06	66.9 64.0 0.17	96.9 23.3 -0.11	136.7 40.6 -0.55	145.3 83.1 -0.10	108.0 111.1 0.11
ST16	40.0 137.2 -0.56	28.9 94.4 -0.07	66.9 64.0 0.17	101.5 72.8 -0.06	108.0 111.1 0.11	**80.0** **151.1** **-0.15**

The strain accommodation between twinning and dislocation glide has been assessed by tensor analysis. As reported in literature [4], the displacement gradient tensor for ET in Mg alloy is $\begin{bmatrix} 0 & 0 & 0.129 \\ 0 & 0 & 0 \\ 0 & 0 & 0 \end{bmatrix}$. This is expressed in the coordinate system (CS) formed by the shear direction (SD), shear plane normal (SPN) and habit plane normal (HPN). Knowing the twin variant and the orientations of twins and the matrix, this tensor can be transformed to any twin or matrix CS. Then the required strain accommodation can be analyzed. For the PT^{ing}-PT^{ed}-ST grouping shown in Fig. 3a, the tensors associated with the six potential PT_i are transformed to the CS of PT^{ed} (i.e. twin b). Similarly, those associated with the six potential ST_j are transformed in this CS as well. The calculation results are listed in Table 2. The calculated SFs for various variants are also given. In Table 2, c and d correspond to v1 and e corresponds to v4. It is clear that twin e has a small SF of -0.059, implying its formation is not favored by the applied stress. According to literature [4], the components of e_{13} and e_{23} in the tensor of $\begin{bmatrix} e_{11} & e_{12} & e_{13} \\ e_{21} & e_{22} & e_{23} \\ e_{31} & e_{32} & e_{33} \end{bmatrix}$ can be accommodated primarily by basal slip in the CS considered (here it is PT^{ed}). Instead, the components of e_{21} and e_{31} require the strain accommodation from prismatic slip. The tensors in Table 2 reveal that the local strains generated by c (or d) and e requires the strain accommodation mainly from basal slip. Considering the critical resolved shear stress (CRSS) for basal slip is relatively small, the selection of these variants are preferred. Moreover, the absolute values of $e_{13} + e_{23}$ are close for c (or d) and e, and their signs are of opposite. This implies that the co-activation of PT^{ing} and ST can partly cancel the accommodation strain required from basal slip. This is in agreement with the m' analysis reported in previous work [19].

Table 2. The displacement gradient tensors associated with PT^{ing} and ST expressed in the PT^{ed} crystal frame for the grouping shown in Figure 3b.

	PT-SF	PT-Tensor		ST-SF	ST-Tensor
V1 (c, d)	**0.468**	$\begin{pmatrix} -0.021 & 0.107 & \mathbf{-0.043} \\ -0.007 & 0.035 & \mathbf{-0.014} \\ -0.007 & 0.035 & -0.014 \end{pmatrix}$	V1	-0.068	$\begin{pmatrix} -0.064 & 0 & -0.069 \\ 0 & 0 & 0 \\ 0.06 & 0 & 0.064 \end{pmatrix}$
V2	0.455	$\begin{pmatrix} 0.015 & 0.03 & -0.036 \\ 0.025 & 0.049 & -0.059 \\ 0.028 & 0.053 & -0.064 \end{pmatrix}$	V2	-0.475	$\begin{pmatrix} -0.016 & -0.028 & -0.034 \\ -0.028 & -0.048 & -0.059 \\ 0.03 & 0.052 & 0.064 \end{pmatrix}$
V3	0.458	$\begin{pmatrix} -0.021 & -0.008 & 0.01 \\ 0.101 & 0.039 & -0.049 \\ 0.037 & 0.014 & -0.018 \end{pmatrix}$	V3	-0.164	$\begin{pmatrix} -0.016 & 0.028 & 0.034 \\ 0.028 & -0.048 & -0.059 \\ -0.03 & 0.052 & 0.064 \end{pmatrix}$
V4	0.465	$\begin{pmatrix} -0.027 & -0.011 & -0.008 \\ 0.103 & 0.041 & 0.031 \\ -0.045 & -0.018 & -0.014 \end{pmatrix}$	**V4 (e)**	**-0.059**	$\begin{pmatrix} -0.064 & 0 & \mathbf{0.069} \\ 0 & 0 & \mathbf{0} \\ -0.06 & 0 & 0.064 \end{pmatrix}$
V5	0.427	$\begin{pmatrix} 0.015 & 0.025 & 0.024 \\ 0.029 & 0.048 & 0.046 \\ -0.04 & -0.066 & -0.063 \end{pmatrix}$	V5	-0.454	$\begin{pmatrix} -0.016 & -0.028 & 0.034 \\ -0.028 & -0.048 & 0.059 \\ -0.03 & -0.052 & 0.064 \end{pmatrix}$
V6	0.432	$\begin{pmatrix} -0.015 & 0.105 & 0.035 \\ -0.005 & 0.032 & 0.011 \\ 0.008 & -0.052 & -0.017 \end{pmatrix}$	V6	-0.151	$\begin{pmatrix} -0.016 & 0.028 & -0.034 \\ 0.028 & -0.048 & 0.059 \\ 0.03 & -0.052 & 0.064 \end{pmatrix}$

4. Conclusions

Statistical analysis is performed on the PT^{ing}-PT^{ed}-ST groupings formed in AZ31 Mg alloys after tensile deformation. The following major conclusions can be drawn:

(1) The two primary variants (PT^{ing} and PT^{ed}) in the groupings likely form OP orientation with ψ and κ being 40.0° and 42.8°, respectively. This corresponds to a m' of 0.562, the largest one among the three possible values.

(2) The detwining of PT^{ed} and ST also have OP orientation, indicating the groupings of these variants are beneficial for minimizing the local strain anisotropy.

(3) PT^{ing} and the selected ST almost have opposite shear directions (with the angle of 151.1°), implying that the local strain generated by these two variants can be partly cancelled. The extra accommodation strain might be provided by basal slip.

Acknowledgements

This project was supported financially by the National Natural Science Foundations of China (Project No. 51571045 and 51421001).

References

[1] Barnett M, Keshavarz Z, Beer A and Ma X 2008 *Acta Mater.* **56** 5
[2] Koike J, Sato Y and Ando D 2008 *Mater. Trans.* **49** 2792
[3] Martin É, Capolungo L, Jiang L and Jonas J 2010 *Acta Mater.* **58** 3970
[4] Jonas J, Mu S, Al-Samman T, Gottstein G, Jiang L and Martin E 2011 *Acta Mater.* **59** 2046
[5] Luo J, Godfrey A, Liu W and Liu Q 2012 *Acta Mater.* **60** 1986
[6] Mu S, Jonas J and Gottstein G 2012 *Acta Mater.* **60** 2043

[7] Shi Z, Zhang Y, Wagner F, Juan P, Berbenni S, Capolungo L, Lecomte J and Richeton T 2015 *Acta Mater.* **83** 17
[8] Wang S, Schuman C, Bao L, Lecomte J, Zhang Y, Raulot J, Philippe M, Zhao X and Esling C 2012 *Acta Mater.* **60** 3912
[9] Qin H and Jonas J 2014 *Acta Mater.* **75** 198
[10] Xu S, Toth L, Schuman C, Lecomte J and Barnett M 2017 *Acta Mater.* **124** 59
[11] Bingert J, Mason T, Kaschner G, Maudlin P and Gray G 2002 *Metall. Mater. Trans.* A **33** 955
[12] McCabe R, Proust G, Cerreta E and Misra A 2009 *Int. J. Plast.* **25** 454
[13] Xu S, Liu T, Chen H, Miao Z, Zhang Z and Zeng W 2013 *Mater. Sci. Eng.* A **565** 96
[14] Roberts E and Partridge P 1966 *Acta Metall.* **14** 513
[15] Jain J, Zou J, Sinclair C and Poole W 2011 *J. Microsc.* **242** 26
[16] Bian M and Shin K 2013 *Met. Mater. Int.* **19** 999
[17] Jager A, Ostapovets A, Molnár P and Lejček P 2011 *Philos. Mag. Lett.* **91** 537
[18] Suh B, Shim M, Kim D and Kim N 2013 *Scripta Mater.* **69** 465
[19] Xin R, Guo C, Jonas J, Chen G and Liu Q 2017 *Mater. Sci. Eng.* A **700** 226
[20] Nave M and Barnett M 2004 *Scripta Mater.* **51** 881
[21] Yu Q, Wang J, Jiang Y, McCabe R, Li N and Tomé C 2014 *Acta Mater.* **77** 28
[22] Wang L, Eisenlohr P, Yang Y, Bieler T and Crimp M 2010 *Scripta Mater.* **63** 827
[23] Wang L, Yang Y, Eisenlohr P, Bieler T, Crimp M and Mason D 2010 *Metall. Mater. Trans.* A **41** 421
[24] Barnett M, Nave M and Ghaderi A 2012 *Acta Mater.* **60** 1433
[25] Xin R, Guo C, Xu Z, Liu G, Huang X and Liu Q 2014 *Scripta Mater.* **74** 96
[26] Guo C, Xin R, Ding C, Song B and Liu Q 2014 *Mater. Sci. Eng.* A **609** 92

IOP Conf. Series: Materials Science and Engineering **375** (2018) 012022 doi:10.1088/1757-899X/375/1/012022

Crystallographically resolved damage initiation in advanced high strength steel

B. Shakerifard[1], J. Galan Lopez[2], F. Hisker[3] and L. A. I. Kestens[1,4]

[1] Delft University of Technology, Delft, The Netherlands.
[2] M2i, Materials Innovation Institute, Delft, The Netherlands.
[3] ThyssenKrupp Steel Europe AG, Duisburg, Germany.
[4] Ghent University, Ghent, Belgium.

b.shakerifard@tudelft.nl

Abstract. Recently, the third generation of advanced high strength steels (AHSSs) show promising properties for automotive applications. The improvement of macroscopic mechanical performance is not feasible without a deep understanding of the micromechanical behavior and failure micro-mechanisms involved during its response under various loading conditions. In this study, a uniaxial tensile test is conducted on a low silicon bainitic steel with second phase constituents (martensite and carbides). A comprehensive image processing on SEM micrographs is performed in order to quantify the damage evolution as a function of plastic deformation. A new methodology is examined to address the correlation between crystallographic orientation and damage initiation. In this multiphase steel, it appears that orientation dependence of damage initiation is blurred by the presence of different phases and hence there is not an obvious preferential orientation from where damage has initiated.

1. Introduction

In multiphase steels, it has been well studied that microstructural heterogeneities such as different phases with various mechanical contrast, play a significant role in the partitioning of stress (and/or strain) during plastic deformation [1-4]. Hence, at the microstructure level, these local deformation incompatibilities may induce void formation in order to dissipate the plastic work.

The morphological effect (size, shape and distribution) of the second phase (*i.e* martensite) on ductile damage initiation in dual phase steels has been extensively studied by advanced experiments and crystal plasticity models [5-8]. In addition, the role of crystallographic orientation in stress (and/or strain) partitioning is another factor that needs to be considered. Jia *et al* have shown in a duplex stainless steel that at high macroscopic stresses when all grains in both constituent phases deform plastically, the role of the crystallographic orientation anisotropy on stress-strain partitioning is more significant than the phase contrast stresses [9]. The role of crystallographic orientation on damage initiation at relatively high levels of plastic deformations was also observed on metal-matrix [10] and rigid fiber composites [11] or alternatively in an austenitic stainless steel under the cyclic loading condition [12]. Eventually, the competing effect of phase contrast, with various topologies, and crystallographic orientation on damage initiation is a frequent topic of investigation [6, 12, 13].

In the current work, first a complete quantitative analysis on distribution of voids and its correlation to the strain is performed. This allows deriving a criterion for the initiation of the damage within the microstructure. In order to study the role of crystallographic orientation on damage initiation, an electron backscatter diffraction (EBSD) based analysis is used to capture the orientations around the voids. Afterwards, textures are compared in order to understand the correlation between damage and crystallographic orientation.

2. Methodology

2.1. Material

The bainitic seel used in the current study is a low silicon bainitic steel. The chemical composition is shown in table 1. The 1 mm thick sheet of cold rolled ferritic-pearlitic steel is annealed above the Ac3 temperature and quenched to the bainite holding temperature for a particular holding time, which results in a microstructure that is composed of a predominant bainite phase and a small fraction of martensite (~3%). Following, bainite transformation, the material is quenched to room temperature.

Table 1. Chemical composition of the studied bainitic steel (wt.%).

Elements	C%	Mn%	Si%	Cr%	Al%	P%	S%	N%
Bainitic steel	0.215	1.96	0.10	0.61	0.020	0.007	0.0027	0.005

2.2. Mechanical test

Two uniaxial tensile tests were conducted at room temperature with strain rates of 6.10^{-3} and 8.10^{-1} s^{-1} in elastic and plastic regions of flow curve, respectively. The tensile tests were carried out according to the DIN EN 6892-1 standard using a Zwick Z100 tensile machine. The flat specimens with gauge length of 25 mm are loaded along the rolling direction (RD). Two interrupted tensile tests were performed in the uniform and non-uniform regions of the plastic domain in order to capture the evolution of the voids within the microstructure.

2.3. Microstructure characterization

Specimen surfaces are prepared by the conventional metallographic procedure of grinding, diamond polishing and eventually polishing with colloidal SiO2. A scanning electron microscope FEI® (Quanta FEG 450) equipped with EBSD detector is used to characterize the microstructure of the samples prior to and after the tensile test. EBSD measurements are carried out with the step size of 0.05 μm in the conventional configuration of 70° inclined samples. Results obtained from EBSD analysis are post processed in the EDAX-OIM Analysis™ software.

In order to quantify the area fraction and number density of voids within the microstructure after the tensile test, a high resolution FEG-SEM (FEI XL30) was used. An area of approximately 1x1 mm^2 was scanned by running an image acquisition script to automatically capture 1000 secondary electron SEM micrographs. The contiguous collection of micrographs with pixel size of 31 nm could enable to detect all the micro-voids within the microstructure. Eventually, all the images were merged and processed to quantify voids by image processing using Matlab@ software. A void size criterion was considered in which all the voids larger than 0.1μm^2 (≈33 pixels) were counted.

3. Results and discussions

3.1. Mechanical response

The mechanical response of the material is shown in figure 1. Considering the engineering flow curve of the material, small non-uniform region is observed. Therefore, the material could not accommodate considerable plastic deformation right after the uniform region of the flow curve. The localized necking behaviour of the material also confirms the small post-necking region of the flow curve.

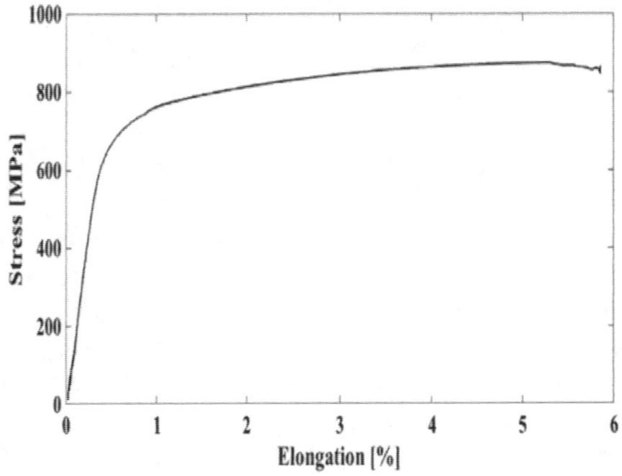

Mechanical properties	average
Yield stress (Y_s) [MPa]	743.5 ±9
Ultimate Tensile Strength (UTS) [MPa]	874±10
Uniform elongation (e_u) [%]	5±0.14
Fracture elongation (e_f) [%]	5.3±0.5
Fracture strain (%)	107.8±0.14

Figure 1. Engineering flow curve of the material under quasi-static loading condition. Mechanical properties calculated from the tensile curve.

3.2. Microstructure characterization

3.2.1. Void mapping. The void map shown in figure 2(a), reveals the accumulation of voids adjacent to the fracture surface. Voids are only initiated after uniform plastic deformation at the finishing stage of necking, prior to final failure of the material. Indeed, the interrupted tensile sample at the elongation of 5% confirmed the absence of void formation within the microstructure prior to necking. However, in dual phase (DP) ferritic and martensitic steels, it has been observed that voids started to nucleate at earlier stages of plastic deformation [14]. The retardation of the void initiation in this bainitic steel is a consequence of the less mechanical phase contrast between bainite and martensite compared to ferrite and martensite phases in DP steels [15].

The voids are quantified by two parameters: the void number density and the void area fraction. The void number density is related to void nucleation activity within the microstructure during deformation. The void area fraction reveals information with regard to subsequent steps of the micro-mechanisms of ductile damage initiation, growth and coalescence (see figure 2b). True plastic strain values are calculated in equally distanced intervals from the fracture surface based on the transversal and normal strains. The strain gradient, as function of distance from the fracture surface, is remarkably high close to the fracture surface. It is observed that both factors increase with plastic deformation. These two parameters are relatively low until almost 72% of plastic deformation. However, upon reaching this strain level, a pronounced increase in the damage incidents is observed. The localization of the deformation (localized necking) close to fracture surface leads to higher triaxiality of the stress mode and, consequently, damage is accumulated in this region.

3.2.2. Correlation between crystallographic orientation and damage. The role of crystallographic orientation on damage initiation was investigated. Scanning electron micrographs of the etched microstructure showed that damage has occurred at the interface of bainite and martensite (B/M). This reveals the presence of mechanical phase contrast between B and M, which causes stress (and/or strain) partitioning. This incompatibility at some critical moment when the imposed loads cannot be relaxed anymore by plastic glide, leads to interface tearing or void initiation. In addition, the role of crystallographic orientation on strain (and/or stress) partitioning is inevitable. In this effort, 33 voids have been studied locally by EBSD in order to capture the local surrounding orientations. The selection procedure is schematically represented in figure 3. As it is shown in figure 3b, an area with sides of approximately 3 times the void diameter was considered as the local environment surrounding a specific void.

Due to severe plastic deformation and distortion in the crystal lattices in the vicinity of the voids, a poor pattern quality was observed. Only orientations with confidence index above 0.1 were retained. Later, the ODF was derived from approximately 131,000 void neighbouring pixels, represented in the $\varphi_2 = 45°$ section of Euler space and denominated the VODF (Void ODF), cf. Fig. 4b. Because on the level of individual voids sample symmetry has to be ignored and thus the Euler space limits are set to $0 < \varphi_1 < 360°$ and $0 < \varphi_2, \Phi < 90$. Similarly, also the texture was considered of the regions with no voids, which produces the No-Void ODF (NVODF), cf. Figure 4c. It is expected that regions with no voids have more dispersed orientations compared to regions selected around the voids. Therefore, after merging all of these orientations away from the voids, the No-Void ODF (NVODF) plot reveals smoother distribution.

As it was mentioned earlier in section 3.2.1, during uniform plastic deformation no microstructural damage has appeared, thus, the deformation texture of the material at the UTS point, cf. Figure 4a, can be considered as a reference texture from which voids will be initiated. By merging the textures from the void neighbouring grains and the grains away from the voids, the entire texture of the necked region is obtained, i.e. the Neck-ODF, cf. Figure 4d. This study is a post-mortem analysis, which implies that all orientations regarding VODF and NVODF were captured after final failure. Moreover, it is not clear which interface was the origin of the damage initiation spot. The rectangular region of interest around the voids includes all possible orientations from which the voids might have initiated.

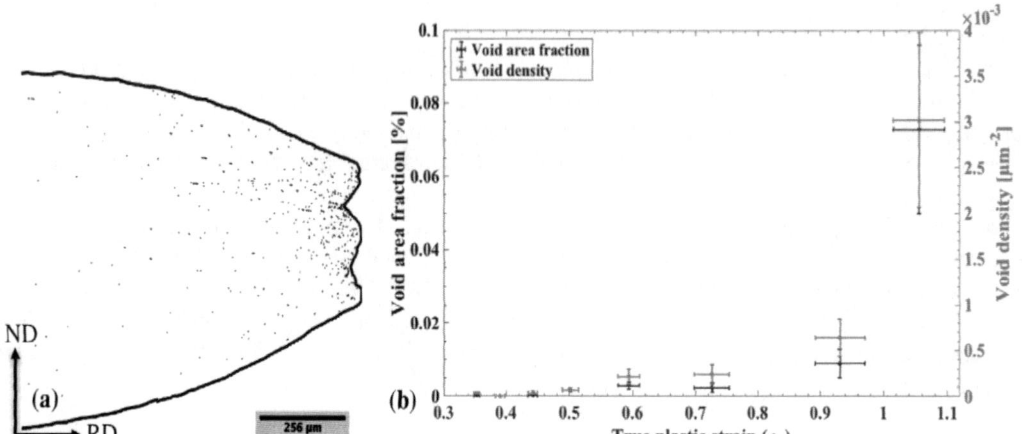

Figure 2. Void mapping analysis: (a) distribution of voids within the microstructure (in order to enhance the visual contrast between voids and the background matrix, voids are enlarged approximately 100x) and (b) void area fraction and void density as a function of true plastic strain.

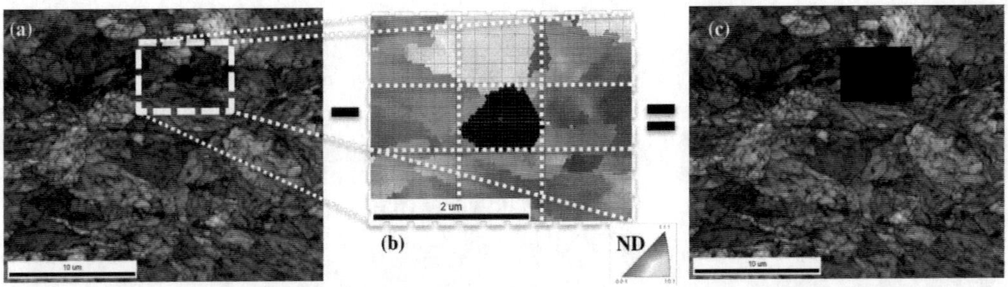

Figure 3. Schematic illustration of orientation selection on an overlapped image quality and inverse pole figure map; (a) complete area of analysis, (b) area selected around the void with faces approximately equal to 3 times of a void diameter and (c) remaining area.

Figure 4 shows the textures in the φ_2=45° section of the material at two different strain levels, uniform and post uniform, respectively. By visually comparing the VODF and NVODF with respect to the Neck-ODF it is observed that the NVODF closer resembles the texture observed in the necked region (Neck-ODF). Conversely, by visually comparing VODF and NVODF with the uniform texture, it appears that the VODF better resembles the uniform ODF. This can be explained by the fact that the local stress relaxation process at a certain moment is occurring by the void initiation rather than by plastic deformation accommodated by dislocation glide. Upon further macroscopic deformation, two

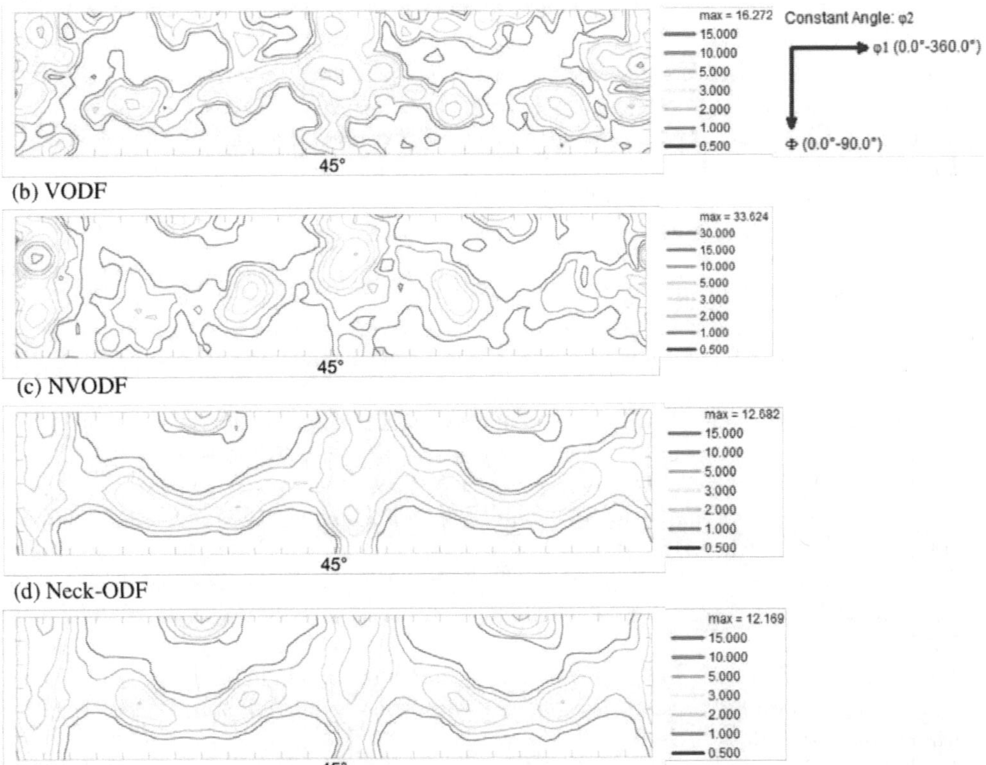

Figure 4. φ2=45° section of ODFs of the material at two different deformation levels; (a) Uniform deformation UODF (calculated from 796000 data points), (b) post uniform VODF (131000 data points), (c) post uniform NVODF (460000 data points) and (d) Necking or post uniform ODF consist of the merged ODFs in b and c.

micro mechanisms can dissipate the local stress concentration produced by the presence of a void, i.e. void growth and/or local further plasticity surrounding the initiated void. The results from the comparison between UODF, VODF and Neck-ODF demonstrate that after voids were formed, the local stress dissipation process around voids is rather governed by void growth than local plasticity as the Void-ODF clearly resembles the uniform ODF. Nevertheless, a void is nothing but a free surface within the microstructure, which may relax the local constraint of the surrounding grains and thus can potentially ease the local plastic deformation. However, the competing role of the locally hardening and free surface lead to the final contribution of plasticity and/or void growth on the stress relaxation.

3.3. Conclusion

Two characterization methods were used to study the initiation of damage within the microstructure of low silicon bainitic steel. It was found that damage initiation and extent of damage could be accurately characterized by void mapping. It was observed that voids are only initiated at terminal stage of deformation prior to final fracture. At approximately 55% of true plastic strain, voids start to initiate.
A new methodology was examined to address the role of crystallographic orientation on damage initiation. In the necked region of the sample, the microstructure was divided in two parts: (i) the crystallographic orientations of the grains surrounding 33 voids were selected and represented by the Void-ODF, (ii) whereas the regions away from the voids represented the no voids ODF (NVODF). It was observed that the VODF more resembles the texture of the uniformly deformed region, outside the necked region. The post mortem crystallographic orientation analysis around voids revealed that after void initiation, the local stress relaxation process is rather controlled by void growth than by further local plastic deformation.

References

1. Ahmad, E., et al., *Effect of microvoid formation on the tensile properties of dual-phase steel.* Journal of Materials Engineering and Performance, 2000. **9**(3): p. 306-310.
2. Avramovic-Cingara, G., et al., *Effect of martensite distribution on damage behaviour in DP600 dual phase steels.* Materials Science and Engineering a-Structural Materials Properties Microstructure and Processing, 2009. **516**(1-2): p. 7-16.
3. He, X.J., N. Terao, and A. Berghezan, *Influence of Martensite Morphology and Its Dispersion on Mechanical-Properties and Fracture Mechanisms of Fe-Mn-C Dual Phase Steels.* Metal Science, 1984. **18**(7): p. 367-373.
4. Steinbrunner, D.L., D.K. Matlock, and G. Krauss, *Void Formation during Tensile Testing of Dual Phase Steels.* Metallurgical Transactions a-Physical Metallurgy and Materials Science, 1988. **19**(3): p. 579-589.
5. Tasan, C.C., et al., *Integrated experimental-simulation analysis of stress and strain partitioning in multiphase alloys.* Acta Materialia, 2014. **81**: p. 386-400.
6. Tasan, C.C., et al., *Strain localization and damage in dual phase steels investigated by coupled in-situ deformation experiments and crystal plasticity simulations.* International Journal of Plasticity, 2014. **63**: p. 198-210.
7. Tasan, C.C., J.P.M. Hoefnagels, and M.G.D. Geers, *Identification of the continuum damage parameter: An experimental challenge in modeling damage evolution.* Acta Materialia, 2012. **60**(8): p. 3581-3589.
8. Yan, D.S., C.C. Tasan, and D. Raabe, *High resolution in situ mapping of microstrain and microstructure evolution reveals damage resistance criteria in dual phase steels.* Acta Materialia, 2015. **96**: p. 399-409.
9. Jia, N., et al., *Interactions between the phase stress and the grain-orientation-dependent stress in duplex stainless steel during deformation.* Acta Materialia, 2006. **54**(15): p. 3907-3916.

10. Needleman, A. and V. Tvergaard, *Comparison of Crystal Plasticity and Isotropic Hardening Predictions for Metal-Matrix Composites.* Journal of Applied Mechanics-Transactions of the Asme, 1993. **60**(1): p. 70-76.
11. Nugent, E.E., R.B. Calhoun, and A. Mortensen, *Experimental investigation of stress and strain fields in a ductile matrix surrounding an elastic inclusion.* Acta Materialia, 2000. **48**(7): p. 1451-1467.
12. Li, R.G., et al., *Unraveling submicron-scale mechanical heterogeneity by three-dimensional X-ray microdiffraction.* Proceedings of the National Academy of Sciences of the United States of America, 2018. **115**(3): p. 483-488.
13. de Geus, T.W.J., et al., *Microscopic plasticity and damage in two-phase steels: On the competing role of crystallography and phase contrast.* Mechanics of Materials, 2016. **101**: p. 147-159.
14. Han, S.K. and H. Margolin, *Void Formation, Void Growth and Tensile Fracture of Plain Carbon-Steel and a Dual-Phase Steel.* Materials Science and Engineering a-Structural Materials Properties Microstructure and Processing, 1989. **112**: p. 133-141.
15. de Geus, T.W.J., R.H.J. Peerlings, and M.G.D. Geers, *Microstructural topology effects on the onset of ductile failure in multi-phase materials - A systematic computational approach.* International Journal of Solids and Structures, 2015. **67-68**: p. 326-339.

Martensitic transformation, twin boundary and phase interface mobility of directionally solidified Ni-Mn-Ga alloys during compression by EBSD tracing

Yanchao Dai[1], Long Hou[1], Yves Fautrelle[2], Zongbin Li[3], Claude Esling[4], Zhongming Ren[1], Xi Li [1,2,*]

[1]State Key Laboratory of Advanced Special Steels, Shanghai University, Shanghai, 200072, P. R. China
[2]EPM-Madylam, ENSHMG BP 38402 St Martin d'HeresCedex, France
[3]Key Laboratory for Anisotropy and Texture of Materials, Northeastern University, Shenyang 110819, P. R. China
[4]Laboratoire d'Étude des Microstructures et de Mécanique des Matériaux (LEM3), CNRS UMR 7239, Université de Lorraine, 57045 Metz, France

Abstract. Directionally solidified Ni-Mn-Ga alloy with orderly arrayed austenite and non-modulated martensite (NM) was formed due to microsegregation between dendrite arm and interdendritic region, leading to preferred orientation of $<001>_A$ and $<110>_M$ coexistence at ambient temperature. The stress-induced martensitic transformation took place due to the increase of martensitic transformation temperature under step-wise uniaxial compression. The detwinning process accompanied with dislocation mobility was easy to carry out on the twin planes 45° inclined to the compression axis, compared with the twin planes parallel to the compression axis in the dominant twinned groups. The martensitic transformation and reorientation of variants were investigated by electron backscatter diffraction.

1. Introduction

Ni-Mn-Ga ferromagnetic shape memory alloys (FSMAs) with giant magnetic field-induced strains (MFIS) in the martensitic phase have been widely investigated as a kind of promising materials for actuator and sensor [1-4]. A 12% MFIS in non-modulated (NM) tetragonal crystal structure has been achieved by modifying the Ni-Mn-Ga alloy's composition with Co and Cu additives, which considerably decrease the twinning stress of the alloy [5]. Therefore, the key requirements to obtain considerable MFIS are high magnetocrystalline anisotropy and low twinning stress [3, 6]. It is relatively easy to obtain Ni-Mn-Ga single crystals with microstructure of multiple martensite variants, provided that they are subjected to a necessary training process [7]. The uniaxial compression tests are usually applied along $<001>_A$ directions referred to high temperature austenite phase. As the shape change is anisotropic, imposition of a unidirectional constraint (tension or compression) can promote the formation of some favorable variants [8]. Therefore, the application of an external load during the martensitic transformation is an effective method to modify the variant distribution [9, 10].

The characteristics of the final martensite phase are closely related to the crystallographic orientation and microstructure of the parent austenite [11]. An austenite with preferred orientation is generally easy to obtain by directional solidification (DS) [12-14]. Moreover, the samples prepared by DS with high growth speed will produce microsegregation based on our previous research [15]. Considering that the transformation temperature is sensitive to the composition in Ni-Mn-Ga alloys, the characteristic in DS can be used to prepare multi-phase mixed microstructures. In this paper, orderly arrayed two-phase Ni-Mn-Ga alloy prepared by DS makes it possible that the coupling of stress-induced martensitic transformation and variant reorientation can be shown and analyzed in detail at ambient temperature. It simplifies the measured difficulty of in-situ observation because there is no thermally-induced process. The electron backscatter diffraction (EBSD) measurement used in this work is highly visualized and suitable for observing the stress-induced processes at different stages. The martensitic transformation processes under step-wise uniaxial compression were traced and direct evidence of the detwinning process induced by stress was captured.

2. Experimental details

The alloy with nominal composition $Ni_{46.7}Mn_{33.9}Ga_{19.4}$ (at.%) was prepared in an arc-melting furnace under an argon atmosphere from high-purity Ni (99.99 wt.%), Mn (99.9 wt.%) and Ga (99.99 wt.%). 2 wt.% excess Mn was added in order to compensate for the Mn loss during melting. The ingots were remelted five times for homogeneity. The experimental apparatus consisted of a Bridgman-Stockbarger-type furnace equipped with a pulling system and a temperature controller. The furnace was heated to 1450°C and held for an hour. Then the specimen was directionally solidified in the Bridgman apparatus by drawing the crucible assembly at a constant speed of 20 μm/s into the liquid Ga-In-Sn metal (LMC) cylinder. A sample with ϕ 3 mm × 6 mm was cut from a stable growth zone in the directionally solidified rod. The compression process was performed with a MTS809 material testing system. Uniaxial compression tests were conducted along the direction of DS with a constant speed of 0.15 mm/min at room temperature.

The compositional variation of the ternary alloy was measured by energy dispersive spectroscopy (EDS) (Octane Plus, EDAX), working at 25 kV. The crystal orientations of austenite and martensite at ambient temperature were characterized in a Hitachi SU70 scanning electron microscope with EBSD acquisition camera and OIM software provided by EDAX Instruments. The identical cross section was measured by EBSD before and after each compression step. The sample after each compression step needed a repolish in order to obtain a high indexing rate. The loss as little as possible was accomplished when using 3000 grit sandpaper. After that the sample was prepared using electropolished (Lectropol-5, Struers^{TM}) with a solution of nitric acid (20 vol.%) and methanol (80 vol.%). The EBSD scanning was carried out at 25 kV, over 1/4 cross section and the detailed maps were detected with a step size of 2 μm and 0.5 μm. The raw data obtained in this study were processed using OIM software. Grain confidence index (CI) standardization and Grain dilation were selected to clear up the noise. The data with high CI value (CI > 0.3) were selected.

3. Results and discussion

Fig. 1 shows the distribution of austenite and martensite variants at the steady solid phase zone of directionally solidified Ni-Mn-Ga alloy rod. It is worth noting that the orientations of non-modulated martensite are all indexed according to the tetragonal $L1_0$ unit cell [16, 17]. Considering that the lattice parameter of the two phases in directionally solidified alloys varies due to the microsegregation, the Bain model is used to give a simplified but competent description of the variant distribution. The transformation from cubic parent phase to tetragonal martensitic phase based on the Bain relationship would result in three different martensite variants [18]. The c-axis of each variant is parallel to one principle axis of the cubic parent phase. The representative relationship is $(001)_A//(001)_M$ and $[100]_A//[1\text{-}10]_M$. Fig. 1(a) shows the IPF map of the cross section at the steady solid phase zone. The

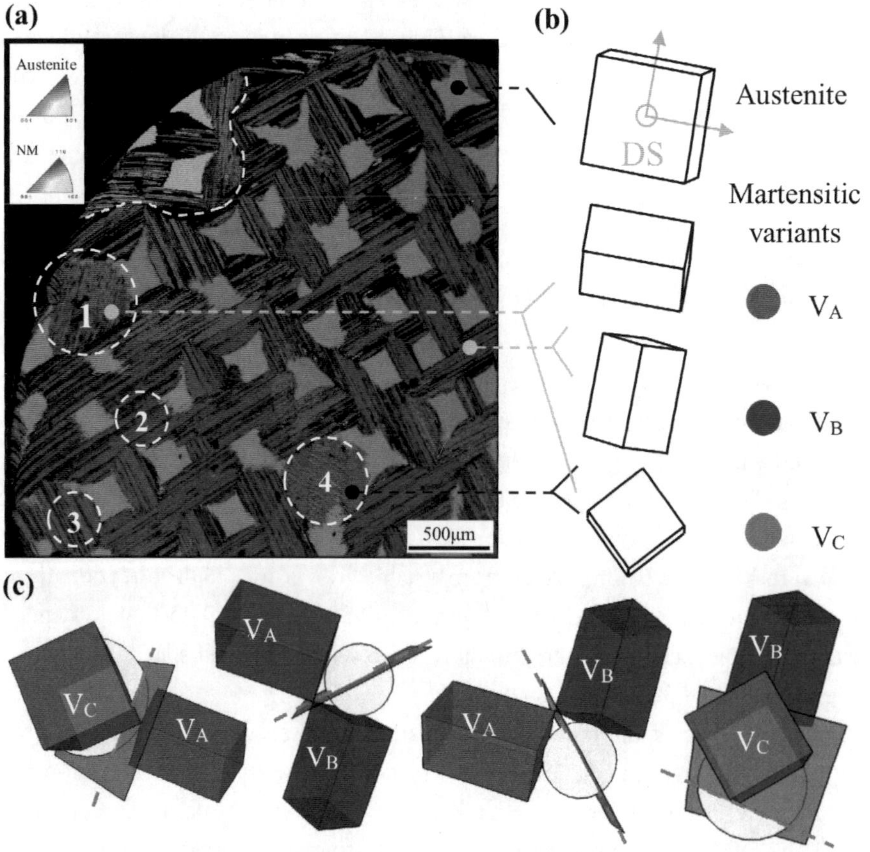

Figure 1. The distribution of austenite and martensite variants at the steady solid phase zone of directionally solidified Ni-Mn-Ga alloy rod. (a) IPF map of the cross section; (b) Orientations of unit cell for twin variants. (c) Geometric perspective views corresponding to the twin orientation relationships of the four variant groups. The unit cell orientations agree with the real ones from the EBSD data. Intersection lines (red) between the twin planes (green) and the sample surface planes (yellow) correspond to the IPBs within the white dotted circles of the EBSD map.

austenite shaped like a star or square is uniformly distributed in the martensitic matrix. The orientation of the residual austenite was verified as <001>$_A$ preferred orientation along the solidification direction for the whole grain before martensitic transformation. The martensite phase has a <110>$_M$ preferred orientation in the cross section. Orientations of the unit cell for twin variants are displayed in Fig. 1(b). It shows that all three variants are all generated in the cross section and each pair of variants creates a colony separated by an intercolony boundary (ICB) [19]. Four representative regions are selected from the colonies and marked with white circles. Within each colony, martensite variants are separated by an interplate boundary (IPB) [19]. The twin orientation relationships for the four variant groups are illustrated by geometric perspective views, which contain three oriented variants named V_A, V_B and V_C (see Fig. 1(c)). Variant groups 2 and 3 both composed of V_A and V_B take up the major parts of the martensite phase. The IPBs in the two main colonies are crisscross and orthogonal. The IPBs in variant groups 1 and 4 are inclined at about ~45° regarding the ones in variant groups 2 and 3. Variant group 1 (consisting of V_A and V_C) and variant group 4 (consisting of V_B and V_C) are not the main twinned structures in the preferential oriented DS rod. It should be pointed out that different martensite variants arrange themselves into variant groups in order to minimize the internal stress [20]. All transformations, free from external stress, tend to be self-accommodating since the shape change in any group will produce a stress field favoring the formation of some particular variants [21].

The mixed austenitic-martensitic zone is derived from the microsegregation between dendrite core and interdendritic region. The microstructure of the longitudinal section of directionally solidified Ni-Mn-Ga alloy is shown in Fig. 2. The strip-shaped areas at the stable solid phase zone represent

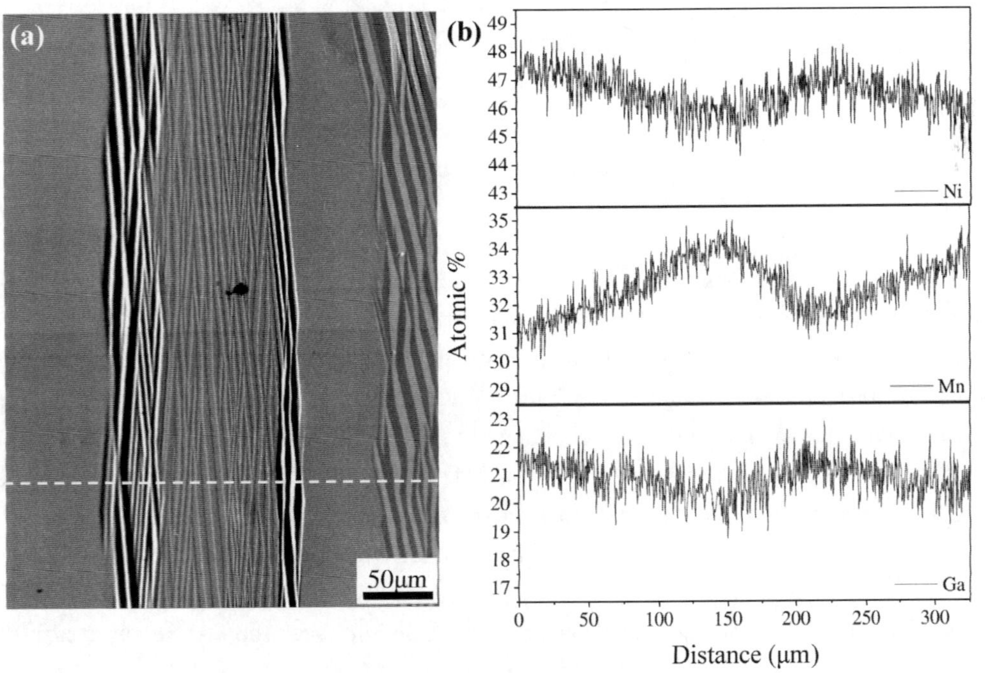

Figure 2. (a) SEM micrograph of the mixed austenitic-martensitic zone in longitudinal section. (b) Distribution of three elements the dashed line measured by EDS.

martensite and the surrounded parts signify austenite. They are also corresponding to the interdendritic region and dendrite core. The distribution of three elements along the dashed line is displayed in Fig. 2(b). It is found that the fluctuation of Mn is more obvious than that of Ni and Ga. The Ni-rich, Mn-poor and Ga-rich composition corresponds to the dendrite core region, while Ni-poor, Mn-rich and Ga-poor composition is found in the interdendritic region. The composition variation results in a change of the e/a value. Generally speaking, the martensitic transformation temperature is linearly depending on e/a values [22]. The martensite embryos initially form near the e/a peak with a high martensitic transformation temperature. As a result, they are constrained by the surrounding austenite with a lower valence electron concentration in the dendrite core and show a preferential variant distribution due to the constraints by their surrounding with lower e/a value.

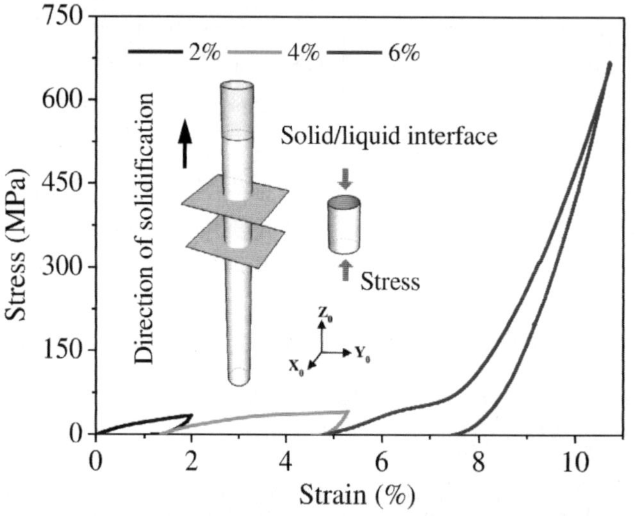

Figure 3. The stress-strain curves obtained from step-wise compression and unloading experiments along the direction of directional solidification.

Fig. 3 shows the stress-strain curves obtained from step-wise uniaxial compression loading and unloading. The hierarchic stress-strain curves manifest the step-wise compressive deformations to strains of 2%, 4% and 6%. The strains are all based on the original length of the column. A conspicuous stress plateau means the twin boundaries become mobile and the reorientation process of the variants starts. The microstructural changes in the directionally solidified two-phase Ni-Mn-Ga alloy subjected to interrupted straining tests were monitored by EBSD, as shown in Fig. 4. It should be noted that the EBSD data were taken from the identical cross section after each cycle of loading and unloading. It is found that the volume fraction of austenite begins to reduce when the strain increased from 2% to 4% and utterly disappears at a strain of 6%.

The detailed transformations of twin boundary and phase interface mobility are depicted in Fig. 4. IPF maps are displayed in the same area before compression and after 4% and 6% strain. The twin and phase boundaries moved at the same time under the uniaxial compression. The reorientation of

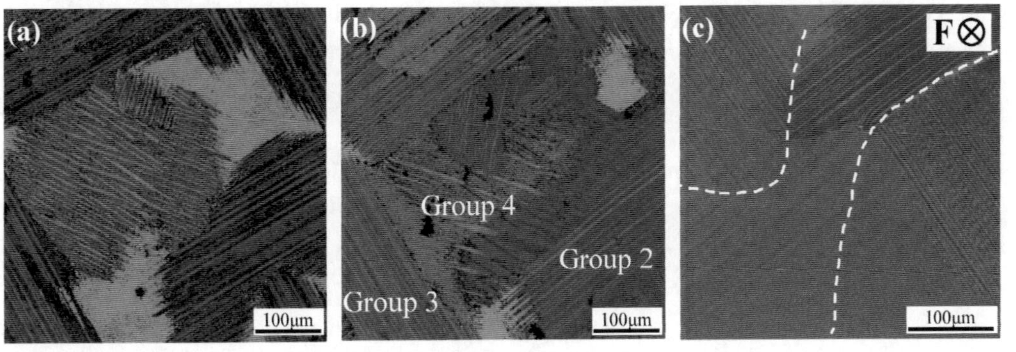

Figure 4. EBSD IPF maps of a magnified region in Ni-Mn-Ga alloy during step-wise compression: (a) before compression; (b) after 4%; (c) after 6%. F means the direction of applied stress. The white dashed lines represent the ICBs between residual group 2 and 3.

martensite variants was accompanied by detwinning. The twin boundaries in group 4 of Fig. 4(b) are curved and irregular. The movement of the twin boundaries in this area can be observed after the first two cycles (2% and 4% strain). In addition, a small dominance of V_B in group 4 (blue color in the middle area) can be noticed. The non-uniform distribution of martensite variants during mechanical training has also been justified [23, 24]. The twin plane in group 4 is at ~45° with respect to the compression axis. The corresponding Schmid factor is close to 0.5. However, V_A and V_B in surrounding of group 2 and 3 keep an uniform distribution after three cycles uniaxial compression along $[001]_A$. As the twin plane is aligned almost parallel to the compression axis, the Schmid factor for any detwinning system potentially operating within this plane is close to zero. The critical detwinning stress cannot be reached to reorient one variant to the other and drive twin boundary motion. After the third cycle in Fig. 4(c), the austenite and variant V_C disappear completely. The group 4 is consumed by the combination of V_A and V_B and the IPBs in adjacent colonies are orthogonal to each other.

4. Conclusions

The influence of step-wise uniaxial compression on the martensitic transformation and detwinning process was investigated in a directionally solidified Ni-Mn-Ga alloy by EBSD. Because of composition segregation, the martensitic transformation temperature gradually increases from the core to the edge of the dendrite, leading to the coexistence of martensite and austenite at ambient temperature. The alloy undergoes a coupling of martensitic transformation and variant reorientation under uniaxial compression. The martensite groups composed of V_C and V_A or V_B with twinning planes inclined to the compression axis at ~45° are detwinned under uniaxial compression, which agrees with the calculated Schmid factor (~0.5).

Acknowledgements

We are grateful to the demo Lab of EDAX China providing the support for EBSD measurement. The authors would like to thank for the financial support from National foundation of Science (Nos. 51690164, 51571056, 51431005), "Shuguang Program" from Shanghai Municipal Education

Commission and the Program for Professor of Special Appointment (Eastern Scholar) at Shanghai Institutions of Higher Learning and Science and Technology Commission of Shanghai Municipality (No. 11JC1413600) and Shanghai Science and Technology Committee Grant (13DZ1108200, 13521101102) and United Innovation Program of Shanghai Commercial Aircraft Engine (AR910, AR911).

[1] Ullakko K, Huang J, Kantner C, OHandley R, Kokorin V. Large magnetic - field - induced strains in Ni2MnGa single crystals. Applied Physics Letters 1996;69:1966-8.
[2] O'Handley RC. Model for strain and magnetization in magnetic shape-memory alloys. Journal of Applied Physics 1998;83:3263.
[3] Sozinov A, Likhachev A, Lanska N, Ullakko K. Giant magnetic-field-induced strain in NiMnGa seven-layered martensitic phase. Applied Physics Letters 2002;80:1746.
[4] Chernenko VA, Anton RL, Kohl M, Barandiaran JM, Ohtsuka M, Orue I, et al. Structural and magnetic characterization of martensitic Ni–Mn–Ga thin films deposited on Mo foil. Acta Materialia 2006;54:5461-7.
[5] Sozinov A, Lanska N, Soroka A, Zou W. 12% magnetic field-induced strain in Ni-Mn-Ga-based non-modulated martensite. Applied Physics Letters 2013;102:021902.
[6] James RD, Tickle R, Wuttig M. Large field-induced strains in ferromagnetic shape memory materials. Materials Science and Engineering: A 1999;273–275:320-5.
[7] Sozinov A, Likhachev AA, Lanska N, Soderberg O, Ullakko K, Lindroos VK. Stress- and magnetic-field-induced variant rearrangement in Ni-Mn-Ga single crystals with seven-layered martensitic structure. Materials Science and Engineering a-Structural Materials Properties Microstructure and Processing 2004;378:399-402.
[8] Wang YD, Brown DW, Choo H, Liaw PK, Cong DY, Benson ML, et al. Experimental evidence of stress-field-induced selection of variants in Ni-Mn-Ga ferromagnetic shape-memory alloys. Physical Review B 2007;75.
[9] Li ZB, Zhang YD, Esling C, Gan WM, Zou NF, Zhao X, et al. In-situ neutron diffraction study of martensitic variant redistribution in polycrystalline Ni-Mn-Ga alloy under cyclic thermo-mechanical treatment. Applied Physics Letters 2014;105.
[10] Li Z, Zou N, Yang B, Gan W, Hou L, Li X, et al. Effect of compressive load on the martensitic transformation from austenite to 5M martensite in a polycrystalline Ni–Mn-Ga alloy studied by in-situ neutron diffraction. Journal of Alloys and Compounds 2016;666:1-9.
[11] Liu J, Zheng H, Huang Y, Xia M, Li J. Microstructure and magnetic field induced strain of directionally solidified ferromagnetic shape memory CoNiAl alloys. Scripta Materialia 2005;53:29-33.
[12] Jiang C, Liu J, Wang J, Xu L, Xu H. Solid–liquid interface morphology and crystal growth of NiMnGa magnetic shape memory alloys. Acta Materialia 2005;53:1111-20.
[13] Liu J, Zheng HX, Li JG. Effect of solidification rate on microstructure and crystal orientation of ferromagnetic shape memory alloys Co–Ni–Al. Materials Science and Engineering: A 2006;438-440:1061-4.
[14] Wang J, Jiang C. Single-crystal growth of NiMnGa magnetic shape memory alloys. Journal of Crystal Growth 2008;310:865-9.
[15] Li X, Gagnoud A, Ren Z, Fautrelle Y, Debray F. Effect of strong magnetic field on solid solubility and microsegregation during directional solidification of Al-Cu alloy. Journal of Materials Research 2013;28:2810-8.
[16] Niemann R, Fähler S. Geometry of adaptive martensite in Ni-Mn-based Heusler alloys. Journal of Alloys and Compounds 2017;703:280-8.
[17] Pons J, Chernenko VA, Santamarta R, Cesari E. Crystal structure of martensitic phases in Ni–Mn–Ga shape

IOP Conf. Series: Materials Science and Engineering **375** (2018) 012023 doi:10.1088/1757-899X/375/1/012023

memory alloys. Acta Materialia 2000;48:3027-38.
[18] Bain EC. Nature of martensite. Trans AIME 1924;70:25–46.
[19] Szczerba MJ, Chulist R. Detwinning of a non-modulated Ni–Mn–Ga martensite: From self-accommodated microstructure to single crystal. Acta Materialia 2015;85:67-73.
[20] Christian JW. Deformation by moving interfaces. Metallurgical Transactions A 1982;13:509-38.
[21] Zhu JJ, Huang WM, Liew KM. Deformation energy in martensitic transformation. J Phys IV 2003;112:179-82.
[22] Chulist R, Sozinov A, Straka L, Lanska N, Soroka A, Lippmann T, et al. Diffraction study of bending-induced polysynthetic twins in 10M modulated Ni-Mn-Ga martensite. Journal of Applied Physics 2012;112:063517.
[23] Zhou L, Giri A, Cho K, Heinrich H, Majumdar BS, Sohn Y. Microstructural and Crystallographic Characterization of Ni2+x Mn1−x Ga Alloys (x = 0.14, 0.16, 0.19, 0.22, and 0.24) by Transmission Electron Microscopy. Metallurgical and Materials Transactions E 2014;1:239-46.
[24] Szczerba MJ, Chulist R, Kopacz S, Szczerba MS. Effect of initial plastic strain on mechanical training of non-modulated Ni–Mn–Ga martensite structure. Materials Science and Engineering: A 2014;611:313-9.

IOP Conf. Series: Materials Science and Engineering **375** (2018) 012024 doi:10.1088/1757-899X/375/1/012024

Rolling Texture Development in Aluminum-Zinc Solid Solutions

B.J. Diak, B. Merkley, O. Gopkalo

Mechanical and Materials Engineering, Queen's University, Kingston, Ontario, Canada, K7L3N6

diak@queensu.ca

Abstract. Rolling textures in face-centered cubic polycrystalline metals tend to transform from so-called pure metal to alloy textures with increasing stacking fault energy, so deformation texture correlates directly to stacking fault energy. As a 30th anniversary tribute to the classic copper-brass study by Hirsch and Lücke [1988], we recast the problem for binary aluminium alloys with 1, 10, 34 and 62wt.%Zn. Zinc (Zn) has a very high solid solubility in Al of up to 65at.% at 381°C, but at lower temperatures the Zn solid solution is unstable and can precipitate out or decompose spinodally. Current understanding is that Zn has no effect on the stacking fault energy of aluminum, so a priori, rolling of different Al-xZn alloys should result in no variation in deformation texture. Cast ingots were solutionized and promptly rolled 75 and 90% by multiple passes. X-ray texture pole figures were measured at the mid-plane of the rolled sheet, and the texture components calculated from the crystallographic orientation distribution function. Results show a decrease in {112}<11-1> and increase in {011}<21-1> with Zn content but uncorrelated changes in {123}<63-4> components. Bird's eye - like microstructural features observed in the high Zn containing alloys at 90% reduction were {011}<21-1> and resemble those observed by other researchers in severely rolled brass, Al-4.5Mg and 316L.

1. Introduction

Aluminum (Al) has a relatively high stacking fault energy compared to other face-centered cubic (f.c.c.) crystalline elements, and so its deformation is often characterized by greater potential for cross-slip and recovery at higher temperatures and organization of a more cellular work-hardened dislocation structure. It has been observed that the rolling textures in face-centered cubic polycrystalline metals tend to transform from so-called pure metal to alloy textures with decreasing stacking fault energy, so deformation texture correlates with stacking fault energy. In one of the first quantitative studies of deformation textures Hirsch and Lücke [1] used the calculated crystallographic orientation distribution function to quantify the role of rolling reduction and Zn content on the texture evolution of copper and α-brass. As a 30th anniversary tribute to this study, we recast the problem for binary Al alloys with 1, 10, 30 and 60wt.%zinc. Zinc (Zn) has a very high solid solubility in Al of up to 83wt.% (65at.%) at 381°C, but at lower temperatures the Zn solid solution is unstable and can precipitate out or decompose spinodally [2-5]. It is mutedly accepted that Zn has no effect on the stacking fault energy of aluminum [5, 6], so a priori, rolling of solutionized Al-xZn alloys should result in no variation in deformation texture due to slip compared to pure Al.

The objective of the present work is to study the rolling texture development in Al-xZn alloys aux Hirsch and Lücke [1] to indirectly test if Zn modifies the intrinsic stacking fault energy of aluminum.

2. Experimental Procedure

2.1 Alloy preparation

Five aluminum alloys with target compositions of 0, 1, 10, 30 and 60wt.%Zn were prepared from 99.99 or 99.97wt.%Al, and 99.99wt.%Zn starting materials. For each melt composition Al was place in a refractory crucible coated with red mud and gas fired to 800°C after which TiB_2 wafer was added for grain refining, followed by addition of the Zn and stirred with a graphite rod. To reduce Zn loss to vaporization, excess Zn was added according to past practice and the temperature lowered to about 70°C above the liquidus for as superheat for each composition. Dross was skimmed from the melt surface before gravity pouring into a graphite-coated steel book mold at the temperatures listed in Table 1 and air cooled with no chill. The ingots were scalped, wrapped in aluminum foil and then homogenized at 428°C for 96 hours in a salt bath and water quenched to room temperature. Chemical analysis of the homogenized castings listed in Table 2 indicates compositions close to the targets and are subsequently identified by these chemistries. The differences in Fe and Si content between the pure Al and Al-1wt.%Zn, and other alloys is due to the use of a 99.99 Al master alloy for the former two.

Table 1: Pouring temperatures used in the casting process of each alloy.

Sample ID	Pure Al	Al-1wt%Zn	Al-10wt.%Zn	Al-30wt.%Zn	Al-60wt.%Zn
Temperature (°C)	730	730	720	690	640

Table 2: Alloy composition (wt.%) of the cast alloys in addition to aluminum determined according to ASTM E1097-12 and E1470-99(2011).

Alloy	Zn	Ti	B	Mg	Si	Cu	Mn	Fe	Cr
Pure Al	<0.01	0.10	0.010	<0.01	<0.01	<0.01	<0.01	<0.01	<0.01
1wt%Zn	1.06	0.13	0.019	<0.01	<0.01	<0.01	<0.01	0.01	<0.01
10wt.%Zn	10.7	0.11	0.014	<0.01	0.02	<0.01	<0.01	0.07	0.01
30wt.%Zn	33.8	0.08	0.009	<0.01	0.04	<0.01	<0.01	0.08	<0.01
60wt.%Zn	62.2	0.05	0.007	<0.01	0.03	<0.01	<0.01	0.03	<0.01

2.2 Solutionizing and plane strain deformation

Samples with final dimensions of 57mm (l) x 12.7mm(w) x 8.2mm (t) were extracted from the casting, and individually re-solutionized at 428°C for 10 minutes in the salt bath and quenched into ice water. Within 5 minutes of quenching, reductions to 75% (2.4mm) and 90% (0.75mm) were started at 25°C using a channel-die for the pure aluminium, and a Stanat 2-high rolling mill for the other alloys. 75% and higher is beyond the texture transition issue in brass [7]. After the target reduction, half of the specimen was sectioned from the front end and put aside for characterization, with the remaining sample being further reduced. Note that the rolling deformation used approximately 5% reduction steps towards each target, while the channel die deformation was done continuously at a rate of 0.1mm/sec.

2.3 Texture and microstructure characterization

After deformation the specimens were mechanically ground using SiC papers, polished with diamond and colloidal silica to a center thickness on the RD-TD section, and electropolished using a by volume 70% ethanol – 20% distilled water – 8% perchloric acid – 2% butyl cellusolve solution at -20°C and 1.5 A/ cm^2 for 60 seconds Four incomplete (111), (200), (220) and (311) pole figures were measured for each deformation condition using 50kV 190mA Cr-Kα x-rays produced from a Rigaku DMAX rotating anode system with a 4-axis Huber cradle set-up with a Bragg-Brentano diffraction geometry. A vanadium filter was used in the measurement of the 10, 30 and 60%Zn alloys. Crystallographic orientation distribution functions were calculated from the pole figure data using the MTM-FHM analysis software to an LMAX of 22 [8]. All diffracted intensities were referenced to a random powder standard. Scanning electron microscopy imaging and orientation observations were made of select RD-ND sections using an FEI nano-SEM 450 equipped with Bruker e-FLASH detector.

IOP Conf. Series: Materials Science and Engineering **375** (2018) 012024 doi:10.1088/1757-899X/375/1/012024

3. Results

Figure 1 shows the incomplete {111} pole figures measured for all the alloys at 75 and 90% reductions. From this data it is evident that the textures intensify with rolling reduction but weaken with alloying. Integrating all data, Fig. 2 plots the ϕ_2=90, 60 and 45° sections for the same cases. Also shown for visual reference in overlay are some ideal texture components: cube, C {001}<100>; goss, G {011}<100>; brass, Br {011}<21-1>; S {123}<63-4>; and copper, Cu {11-2}<111>.

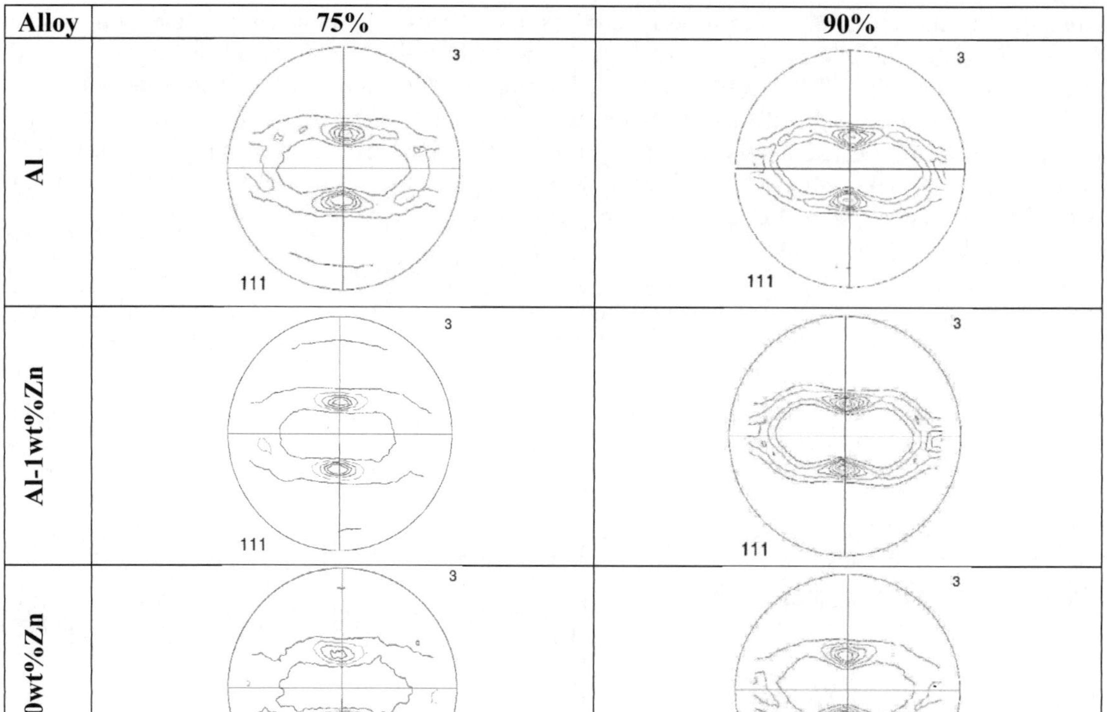

Fig. 1: As-measured incomplete {111} pole figures for the Al-xZn alloys after 75 and 90% reductions in RD-TD sections of the center-plane. Contour lines are 1, 2, 3, 4, 5, 7.5 times random.

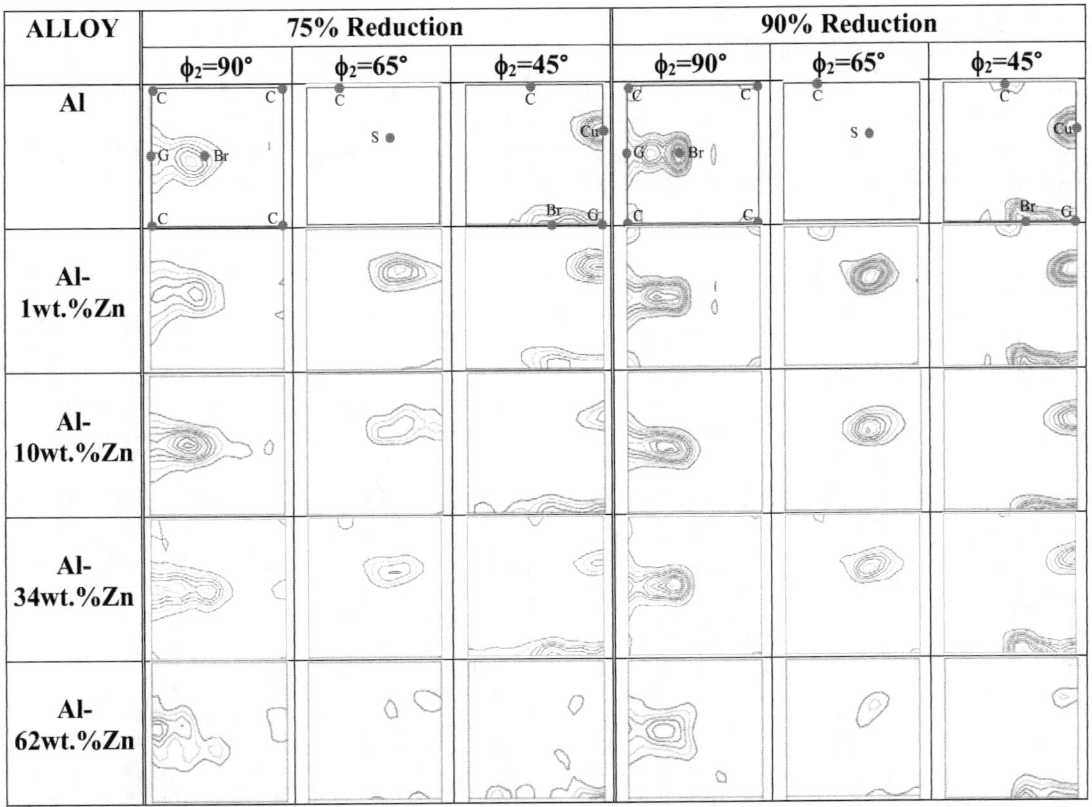

Fig. 2: CODF ϕ_2=90, 65 and 45° sections for the same conditions as Fig. 1. Overlaid for comparison are ideal texture components cube (C), brass (Br), goss (G), copper (Cu) and S. Contour lines are 2, 3, 4, 5, 6, 7, 10, 12 times random.

4. Discussion

The rolling deformation of Al is similar yet quite different from the classic fcc textures of pure metal copper and the alloy brass, so that Al has its own texture characteristics. In the present work, plane strain deformation of pure Al from a random cast texture to 70 and 90% reductions results in the typical texture components of increasing Cu, Br and S components observed in many other rolling studies. Figure 3 shows the volume fractions of the main components in pure Al increase with deformation, but does not change for C and G. This trend for all the components remains in the Zn containing alloys.

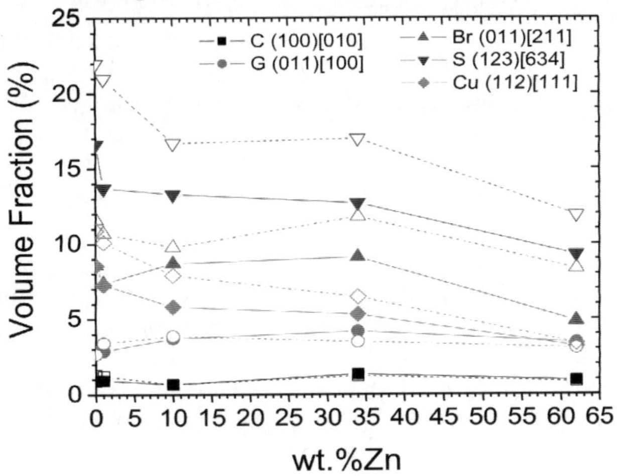

Fig. 3: Calculated volume fraction of select ideal texture components with Zn alloying and rolling reduction, where closed-symbols/solid lines are 70% and open-symbols/dashed lines are 90% reductions. The normalizing spread, Φ_o, about the ideal orientation in Euler space is 11°.

Figure 3 also shows that alloying with Zn has some notable effects on the fraction of texture components. Firstly, C and G do not measurably change with alloying. Secondly, S, the largest component, and Cu decrease with Zn content. In contrast, Br decreases for 1%Zn, but then increases with alloying peaking at 12% for the Al-34wt.%Zn alloy rolled 90%. The S texture too shows a notable increase from 70 to 90% reduction. For the Al-62wt.%Zn alloy, all components decreased from the peak at 34wt.%Zn except the C and G, which did not change. The texture intensification at 90% reduction warranted a closer examination of the microstructure.

Figure 4 shows backscattered scanning electron images of the transverse sections of the Zn containing alloys after 90% reduction and sitting at room temperature for more than a year. All microstructures show lamellar banded structures aligned with the rolling direction with some particle decoration at the original boundaries. The pure Al microstructure (not shown) is very similar to the Al-1wt.%Zn. With alloying there is a greater propensity of heterogeneous deformation characterized by shear banding with rhombic-shaped bird's eye features prevalent in the 34 and 62wt.%Zn containing alloys. The eyes were more rounded, but less clear in the 60wt%Zn alloy due to etching artefacts from the electropolishing of this reactive alloy.

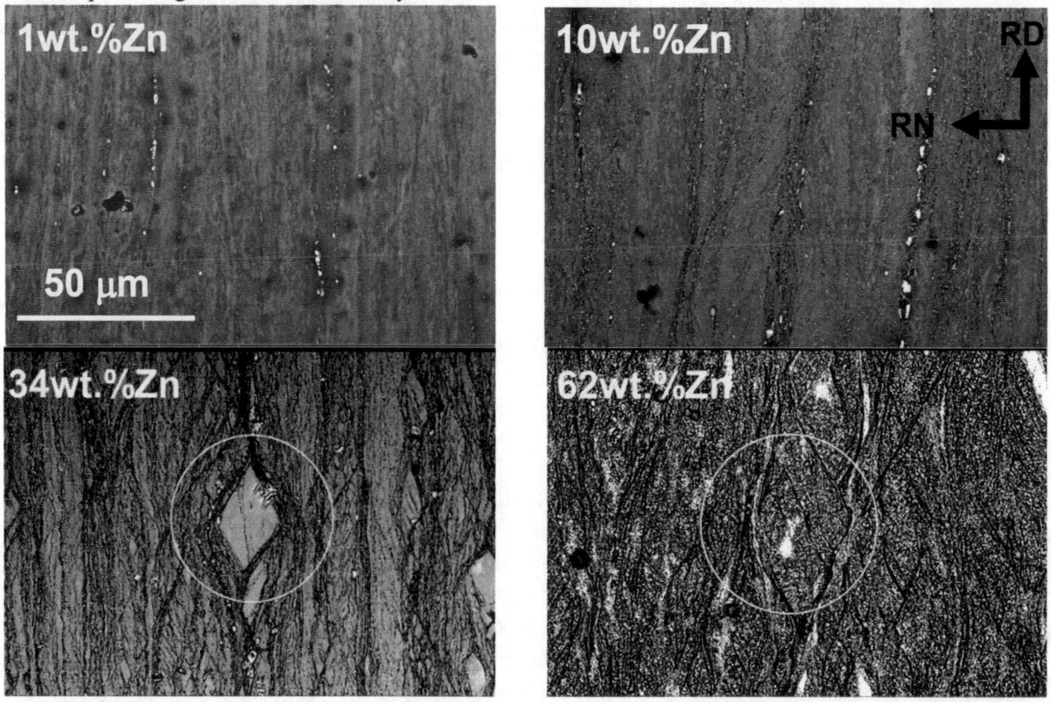

Fig. 4: Backscattered electron images of transverse sections of Al-xZn alloys after 90% reductions for (a) Al-1wt.%Zn, (b) Al-10%Zn, Al-34wt.%Zn and 62wt.%Zn. Circled regions show bird's eye features which are prevalent in the high Zn containing alloys. Imaging conditions were 10kV and 8mm WD using a CBS detector.

The orientation of the eye in the Al-34wt.% specimen is quite different from the surrounding orientations as indicated by the electron channelling effect of the backscattered image. The eye orientation was checked by electron backscattered diffraction, and the orientations plotted in {111} and {220} pole figures of Fig. 5. The feature is clearly a {110}<1-12> oriented grain and there is a splitting in the orientation data that matches an apparent slip band in the feature. The overall shape of the bird's eye is symmetric to the external strain coordinates and with an outline close to the {111} slip intersections with the sectioning plane. Rhombic, lens-shaped domains have been reported for a range of rolled f.c.c. alloys since the 1970's: Cu-30Zn [10], Al-4.5Mg [11], and 316L stainless steel [12], but not for Al-xZn alloys to our knowledge. We will not speculate on the origin of the birds' eyes in this work except to restate that it has a stable Br orientation and its presence is accompanied by surrounding slip and shear banding. Is the texture and microstructure development change with Zn a solid solution effect, or an effect of the deformation temperature? Even with a rapid quench the solid solution is unstable at 25°C for alloys with greater than 10wt.%Zn. At 298K, the T/T_m for the 30 and 60wt.%Zn alloys is greater than 0.35, which is the transition to non-octahedral slip [13]. The observed results are unlikely a stacking fault energy effect unless direct evidence of faults can be observed. See Leffers [14, 15] for a full review of the many nuances about the origin of the Br texture.

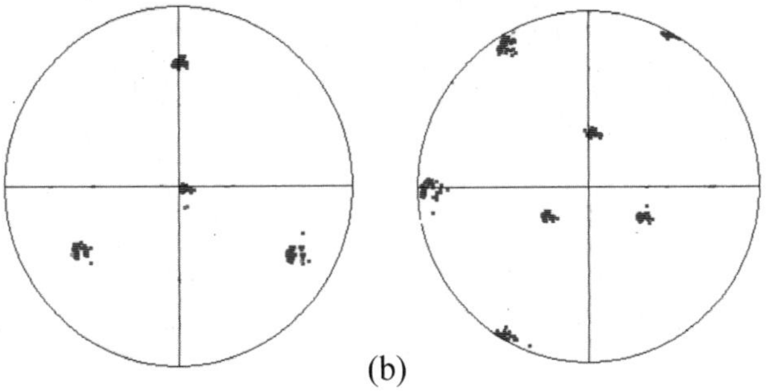
(a) (b)

Fig. 5: (a) {111} and (b) {220} pole figures of bird's eye feature in Fig. 4(c) determined by electron backscattered diffraction. Vertical direction is the rolling direction and matches the orientation of the image in Fig. 4(c).

5. Conclusions

The evolution in rolling texture and microstructure of Al-xZn alloys after 70 and 90% reductions was studied. The Br-S-Cu intensity increased with rolling reduction, but increasing Zn weakened this effect except for some stabilization of Br. {110}<1-12> microstructural "bird's eye" features were observed in 34 and 62wt.% Zn containing alloys at 90% reduction. No evidence for intrinsic stacking fault energy effects was observed, but the fundamental role of Zn in Al solid solution remains unclear and requires more detailed work hardening and microstructure-defect studies at higher reductions.

Acknowledgments

One of the authors (bjd) would like to thank M. Gallerneault for advice on using TiB_2 in casting the alloys, and acknowledges patient discussions with B. Hutchison, D.J. Lloyd and S. Saimoto about birds' eyes. The work was funded by an NSERC Discovery grant.

References

[1] Hirsch J and Lücke K 1988 *Acta metall.* **36** 2863.
[2] Douglass DL and Barbee TW 1969 *J. Matls. Sci.* **4** 121.
[3] Douglass DL and Barbee TW 1969 *J. Matls. Sci.* **4** 138.
[4] Hoyt JJ, Clark B, de Fontaine D, Simon JP, Lyon O 1989 *Acta metall.* **37** 1597.
[5] Lloyd DJ, private communication, Dec. 2, 2015.
[6] Karmakar G, Sen S, Chattopadhyay SK, Meikap AK, Chaterjee SK 2002 *Bull. Mater. Sci.* **25** 315.
[7] Yeung WY, Hirsch J, Hatherley M 1988 *8th International Conference on Textures of Materials* (Santa Fe) (Warrendale: TMS) p 467.
[8] Van Houtte P. MTM-FHM, Leuven, Belgium 2000.
[9] Yeung WY, Hirsch J, Hatherley M 1988 *8th International Conference on Textures of Materials* (Santa Fe) (Warrendale: TMS) p 467.
[10] Hutchinson WB, Duggan BJ, Hatherly M 1979 *Metals Tech.* Oct 398.
[11] Lloyd DJ, Burtryn EF, Ryvola M 1982 Microstructural Science vol. 10 (Elsevier) p 373.
[12] Donadille C, Valle R, Dervin P, Penelle R 1989 *Acta metall.* **37** 1547.
[13] Cortez-Regaldo H and Saimoto S 1984 7th *International Conference on Textures and Materials* (Noordwijkerhout) (Netherlands: The Netherlands Society for Materials Science) p 39.
[14] Leffers T 1996 *11th International Conference on Textures and Materials* (Bejing) vol. 1 (Bejing: Intl. Academic Publishers) p 299.
[15] Leffers T, Ray RK 2009 *Prog. Matls. Sci.* **54** 351.

Comparison of rolling texture of austenite and ferrite phases of duplex steel with single-phase austenitic and ferritic steel

J Capek[1], M Cernik[2], N Ganev[1], K Trojan[1], J Nemecek[1], K Kolarik[1]
[1] Department of Solid State Engineering, Faculty of Nuclear Sciences and Physical Engineering, Czech Technical University in Prague, Trojanova 13, Prague, Czech Republic
[2] U. S. Steel Košice, s.r.o., Košice, Slovak Republic

E-mail: Jiri.Capek@fjfi.cvut.cz

Abstract. In this contribution, the behavior of three types of steel after rolling was investigated. Ferritic and austenitic steels or phases have different mechanical and thermal properties. Due to their mutual influence during plastic deformation, it is possible to suppose the differences between preferred orientations of austenite and ferrite phases in duplex steel and single-phase austenitic and ferritic steels. For this reason, the preferred orientations of austenite and ferrite phases of rolled duplex steel were compared with rolled single-phase austenitic and ferritic steel. The 0–50% reductions of rolled steel plates were selected. Mainly, the strength and type of preferred orientation were compared in the relation to reduction and material. For this purpose, the rolled surfaces were measured by X-ray diffraction. The significant differences between intensities and types of texture components were investigated. For austenite-ferrite interaction during plastic deformation and phase transformation, samples were analysed by electron back-scattered diffraction. The Kurdjumov-Sachs orientation relationship was investigated.

1. Introduction

Duplex stainless steels or duplex steels are dual-phases steels with ferrite and austenite phases with roughly 50/50 proportion. Due to the combination of properties of both phases and two phase microstructure, duplex steels exhibit high corrosion and wear resistance in many environments, where the standard austenitic and ferritic steels are usually used [1]. For this reason, it is possible to manufacture lighter functional components. However, disadvantages are combined too. Due to austenite phase, duplex steels are susceptible to mechanical hardening and, because of ferrite phase, to hydrogen embrittlement.

The importance of study of preferred orientation (texture) originates in the anisotropy of most material properties. Consequently, the analysis and subsequent interpretation of the texture is crucial, especially in material engineering. Moreover, texture analysis during the thermo-mechanical processing of materials provides information on the basic mechanism. The properties, which are influenced, are the Young's modulus of elasticity, Poisson number, hardness, strength, ductility, abrasion resistance, magnetic permeability, electrical conductivity etc. Therefore, materials with a strong texture are used to produce components with specific properties [2]. Major deformation mechanisms responsible for the formation of ferrite and austenite rolling textures in duplex steels should be similar as in the single-phase steels. However, austenite and ferrite phases are generally different in plastic and elastic properties. Therefore, their behavior during deformation can be different in comparison with single-phase steels [3].

2. Theory

Texture is interpreted in the form of orientation distribution function (ODF), pole figures, and inverse pole figure (IPF) maps. ODF is usually presented using Euler space, which is defined by three rotations of crystallites $\mathbf{g} = (\varphi_1, \Phi, \varphi_2)$, so called Euler angles in Bunge notation [2]. ODF or $f(\mathbf{g})$ function determines the bulk density of crystals with the **g** orientation. The unit of $f(\mathbf{g})$ is multiple of a random distribution, i.e. the function is normalized to one, which corresponds to the random orientation of crystals. According to crystal and sample symmetries, the Euler angles of materials with cubic crystal and orthorhombic sample symmetries are in the range $(\varphi_1, \Phi, \varphi_2) \in \langle 0°; 90° \rangle$. For materials with a face centered cubic (fcc) and a body centered cubic (bcc) lattice, the ODF is usually interpreted using different texture components and textures with fibre components, shortly fibres, [4], see Fig. 1. These fibres are usually interpreted according to rolling (RD), transversal (TD) and normal (ND) directions. The type and strength of texture are significantly dependent on the material structure (especially on the grain size and initial texture), chemical and phase composition, and mainly the thermo-mechanical properties of presented phases [2, 3, 5].

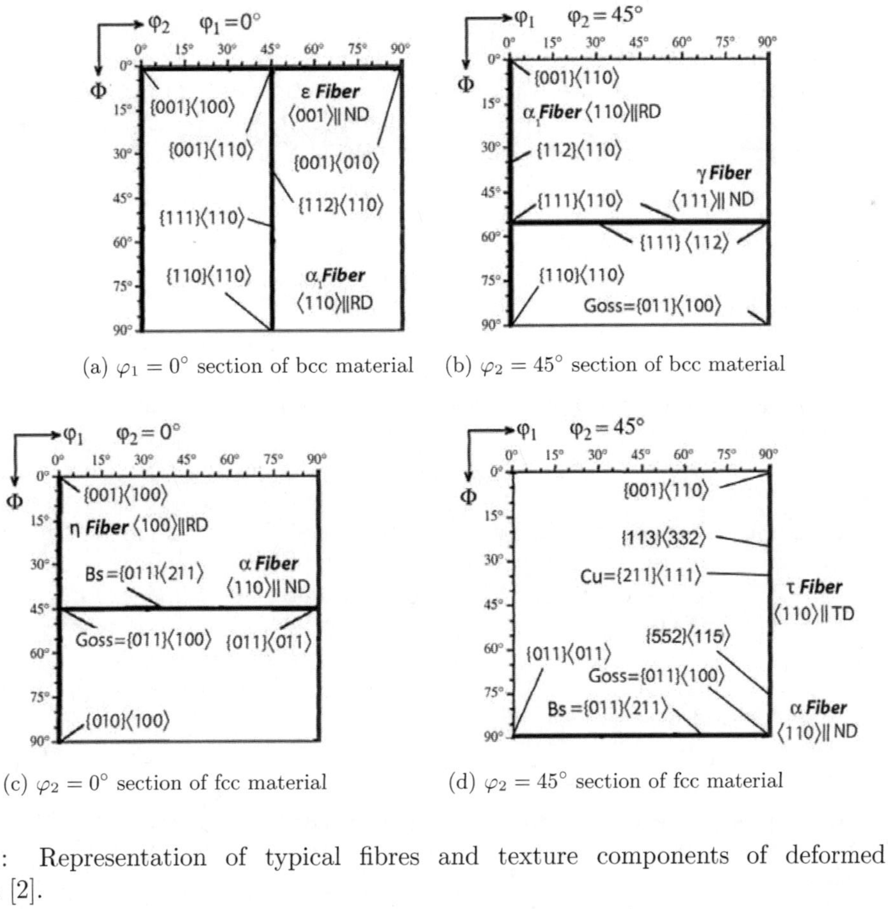

(a) $\varphi_1 = 0°$ section of bcc material (b) $\varphi_2 = 45°$ section of bcc material

(c) $\varphi_2 = 0°$ section of fcc material (d) $\varphi_2 = 45°$ section of fcc material

Figure 1: Representation of typical fibres and texture components of deformed cubic materials [2].

Typical orientations in deformed bcc materials are formed into three characteristic fibres. During cold rolling, primarily the α_1, ε and γ-fibres are created. The α_1-fibre is characterized by $\langle 110\rangle\|$RD and includes e.g. the rotated cubic component $\{001\}\langle 110\rangle$. The γ-fibre includes crystallographic directions with $\{111\}$ planes which are parallel to ND. Rotated cubic component (or orientation) $\{001\}\langle 110\rangle$ and cubic component $\{001\}\langle 100\rangle$ are part of ε-fibre which is characterized by $\langle 100\rangle\|$ND.

The most important texture components for fcc materials of Brass type (materials with small stacking fault energy — SFE) are Goss (G) $\{011\}\langle 100\rangle$ and Brass (Bs) $\{\bar{1}10\}\langle 11\bar{2}\rangle$, see Fig. 1. In the paper [5], this Brass type is called Silver type (Ag). Less common texture component is Copper (Cu) $\{\bar{2}11\}\langle 111\rangle$, mostly for relatively high SFE. These texture components are part of three main texture fibres: α — $\langle 110\rangle\|$ND, τ — $\langle 110\rangle\|$TD and η — $\langle 100\rangle\|$RD. The potentional difference between duplex and austenite steel may be caused by the difference in SFE, which depends on chemical composition. According to [6], the SFE of austenite steel is arround 18 mJ/m^2 and for γ phase of duplex steel, the SFE is 10 mJ/m^2.

The orientation relationship (OR) of all phases is necessary to known for interpretation of austenite-ferrite interaction. During the manufacturing of duplex steels, the samples are usually quenched. Therefore, the orientations of precipitated/transformed grains are not random. According to [7], solidification of duplex steels is controlled by chemical composition. So initially, the solid phase is completely ferritic (α). Subsequent cooling results in austenite phase (γ) generation in diffusion reaction $\alpha \rightarrow \alpha+\gamma$. According to temperature gradient during quenching and diffusion, four basic types of dual-phase structure can exist [7]. The most common structures are duplex and dispersion. The duplex structure is created by austenite grains precipitated on the grain boundaries of α phase. In the case of dispersion structure, not all dissolved elements, which stabilize the austenitic structure, diffuse on grain boundaries. From these elements, γ phase precipitate intergranular [7]. For this reason, the microstructure of typical duplex steel consists of α matrix with coarse intragranular (islands creating) and fine intergranular γ grains.

The orientation relationships (ORs) are usually represented in Rodrigues-Frank (R-F) space which is a map of all corresponded orientations and variants [8]. From R-F space is possible to reach information about differences between ORs, see Tab. 1. During most phase transformations, certain favoured orientation relationship exists between the parent and the daughter phases. The OR can be described according several models: Bain, Kurdjumov-Sachs (K-S), Nishiyama-Wasserman (N-W) etc., see Tab. 1. The Greninger-Troiano is called an irrational OR because of high Miller index numbers. The rest of ORs are called rational ORs due to low Miller indices. From a single-austenite crystal, 12 and 24 crystallographic variants can be formed with a N-W and K-S/Pitsch orientations relationship, respectively, due to the symmetry of cubic systems. By comparing all 12 variants, only five possible misorientation angles (MAs) (13.76°, 19.47°, 50.05°, 53.69° and 60°) between variants are maintained in the case of the N-W relationship [9]. For K-S orientations relationship, the MAs are about 4° smaller, i.e. 10.53° and 55.53° [8].

Each material may show its specific combination of nucleation OR and OR continuum [10]. In low-carbon steels the K-S OR nucleation is observed and the orientation gradually changes to a Pitsch variant [10]. Therefore, duplex steel proved to be useful for understanding ORs in steel, in which γ precipitations nucleate in a α matrix [11] and the K-S OR was predominantly found.

3. Experiment

The plate shape samples of size 19×120 mm^2 and different thicknesses were made of AISI 420 (ferritic), AISI 304 (austenitic) and AISI 318LN (duplex) type of stainless steel using grinding and milling. Before deformation, the austenitic and ferritic with duplex samples were annealed in air laboratory furnace for 4 hours at 840 °C and 7 hours at 650 °C, respectively, in order to

Table 1: Common orientation relationship models: Bain, Kurdjumov-Sachs, Nishiyama-Wassermann, Pitsch and Greninger-Troiano, where α and γ are ferrite and austenite phases, respectively [7].

OR	crystalographic planes	crystalographic directions	MA of α -γ
Bain	$\{001\}_\gamma \| \{001\}_\alpha$	$\langle 100\rangle_\gamma \| \langle 110\rangle_\alpha$	45°
Kurdjumov-Sachs	$\{111\}_\gamma \| \{011\}_\alpha$	$\langle 011\rangle_\gamma \| \langle 111\rangle_\alpha$	42.8°
Nishiyama-Wassermann	$\{111\}_\gamma \| \{011\}_\alpha$	$\langle 112\rangle_\gamma \| \langle 011\rangle_\alpha$	45.9°
Pitsch	$\{100\}_\gamma \| \{110\}_\alpha$	$\langle 110\rangle_\gamma \| \langle 111\rangle_\alpha$	45.9°
Greninger-Troiano	$\{111\}_\gamma \| \{110\}_\alpha$	$\langle 123\rangle_\gamma \| \langle 133\rangle_\alpha$	44.2°

reduce residual stresses. In the end, the samples were cold rolled with 0, 10, 20, 30, 40, and 50% reductions of thickness, thus the final thickness of all samples was 1.5 mm. Therefore, samples made of ferritic, austenitic and duplex steel were marked as F0–F50, A0–A50 and D0–D50, respectively.

X'Pert PRO MPD diffractometer with Co$K\alpha$ radiation was used to the analyses of RD-TD plane samples by X-ray diffraction (XRD). Texture analysis was performed based on ODF calculation from experimental pole figures which were obtained on three diffraction lines $\{110\}$ ($\{220\}$ for α phase in duplex steel), $\{200\}$, $\{211\}$ of ferrite phase and $\{111\}$ ($\{311\}$ for γ phase in duplex steel), $\{200\}$, $\{220\}$ of austenite phases. Irradiated volume of samples was based on size of primary slits (1×1 mm^2), geometry of experiment and penetration depth of the used radiation into given material, see [12].

Contrarily, the TD-ND plane of samples was investigated by electron backscatter diffraction (EBSD). The *TESCAN VEGA3* electron microscope with *NordlysMax3* EBSD camera was used for measurement of IPF maps. IPF maps were measured at magnification 250× and 1μm step due to a coarse-grain structure in duplex steels.

XRD and EBSD analysis was not done for mutual comparison, but served as complementary data. XRD data was primarily used for texture analysis and EBSD data for austenite-ferrite interaction. The MATLAB™ toolbox MTEX software was used for processing of all experimental data [13].

4. Results and discussions

4.1. X-ray diffraction

Dependence of relative intensities of texture fibres $f(\mathbf{g})$ on the rolling reduction is shown in Figs. 2. In view of the fact that the typical texture components of rolled materials are part of texture fibres, only 2-D ODF sections of samples with 50% deformation are presented, see Figs. 3. For samples without deformation, it is possible to assume that the initial texture is composed of grinding texture components. The rolling texture was generated during rolling of the samples.

In the ferritic samples F0–F50, see Fig. 2a, the typical texture components and fibres were generated. It is evident that strength of the texture components and fibres are increasing with higher deformation. During cold rolling, the strong α_1 and γ-fibres were generated. The dominant texture component are $\{001\}\langle 110\rangle$ (rotated cubic), $\{\bar{1}12\}\langle 110\rangle$ and $\{\bar{1}11\}\langle 110\rangle$. The ε-fibre contains only representative rotated cubic component.

For the austenitic samples A0–A50, see Fig. 2c, the transition between grinding and rolling texture is evident. The rotated cubic component $\{001\}\langle 110\rangle$ is part of grinding texture. During deformation (rolling), grains rotate to the cubic orientation $\{001\}\langle 100\rangle$. From 30% deformation

IOP Conf. Series: Materials Science and Engineering **375** (2018) 012025 doi:10.1088/1757-899X/375/1/012025

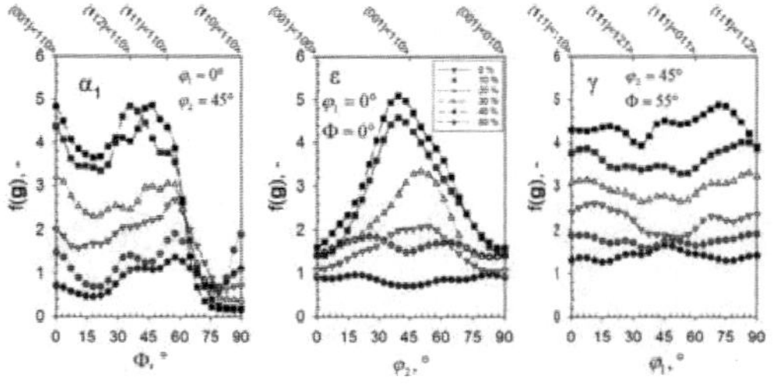
(a) Ferritic steel

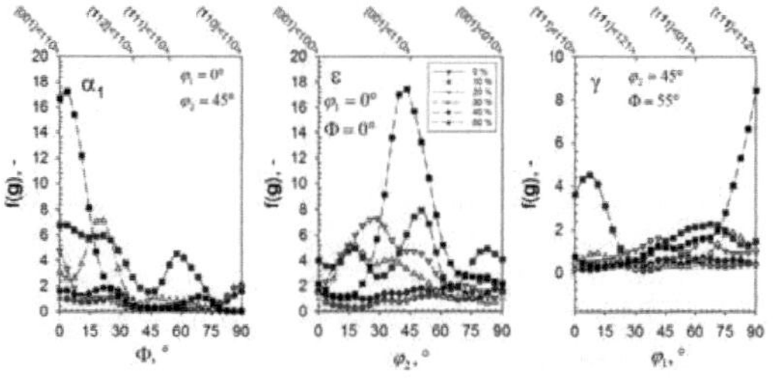
(b) Duplex steel — α phase

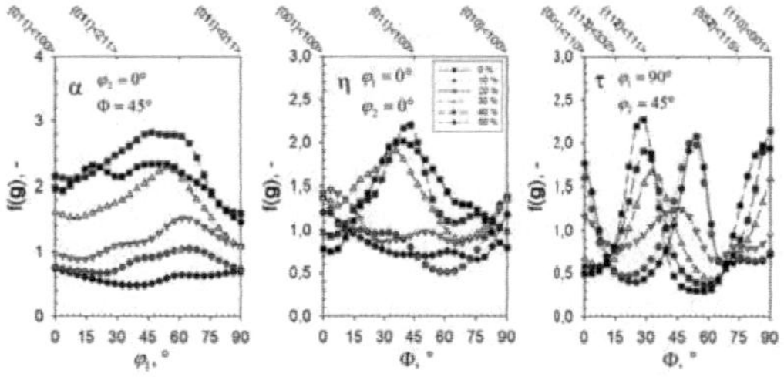
(c) Austenitic steel

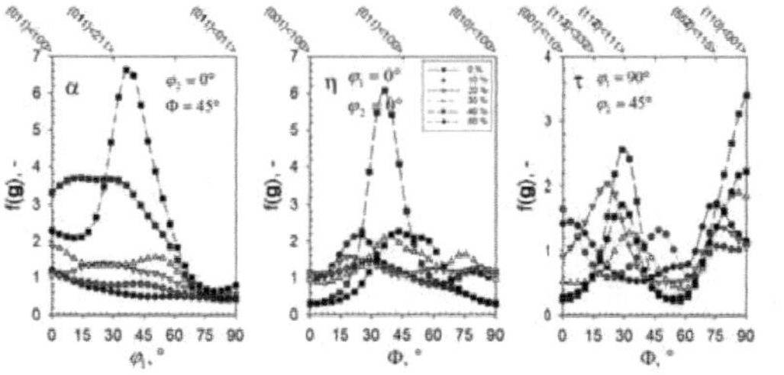
(d) Duplex steel — γ phase

Figure 2: ODFs along individual fibres (α_1, ε and γ) and (α, η and τ) of α and γ phases, respectively.

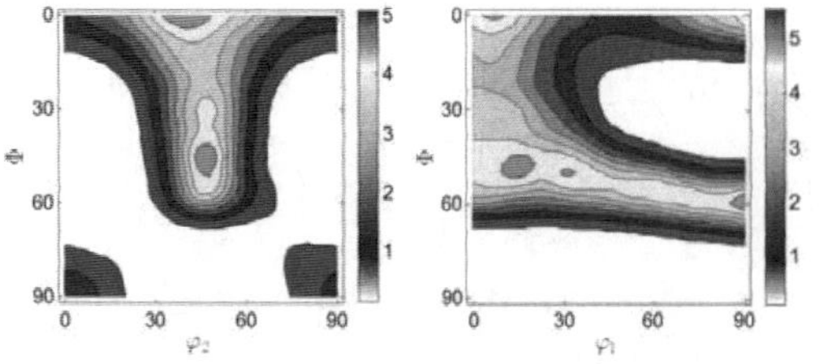
(a) Sample F50

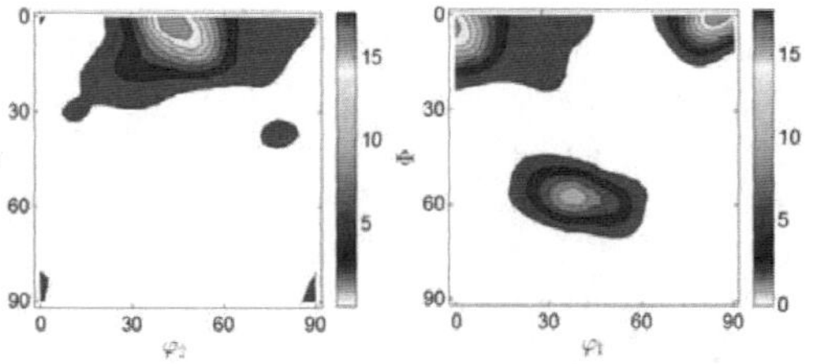
(b) Sample D50 — α phase

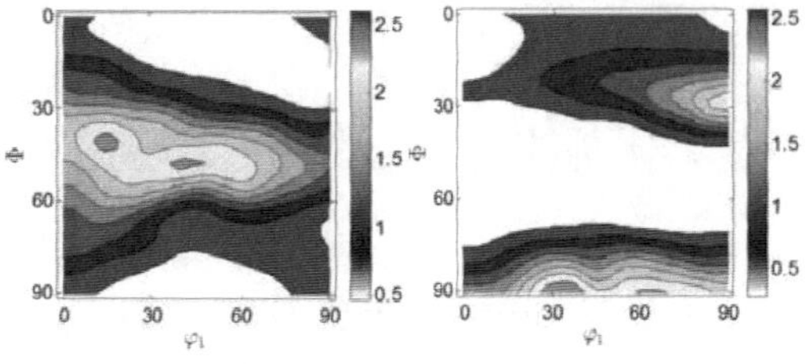
(c) Sample A50

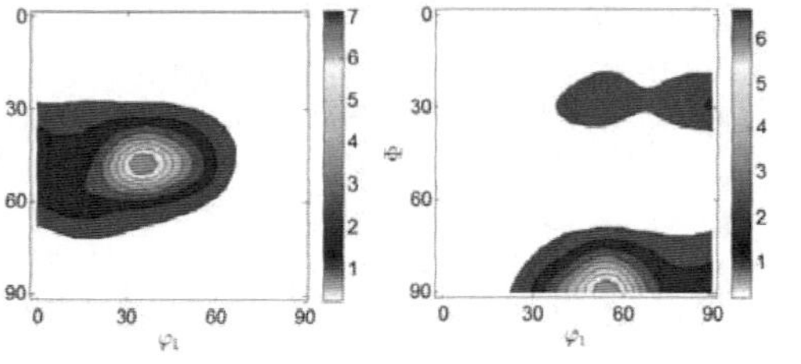
(d) Sample D50 — γ phase

Figure 3: Orientation distribution functions in the $\varphi_1 = 0°$ (left) and $\varphi_2 = 45°$ (right) sections for α phases, and $\varphi_2 = 0°$ (left) and $\varphi_2 = 45°$ (right) sections for γ phases.

the G component $\{110\}\langle 001\rangle$ is generated. Rotating these grains around $\langle 100\rangle\|$RD axis, the η-fibre is formed. The cubic component is a part of η-fibre too. If grains with G orientation rotate around $\langle 110\rangle\|$ND axis, the α-fibre and Bs texture component are generated. The τ-fibre is only composed of G and Cu texture components. Cu texture component is more common for materials with high SFE.

The differences in type and strength of texture of duplex steel are evident in comparison with single-phase steels from Figs. 2. In the case of ferrite phase (Fig. 2b), the mutual influence of α and γ phases and mainly different mechanical properties of both phases results in limitation of rotation of α grains during deformation. During rolling, the harder phase (α) keeps energy until deformation exceeds a certain limit. This released energy results in sudden α grains rotation to another discrete orientation. From this reason, the texture components are very strong and are not typical for this deformation for bcc materials or phases. However, typical texture component is generated in the sample with higher deformation, i.e. rotated cubic component $\{001\}\langle 110\rangle$, see Fig. 3b.

Texture components of γ phase of duplex steel (Fig. 2d) are stronger in comparison with single-phase steel which are caused by mutual interaction during plastic deformation. With increasing of deformation, only Bs, G and Cu texture components are generated. For the sample with 50% deformation, majority of grains have Bs orientation.

4.2. Electron backscatter diffraction

Using EBSD, the austenite-ferrite interaction was investigated. In Figs. 4 IPF maps of D0 and D50 samples are presented. The MAs of both phases are depicted in Figs. 5.

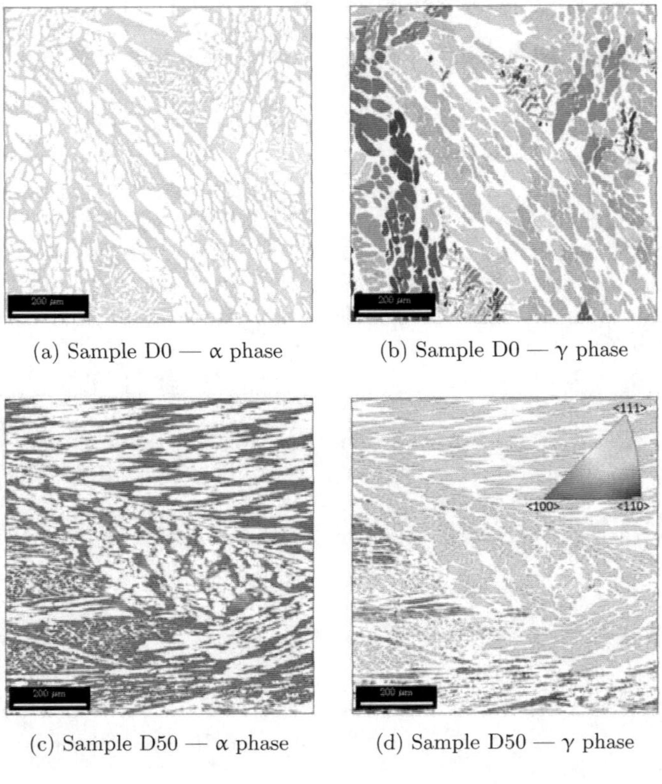

(a) Sample D0 — α phase

(b) Sample D0 — γ phase

(c) Sample D50 — α phase

(d) Sample D50 — γ phase

Figure 4: Inverse pole figure maps of α and γ phases of D0 and D50 samples.

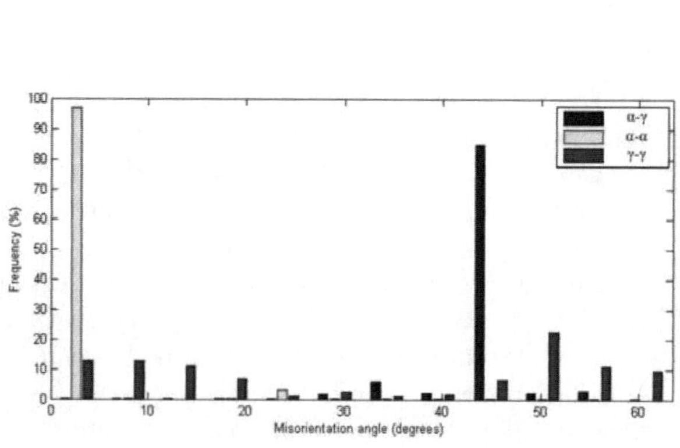

(a) Sample D0

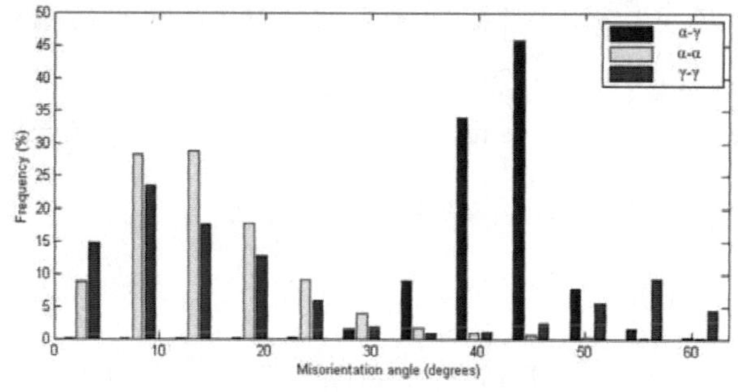

(b) Sample D50

Figure 5: Misorientation angles between γ -γ , α -γ and α -α phases of D0 and D50 samples.

From IPF maps in Figs. 4 and MA in Figs. 5, the solidification history of duplex steels is evident. According to the theory, see Section 2, initially, the solid is completely ferritic. This microstructure is composed of ferritic coarse-grain with small MA which is under 10°. During further cooling, the austenite grains are precipitated on the grain boundaries and in the grains and create duplex and disperse structure, respectively. Generally, the phase transformation is a solid-solid transformation with minimum diffusion, as supported by the high cooling rates. The orientation of γ grains is not random, but MAs between α and γ grains is around 43°, where the best fit is with K-S OR, see Tab. 1. According to K-S or N-W OR, the MAs between γ and γ grains are around 10° and 55°. Rest of MAs, i.e. mainly 20–40°, represents random orientations.

During deformation, the grains are re-oriented. The MAs between α and γ grains is not strictly 42.8°, but it creates Gaussian distribution. For high deformation, the MAs between α and α is similar to Gaussian distribution too with maximum at 10°. The MAs between γ and γ grains has the equal trend as the D0 sample. The MA function has maxima around 10° and 55°. These trends are typical for cubic materials and correspond with K-S (10.53° and 55.53°) or N-W (13.76° and 60°) OR.

5. Conclusions

Results obtained using XRD showed:

- The intensity and development of texture components are different for single and dual-phase types of steels.
- With increasing degree of deformation of single-phase samples, a considerable increase in the texture intensity was observed.

- For ferritic sample, there are four main texture components $\{001\}\langle 110\rangle$ (rotated cubic), $\{\bar{1}12\}\langle 110\rangle$, $\{\bar{1}11\}\langle 110\rangle$ and $\{11\bar{1}\}\langle 112\rangle$ which are parts of α_1, ε and γ-fibres.
- For α phase of duplex steel, only rotated cube component $\{001\}\langle 110\rangle$ is dominant after 50% deformation. The texture for α phase of duplex steel is stronger compared with ferritic samples.
- Texture of austenitic steel is primarily composed of expected Brass $\{110\}\langle\bar{1}12\rangle$, Goss $\{110\}\langle 001\rangle$ components and less common Copper $\{11\bar{2}\}\langle 111\rangle$ component which are parts of α, τ and η-fibres.
- In the case of γ phase of duplex steel, Goss, Brass and Copper texture components are also presented. But for high deformation, only Brass component develops texture, which is typical for steels with low SFE. These texture fibre components are not conspicuous.

Using EBSD analyses, further results were acquired:

- Microstructure of duplex steel is composed of α grains which have MA approx. 10°. During solidification, two types of γ grains precipitate. Firstly, on the grain boundaries (coarse-grains) and in the grains (disperse fine-grains).
- The MA of α -γ is 42.8° according to K-S OR.
- Because of plastic deformation during rolling, the MAs between α and γ grains after rolling are not strictly 42.8°, but it creates a Gaussian distribution.
- Behavior of MA of α -α and γ -γ grains is according to K-S or N-W OR with maxima at around 10° and 55°.

References

[1] Dakhlaoui R, Braham C and Baczmanński A 2007 Mechanical properties of phases in austeno-ferritic duplex stainless steel-Surface stresses studied by X-ray diffraction *Mater. Sci. Eng.: A* **444** 6–17
[2] Bunge H J 1982 *Texture Analysis in Materials Science* (London: Butterworth)
[3] Ryś J, Ratuszek W and Witkowska M 2006 Comparison of the Rolling Texture Formation in Duplex Steels with Various Initial Textures *Arch. Metall. Mater.* **51** 495–502
[4] Suwas S and Ray R K 2014 *Crystallographic Texture of Materials* (Springer-Verlag)
[5] Hu H 1974 Texture of metals *Texture* **1** 233–58
[6] Reick W, Pohl M and Padilha A F 1996 Determination of stacking fault energy of austenite in aduplex stainless steel *steel res. int.* **67** 253–6
[7] Knyazeva M and Pohl M 2013 Duplex Steels: Part I: Genesis, Formation, Structure *Metallogr. Microstruct. Anal.* **2** 113–21
[8] He Y, Godet S, Jonas J J 2004 Representation of misorientations in Rodrigues-Frank space: application to the Bain, Kurdjumov-Sachs, Nishiyama-Wassermann and Pitsch orientation relationships in the Gibeon meteorite *Acta Mater.* **55** 1179–90
[9] Beladi H, Adachi Y, Timokhina I and Hodgson P D 2009 Crystallographic analysis of nanobainitic steels *Scripta Mater.* **60** 455–8
[10] de Jeer L T, Ocelík V and De Hosson J T 2017 Orientation Relationships in $Al_{0.7}$CoCrFeNi High-Entropy Alloy *Microsc. Microanal.* **23** 905–15
[11] Jiao H, Aindow M and Pond R C 2003 Precipitate orientation relationships and interfacial structures in duplex stainless steel Zeron-100 *Philos. Mag.* **83** 1867–87
[12] Čapek J and Pala Z 2014 Programs for Visualizing Results of Diffraction Experiments *Mater. Struct.* **21** 93–4
[13] Bachmann F, Hielscher R and Schaeben H 2010 Texture Analysis with MTEX — Free and Open Source Software Toolbox *Solid State Phenomen.* **60** 63–8

Acknowledgments

This work was supported by the Grant Agency of the Czech Technical University in Prague, grant No. SGS16/245/OHK4/3T/14, Czech Science Foundation, grant No. 14-36566G, and the "Nadání Josefa, Marie a Zdeňky Hlávkových" Foundation.

IOP Conf. Series: Materials Science and Engineering **375** (2018) 012026 doi:10.1088/1757-899X/375/1/012026

Development of crystallographic texture under plane and shear strain in ultrahigh-strength strip steels

A. Kaijalainen[1], V. Javaheri[1], D. Lindell[2] and D.A. Porter[1]

[1] Materials and Production Engineering, University of Oulu, Finland.
[2] Swerea KIMAB, Stockholm, Sweden.

antti.kaijalainen@oulu.fi, vahid.javaheri@oulu.fi, david.lindell@swerea.se, david.porter@oulu.fi

Abstract. The effect of centerline and subsurface microstructures on the crystallographic texture of three 8 mm thick low-alloyed hot-rolled and direct-quenched ultrahigh-strength strip steels with yield strengths in the range 800 – 1100 MPa has been investigated. Detailed microstructural features were studied using LCSM, FESEM, FESEM-EBSD. In addition textures and crystallographic features were analyzed using Matlab combined with MTEX software. Rolling to lower finish rolling temperatures increased austenite pancaking leading to the formation of ferritic/granular bainitic and the upper bainitic microstructures at the subsurface. In addition, increased austenite pancaking was found to increase the intensities of ~{554}<225>$_\alpha$, ~{112}<110>$_\alpha$ and ~{112}<131>$_\alpha$ texture components at the centerline and ~{112}<111>$_\alpha$ and ~{110}<112>$_\alpha$/<111>$_\alpha$ texture components in the surface layers, especially in upper bainitic microstructures. Parent austenite reconstruction shows that crystallographic texture at the centerline derived from {112}<111>$_\gamma$ and {110}<112>$_\gamma$ and the subsurface the shear texture components derived from the {112}<110>$_\gamma$ and {111}<112>$_\gamma$ components, as expected. The Matlab reconstruction code was found to work well for martensitic and upper bainitic morphologies even with the highly pancaked prior austenite structure. However, it was less precise for granular bainite and ferrite.

1. Introduction

As previous studies are shown [1,2], the microstructure and crystallographic texture of steel products is greatly influenced by the austenite morphology before phase transformation. Therefore, an understanding of prior austenite morphology is very important for microstructural control in high-strength steels.

The grain reconstruction method is commonly used with martensitic and bainite microstructures [3,4] that have transformed from equiaxed prior austenite with a Kurdjumov–Sachs (K–S) [5] or Nishiyama-Wasserman orientation relationship [6,7]. Furthermore, it is well-known that when austenite is deformed under plane strain conditions below its non-recrystallization temperature, a strong deformation texture develops, the main components of which are {112}<111>$_\gamma$ (Cu), {110}<112>$_\gamma$ (Br), and {110}<001>$_\gamma$ (Goss) as shown in Fig. 1a. The Cu component transforms primarily into {112}<110>$_\alpha$, also known as "transformed Cu", with a much smaller fraction appearing in the vicinity of the "rotated Goss" ({110}<110>$_\alpha$) in Fig. 1a. Finally, the Br component transforms primarily into "transformed Br" located in the vicinity of {111}<112>$_\alpha$ and {554}<225>$_\alpha$, with minor amounts rotating to {112}<131>$_\alpha$ and {100}<011>$_\alpha$ in Fig. 1a [8,9]. Similarly, when austenite is deformed by rolling below its non-recrystallization temperature, there are shear texture components in the austenite close to the rolled surfaces, i.e., {111}<211>$_\gamma$ and {112}<110>$_\gamma$. After cooling and transformation to ferrite these lead to the formation of shear components {112}<111>$_\alpha$, {110}<112>$_\alpha$ and {110}<111>$_\alpha$, respectively [10] as shown in Fig. 1b.

Therefore, the aim of this study is to use EBSD measurements to reconstruct the prior austenite grain morphology and crystallographic texture in the case of highly pancaked prior austenite to help understand the microstructural transformations that occur during cooling after hot rolling. Such studies can help elucidate the specific reason for the higher intensity of the ~{112}<111>$_\alpha$ texture component found in upper bainite in direct quenched steels.

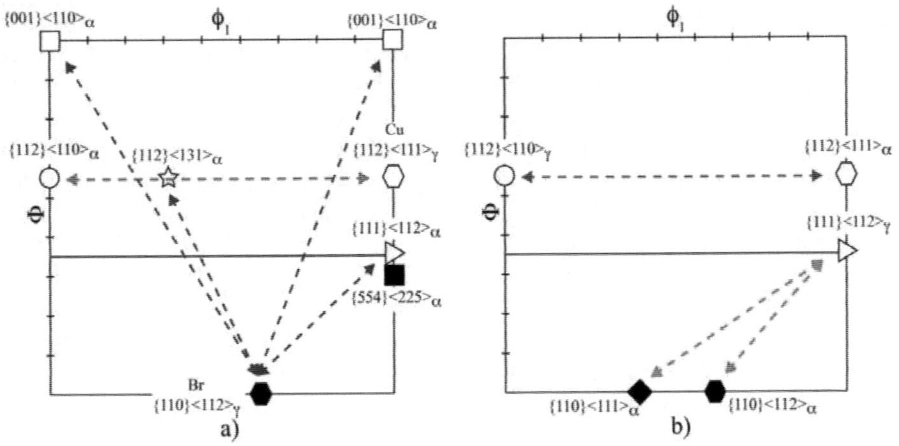

Figure 1. Texture changes resulting from γ to α transformation in pancaked austenite as seen in the ϕ_2 = 45° ODF section. (a) Austenite deformed under plane strain conditions, (b) shear-strained austenite.

2. Experimental

The influence of composition and hot rolling parameters on the subsurface microstructure, texture and bendability in various Nb-microalloyed steels having yield strengths in the range 800 – 1100 MPa containing was investigated using hot rolling experiments. The chemical compositions of the steels are given in Table 1. In these experiments, a 210 mm thick continuously cast slab was reheated at 1250 °C and thermomechanically controlled rolled to a final thickness (t) of 8 mm and direct quenched to room temperature at a rate of about 50-70 °C/s. In all, 14 strips were produced with different subsurface microstructures by varying the finish rolling temperatures (FRT) to 920 °C, 880 °C, 840 °C, 820 °C and 800 °C. Each strip was coded according to its chemical composition (A-C) and the finish rolling temperature (920-800).

Table 1. Chemical composition of the investigated steels in weight percent.

Material	C	Mn	Si	Nb	Cr	Mo	V	B	Ti	Al
Steel A	0.065	1.4	0.2	0.04	1.0	0.02	0.011	0.0013	0.023	0.03
Steel B	0.076	1.8	0.2	0.04	1.0	0.01	0.017	0.0015	0.021	0.04
Steel C	0.093	1.1	0.2	0.04	1.1	0.15	0.012	0.0013	0.021	0.04

The prior austenite grain structure was studied using a laser scanning confocal microscope (LSCM) after picric acid etching. General characterization of the transformation microstructures was performed on nital-etched specimens with a field emission scanning electron microscope (FESEM) (Ultra plus, Zeiss). Electron backscatter diffraction (EBSD) measurements (ODF texture) and analysis were performed using the Oxford-HKL acquisition and analysis software. The FESEM for the EBSD measurements was operated at 10 kV and the step size was 0.2 μm.

In order to reveal the parent austenite grain structure and texture, the EBSD data was subjected to a reconstruction technique that employed Matlab supplemented with the MTEX texture and crystallography analysis toolbox [11]. Briefly, grain maps were initially assembled from the data sets with a grain boundary tolerance of 3-5 degrees. Subsequently, the parent austenite orientation map was reconstructed from this data with a two-step reconstruction algorithm. In the first step of the process, the orientation relationship between austenite and martensite was determined using Kurdjumov-Sachs orientation relationship (K-S OR) i.e. $\{111\}_{\gamma}//\{110\}_{\alpha}$, $<110>_{\gamma}//<111>_{\alpha}$. There are 4 $\{111\}_{\gamma}$ planes which are parallel to a $\{110\}_{\alpha}$ plane and each $\{111\}_{\gamma}$ plane contains 3 $<110>_{\gamma}$ directions each of which is parallel to 2 $<111>_{\alpha}$ directions. Therefore, 24 different crystallographic variants of bcc structure (bainite/martensite) can form within one former fcc structure (parent austenite). Regarding the variant selection, according to the Morito et al. [12], 24 variants generated by the K-S OR have been classified into four packet groups. In a packet, all variants share the parallelism of the same close-packed planes. Each packet contains six variants with parallel direction relationships on the close-packed parallel planes. This approach has been done for all samples. Although the exact parallelism of planes and directions of the K-S OR cannot be observed in actual crystallographic measurements of martensite, the consistent variant labeling system by Morito et al. [12] has been used to successfully characterize a number of martensitic/bainitic microstructures with varying orientation relationships between fcc and bcc structure. In the second step, the grain map was divided into discrete clusters using the Markov clustering method [13] proposed by Gomes and Kestens [14]. The parent austenite orientation was then calculated for each cluster separately, resulting in a reconstructed orientation map. The average misorientation between the reconstructed orientation for each cluster and the best fit for each individual grain was approximately 2 degrees, indicating a good fit for the reconstructed result. The full details for the reconstructed procedure are available in Ref. [15].

3. Results and discussion

The mechanical properties of the steels are given in Ref. [1]. The longitudinal yield stress ($R_{p0.2}$) and tensile strength (R_m) of studied steels vary in the ranges 793-1102 MPa and 933-1247 MPa, respectively.

General findings from the centerline microstructural characterization of the 14 strips can be summarized as follows: (1) a decrease in FRT increased the total reduction (R_{tot}) in the non-recrystallization (NRX) regime of austenite (i.e. pancaking) (in Fig. 2), (2) a decrease of FRT increased the incidence of granular bainite and ferrite at the expense of auto-tempered martensite and upper bainite. However, the microstructures at the centerline consisted of mostly auto-tempered martensite with some bainite. (3) Prior austenite pancaking increases the intensities of the ~$\{554\}<225>_{\alpha}$, ~$\{112\}<110>_{\alpha}$ and ~$\{112\}<131>_{\alpha}$ texture components (Figs. 4-6). More details of the effect of the thermomechanical rolling on prior austenite grain structure, transformation microstructure, texture, mechanical properties and microhardness are discussed in earlier papers [1,16]. The importance of the subsurface observations is discussed in more detail below.

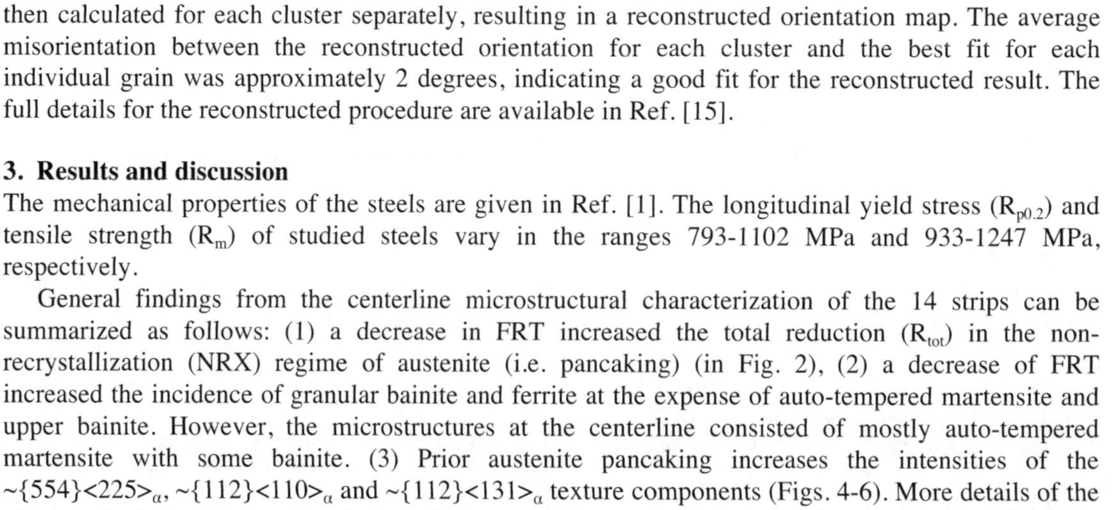

Figure 2. LSCM images of prior austenite morphologies following etching with picric acid: Steel B with (a) 920 °C, (b) 820 °C FRT and (c) Steel C with 820 °C FRT at the centerline.

Specimens consisted of polygonal ferrite interspersed among other carbon-rich microstructural components down to ~50 μm from the surface. This microstructure is seen in Figs. 8 and 9 (bcc EBSD). In the layer from 50 μm to 400 μm, specimens with high FRT (above 880 °C) and Steel B820 mainly comprised upper bainite and martensite with a small fraction of granular bainite. In Steels A and C, with FRT below 840 °C, the subsurface was mainly granular bainite with some upper bainite and ferrite. A decrease of FRT increased the incidence of softer microstructures like ferrite and granular bainite. In addition, increased austenite pancaking increases the intensities of ~{112}<111>$_\alpha$ and ~{110}<112>$_\alpha$ - {110}<111>$_\alpha$ texture components in the surface layers. Fig. 3 summarizes the effect FRT on the intensity of the main texture components and microstructures at the centerline and the subsurface. In Fig. 3 the main microstructural components constituting more that 50 % of the microstructure are indicated.

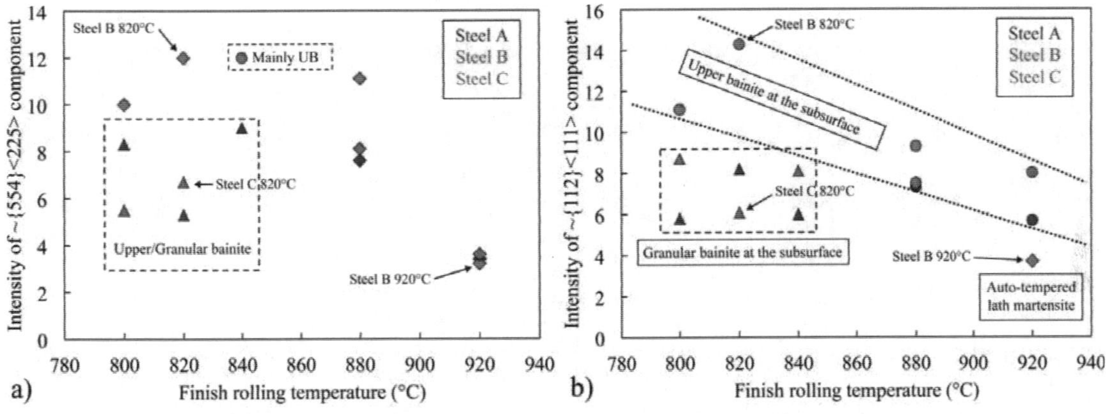

Figure 3. Effect of the finish rolling temperature on the intensity of the a) ~{554}<225>$_\alpha$ texture component at the centerline and b) ~{112}<111>$_\alpha$ texture component near the sheet surface in various microstructures. Symbols: ◆lath martensite, ●upper bainite, ▲granular bainite.

Figs. 4-6 and Figs. 7-9 present original EBSD maps (above) and reconstructed parent austenite maps (below) together with ODFs for the centerline and subsurface layer, respectively. The parent austenite reconstruction works well for martensite and upper bainite even in the case of highly pancaked prior austenite and a relatively large scanned area (Figs. 4, 5 and 7). However, in the case of granular bainite or even ferrite, the reconstructed grain map (Fig. 6) does not appear feasible due to the presence of intermittent grains and boundaries. Even the change of orientation relationship from K-S to Bain [17], Nishiyama-Wassermann [6,7] or Greninger-Troiano [18] for the austenite to granular bainite and ferrite transformations, cannot be reason for the intermittent grains, because the orientation relationships only differ from each other by a few degrees [8]. In the case of granular bainite, the problems in the reconstruction could be caused by errors in the Matlab code. In the case of polygonal ferrite, the reconstruction algorithm will not work as polygonal ferrite grows into the austenite grains with which it has no orientation relationship. Since, the similar the intermittent grain boundaries and unreconstructed polygonal ferrite can be seen in near surface (50 μm below from surface) in Fig. 8 and even lower from the sheet surface in Fig. 9.

The crystallographic texture at the centerline transformed from {111}<112>$_\alpha$, {554}<225>$_\alpha$, {112}<131>$_\alpha$ and {100}<011>$_\alpha$ to {112}<111>$_\gamma$ and {110}<112>$_\gamma$, as was presented schematically in Fig. 1a. Steel B with 820 °C FRT (Fig. 5) shows irregular transformation between ferrite to austenite texture, while the intense additional {112}<110>$_\gamma$ texture component developed at the centerline. In subsurface EBSD measurements, the shear texture components {112}<111>$_\alpha$ and {110}<112>$_\alpha$/<111>$_\alpha$ transformed to the {112}<110>$_\gamma$ and {111}<112>$_\gamma$ components in the parent austenite. Additionally, the {111}<112>$_\gamma$ component can be seen in two locations, (111)[1-21] and (111)[-1-12], for symmetry reasons in the fcc ϕ_2=45° ODF.

After austenite reconstruction from mainly martensitic microstructures (in Steel B 920 °C), the maximum texture intensities increased from 5.4 to 14.0 at the centerline and 3.0 to 3.7 in the subsurface layer. For upper and granular bainitic starting microstructures, on the other hand, reconstructed austenite texture intensities decreased. This phenomenon needs to be studied further with different types of orientation relationships as the basis of the reconstructions. Hence, the reasons for the extremely intense bcc texture seen in the mainly upper bainitic microstructures (Steel B 820 °C, in Figs. 5 and 8) cannot be explained by the present parent austenite reconstruction method. Even with the same finish rolling temperature and prior austenite pancaking, for Steel C FRT 820 °C, which consists of granular bainite, the texture intensities are much lower than those of the upper bainite in Steel B.

Figure 4. Centerline of Steel B with 920 °C FRT (martensitic). (above) Original EBSD band contrast map with high-angle boundaries (>15°) and IPF colored in the RD direction and bcc ϕ_2=45° ODF and (below) reconstructed parent austenite grain map in Euler coloring and fcc ϕ_2=45° ODF.

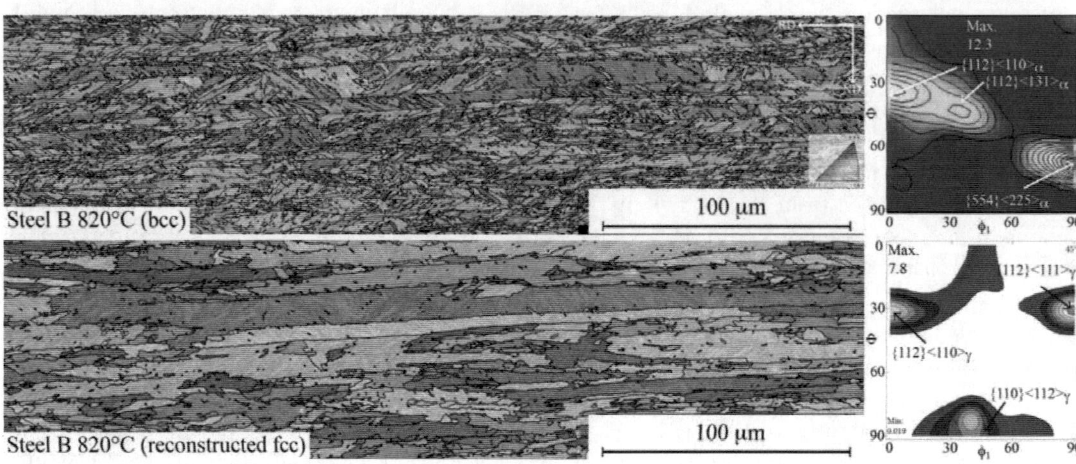

Figure 5. Centerline of Steel B with 820 °C FRT (martensitic). (above) Original EBSD band contrast map with high-angle boundaries (>15°) and IPF colored in the RD direction and bcc ϕ_2=45° ODF and (below) reconstructed parent austenite grain map in Euler coloring and fcc ϕ_2=45° ODF.

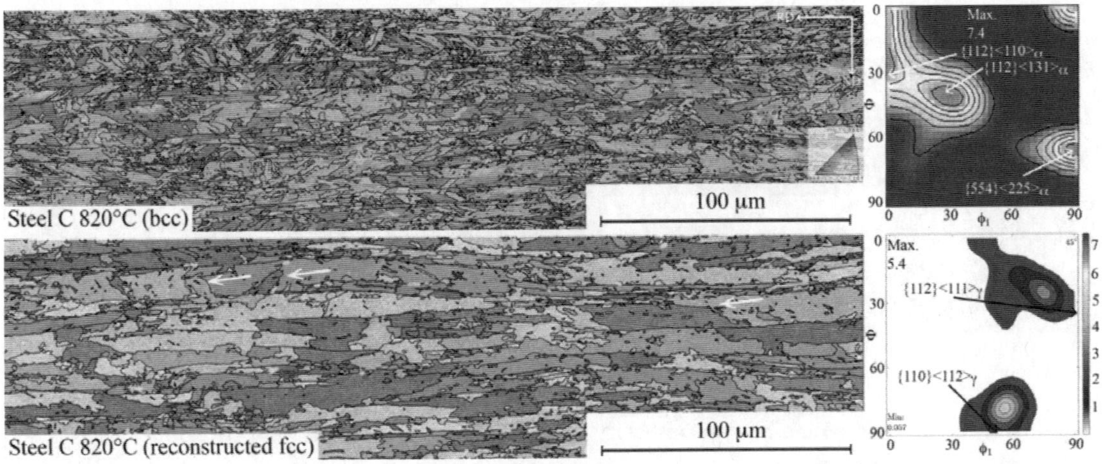

Figure 6. Centerline of Steel C with 820 °C FRT (upper/granular bainite). (above) Original EBSD band contrast map with high-angle boundaries and IPF colored in the RD direction and bcc ϕ_2=45° ODF and (below) reconstructed parent austenite grain map in Euler coloring and fcc ϕ_2=45° ODF.

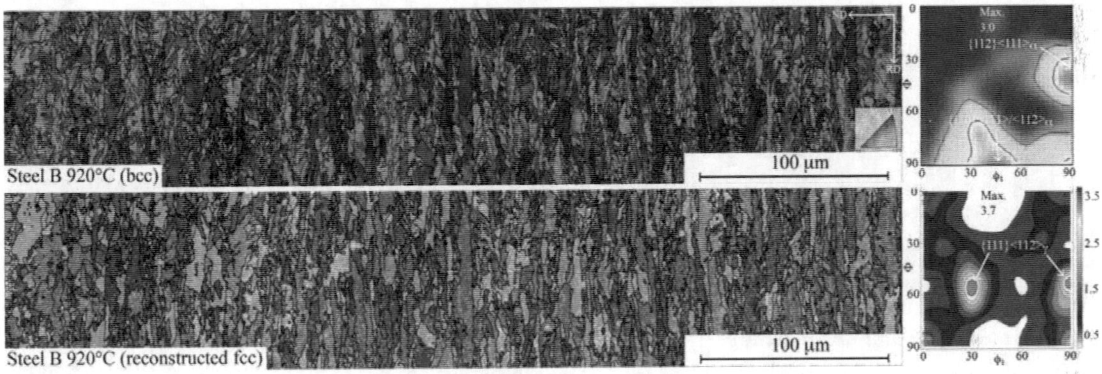

Figure 7. Subsurface of Steel B with 920 °C FRT (martensitic). (above) Original EBSD band contrast map with high-angle boundaries (>15°) and IPF colored in the RD direction and bcc ϕ_2=45° ODF and (below) reconstructed parent austenite grain map in Euler coloring and fcc ϕ_2=45° ODF. On the left hand side is located surface of the strip.

Figure 8. Subsurface of Steel B with 820 °C FRT (upper bainitic). (above) Original EBSD band contrast map with high-angle boundaries (>15°) and IPF colored in the RD direction and bcc ϕ_2=45° ODF and (below) reconstructed parent austenite grain map in Euler coloring and fcc ϕ_2=45° ODF. On the left hand side is located surface of the strip.

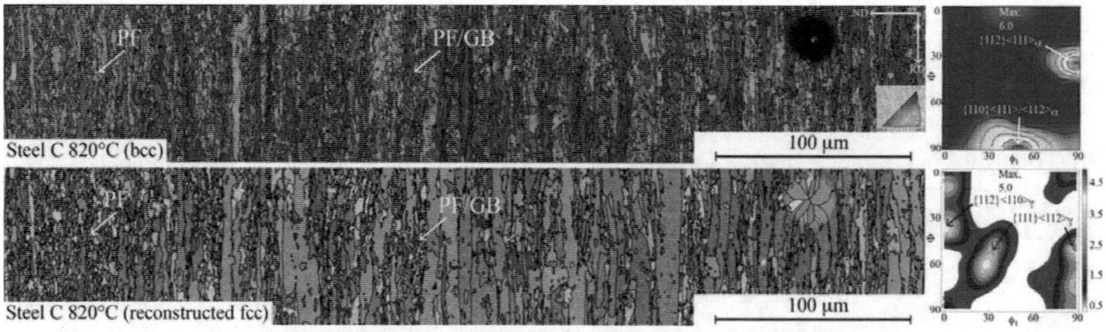

Figure 9. Subsurface of Steel C with 820 °C FRT (granular bainitic). (above) Original EBSD band contrast map with high-angle boundaries (>15°) and IPF colored in the RD direction and bcc ϕ_2=45° ODF and (below) reconstructed parent austenite grain map in Euler coloring and fcc ϕ_2=45° ODF. On the left hand side is located surface of the strip.

4. Summary

The purpose of this research was to investigate the effect of various microstructures on the parent austenite reconstruction and crystallographic texture at the centerline and subsurface of hot-rolled and direct-quenched ultrahigh-strength steels. The main observations and conclusions of the work can be summarized as follows:

- An increase in the total reduction in the non-recrystallization temperature region in conjunction with a lowering of the finishing rolling temperature (FRT) increased austenite pancaking.
- A decrease of FRT increased the formation of softer microstructures such as ferrite and granular bainite in the subsurface layers. At high FRTs, the microstructures at the centerline consisted mainly of auto-tempered martensite especially in the case of higher manganese content. Lowering FRT increased the fractions of granular bainite and ferrite at the expense of martensite and upper bainite in both the central and subsurface parts of the strip thickness.
- Prior austenite pancaking increases the intensities of the plane strain ~{554}$<225>_\alpha$, ~{112}$<110>_\alpha$ and ~{112}$<131>_\alpha$ texture components at the centerline and shear strain ~{112}$<111>_\alpha$ and ~{110}$<112>_\alpha$ - {110}$<111>_\alpha$ texture components in the surface layers.
- The current Matlab reconstruction code works in martensitic and upper bainitic microstructures even with the highly pancaked prior austenite morphology, although not in granular bainite or ferritic microstructures. The code also generated the expected austenite crystallographic texture from the bcc textures.
- For martensitic microstructures, the maximum texture intensities increased after austenite reconstruction while the austenite reconstructed from upper and granular bainite had lower maximum texture intensities than the original bcc microstructures.
- No explanation has been found for why upper bainite shows more intense texture than granular bainite even with same prior austenite pancaking.

References

[1] Kaijalainen A, Liimatainen M, Kesti V, Heikkala J, Liimatainen T and Porter D 2016 *Metall. Mater. Trans. A.* **47** 4175
[2] Kaijalainen A, Suikkanen P, Karjalainen LP and Jonas JJ 2014 *Metall. Mater. Trans. A.* **45** 1273
[3] Cayron C, Artaud B and Briottet L 2006 *Mater. Charact.* **57** 386
[4] Miyamoto G, Iwata N, Takayama N and Furuhara T 2011 *ISIJ Int.* **51** 1174
[5] Kurdjumow V and Sachs G 1930 *Zeitschrift Für Phys.* **64** 325
[6] Nishiyama Z 1934 *Sci. Rep. Tohoku Imp. Univ.* **23** 637

[7] Wassermann G 1933 *Arch. Eisenhuttenw.* **16** 647
[8] Ray RK and Jonas JJ 1990 *Int. Mater. Rev.* **35** 1
[9] Kestens L and Jonas JJ 2005 *ASM Handbook vol 14A Metalworking: Bulk Forming*, ed SL Semiatin (Materials Park: ASM International) p 685
[10] Wittridge N and Jonas JJ 2000 *Acta Mater.* **48** 2737
[11] Nyyssönen T, Isakov M, Peura O and Kuokkala VT 2016 *Metall. Mater. Trans. A.* **47** 2587
[12] Morito S, Huang X, Furuhara T, Maki T, Hansen N, 2006 *Acta Mater.* **54** 5323
[13] van Dongen S 2000 *Graph Clustering by Flow Simulation* University of Utrecht
[14] Gomes E and Kestens L 2015 *IOP Conf. Ser. Mater. Sci. Eng.* **82** 12059
[15] Nyyssönen T 2017 *Quenching and Partitioning of High-Aluminum Steels* Tampere University of Technology
[16] Kaijalainen A, Suikkanen P, Karjalainen LP and Porter D 2016 *Mater. Sci. Eng. A.* **654** 151
[17] Bain E and Dunkirk N 1924 *Trans. AIME* **70** 25
[18] Greninger A and Troiano A 1949 *Trans. AIME* **185** 590

IOP Conf. Series: Materials Science and Engineering **375** (2018) 012027 doi:10.1088/1757-899X/375/1/012027

Parameterization of rotations in reference frames with redundant crystallographic axes

Adam Morawiec

Polish Academy of Sciences, Institute of Metallurgy and Materials Science, Reymonta 25, 30-059, Krakow, Poland

E-mail: `nmmorawi@cyf-kr.edu.pl`

Abstract. Parameterizations of rotations in three-dimensional space are fundamental for description of crystallographic textures. Such parameterizations (e.g., Rodrigues parameters) are usually specified using orthonormal coordinate systems, whereas bases of crystal lattices are generally non-orthogonal. In the case of crystals with hexagonal or rhombohedral lattices, the reference frames involve redundant crystallographic axes, and hence a question arises about feasibility of the generalization of the rotation parameterizations to such frames. The definition of Rodrigues parameters can be extended so they are directly linked to non-Cartesian bases of crystal lattices. The new Rodrigues parameters are contra- or covariant components of vectors specified with respect to exactly the same lattice basis as atomic positions in a unit cell. The generalized formalism allows for using redundant crystallographic axes. Also the orientation matrices can be represented in such frames. The Rodrigues parameterization in non-Cartesian coordinate frames is convenient for crystallographic applications because the generalized parameters are directly related to indices of rotation-invariant lattice directions and to Miller indices of rotation-invariant lattice planes. In the case of the hexagonal and rhombohedral lattices, the frames with redundant axes are used to account for lattice symmetry, but one may apply such frames for other reasons. They can be convenient for handling arbitrary symmetries, in particular symmetries arising in description of some phenomena or symmetries of physical processes. The practicality of frames is illustrated by an alternative description of body-centered lattices, a formula for lattice rotation in deformation by slip, and a new interpretation of indexing of single crystal diffraction patterns.

1. Introduction

Texture analysis relies heavily on the notions of reference frames and rotations in three-dimensional space [1, 2]. The usual approach is to represent proper rotations by special orthogonal matrices. Also rotation vectors are a commonly used representation. In particular, the so-called Rodrigues vector is convenient in analysis of crystal misorientations.

Special orthogonal matrices and Rodrigues vectors are defined in reference to Cartesian systems [1, 3]. On the other hand, bases of crystalline lattices are often non-orthogonal. Rotation matrix formalism in non-orthogonal bases has been used in crystallography for a long time; cf. [4]. The issue of referring Rodrigues vectors directly to oblique frames was considered only recently in [5]. The latter paper describes a more general formalism which allows to deal with rotations not only in arbitrary lattice bases comprised of linearly independent vectors but also in frames with redundant crystallographic axes. This formalism encompasses the description of rotations in the hexagonal and rhombohedral cases with four axes of the frame. It can be

IOP Conf. Series: Materials Science and Engineering **375** (2018) 012027 doi:10.1088/1757-899X/375/1/012027

applied to any finite number (≥ 3) of axes, and it can be seen as a generalization of the classic schemes of indexing of lattice planes and directions. The concept of indexing in systems with redundant crystallographic axes has been considered before [6, 7, 8]. For complete understanding of indexing in systems with redundant axes, one needs to establish a link between the frame and its dual on one side, and the lattice and its reciprocal on the other, and this issue has been clarified in [5]. The formalism relies on the equivalence between the indices of nodes of direct and reciprocal lattices and contra- and co-variant vector coordinates [4, 9]. In particular, the generalized Rodrigues parameters are defined as contra- or co-variant components of vectors in the physical space of the crystal.

The paper illustrates some simple but quite different applications of the frames: these are a symmetric description of body-centered lattices, a new formula for lattice rotation in deformation by slip, and an original interpretation of ab initio indexing of single crystal diffraction patterns. The nomenclature and notation of the paper are close to those of [5] and some chapters of the International Tables for Crystallography [4, 9]. The summation convention is assumed, i.e., there is a summation over each index which appears in a term twice, as a subscript and a superscript. At the outset, the concepts of [5] are briefly reformulated so one can easily grasp the essence of the formalism.

2. Rotations in three dimensions and frames

Rotation about $\mathbf{n}$ by the angle θ is frequently represented by the matrix

$$\mathbf{R} = \mathbf{g}\,\cos\theta + \mathbf{n}\otimes\mathbf{n}\,(1-\cos\theta) + \boldsymbol{\epsilon}\cdot\mathbf{n}\,\sin\theta \tag{1}$$

or Rodrigues vector $\mathbf{r} = \mathbf{n}\tan(\theta/2)$. Composition of rotations represented by matrices $\mathbf{S}$ and $\mathbf{P}$ is represented by matrix product of $\mathbf{P}$ and $\mathbf{S}$. Composition of rotations parametrized by Rodrigues vectors $\mathbf{p}$ and $\mathbf{s}$ is given by $\mathbf{p}\circ\mathbf{s} = (\mathbf{p}+\mathbf{s}+\mathbf{p}\times\mathbf{s})/(1-\mathbf{p}\cdot\mathbf{s})$. The vector $\mathbf{w}$ resulting form the rotation of $\mathbf{v}$ by a rotation represented by matrix $\mathbf{R}$ and Rodrigues vector $\mathbf{r}$ is $\mathbf{w} = \mathbf{R}\cdot\mathbf{v} = (-\mathbf{r})\circ\mathbf{v}\circ\mathbf{r}$. These two representations are linked by

$$\mathbf{R} = \frac{1}{1+\mathbf{r}\cdot\mathbf{r}}\left(\mathbf{g}\,(1-\mathbf{r}\cdot\mathbf{r}) + 2\,\mathbf{r}\otimes\mathbf{r} + 2\boldsymbol{\epsilon}\cdot\mathbf{r}\right) \quad \text{and} \quad \mathbf{r} = \boldsymbol{\epsilon} : \mathbf{R}/(1+\mathrm{tr}\,\mathbf{R})\ .$$

For small angle rotations $\mathbf{R} \approx \mathbf{g} + \boldsymbol{\Omega}$, where $\boldsymbol{\Omega}$ is linked to Rodrigues vector via $\boldsymbol{\Omega} = 2\boldsymbol{\epsilon}\cdot\mathbf{r}$ and $\mathbf{r} = \boldsymbol{\epsilon} : \boldsymbol{\Omega}/4$. The formulas listed above are well known. They are usually interpreted in Cartesian reference systems with $\mathbf{g}$ being the identity matrix and $\boldsymbol{\epsilon}$ representing the permutation symbol. It is interesting, however, that they can also be considered in more general terms. First, the vectors and tensors can be given in rectilinear non-orthonormal systems, in particular, in a basis of a trigonal crystal lattice. Second, the above expressions can be applied using frames with redundant axes. In particular, they are valid in the conventional four-axis frames used for describing hexagonal and rhombohedral crystals. Clearly, this general approach requires a proper interpretation of $\mathbf{g}$ and $\boldsymbol{\epsilon}$.

2.1. Overcomplete basis sets

For our purpose, a frame is a set of $M \geq 3$ non-coplanar vectors $\mathbf{a}_\mu$, where $\mu = 1, 2, ..., M$, and no two vectors are collinear. The metric tensor is defined as

$$g_{\mu\nu} = \mathbf{a}_\mu \cdot \mathbf{a}_\nu\ .$$

The rank of the corresponding matrix is 3. Let $g^{\mu\nu}$ be the components of the Moore-Penrose pseudoinverse [10] of the matrix $[g_{\mu\nu}]$. The vectors

$$\mathbf{a}^\mu = g^{\mu\nu}\mathbf{a}_\nu$$

constitute the frame canonically dual to $\mathbf{a}_\mu$. They satisfy the relationships $\mathbf{a}^\mu \cdot \mathbf{a}^\nu = g^{\mu\nu}$ and $g_{\mu\nu}\mathbf{a}^\nu = \mathbf{a}_\mu$. With $g_\mu{}^\nu = \mathbf{a}_\mu \cdot \mathbf{a}^\nu = g_{\mu\rho}\, g^{\rho\nu}$, one has $g_\mu{}^\nu = g^\nu{}_\mu$. The idempotent matrix corresponding to $g_\mu{}^\nu$ is equal to its pseudoinverse, its trace $g_\mu{}^\mu$ equals 3, and there occurs $g_{\mu\kappa}g^\kappa{}_\nu = g_{\mu\nu}$ and $g^{\mu\kappa}g_\kappa{}^\nu = g^{\mu\nu}$. Moreover, one has

$$g^\mu{}_\nu \mathbf{a}^\nu = \mathbf{a}^\mu \ , \quad g_\mu{}^\nu \mathbf{a}_\nu = \mathbf{a}_\mu \ . \tag{2}$$

With $M > 3$, the decomposition of a given vector, say $\mathbf{v}$, into linear combinations

$$v^\mu \mathbf{a}_\mu = \mathbf{v} = v_\mu \mathbf{a}^\mu$$

of $\mathbf{a}_\mu$ or $\mathbf{a}^\mu$ is ambiguous. It is made unique by imposing the constraints

$$g^\kappa{}_\nu v^\nu = v^\kappa \ , \quad g_\kappa{}^\nu v_\nu = v_\kappa \ . \tag{3}$$

With these conditions satisfied, one has

$$v^\mu = \mathbf{v} \cdot \mathbf{a}^\mu \ , \quad v_\mu = \mathbf{v} \cdot \mathbf{a}_\mu \ ,$$

$$g_{\mu\nu} v^\nu = v_\mu \ , \quad g^{\mu\nu} v_\nu = v^\mu \ .$$

This formalism belongs to the well developed theory of frames; see, e.g., [11] and references therein. The principal applications of frames are in signal and image processing.

The frames of particular type described above (non-orthogonal overcomplete sets of vectors in finite-dimensional inner-product space) can also be applied in crystallography. The redundancy of frames has some advantages. First, frames improve computational robustness to random errors as information is spread over a larger number of vector components. More importantly, the redundancy of frames provides some flexibility; for instance, a frame can match a given symmetry, as it is in the case of hexagonal and rhombohedral lattices. Finally, from the crystallographic standpoint, the use of finite frames provides a panoramic view on indexing of lattice planes and directions.

Going back to expressions opening this section and the description of rotations in frames, the tensors $\mathbf{g}$ and $\boldsymbol{\epsilon}$ are shorthand symbols for the metric tensor and Levi–Civita tensor with components $\epsilon_{\kappa\mu\nu} = (\mathbf{a}_\kappa \times \mathbf{a}_\mu) \cdot \mathbf{a}_\nu$ and $\epsilon^{\kappa\mu\nu} = g^{\kappa\alpha}g^{\mu\beta}g^{\nu\gamma}\epsilon_{\alpha\beta\gamma} = (\mathbf{a}^\kappa \times \mathbf{a}^\mu) \cdot \mathbf{a}^\nu$. For instance, the explicit form of eq.(1) is $R^\mu{}_\nu = g^\mu{}_\nu \cos\theta + n^\mu n_\nu (1 - \cos\theta) + g^{\mu\kappa}\epsilon_{\kappa\nu\rho}n^\rho \sin\theta$. The entries $R^\mu{}_\nu$ satisfy the orthogonality conditions $g_{\kappa\rho}R^\kappa{}_\mu R^\rho{}_\nu = g_{\mu\nu}$ with the pseudodeterminant of $[R^\mu{}_\nu]$ equal to +1. Similarly, for small angle rotations, one can write

$$\Omega_{\mu\nu} = 2\epsilon_{\mu\nu\kappa}r^\kappa \quad \text{and} \quad r^\kappa = \epsilon^{\kappa\mu\nu}\Omega_{\mu\nu}/4 \ . \tag{4}$$

The explicit forms of other equations are available in [5]. They easily follow from the covariance principle and mechanical positioning of sub- and super-scripts.

3. Frames and crystallographic lattices

The representations of rotations in frames can be used in various fields, but our considerations are confined to the realm of classic crystallography in order to conform to the subject matter of ICOTOM. In crystallographic applications, the bases of reference systems are directly linked to crystal lattice bases. Also frames can be used for determining lattices. This has been unknowingly done for a long time in the case of symmetry-based four-index schemes of Miller-Bravais [12] and Weber [13].

The crucial point for using frames in description of a crystal lattice is to specify the link between the frame vectors $\mathbf{a}_\mu$ and the lattice nodes. Clearly, not every frame naturally induces

IOP Conf. Series: Materials Science and Engineering **375** (2018) 012027 doi:10.1088/1757-899X/375/1/012027

a lattice (as for some sets of vectors $\mathbf{a}_\mu$, integer combinations of these vectors can be arbitrarily close to each other), and it is easier to consider the link from the perspective of a given lattice. Two types of the relationship between the lattice nodes and the vectors $\mathbf{a}_\mu$ are more natural than other.

First, the standard or canonical case embodying the classic Miller-Bravais and Weber indexing is based on the following rule: $\mathbf{a}_\mu$ are direct lattice vectors, and *arbitrary* integer combinations of $\mathbf{a}_\mu$ constitute the lattice. The integer coefficients may fail to satisfy conditions (3). The lattice vectors, when expressed in the frame $\mathbf{a}_\mu$ as $q^\mu \mathbf{a}_\mu$ with $q^\mu = g^\mu{}_\nu q^\nu$, have rational (i.e., not necessarily integer) components q^μ. The above arrangement between the direct lattice nodes and $\mathbf{a}_\mu$ determines the relationship between the reciprocal lattice and the dual frame $\mathbf{a}^\mu$: the reciprocal lattice consists of linear combinations of $\mathbf{a}^\mu$ with integer coefficients satisfying conditions (3). One must be aware, however, that an arbitrary integer combination of $\mathbf{a}^\mu$ may not be a vector of the reciprocal lattice. In particular, the vector $\mathbf{a}^\mu$ of the dual frame may not be a vector of the reciprocal lattice.

Clearly, the above scheme is not the only possible. The other natural approach would be to exchange the roles of the direct and the reciprocal lattices. In such a 'dual-canonical' case, the direct lattice vectors would be linear combinations of $\mathbf{a}_\mu$ with integer coefficients satisfying eqs.(3), and the reciprocal lattice vectors would be linear combinations of $\mathbf{a}^\mu$ with arbitrary integer coefficients.

In what follows, the canonical scheme is assumed. Clearly, a frame determines a lattice, and the lattice can be based on numerous different frames. Classic lattice bases constitute particular types of frames. Having a given lattice frame $\mathbf{a}_\mu$, one can specify the lattice direction $q^\mu \mathbf{a}_\mu$ using the generalized Weber indices

$$[q^1\, q^2\, \ldots\, q^M] \propto [u\, v\, \ldots\, w]\ .$$

Similarily, the generalized Miller-Bravais indices

$$[k_1\, k_2\, \ldots\, k_M] \propto (h\, k\, \ldots\, l)$$

specify the plane normal to the vector $k_\mu \mathbf{a}^\mu$. Both q^μ and k_μ are expected to satisfy eqs.(3).

Now, assuming a crystal lattice is based on the frame $\mathbf{a}_\mu$, the Rodrigues parameters of a rotation have a simple crystallographic interpretation. Since covariant vector components are proportional to indices of a lattice plane, the covariant components r_μ are proportional to generalized Miller-Bravais indices $(h\, k\, \ldots\, l)$ of the rotation-invariant lattice plane. Similarly, the contravariant components r^μ are proportional to generalized Weber indices $[u\, v\, \ldots\, w]$ of the rotation-invariant lattice direction. Calculating the components of $\mathbf{r}$ in the reference system constituting the lattice frame is a natural way to get indices of the rotation-invariant lattice planes and directions in low-symmetry crystals.

4. Applications

With the use of frames, some old lattice-related issues can be viewed in a new light. Three example applications are presented below. We begin with a note on using the tetrahedral frame for describing body centered lattices.

4.1. Tetrahedral frame

Frames can be used to account for lattice centering in a 'symmetric' manner. Take the simplest case of the body-centered cubic (bcc) lattice. It is naturally described using the 'tetrahedral' frame. Let the vectors $\mathbf{a}_\mu$ ($\mu = 1, 2, 3, 4$) be related to the conventional orthonormal basis $\mathbf{e}_J$ by

$$\mathbf{a}_1 = a(\mathbf{e}_1+\mathbf{e}_2+\mathbf{e}_3)/2,\ \mathbf{a}_2 = a(\mathbf{e}_1-\mathbf{e}_2-\mathbf{e}_3)/2,\ \mathbf{a}_3 = a(-\mathbf{e}_1+\mathbf{e}_2-\mathbf{e}_3)/2,\ \mathbf{a}_4 = a(-\mathbf{e}_1-\mathbf{e}_2+\mathbf{e}_3)/2; \quad (5)$$

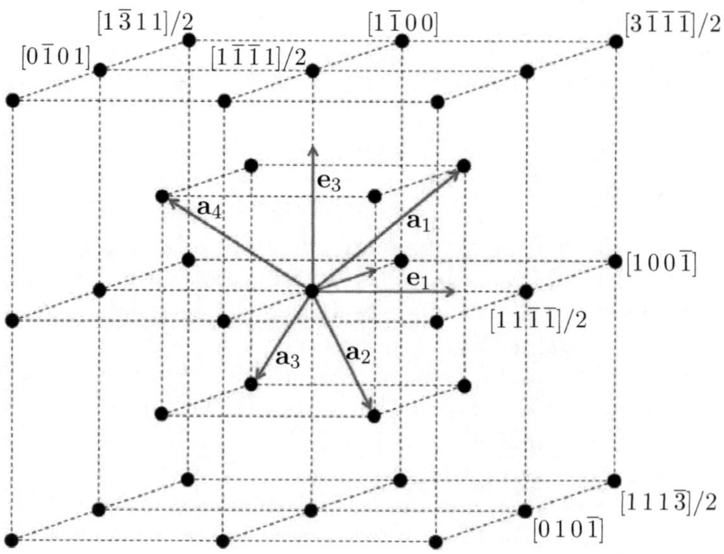

Figure 1. Schematic of body centered cubic lattice and the tetrahedral frame $\mathbf{a}_\mu$. The red and blue vectors constitute the tetrahedral frame and the basis of the Cartesian reference system, respectively. Indices of some lattice nodes in the tetrahedral frame are shown.

see Fig.1. Integer combinations of these vectors indicate nodes of a bcc lattice with the conventional lattice parameter a.[1] The metric corresponding to the above frame is

$$g_{\mu\nu}/a^2 = a^2 g^{\mu\nu} = g_\mu{}^\nu = \delta_\mu{}^\nu - 1/4 \,, \tag{6}$$

and eqs.(3) take the explicit form $v^1 + v^2 + v^3 + v^4 = 0 = v_1 + v_2 + v_3 + v_4$. With the canonical approach, the direct lattice is built of all integer combinations of $\mathbf{a}_\mu$, but the integer coefficients may violate conditions (3). For instance, the direct lattice node at $\mathbf{v} = a\,\mathbf{e}_1 + a\,\mathbf{e}_2 + a\,\mathbf{e}_3$ can be expressed as the integer combination $2\mathbf{a}_1 + 0\mathbf{a}_2 + 0\mathbf{a}_3 + 0\mathbf{a}_4$, but the integer coefficients do not satisfy eqs.(3). The coordinates of $\mathbf{v}$ satisfying (3) are rational: $\mathbf{v} = (3/2)\mathbf{a}_1 - (1/2)\mathbf{a}_2 - (1/2)\mathbf{a}_3 - (1/2)\mathbf{a}_4$.

The frame dual to (5) is given by $\mathbf{a}^\mu = \mathbf{a}_\mu/a^2$. The reciprocal lattice is built of combinations of vectors $\mathbf{a}^\mu$ with integer coefficients satisfying conditions (3). This lattice corresponds to linear combinations of vectors $\mathbf{e}^J/a$ with integer coefficients such that their sum is even. Thus, as expected, it is the face-centered cubic (fcc) lattice with the lattice parameter of $2/a$.[2] With the four-index labeling of the reciprocal lattice nodes, there are no diffraction selection rules (i.e., no centering-caused systematic extinctions); an arbitrary quadruple of integers summing up to zero corresponds to a diffraction reflection and vice versa. For illustration, here are some corresponding three- and four-index designations of reflecting planes:

Orthogonal basis	(000)	(002)	(011)	(112)	(022)	(013)	(222)
Tetrahedral frame	(0000)	$(1\bar{1}\bar{1}1)$	$(1\bar{1}00)$	$(2\bar{1}\bar{1}0)$	$(2\bar{2}00)$	$(2\bar{2}\bar{1}1)$	$(3\bar{1}\bar{1}\bar{1})$.

The modification of the above considerations to body-centered tetragonal or body-centered orthorhombic lattices is straightforward, and the indexing scheme of all three body-centered cases is the same.

[1] Such reference system was discussed by Rogers and Klyne [14] in relation to tetrahedrally arranged bonds, i.e., a network of atoms, not a crystal lattice. Moreover, the coordinates v^μ are related to but different from tetrahedral coordinates used in applications of finite element method (e.g., [15], p.106).

[2] If $a = 1$, the frame $\mathbf{a}_\mu$ is self-dual ($\mathbf{a}^\mu = \mathbf{a}_\mu$), but obviously the lattice is not self-reciprocal.

IOP Conf. Series: Materials Science and Engineering **375** (2018) 012027 doi:10.1088/1757-899X/375/1/012027

4.2. Lattice rotation in deformation by slip

Frames can be applied to accommodate other symmetries. In particular, these can be symmetries arising in characterization of phenomena. For instance, in description of orientation changes during plastic deformation, convenient frames can be devised based on characteristic directions and/or planes of slip or twinning systems. The tetrahedral frame (5) spans the Thompson tetrahedron illustrating the fcc $\{111\}\langle 011\rangle$ slip systems [16], and these slip systems could be described using tetrahedral coordinates, but also other frames can be suitable.

Let us consider the general case without the limitation to the fcc lattice. Single slip results in a rotation of the lattice around certain axis, and this axis identifies the slip system. The unit vector along the axis will be taken as $\mathbf{a}_\mu$. The total rotation caused by slip on multiple systems is given in the Cartesian reference frame by the well known formula (see, e.g., [17])

$$\Omega_{IJ} = \sum_\mu \gamma^\mu \left(n_I^\mu b_J^\mu - b_I^\mu n_J^\mu\right)/2 \,, \tag{7}$$

where b_I^μ and n_I^μ are Cartesian components of the unit vectors along the slip direction and slip plane normal, respectively, and γ^μ is the shear on the μ-th system. The (small) angle of rotation is $\gamma^\mu/2$. Based on eq.(4), the relationship (7) can be written in the form $\Omega_{IJ} = 2\varepsilon_{IJK} r^K$, where

$$r^K = \sum_\mu (\varepsilon^{KIJ} n_I^\mu b_J^\mu)\,\gamma^\mu/4 = (\mathbf{a}_\mu \cdot \mathbf{e}^K)\,\gamma^\mu/4 = (\mathbf{a}_\mu\,\gamma^\mu/4)\cdot \mathbf{e}^K$$

are Cartesian components of the Rodrigues vector $\mathbf{r} = r^K \mathbf{e}_K$. The vector $\mathbf{r}$ can be expressed in the frame $\mathbf{a}_\mu$ as $\mathbf{r} = r^\mu \mathbf{a}_\mu$ with $r^\mu = \mathbf{r}\cdot\mathbf{a}^\mu$. Now, in this representation, equation (7) takes the simple form[3]

$$r^\mu = g^\mu{}_\nu \gamma^\nu/4 \,. \tag{8}$$

The components r^μ satisfy eqs.(3), whereas γ^μ are unconstrained. With given shears γ^μ, one can obtain the resultant rotation via $\mathbf{r} = r^\mu \mathbf{a}_\mu = (g^\mu{}_\nu \gamma^\nu/4)\mathbf{a}_\mu = \gamma^\nu(\mathbf{a}_\mu g^\mu{}_\nu)/4 = \gamma^\mu \mathbf{a}_\mu/4$. This relationship can be seen as a compact form of eq.(4) with $\mathbf{r} = \boldsymbol{\epsilon} : \boldsymbol{\Omega}/4 = \sum_\mu (\gamma^\mu/4)\boldsymbol{\epsilon} : (\mathbf{n}_\mu \times \mathbf{b}_\mu) = \gamma^\mu \mathbf{a}_\mu/4$ and $\mathbf{a}_\mu = \boldsymbol{\epsilon} : (\mathbf{n}_\mu \times \mathbf{b}_\mu)$; compare, e.g., eqs.(1–2) of [18]. The convenience of using frames lies in the presence of $g^\mu{}_\nu$ on the right hand side of eq.(8); by the properties of the pseudoinverse [10], if the resultant rotation vector $\mathbf{r}$ is known, the shears are given by

$$\gamma^\mu = 4\,\mathbf{r}\cdot\mathbf{a}^\mu + \xi^\mu \,,$$

where the terms ξ^μ satisfy $g^\mu{}_\nu \xi^\nu = 0$, or $\xi^\mu = x^\mu - g^\mu{}_\nu x^\nu$ with x^μ being arbitrary numbers. In the case of $\xi^\mu = 0$, the shears $\gamma^\mu = 4\,\mathbf{r}\cdot\mathbf{a}^\mu$ are optimally distributed over the slip systems in the sense that $\sum_\mu (\gamma^\mu)^2$ is the smallest possible.

As a concrete example, let us consider the aforementioned case of fcc metals with slip on $\{111\}$ planes in $\langle 011\rangle$ directions leading to rotations about $\langle 112\rangle$ rotation axes. The frame consists of twelve vectors

$$\begin{array}{llll}
\mathbf{a}_1 = [2\,1\,1]a/2 & \mathbf{a}_4 = [2\,\bar{1}\,\bar{1}]a/2 & \mathbf{a}_7 = [\bar{2}\,1\,\bar{1}]a/2 & \mathbf{a}_{10} = [\bar{2}\,\bar{1}\,1]a/2 \\
\mathbf{a}_2 = [1\,2\,1]a/2 & \mathbf{a}_5 = [1\,\bar{2}\,\bar{1}]a/2 & \mathbf{a}_8 = [\bar{1}\,2\,\bar{1}]a/2 & \mathbf{a}_{11} = [\bar{1}\,\bar{2}\,1]a/2 \\
\mathbf{a}_3 = [1\,1\,2]a/2 & \mathbf{a}_6 = [1\,\bar{1}\,\bar{2}]a/2 & \mathbf{a}_9 = [\bar{1}\,1\,\bar{2}]a/2 & \mathbf{a}_{12} = [\bar{1}\,\bar{1}\,2]a/2
\end{array}$$

with $a = 2\times 6^{-1/2}$ so the vectors are of unit magnitude.[4] The tensors characterizing the frame have the form

$$\frac{1}{6\,a^2}\,[g_{\mu\nu}] = 6\,a^2\,[g^{\mu\nu}] = [g_\mu{}^\nu] = \frac{1}{24}\begin{bmatrix} l_0 & l_1 & l_2 & l_3 \\ & l_0 & l_3 & l_2 \\ & & l_0 & l_1 \\ \text{sym} & & & l_0 \end{bmatrix} ,$$

[3] This expression follows from $\mathbf{r}\cdot\mathbf{a}^\mu = (r^K\mathbf{e}_K)\cdot\mathbf{a}^\mu = (\mathbf{a}_\nu\cdot\mathbf{e}^K)(\mathbf{e}_K\cdot\mathbf{a}^\mu)(\gamma^\nu/4) = (\mathbf{a}^\mu\cdot\mathbf{a}_\nu)(\gamma^\nu/4)$ and the definition of $g^\mu{}_\nu$.

[4] This frame is linked to the fcc (direct) lattice with lattice parameter a.

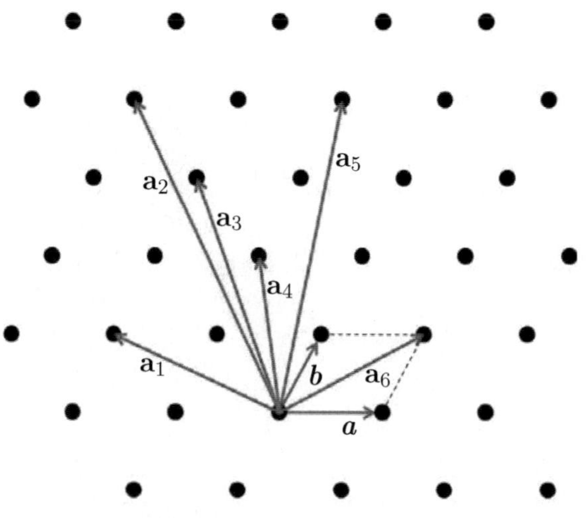

Figure 2. Schematic two-dimensional illustration of the indexing problem. Given positions of some (reciprocal) lattice nodes, determine the the parameters of the lattice, or given a frame (red vectors $\mathbf{a}_\mu$), get the (blue) vectors $\boldsymbol{a}$ and $\boldsymbol{b}$ of the lattice basis.

where

$$l_0 = \begin{bmatrix} 6 & 5 & 5 \\ & 6 & 5 \\ \text{sym} & & 6 \end{bmatrix}, \; l_1 = \begin{bmatrix} 2 & -1 & -1 \\ & -4 & -3 \\ \text{sym} & & -4 \end{bmatrix}, \; l_2 = \begin{bmatrix} -4 & -1 & -3 \\ & 2 & -1 \\ \text{sym} & & -4 \end{bmatrix}, \; l_3 = \begin{bmatrix} -4 & -3 & -1 \\ & -4 & -1 \\ \text{sym} & & 2 \end{bmatrix}.$$

With this $[g^{\mu\nu}]$, the dual frame is $\mathbf{a}^\mu = \mathbf{a}_\mu/(6a^2) = \mathbf{a}_\mu/4$. Thus, the principal part $(4\,\mathbf{r}\cdot\mathbf{a}^\mu)$ of the shear on the μ-th system is given simply by the scalar product of the rotation vector $\mathbf{r}$ and the unit vector $\mathbf{a}_\mu$ along the rotation axis of the system.

Clearly, other tensor quantities used for describing the slip-based deformation can also be expressed using frames, but such broader considerations are beyond the scope of this short communication.

4.3. Indexing of single crystal diffraction patterns

The ab initio indexing problem can be stated as follows: having measured positions of some diffraction reflections, get the parameters of the crystal lattice, and ascribe Miller indices to the reflections. One can formulate it more formally: having a number of vectors of the crystal reciprocal lattice, determine the basis of the crystal lattice; see, e.g., [19] and references therein. It is easy to notice that the input data (vectors of the reciprocal lattice) constitute a dual frame (an overcomplete set of basis vectors), and the problem is to reduce this frame to a conventional basis of the reciprocal lattice built of three linearly independent vectors; see Fig.2. This is similar to the (Pohst) problem of computer science: given some lattice vectors which exceed in number the space dimension, find a lattice basis made of short vectors, or in other words, given a frame, find a basis made of short vectors. This problem is solved by a modified version of LLL lattice reduction algorithm (sometimes referred to as MLLL) by Pohst [20]. With this solution, the center of the indexing problem is shifted to accounting for experimental inaccuracies in input data. This explicitely validates the frequently postulated property that indexing is simple for accurate data.

IOP Conf. Series: Materials Science and Engineering **375** (2018) 012027 doi:10.1088/1757-899X/375/1/012027

5. Final remarks

Summarizing, frames open an opportunity to take a new look at the subject of reference systems and rotation parameterizations. Since the restrictions on the vectors $\mathbf{a}_\mu$ are week, frames can match various geometries. They allow for general but also very concise formulation of problems. Therefore, one can expect numerous texture-related applications of frames. The most obvious use is in crystallographic software involving plane and direction indices. This includes software for analysis of orientation relationships, coincident lattices, slip systems, twinning systems et cetera. The most basic advantage of using frames is that the four-index (Miller-Bravais + Weber) system standardly used for crystals of the hexagonal family can be treated on an equal footing with the three-index system used for other crystal families.

References

[1] Bunge H J 1982 *Texture Analysis in Materials Science* (Butterworth's, London)
[2] Morawiec A 2004 *Orientations and Rotations. Computations in Crystallographic Textures* (Springer–Verlag, Berlin)
[3] Frank F C 1988 *Metall. Trans. A* **19** 403–408
[4] Shmueli U 2006 *International Tables for Crystallography, Vol. B, Section 1.1* ed Shmueli U (Springer, Dodrecht) pp 2–9
[5] Morawiec A 2016 *Acta Cryst. A* **72** 548–556
[6] Frank F C 1965 *Acta Cryst.* **18** 862–866
[7] Nicholas J F and Segall R L 1970 *Acta Cryst. A* **26** 522–527
[8] Mackay A L 1977 *Acta Cryst. A* **33** 212–215
[9] Sands D E 2006 *International Tables for Crystallography, Vol. B, Section 3.1* ed Shmueli U (Springer, Dodrecht) pp 348–352
[10] Ben-Israel A and Greville T N E 2003 *Generalized Inverses, Theory and Applications* (Springer, New York)
[11] Casazza P G, Kutyniok G and Philipp F 2013 *Finite Frames. Theory and Applications* ed Casazza P G and Kutyniok G (Birkhuser, Basel) pp 1–53
[12] Bravais M 1866 *Études cristallographiques* (Gauthier-Villars, Paris)
[13] Weber L 1922 *Z. Kristallogr.* **57** 200–203
[14] Rogers D and Klyne W 1972 *Tetrahedron Lett.* **13** 1441–1444
[15] Rao S S (ed) 2005 *The Finite Element Method in Engineering* fourth edition ed (Burlington: Butterworth-Heinemann)
[16] Thompson N 1953 *Proc. Phys. Soc. B,* **66** 481–492
[17] Winther G 2008 *Acta Mater.* **56** 1919–1932
[18] Kashihara K and Shibayanagi T 2009 *Mater. Trans.* **50** 2192–2200
[19] Morawiec A 2017 *J. Appl. Cryst.* **50** 647–650
[20] Pohst M 1987 *J. Symb. Comput.* **4** 123–127

Crystal plasticity and continuum mechanics-based modelling of deformation and recrystallization textures in aluminum alloys

Jurij J. SIDOR

Eötvös Loránd University (ELTE), Faculty of Informatics, Savaria Institute of Technology, Károlyi Gáspár tér 4, 9700 Szombathely, Hungary

Corresponding author's e-mail: js@inf.elte.hu

Abstract. This contribution provides a brief overview on modeling the deformation textures by combining both crystal plasticity approaches and models based on principles of continuum mechanics, dealing with a flow of a material during deformation. The behavior of Al alloys during deformation is analyzed by means of well-established crystal plasticity theories such as full constraints Taylor, advanced Lamel, visco-plastic self-consistent and Cluster V model. As a first approximation, the deformation flow in rolling is approximated by plane strain compression, however, it is shown that an improvement in texture prediction is reached by considering strain heterogeneities, evolved across the thickness of a rolled sheet. The effective modeling strategies employed in texture simulations are discussed in the current contribution. It is analyzed how microstructural heterogeneities might influence the evolution of recrystallization texture and plastic strain ratio in aluminum alloys.

Keywords: texture, Al alloys, deformation, recrystallization, crystal plasticity, continuum mechanics.

1. Introduction

Thermo-mechanical processing of flat aluminum products involves both deformation and recrystallization processes, which cause specific texture evolution. The crystallographic texture, evolved during thermo-mechanical processing, is mainly affected by technological parameters such as the degree of deformation, strain mode, recrystallization temperature e.t.c. The influence of technological parameters on microstructure and texture evolution can be revealed by means of numerical approaches, dealing with (i) strain distribution in deformed material, (ii) strain partitioning in polycrystalline aggregate and (iii) stress relaxation in a given crystal subjected to internal or external influence. Over the past several decades, extensive research endeavors in the field of plasticity, micro- and nano-mechanics have led to the development of various models enabling understanding of material's behavior during plastic deformation on different length scales. The so-called engineering scale, dealing with macroscopic properties of a material such as stress, strain and hardening parameters, considers each material as a continuum and therefore the crystallographic aspects of microstructure evolution during deformation are neglected. Various aspects of continuum mechanics (CM) are incorporated for instance in finite element (FEM) or flow-line (FLM) models. Contrarily, Taylor-type crystal plasticity (CP) homogenization models [1-5], dealing with partitioning of macroscopic load in a polycrystalline aggregate, are not capable of explaining the effect of various technological parameters on the deformation flow. Basic principles of both continuum mechanics and crystal plasticity theories extend our understanding on deformation process and enable explaining mesoscopic scale's phenomena involved in recrystallization (RX).

Controlling the evolution of crystallographic texture during thermo-mechanical processing is of crucial importance since the texture is one of the main sources of plastic anisotropy. The main sources of anisotropy in polycrystalline systems are discussed in detail elsewhere [6]. Employing Taylor-type homogenization approaches enables modeling the anisotropy of plastic strain ratio, which in turn provides information on material's behavior during deep drawing operation. This aspect of mesoscopic performance is of importance particularly in deep-drawing qualities of Al alloys used for instance in the automotive industry. Although many efforts have been made to model the texture development during processing chain, but there are many aspects of crystallographic texture evolution, which are still not explained by existing models. In this view, the current contribution provides a brief overview on modeling the deformation textures by combining both CP and CM approaches and employing various boundary conditions. The influence of grain interaction phenomena, incorporated in particular CP model, on the quality of deformation texture prediction is analyzed and the effective strategies employed in texture simulations are presented. It is evaluated how microstructural heterogeneities might influence the evolution of recrystallization and plastic strain ratio in aluminum alloys.

2. Experimental and Computational Procedure

Al-Mg-Si aluminum alloy from 6016 series was investigated in the current study. In order to ensure a high degree of rolling reduction in one pass, the investigated material was subjected to heat treatment at 550°C prior to cold rolling. The annealed material with fully recrystallized microstructure and the initial thickness of 1.125 mm was cold rolled to a final thickness in one pass with 47% reduction. The roll diameter of 129 mm ensured a high contact length to mean thickness ratio ($l/h = 5.86$).

The through-thickness pre-rolling, cold rolling and recrystallization textures were measured by means of electron back scattering diffraction (EBSD) detector attached to scanning electron microscope. EBSD data were collected with Hikari type detector. The orientation contrast scans were collected and analyzed by the commercial OIM-TSL-6® software. In the case of recrystallized microstructures, the acceleration voltage of SEM was kept at 20kV, while acquisition of EBSD patterns in the deformed sample was realized at a lower acceleration voltage of 15kV in order to avoid overlapping of acquired pattern with those originating from deeper layers. The investigated samples were 70° tilted with respect to the EBSD detector, whereas the data were acquired on a hexagonal scan grid. The EBSD measurements were carried out in the plane perpendicular to the sample transverse direction (TD-plane) extending over the entire thickness of the investigated sample. The EBSD dataset of the deformed sample consists of $\approx 5.6\times10^5$ indexed points and the total number of grains measured is $\approx$ 5000, whereas the number of grains sampled prior to rolling is $\approx$ 8000 with $\approx 4.5\times10^5$ indexed points.

Samples for EBSD examination were prepared by mechanical grinding and polishing according to the typical procedure, finishing with 1 μm diamond paste. The final preparation step was electrolytic polishing for ~1 min at a voltage of 18V with A2 Struers® electrolyte, cooled to temperatures ranging between -5 and 0°C.

Both measured orientation data and simulated ones were post-processed with the MTM-FHM software, developed by Van Houtte [7], to the purpose of meaningful comparison. The discrete sets of orientations obtained both from the experimental measurements and texture simulations were converted to continuous ODFs by superimposing a Gaussian peak for each individual orientation with a spread of 7° around the exact orientation. The ODFs are presented in the φ_2=const sections.

The texture evolution during cold rolling was predicted by taking into account only the {111}<110> octahedral slip systems with the following crystal plasticity (CP) approaches: the full constraints Taylor model (FCT) [1, 2], the visco-plastic self-consistent (VPSC) model [4], the Alamel model [2, 3] and Cluster-V model [5]. For modeling purposes, the experimentally measured textures were converted to continuous ODFs and subsequently discretized according to a method proposed by Tóth and Van Houtte [8]. According to this procedure, the total number of experimentally measured points is reduced to a set of 8000 equally weighted orientations. As a first approximation, the rolling was approximated by plane strain compression (PSC). However, in order to obtain more accurate displacement fields across the thickness of rolled materials, two-dimensional finite element calculations and flow-line modeling [9] were performed. In the FEM calculations, the isotropic material's plastic behavior is described with the stress-strain curve fitted by piecewise linear segments, whereas the rolls were considered as fully rigid objects. The following material parameters were used in simulations for isotropic aluminum matrix: Young's Modulus E = 68.9 GPa, Poisson's ratio ν = 0.33 and the yield strength σ_y = 80 MPa. In the case of FEM and FLM simulations, the displacement fields were extracted from the surface, 1/10th, 1/5th, 3/10th, 2/5th of the half thickness and mid-thickness plane. The extracted outputs were fed to the CP codes for calculation of through-thickness textures. Results obtained by FEM-CP and FLM-CP simulations were compared with the textures computed by assuming homogeneous plane strain deformation across the entire thickness in rolled materials.

3. Brief description of crystal plasticity models

Simulation of crystallographic texture evolution in a polycrystalline aggregate, subjected to an arbitrary strain mode, is generally completed by means of crystal plasticity approaches. These theories enable explaining how the macroscopic strain is translated to the mesoscopic level. Crystal plasticity models employed in the current investigation (FC Taylor [1], Alamel [2, 3], VPSC [4] and Cluster V [5]) employ diverse interaction phenomena on the grain level while considering each grain as a perfect crystal. In the full constraints approach, each grain behaves individually independently of neighbors, while the macroscopic strain velocity field L_{ij} and the microscopic counterpart of individual crystal l_{ij} are identical. In contrast to the FC Taylor theory, the behavior of a number of neighbors is accounted for in the grain interaction CP models [2-5]. A short-range Alamel, an intermediate-range Cluster V and a long-range VPSC models were employed for calculation the texture evolution during rolling. In the grain interaction approach such as Alamel, the polycrystalline aggregate is subdivided into a number of pairs. The equilibrium of stresses and the strain compatibility are fulfilled in the considered pair of crystals. Additionally, the interaction between grains is conditioned by the orientation of grain boundary, which is defined via grain boundary segment orientation distribution function.

In the visco-plastic self-consistent model one particular grain, approximated by an ellipsoidal visco-plastic inclusion, interacts with its surroundings represented by a hypothetical homogeneous effective medium (HEM). The homogeneous effective medium represents an averaged constitutive behavior of polycrystalline aggregate surrounding a given crystal. In the self-consistent approach, the stress applied to the outer boundary of the HEM gives rise to local deviations of strain rates in its vicinity. In the constitutive law, the degree of interaction is expressed by means of so-called interaction parameter n_{eff}, which represents the deviation of the strain rate in particular grain with respect to the macroscopic average one. Assumption of $n_{eff}=0$ assures identical strain-rate state in each grain of a polycrystalline aggregate and this is known as full constraints Taylor or upper-bound approximation. Self-evidently, the secant formulation with $n_{eff} = 1$ is fairly stiff and resembles many features of the full constraints model. By way of contrast to linearization schemes with a low value of n_{eff}, the tangent approximation with $n_{eff} = 20$ assumes relatively uniform stress state in the polycrystalline material. As soon as $n_{eff} \rightarrow \infty$ a stress deviation diminishes to zero (Sachs model) and therefore a lower-bound approximation is maintained. Obviously, the choice of linearization scheme has a great impact on the results of texture simulation. In the current calculations, n_{eff} is considered to be 10, corresponding to an intermediate grain interaction level.

In the cluster V model, the polycrystalline aggregate is subdivided into a number of clusters consisting of five grains, however, each particular grain assembly is exposed to the Taylor type interaction. The grain interaction inside individual clusters is approximated by the VPSC linearization scheme with n_{eff} =10.

4. Modeling the material's flow by continuum mechanics-based approaches

4.1 Geometrical Approaches

Plane strain compression (PSC) is the simplest geometrical approach, which is often used in approximating the rolling process. This approach primarily accounts for a change of material's shape and enables the estimation of an important technological parameter such as strain *e*. This strain mode is justified to a large extent due to the fact that the ratio between the compressive strains (e_{33}=-e_{11}) and other components of strain tensor e_{ij} ($i \neq j$) is large. Assuming a linear dependence of elongation change with time, the deformation gradient tensors (F) gains the following form:

$$F = \begin{bmatrix} \frac{\Delta ht}{h_f} + 1 & 0 & 0 \\ 0 & 0 & 0 \\ 0 & 0 & -\frac{h_f}{\Delta ht + h_f} \end{bmatrix} \quad (1)$$

where h_i is the initial sheet thickness, h_f is the final sheet thickness, t is time (t is normalized between 0 and 1) and $\Delta h = (h_i - h_f)$.

Knowing the deformation gradient tensor enables calculating the strain velocity gradient tensor L, which is used in crystal plasticity calculations.

$$L = \dot{F}F^{-1} = \begin{bmatrix} \frac{\Delta h}{\Delta ht + h_f} & 0 & 0 \\ 0 & 0 & 0 \\ 0 & 0 & -\frac{\Delta h}{\Delta ht + h_f} \end{bmatrix} \quad (2)$$

It should be underlined that this approximation excludes both the effect of roll gap geometry and strain heterogeneity across the thickness of a sheet. In this analysis, the components of L tensor might be approximated by a formula, used for the estimation of a true strain (L_{11}=-$L_{33} \approx \ln(h_i/h_f)$).

Shear strain e_{13} evolving due to the roll gap geometry during rolling can be calculated by [10]:

$$e_{13} = \frac{\Delta h}{(R\Delta h - (\Delta h/2)^2)^{0.5}} \quad (3)$$

where R is a roll radius.

Since for small strain increments $\dot{e}_{13}=0.5(L_{13}+L_{31})$, and in this analysis L_{31} =0, consequently L_{13} = $2e_{13}$. Taking into consideration that e_{13} strain is mainly concentrated within a thin sub-surface layer, it is reasonable to assume an exponential character of the shear strain distribution:

$$e_{13}^s = e_{13} exp\left(\frac{s-1}{s}\right) \quad (4)$$

here, the superscript s indicates the position of a given layer with respect to the surface of a sheet. In the equation s might change between 0 and 1: s =1 for the surface while s =0 for the mid-thickness plane.

Combining equations 2-4 enables calculation of strain velocity gradient tensor L^s for various thickness layers s:

$$L^s = \begin{bmatrix} \frac{\Delta h}{\Delta ht + h_f} & 0 & \frac{2e_{13}^s(\Delta h + h_f)}{\Delta ht + h_f} \\ 0 & 0 & 0 \\ 0 & 0 & -\frac{\Delta h}{\Delta ht + h_f} \end{bmatrix} \quad (5)$$

4.2 Finite element modeling

In order to account for the effect of various rolling parameters such as the reduction level, the initial sheet thickness, the friction conditions, the circumferential velocity of the rolls, and their diameter on the deformation flow, finite element simulations were performed. The behavior of anisotropic material was studied. It is important to note that the FEM outputs depend on the choice of friction condition. Inasmuch as the rolling was performed on well-lubricated surfaces of both sheet and rolling cylinders, the deformation with a constant Coulomb friction coefficient μ, slightly exceeding the minimum value necessary for rolling ($\mu=1.1 \cdot \mu_{min}$), was assumed in this study. The value of μ_{min} was approximated by the following formula [11]:

$$\mu_{min}=\frac{1}{2}\sqrt{\frac{h_f}{R}}\frac{\ln\left(\frac{h_i}{h_f}\right)+\frac{1}{4}\sqrt{\frac{h_i-h_f}{R}}}{\tan^{-1}\sqrt{\frac{h_i}{h_f}-1}} \tag{6}$$

4.3 Flow line modeling

Flow line model is an analytical solution of continuum mechanics-based equations, describing the material's flow in rolling under predefined boundary conditions. In the FLM, developed by Decroos et al [9], a kinematically admissible displacement velocity field fulfills the following boundary conditions: (i) material's velocity is uniform across the thickness at the entrance and the exit of the rolling process, (ii) the incompressibility condition is fulfilled at all points of the rolled material, (iii) material strictly follows the prescribed streamlines in the middle and surface of the plate, while some deviation from the predefined stream-lines is allowed for other layers, (iv) at the surface, the velocity field is prescribed by means of model parameter α that is linked to the friction conditions between the rolls and the plate (α ensures a difference between velocities of surface and mid-thickness layers), (v) the variation of the velocity across the thickness is conditioned by the n-th power law. The comprehensive mathematical description of this approach is described by Decroos et al [9]. In the current contribution the model parameters α and n are set as to resemble the deformation patterns, calculated by FEM.

5. Rolling texture simulation

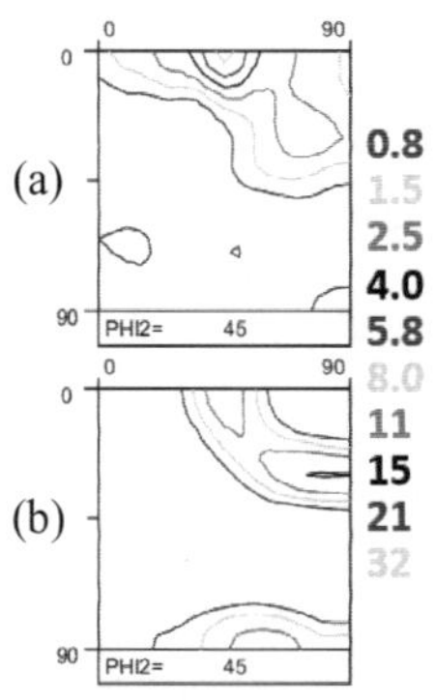

Fig.1. Pre-rolling texture (a) and texture evolved after 47% reduction (b).

Prior to cold rolling, annealing process has ensured a strong recrystallization texture (Fig.1), which tends to transform to the β fiber described analytically by [12]:

$$\{h,\ 1,\ h+1\}\left\langle\frac{h(h+1)}{3/4-h},\ \frac{2h(h+1)}{1/2-h},\ \frac{h^2}{h-3/4}+\frac{2h}{h-1/2}\right\rangle \tag{7}$$

Fig. 2 reveals the distribution of L_{ij} components calculated by equations 2 and 5. It is obvious from Fig. 2 that the shear component L_{13} drastically diminishes from the surface toward the mid-thickness independently of the model used. The shear components L_{13}, calculated by formula 5, are significantly smaller compared to ones computed either by FEM or FLM. Fig. 3 shows results of texture simulations with various both CP and continuum mechanics-based approaches.

The quality of texture simulation is assessed by means of texture index difference $(TID = \int[f_{exp}(g) - f_{sim}(g)]^2 dg)$, which rises with the increase of differences between the compared ODFs. The CP models are capable of reproducing individual texture components along the β-fiber as well as the retained RX orientation with various degree of accuracy. A short-range grain interaction of the Alamel model accounts for the most accurate texture prediction, independently of strain mode approximation. Both, the intermediate and long-range grain interaction schemes, incorporated in the Cluster V and VPSC models respectively, provide the second most successful and at the same time very comparable and texture prediction. Regarding CM approaches, both FEM and FLM are superior over the geometrical approximations. It seems that computationally less heavy flow line model is capable of providing texture prediction comparable to one simulated with the L_{ij} obtained from the finite element model, independently of CP approach employed.

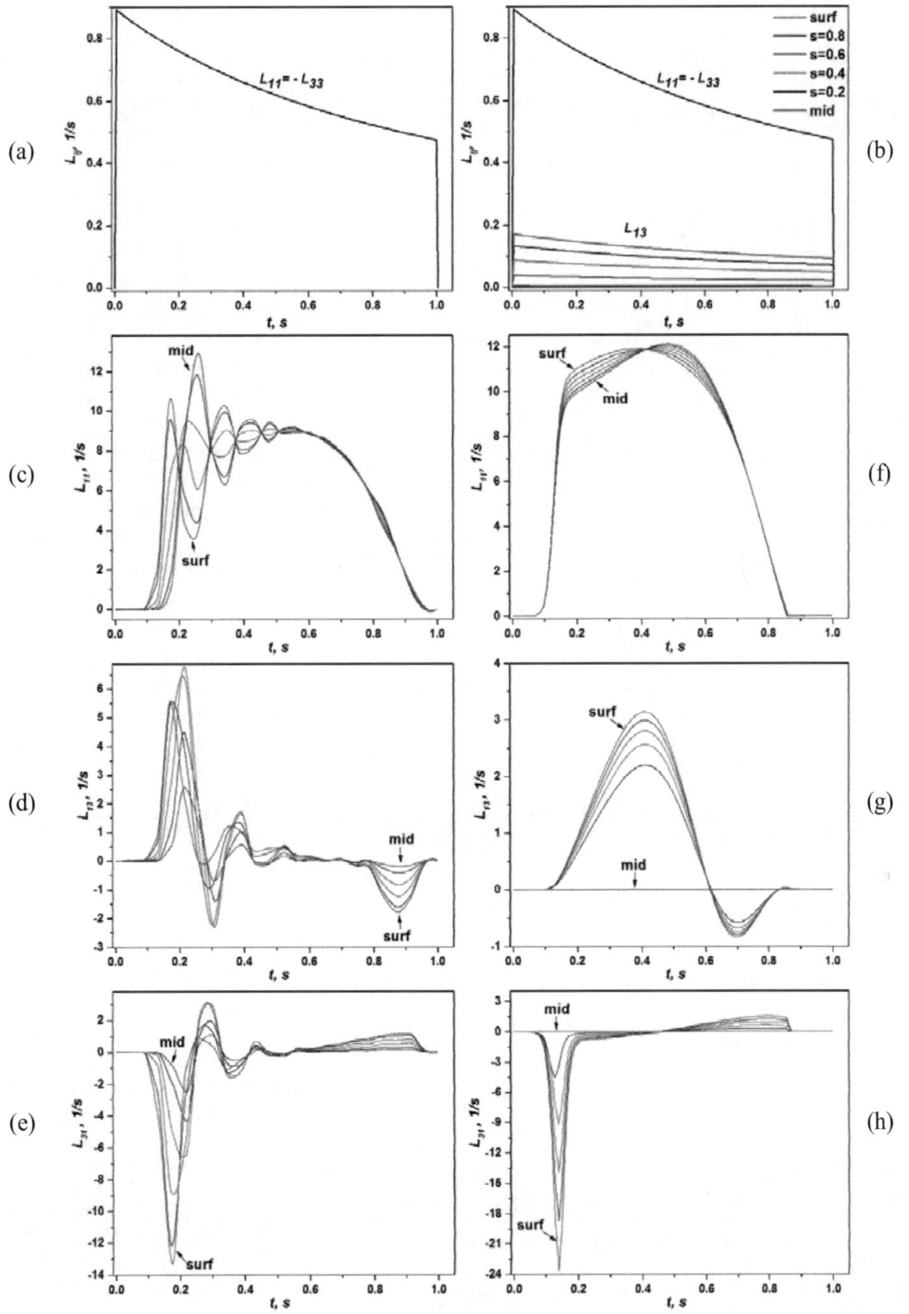

Fig.2. Strain velocity gradient components L_{ij} calculated for various boundary conditions: a) PSC, equation 2; b) PSC+e_{13}, equation 5; (c-e) FEM; (f-h) FLM. For a meaningful comparison, time t is normalized between 0 and 1.

IOP Conf. Series: Materials Science and Engineering **375** (2018) 012028 doi:10.1088/1757-899X/375/1/012028

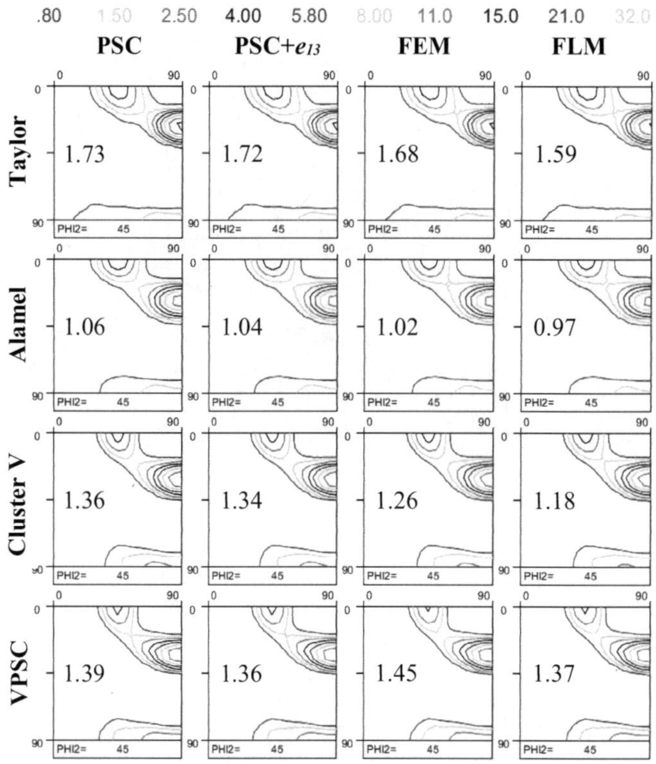

Fig.3. Simulated textures with different both crystal plasticity and continuum mechanics-based approaches. The *TID* number for a given pair of CP and CM models is indicated on the corresponding ODF section.

6. Modeling the influence of recrystallization texture on plastic strain ratio

Many of the principles and concepts addressed by the above-discussed CP models also apply to RX phenomena. Various RX modeling approaches [13-16] are based on orientation selection or determination of nucleation spectra, considering the diversity in dissipation of plastic power in different crystal orientations. This is of particular importance while investigating the effect of microstructural heterogeneities on the evolution of recrystallization texture. The evolution of nucleation spectra in the bulk of a material as well as at micro shear bands or in the particle affected deformation zone might be explained by RX models considering both basic principles of micromechanics and crystal plasticity approaches [15-18]. Accounting for strain path heterogeneities in the vicinity

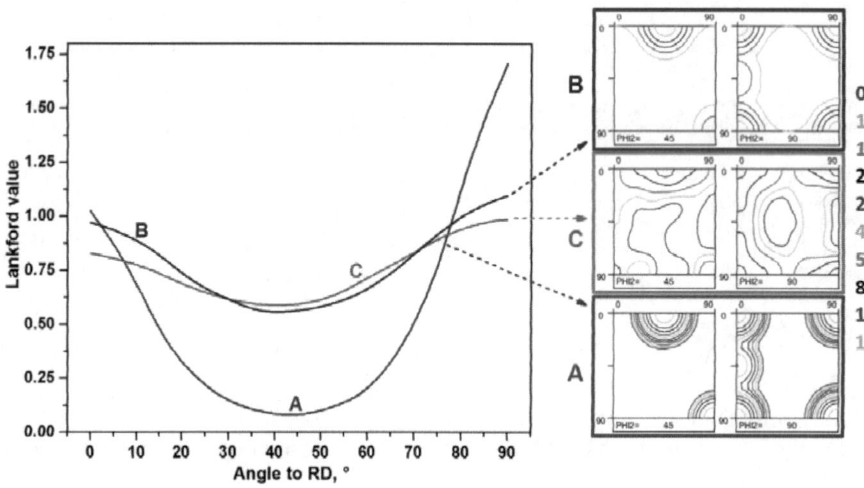

Fig.4. Simulated r-value profiles with the Alamel model.

IOP Conf. Series: Materials Science and Engineering **375** (2018) 012028 doi:10.1088/1757-899X/375/1/012028

of non-deformable inclusions allows explaining the appearance of the P {011}<233>, {100}<130> or weak γ-fibre <111>//ND orientations, which typically evolve in particle containing Al alloys [15, 16]. Likewise, considering strain path deviations inside the copper-type shear bands from the macroscopic one sheds light on the evolution of P, Goss {011}<100> and Q {013}<231> orientations [17, 19]. The evolution of mentioned texture components has a decisive influence on anisotropy of plastic strain ratio, which measured in tension at various angles with respect to rolling direction (RD). The conventional measure of plastic anisotropy (Lankford value or *r*-value) can be modelled by means of CP approaches with reasonable accuracy [18]. The normal anisotropy ($\bar{r}$) is calculated from the ratio of a true strain in the sheet transverse direction to the logarithmic strain in the sheet normal direction at various angles to RD and averaged by a trapezoidal rule, while the degree of planar anisotropy (Δr) is assessed based on the *r*-value profile in the plane of a rolled sheet. Fig. 4 reveals the effect of various hypothetical RX textures on the plastic strain ratio. First, the *r*-value profile of a material, characterized by a strong Cube {001}<100> mixed with weaker Goss texture, is examined (Fig.4, texture A). This type of texture often dominates in Al-alloys [18,19, 20] after recrystallization and tends to produce a V-shaped *r*-profile with a local minimum at ~45° with respect to RD (Fig.4, Lankford profile A). The estimated normal anisotropy for the A-type texture ($\bar{r}$=0.72) might ensure a minimum level of deep drawability, however, the high value of in-plane anisotropy (Δr=1.29) makes this material inappropriate for deep drawing applications. In ideal case, the in-plane anisotropy should be minimized to zero and this can be accomplished by texture randomization. Taking into consideration that it is almost impossible to produce an isotropic material with a random texture, it is assumed here that the strongly textured material A is randomized, and, as a result a B-type texture is evolved, consisting of a weakly developed Cube and Goss components (Fig.4, texture B). The A and B-type textures are qualitatively identical, while the quantitative diversity accounts for significant changes in Lankford profiles. Even if the normal anisotropy of B-type texture ($\bar{r}$=0.8) did not improve drastically as compared to A-type counterpart, the planar anisotropy (Δr=0.47) reveals considerable improvement. It is evident that even after sever texture randomisation the *r*-value is still anisotropic, i.e. dependent on the direction of the tensile sample axis. It should be mentioned, that during thermo-mechanical processing, the evolution of particular texture component occurs at the expense of other orientations. For instance, the evolution of Cube-oriented grains, which dominate in Al alloys after recrystallization, can be supressed by activating nucleation at microstructural heterogeneities (shear bands) or via particle stimulated nucleation (PSN) mechanism. Each nucleation phenomenon produces characteristic orientation spectrum, as noted previously. Since texture weakening caused either by PSN or via nucleation at shear bands is not of random nature, it is assumed that the A-type texture is randomized by the P, Q, weak γ-fibre and components scattered around these orientations. The resulting C-type texture (Fig.4) serve to produce a V-shaped *r*-value profile with $\bar{r}$=0.75 and Δr=0.32. Even though the normal anisotropy remains far below the desired value of $\bar{r}$=1, the planar anisotropy seems to be acceptable for deep drawing.

Fig. 4 clearly demonstrates that the presence of orientations, originating from microstructural heterogeneities (P and Q) which evolve at the expense of components evolved in the bulk of a material, leads to decrease in planar anisotropy while the value of normal anisotropy does not improve significantly. In summary, application of CP models enables explaining the evolution of crystallographic textures and related properties in materials with various crystal structures [18, 21-23].

7. Conclusions

Analyzing the quality of texture simulation indicates that grain interaction phenomena, incorporated in CP calculations, play far more important role than CM-based approximation of deformation. Approximation of strain path by either FEM or FLM is superior over the geometrical approximations. Analytical flow line model combined with a given CP approach is capable of providing texture prediction as accurate as one simulated with the strain path obtained from the FEM.

The improvement of texture prediction by the Alamel model justifies the importance of short-range grain interactions in CP modeling. It seems that the intermediate grain interaction, incorporated in the Cluster V model, is the second most important homogenization scheme, occurring during deformation. Since the quality of simulation by the VPSC model is comparable to the Cluster V approach, this suggests that long-range grain interactions cannot be neglected, while modeling the texture evolution. Modeling the plastic strain ratio with CP models indicates that presence of texture components, originating from microstructural heterogeneities, tends to decrease the planar anisotropy and, thus, enhances the forming characteristics of a material.

Acknowledgement

The financial support of EFOP-3.6.1-16 project: “Innovative processing technologies, applications of energy engineering and implementation of wide range techniques for microstructure investigation” within the Széchenyi 2020 program is gratefully acknowledged.

References

[1] Van Houtte P and Aernoudt E 1975 *Z. Metallkunde* **66** 202.
[2] Van Houtte P, Li S, Seefeldt M and Delannay L *Int. J. Plast.* **21** 589.
[3] Van Houtte P, Kanjarla AK, Van Bael A, Seefeldt M and Delannay L 2006 *European J. Mech. A/Solids* **25** 634.
[4] Lebensohn RA and Tomé CN 1993 *Acta Metal. Mater.* **41** 2611.
[5] Xie Q, Van Bael A, Sidor J, Moerman J and Van Houtte P 2014 *Acta mater.* **69** 175.
[6] Hutchinson B 2015 *Mater. Sc. and Techn.* **31** 1393.
[7] Van Houtte P 1995 *MTM-FHM Software* 2nd ed. (MTM-KU Leuven).
[8] Tóth LS and Van Houtte P *Text. Microstruct.* **19** 229.
[9] Decroos K, Sidor J and Seefeldt M 2014 *Met. and Mater. Trans.* **45A** 948.
[10] Lee CS and Duggan BJ 1991 *Met. and Mater. Trans.* **22A** 2637.
[11] Avitzur B 1980 *Int. J. Mach. Tool. Des. Res.* **20** 197.
[12] Sidor JJ and Kestens LAI 2013 *Scripta Mater.* **68** 273.
[13] Jonas JJ and Tóth LS 1992 *Scripta Metall. Mater.* **27** 1575.
[14] Engler O 1997 *Textures and Microstr.* **28** 197.
[15] Sidor JJ, Decroos K, Petrov RH and Kestens LAI 2015 *Int. J. Plast.* **66** 119.
[16] Sidor JJ, Petrov RH and Kestens LAI 2011 *Acta Mater.* **59** 5735.
[17] Lapeire L, Sidor J, Verleysen P, Verbeken K, De Graeve I, Terryn H and Kestens LAI 2015 *Acta Mater.* **95** 224.
[18] Sidor J, Petrov R, Xie Q, Van Houtte P and Kestens L 2017 *Mater. Sc. Techn.* **33** 667.
[19] Engler O 2001 *Scripta Mater.* **44** 229.
[20] Bennett TA, Sidor J, Petrov RH and Kestens LAI 2010 *Adv. Eng. Mater.* **12** 1018.
[21] Wang XF, Guo MX, Chen Y, Zhu J, Zhang JS and Zhuang LZ 2017 *Met. and Mater. Trans.* **48A** 3540.
[22] Wu Y, Liao H, Zhou K and Yang J 2015 *Mater. and Des.* **57** 416.
[23] Lee KM, Huh MY, Lee HJ, Park JT, Kim JS, Shin EJ and Engler O 2015 *JMMM* **396** 53.

IOP Conf. Series: Materials Science and Engineering **375** (2018) 012029 doi:10.1088/1757-899X/375/1/012029

Orientation-dependent fatigue damage in planar slip metals

R G Li[1], W J Liu[2] and Y D Wang[1, *]

[1]State Key Laboratory for Advanced Metals and Materials, University of Science and Technology Beijing, Beijing 100083, China
[2]Advanced Photon Source, Argonne National Laboratory, Argonne, IL 60439, USA

E-mail address: ydwang@ustb.edu.cn

Abstract. Shear banding as a ubiquitous phenomenon is mostly observed in planar slip metals under complex plastic deformation conditions, especially in cyclic deformation. The deformation substructures and local damages during fatigue vary from grain to grain, which depends greatly on the crystallographic orientation. Despite previous extensive studies, the orientation-dependent deformation substructure is still elusive due to the complexity of damage evolution and the limitation of spatial resolution in both reciprocal and real sample spaces by conventional characterization methods. In this study, we conducted synchrotron-based X-ray micro-diffraction (μXRD) experiments to explore the dependence of localized elastic strains on grain orientation. It is demonstrated that the typical intersection of slip shear bands (SSBs) with Lomer-Cottrell locks (LCLs) are the potential damage locations, which are most likely to form in the near [001]//LD grains. The advanced μXRD with ability of spatially quantify the localized strains/fine orientations opens new opportunities for understanding the microscopic mechanisms of damage and failure in bulk materials.

1. Introduction

Shear bands are the narrow layers of materials undergoing intense shearing strains. Under a fast monotonic or low-strain-amplitude cyclic load, cumulative damage gradually develops within the microscale shear bands such as dislocation slip bands, leading to the eventual material failure. SSBs have been extensively observed in the planar slip alloys subjected to cyclic deformation [1-3], in which damage accumulation and resulting instabilities represent one of the most common modes of material failure. It is still a significant challenge to rationally evaluate the cyclic failure life of structural materials due to a lack of knowledge on the microscale stress heterogeneity across SSBs and the criterion of localized instability.

Deformation substructures and microscale stresses are developed during cyclic deformation, which can be used as an important microstructural indicator to scale the progress of fatigue damage. It is in a close correlation with primary slip systems activated during fatigue of planar slip materials. Detailed investigations on the evolution of deformation substructure and its dependence of crystal orientation are of great value in the understanding of fatigue damage. The traditional characterization techniques such as EBSD can be used to obtain the nanoscale misorientation distribution on the surface of deformed sample. However, the stress relaxing on the sample surface may destroy the configuration of defects and alter the microscale stress field. The synchrotron-based X-ray micro-diffraction (μXRD) technique provides a powerful tool to study the microscale distribution of elastic strain and grain orientation within three-dimensional deformed sample. From the microscale stress field, it is also possible to determine

IOP Conf. Series: Materials Science and Engineering **375** (2018) 012029 doi:10.1088/1757-899X/375/1/012029

the configuration of various defect structures with submicrometer resolution, being the fundamental element of substructures.

AL6XN SS is a typical austenite stainless steel (SS) with low stacking fault energy. As a result, planar slip prevails, yielding rich planar slip bands during fatigue cycling, which makes it an ideal target material for the study of orientation-dependent fatigue damage.

2. Experimental procedure

The AL6XN SS has a single-phase face-centered-cubic (FCC) structure. Samples for mechanical testing were cut from the sheet after 1160 °C solution treatment for 2 hours followed by water quenching and hot rolling with a final thickness of 15 mm. The final grain sizes range from 80 to 120 μm. The standard rod-shaped samples were used in fatigue tests with gauge section of 8 mm in diameter and 20 mm in length. The strain-controlled fatigue test was conducted under a total strain control with the maximum strain amplitude of 0.3%, the cycle asymmetry coefficient R = -1, and the load frequency of 0.5 Hz at room temperature in air.

The μXRD [4-8] was used to explore the strain heterogeneity of grains in a fatigued bulk AL6XN SS. Spatially-resolved 3D μXRD measurements were conducted by using differential-aperture X-ray microscopy (DAXM) at beamline 34-ID-E of the Advanced Photon Source in Argonne National Laboratory. An X-ray microbeam (polychromatic or monochromatic) is focused to about 0.5 μm × 0.5 μm by a pair of elliptical Kirkpatrick-Baez focusing mirrors. Figure 1a is a schematic illustration of the diffraction geometry with the loading direction (LD), transverse direction (TD) and normal direction (ND) labeled in light blue. Diffracted X-ray beams will be recorded on an X-ray area detector.

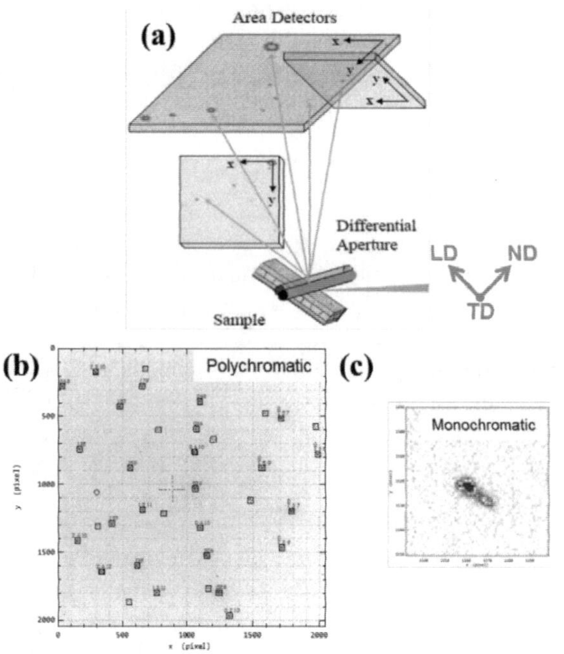

Figure 1. Schematic of μXRD experiments. (**a**) Diffraction geometry. (**b**) The polychromatic Laue pattern. (**c**) The monochromatic Laue pattern.

The polychromatic and monochromatic Laue spots are illustrated in Fig. 1b and 1c, respectively. The point-to-point micro-orientation can be calculated from the polychromatic Laue spots, and the high accuracy lattice strain along ND can be calculated from the energy scanning for monochromatic Laue

IOP Conf. Series: Materials Science and Engineering **375** (2018) 012029 doi:10.1088/1757-899X/375/1/012029

spots. The depth resolution is provided by scanning a 100-μm-diameter platinum wire as diffracted-beam profiler in a knife-edge fashion. By triangulation, the origins of the X-rays can be determined on a pixel-by-pixel basis. A typical scanning step of 1.4 μm gives the depth resolution of ~1.0 μm. The detailed information on μXRD is well described previously [6].

3. Results and discussion

3.1. Orientation-dependent deformation substructure

The micro-orientation with an order of magnitude higher on precision than Electron Back-Scattered Diffraction (EBSD) [9] can be obtained from the analysis on monochromatic Laue spots in the μXRD experiments. Better yet, we can get the point-to-point local strain tensor map from the calculations, which provides valuable date for studying orientation-dependent features.

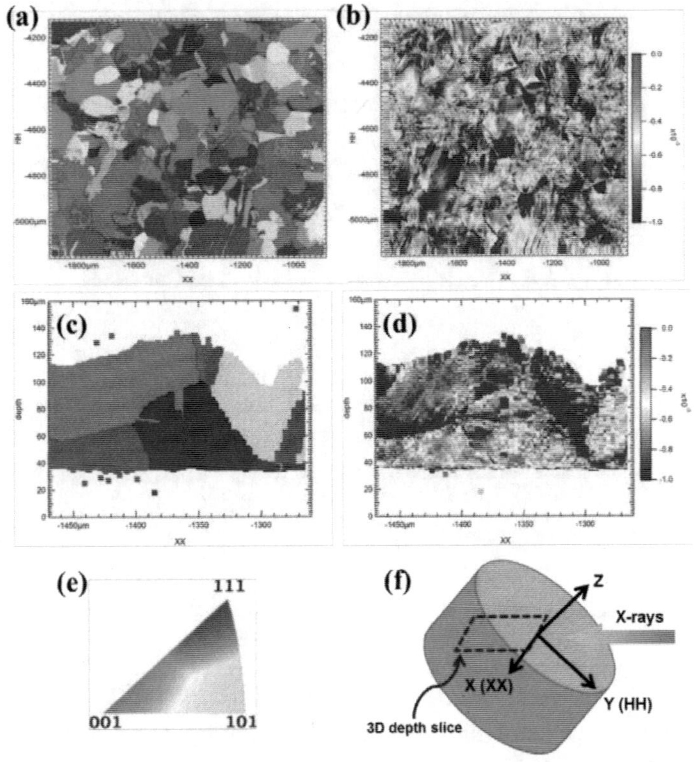

Figure 2. Polychromatic μXRD scanning results. Inverse-pole figure (IPF) map (**a**) and ε_{zz} (**b**) showing grain orientation and corresponding strain distribution of a point-to-point scan area. (**c**) and (**d**) are the local amplifications of (**a**) and (**b**). (**e**) and (**f**) are the IPF (parallel to LD/Y) and coordinate system.

It has been well revealed that the orientation-dependent residual stress is widely existed in fatigued materials. The stress is related to the deformation substructure, which influences the damage of fatigue greatly [10-12]. In this study, the distribution of residual strain is revealed in the monochromatic μXRD experiments. An inverse pole figure (IPF) map and the corresponding residual strain map are shown in Fig. 2a and 2b, respectively. Fig. 2c and 2d are their local amplification focussed on several grains. A distribution of orientation dependent residual strain with fine spatial resolution on a bulk fatigued steel is directly visualized. The red/reddish grains with their [001] axis nearly parallel to LD ([001]//LD) in Fig. 2a are consistent with the blue grains having a larger compressive strain along ND in Fig. 2b. More detailed, the deformation substructures of a large number of grains are displayed in the residual strain maps (Fig. 2b and 2d), from which it can be concluded that the near [001]//LD grains undergo severe heterogeneous deformation, with the formation of obvious shear bands through more than one slip systems. This microstructural future is further confirmed by the monochromatic μXRD scan within a [001]//LD grain and a [110]//LD grain.

The diffraction spots of (480) and (006) crystal planes are used to characterize the local features of the [001]//LD grain and the [110]//LD grain in the monochromatic μXRD. The deformation substructure is clearly revealed from the full width at half maximum (FWHM) maps (Fig. 3a and 3b). Shear bands are merged into background (matrix information) in the [110]//LD grain (Fig. 3b), but they are very obvious in the [001]//LD grain (Fig. 3a). The complex evolution of deformation substructure is important to scale the progress of fatigue damage.

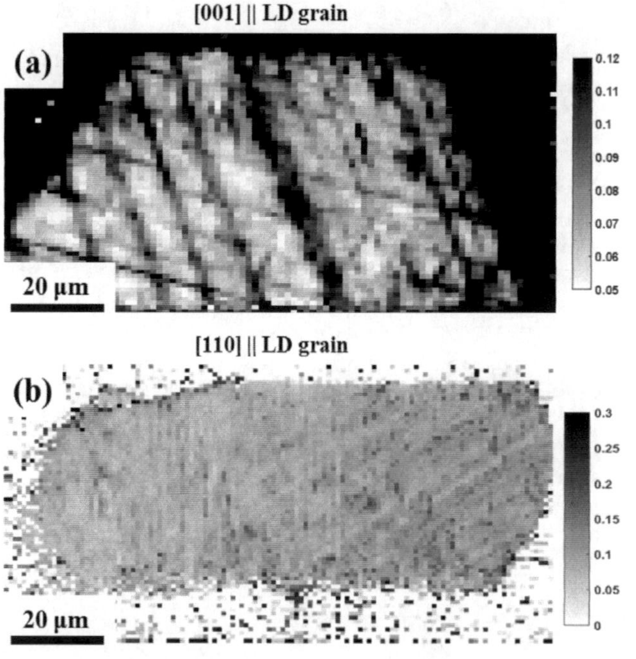

Figure 3. Deformation substructures of AL6XN SS after 29,000 cycles. (**a**) A map of the (480) diffraction peak FWHM of a [001]//LD grain. (**b**) A map of the (006) diffraction peak FWHM of a [110]//LD grain. They were measured by monochromatic X-ray.

It should be emphasized that the orientation-dependent deformation substructure is particularly remarkable in cyclic deformation of planar slip materials, in which no mesoscopic grain reorientation occurs and the SSBs are relatively in favour during the cyclic deformation.

3.2. Micro orientation distribution in a fatigued [001]//LD grain

The micro orientation map of the [001]//LD grain is visualized in Fig. 4a with a local amplification including a wide shear band shown on the right. The colour is based on the coordinates in the [001] pole figure as pictured in Fig. 4c. The misorientation across the grain is no more than 0.2°. A sharp misorientation is observed on the left edge of the shear band, which is consistent with the low-angle grain boundary model with edge dislocations aligned as shown in Fig. 4b. Maps of misorientation showing grain subdivision are consistent with the observed deformation substructure (Fig. 3a).

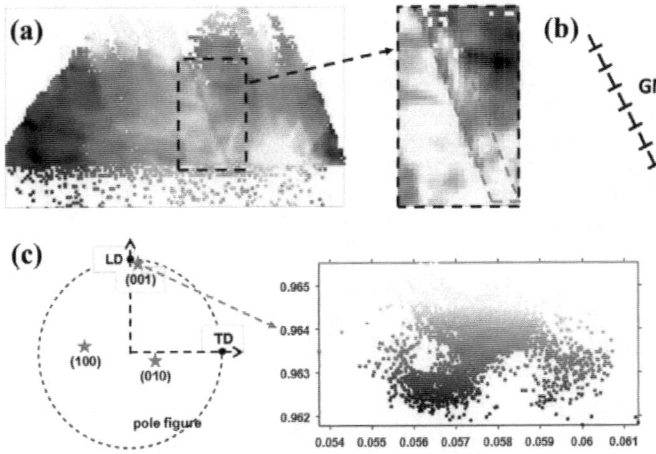

Figure 4. (**a**) Orientation map of a fatigued [001]//LD grain with a local amplification on the right. (**b**) Model of low angle grain boundary with edge dislocations aligned. (**c**) Pole figure of {001} plane with an amplification showing detailed orientation distribution (**d**) on the right.

3.3. Damage source of planar slip stainless steel: Lomer-Cottrell locks

Our former work [13] has confirmed that the interaction of SSBs is the main damage source during cyclic deformation of AL6XN SS. The LCLs are forming by the dislocation reaction in Burger vectors of ½ [10$\bar{1}$] + ½ [011] = ½ [110], in the crossing of SSBs, as shown in Fig. 5. The sessile dislocations deeply promote the dislocation pile-ups, and huge local stress gradients raise around the pile-ups. With the increase in the number of cycles, these locations finally develop into micro cracks [13, 14].

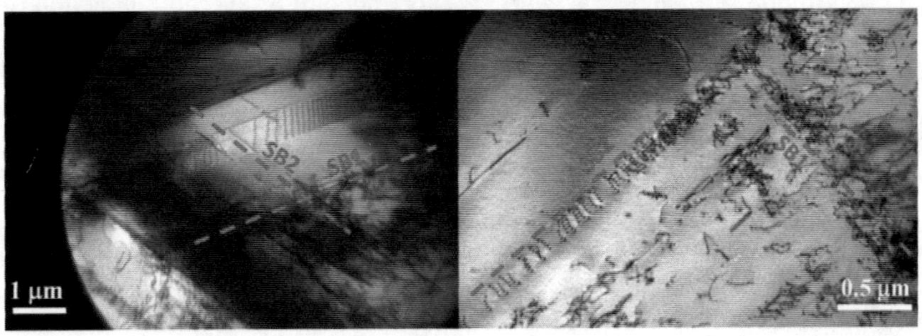

Figure 5. TEM showing the interactions of SSBs in the AL6XN SS after 29,000 cycles.

According to the above characterizations and analyses on the fatigue damage processing, the further development of SSBs through different slip systems within one grain will lead to a quicker local damage and earlier fatigue failure. Analyses on the formation and development of SSBs in various orientation grains are important for predicting the fatigue life. For simplification, the analysis below will expand based on the classical theory of plastic deformation in a qualitative manner due to the complexity of strained conditions of various grains.

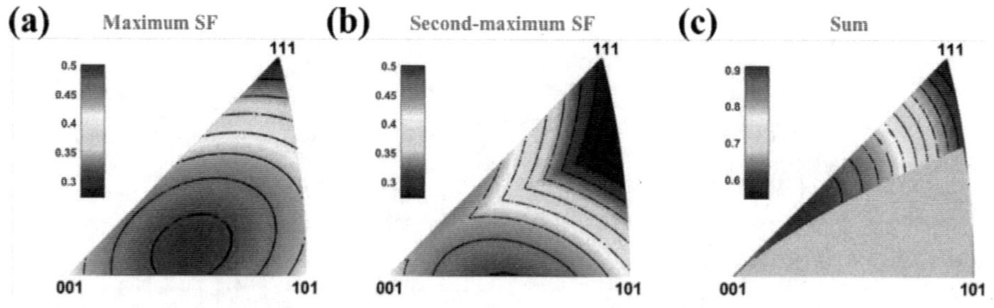

Figure 6. Distributions of Schmidt factors plotted in IPFs along LD. (**a**) Maximum Schmidt factor, (**b**) second-maximum Schmidt factor, (**c**) sum of (**a**) and (**b**) with a 'safe' region coloured in grey.

It is extensively accepted that $(111)[1\bar{1}0]$ slipping is the dominating deformation mode of FCC materials in small/cyclic deformation. The Schmidt factor is often used in the determination of activated slip systems in high cycle fatigue. But focusing on the particular situation of ALXN SS during cyclic deformation that the interaction of mature SSBs is a precursor of grain damage. In another word, grain with the largest maximum Schmidt factor, that generally means it is the easiest one to deform, is not necessarily the weakest grain in fatigue. More factors and LCLs should also be taken into account. In the following, we establish a simplified double SSBs model to analysis the orientation dependent localized deformation substructure.

The model of double SSBs is established on the assumption that mature SSBs are allowed to develop on no more than two (111) planes among grains during cyclic deformation. On the classical theory of slipping deformation, the distributions of maximum Schmidt factor and second-maximum Schmidt factor in IPF along LD are drawn in Fig. 6a and 6b, respectively. Because our motivation is to evaluate the forming probability of LCLs in grains with various orientations and no interaction of SSBs occurs if dislocations slipping on one (111) plane, so what need to be clarified is that the slip systems having the maximum and second-maximum Schmidt factor should not be a same (111) plane.

In Fig. 6a and 6b, the red parts present the orientations easiest and second-easiest to be activated, respectively. Fig. 6c is the sum of Fig. 6a and 6b, from which we can evaluate the forming probability of double SSBs in simple way. The grey part in Fig. 6c is the orientation with no $\frac{1}{2}[10\bar{1}] + \frac{1}{2}[011] = \frac{1}{2}[110]$ LCLs forming on the above assumption that only two slip systems are allowed to be activated. This means that the grains in grey may be in favourable orientations to activate two slip systems simultaneously, but dislocations on these planes cannot be locked on the SSBs via dislocation reactions. The SSBs of these grains will be thinner and distribute in a homogeneous way. The analysis is in good consistent with the above-mentioned observation of deformation substructures. It can be see clearly from the model that double SSBs with Lomer-Cottrell locks are most likely to form in the near [001]//LD grains. It still need further investigations in order to reveal the underlying mechanisms of orientation-dependent crystal damage in a quantitative way, which is now in progress.

4. Conclusions

Our μXRD experiments illustrated the orientation-dependent residual strains in bulk fatigued stainless steel. The orientation-dependent local damage of fatigued planar slip metals is analysed by the classical slip theory in a qualitative manner. The typical double SSBs with LCLs are the potential damage spots, which are most likely to form in the near [001]//LD grains. The experimental findings and theoretical analyses of orientation-dependent properties will contribute to the understanding of fatigue failure mechanism, and further promote the design of materials. As demonstrated in this work, the spatially resolved 3D μXRD mapping of orientation and lattice stress field at submicron scale provides a new opportunity to probe the microscopic damage and failure mechanisms in plastically deformed bulk materials.

Acknowledgements We thank the financial support from The National Key Research and Development Program of China (2017YFA0403804), National Natural Science Foundation of China (NSFC) (Grant number 51471032 and 51527801), the Fundamental Research Funds for the Central Universities (Grant number 06111020), and the fundamental research fund at the State Key Laboratory for Advanced Metals and Materials. Use of APS was supported by the US Department of Energy, Office of Science, Office of Basic Energy Science, under Contract DE-AC02-06CH11357.

References

[1] V Gerold, H Karnthaler 1989 On the origin of planar slip in f.c.c. alloys, *Acta Metal.* **37**, 2177-2183.

[2] K Tanaka, T Mura 1984 Fatigue crack growth along planar slip bands *Acta Metal.* **32**, 1731-1740.

[3] Z Wang 2004 Cyclic deformation response of planar-slip materials and a new criterion for wavy to-planar-slip transition, *Philo. Mag.* **21**, 351-379.

[4] G Ice, B Larson, W Yang, J Budai, J Tischler, J Pang, R Barabash and W Liu 2005 Polychromatic X-ray microdiffraction studies of mesoscale structure and dynamics, *J. Synchrotron Rad.* **12**, 155-162.

[5] B Larson, W Yang, G Ice, J Budai, J Tischler 2002 Three-dimensional X-ray structural microscopy with submicrometre resolution, *Nature* **415**, 887-890.

[6] W Liu, P Zschack, J Tischler, G Ice, B Larson 2011 X-ray Laue Diffraction Microscopy in 3D at the Advanced Photon Source. In I. McNulty, Eyberger, C. & Lai, B. (Eds.), *AIP Conference Proceedings* **1365**, 108-111.

[7] N Li, Y Wang, W Liu, Z An, J Liu, R Su, J Li, P Liaw 2014 In situ X-ray microdiffraction study of deformation-induced phase transformation in 304 austenitic stainless steel, *Acta Mater.* **64**, 12-23.

[8] W Liu, in Strain and Dislocation Gradients from Diffraction (ed. GE Ice, & R Barabash) Ch. **2** (Imperial College Press, London, 2014).

[9] D Dingley, V Randle 1992 Microtexture determination by electron back-scatter diffraction, *J. Mater. Sci.* **27**, 4545-4566.

[10] Y Wang, R Peng, X Wang, R McGreevy 2002 Grain-orientation-dependent residual stress and the effect of annealing in cold-rolled stainless steel, *Acta Mater.* **50**, 1717-1734.

[11] Y Wang, R Peng, R McGreevy 1999 High anisotropy of orientation dependent residual stress in austenite of cold rolled stainless steel, *Scrip. Mater.* **41**, 995-1000.

[12] Y Wang, H Tian, A Stoica, X Wang, P Liaw, J Richardson 2003 Grain-orientation-dependent residual stress and the effect of annealing in cold-rolled stainless steel, *Nat. Mater.* **2**, 101-106.

[13] R Li, Q Xie, Y Wang, W Liu, M Wang, G Wu, X Li, M Zhang, Z Lu, C Geng, T Zhu 2018 Unraveling submicron-scale mechanical heterogeneity by three-dimensional X-ray micro-diffraction, *Proc. Natl. Acad. Sci. U. S. A.* **115**, 483-488.

[14] Kuhlmann-Wilsdorf D in Dislocations in Solids (ed. FRN Nabarro, Duesbery MS) Ch. **59** (Elsevier, Amsterdam, 2002).

IOP Conf. Series: Materials Science and Engineering **375** (2018) 012030 doi:10.1088/1757-899X/375/1/012030

Decoding Pattern Dependence of Micro-indentation on Crystal Orientation of Cube-textured Aluminum Foil Using SAXS and Temperature Changes

S. Saimoto[1], M. A. Singh[2], M. R. Langille[1] and C. Gabryel[1]

[1] Mechanical and Materials Engineering, Queen's University, Kingston, ON. Canada, K7L3N6

[2] Physics, Eng. Physics & Astronomy, Queen's University, Kingston, ON Canada, K7L3N6

Abstract

Recent use of nano-indentation to assess work-hardening of complex microstructure of dual phase steels as a function of deformation together with crystal plasticity finite element method to simulate the bulk properties gives rise to the question of the role of crystallite orientation during indentation. Such studies using nano-indentation on μm-sized grains are difficult. To elucidate this effect, Vickers micro-indentation of cube textured Al capacitor foils were examined. These foils of 106 μm thickness are convenient for small angle X-ray scattering (SAXS) examination and the orientation around the indent can be examined by back-scattered electron diffraction (EBSD). The indentations were carried out with the diagonals parallel to the rolling direction (RD) and 45 ° to it at temperatures from 20 °C (295 K) to 160°C (433 K). The SAXS tests were performed only at 20°C on RD specimens.

Keywords: Aluminum, micro-hardness, SAXS, stacking fault tetrahedral, dislocation intersections

1. Introduction

Indentation testing using Vickers hardness with diamond pyramid head has been the standby materials characterization of mechanical properties. Since the discovery of age-hardenable alloys, the great progress is due to the advent of hardness testing to monitor ageing as function of time and annealing temperatures and has become the standard procedure to perform such tests to delineate the optimum hardness for alloy development. As the microstructure of the alloy becomes more complex with multiple hardening precipitates, the indenter size have decreased to micro- and nano-meter scales. Detailed characterization of texture and stored work in damaged volumes using conical indenter have been performed to determine if a simple correlation between hardness and geometric-necessary-dislocation (GND) density exist [1]. The application of such computational analyses was recently illustrated for dual-phase steels [2] in which nano-indentation result was used to correlate to the bulk stress-strain response. However application of such methods to face-centred cubic (fcc) structures may require more statistical testing to assess the crystal orientation dependence of the hardness response. Hardness has been attributed to two main factors; the yield strength due to the size, density and distribution of the embedded nano-particles and the gradients in GND densities around the indent interface. Brown [3] has simulated indentation using Vickers diamond indenter and has developed a circular flow model to predict lattice rotation which could rise as high as 12 °. Such rotations has been observed in Berkovich-tip indented matrix of cube oriented copper crystal using Kikuchi line analysis of electron diffraction (TEM) images (14 °) [4] and in conical indented (111) surface of copper crystal (15 °) using EBSD [1]. Although the analytical means for modelling micro- and nano-plastic flow seems to be at hand [2], for aluminum (Al) alloys dynamic recovery can occur during indentation at room temperatures.

During the recent work of assessing double Bragg scattering (DBS) giving rise to streaks [5] when performing SAXS of Al alloys, it was convenient to use cube textured Al foils to show that crystal orientations exist which do not satisfy the DBS diffraction conditions. The occurrence of such streaks need to be removed prior to the analyses of isotropic data to reveal nano-void formation. During this study it was discovered that the orientation of the indent diagonal whether it was parallel to the rolling

IOP Conf. Series: Materials Science and Engineering **375** (2018) 012030 doi:10.1088/1757-899X/375/1/012030

direction (RD) <100>, or at 45° to it, <110>, the shape of the indent was noticeably different. The origin of this indent pattern difference is examined in this report as a function of temperature.

2. Background on slip systems and interactions in crystals of multiple slip orientation

2.1 Multiple slip effects in copper single crystals under homogeneous deformation

Many past studies [6 - 11] have shown that shape-change of high symmetry oriented crystals do not necessarily occur according to geometric considerations of resolved shear stress since the strengths of dislocation interactions are not all equal. Furthermore, two different slip systems which occur in different parts of the crystal may meet to form strain-compatible dislocation boundaries [7]. One case is shown in Fig. 1, whereby slip traces on $(0\bar{1}1)$ surface of [011] oriented tensile axis pulled at 78 K [6] correspond to primary slip on one side and critical slip on the other to form {001} boundaries as predicted by Kear [7]. As the strain increase, the two portions rotate in opposite directions resulting in break-up of the crystal. The cause of this observation is due to the instability of [011] under tension. However even under stable compression testing using temperatures ranging from 78 K to 813 K, it was found that the predicted four slip systems reduced to two above 295 K [8]. The pair combination may give rise to either Lomer-Cottrell (primary-conjugate) or Stroh-Hirth (primary-critical) locks. The observation indicate that at low temperatures the sessile jogs form from primary-critical intersections to produce vacancies by jog-dragging but at elevated temperatures like-jogs combine to become sufficiently long to dissociate into sessile Stroh-Hirth locks such that strains from this combination ceases. The formation geometry of Lomer-Cottrell locks are more restrictive and they can also be unzipped such that the primary-conjugate combination can continue to operate to accommodate the imposed strain. Strain rate sensitivity using thermal cycling between 78-162 K shows that [001] which includes both primary-critical and the primary-conjugate combinations show lower strain rate sensitivity than [111] which encompass only the primary-conjugate pairs. On the other hand for the 78-273 K cycling, the response are identical showing that the primary-critical obstacle formation, that is jog-dragging, has ceased due to formation of Stroh-Hirth locks [9] as indicated above.

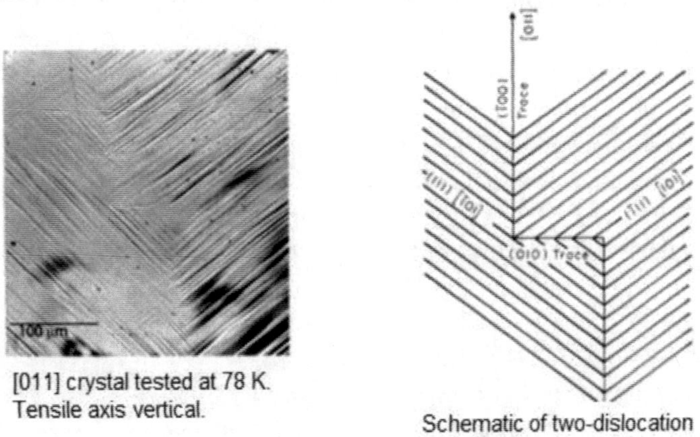

[011] crystal tested at 78 K. Tensile axis vertical.

Schematic of two-dislocation boundaries

Figure 1. Dislocation boundaries formed by primary-critical slip systems according to Kear [7].

2.2 Multiple slip effects in aluminum single crystals

During tensile deformation of [001] Al single crystals, it was found that serrated flow initiated at about 250 K and the critical strain decreased with increasing temperature. These serrations coincided with the appearance of slip lines [9] on the (110) surface wherein three traces are observed with the third and fourth being coincidental. This instability was attributed to strain ageing due to atmosphere formation by vacancies. Since the volume fraction of cube texture in Al sheets has a large effect on its formability, basic studies have been undertaken on plane strain compression of (001) [110] and (001) [010] single crystals at various temperatures [10, 11]. It was found that above strain of 0.20, break-up of the crystals according to the primary-critical intersections were observed with formation of {001} deformation boundaries. Furthermore the increase of cube texture in sheet occurs during recrystallization of rolled sheet wherein the recrystallization nuclei are found in the high GND regions due to strain gradients. To examine this effect, the occurrence of recrystallized grains after Vickers indentation of 12 % pre-rolled sheet of nominally pure aluminum of mm-sized grains were assessed as function of orientation by Xu et al. [12]. For the $(3\overline{1}\overline{1})$ [215] orientation, the indent shape had bulged out; that is sides were convex indicating that the crystal along the indent interface was pushing up along the indenter faces from the original surface plane. These indented grains which were annealed at 310 °C for 1 h showed that grains with (001) parallel to the sheet surface did not form new crystallites whereas most other orientations showed various degrees of recrystallization. This observation is in accordance with that for compression of Al single crystals, in which Nakata [13] found that after about 10 % compression of [001] crystal the amount of stored work was about 5% compared to 25 % for all other orientations including polycrystalline specimen. It should be kept in mind that the primary-critical intersections results in sessile jogs which generate vacancies that would promote dynamic recovery. These aspects of deformation behaviour will be discussed in the slip pattern changes caused by pyramid indenter on (001) with diagonal in RD and 45° orientations at various temperatures.

3. Experimental details

High purity aluminum capacitor foils supplied by Toyo Aluminum of Japan were 106 μm thick sheets that were fully recrystallized with 89.9 % cube texture. Specimens about 25 x 20 mm were bright polished with etch solution of 10 g $NaNO_2$, 100 ml saturated NaOH and 80 ml distilled H_2O. After rinsing in dilute HCl to remove sodium on the surface, electro-polishing was done using in volume % solution of 7 % perchloric acid, 70 % ethanol, 20 % distilled water with balance butyl-cellusolve; the current density was 1.5 A/cm^2. Pre-cooling of apparatus and solution, permitted polishing below -10°C. Vickers micro-hardness indents were made with the rolling direction (RD) either parallel to the diagonal (D) of the indenter, [100] or at 45 ° to it, [110]. For the elevated temperature tests, a programmable hot plate with X-Y control was used as described elsewhere [14] and the top surface was checked by a portable infra-red thermometer. From the indent geometry, not only the diagonals D but also the minimum width W were determined such that D/W can be used as a characterizing parameter with temperature change. For SAXS examination, only the RD type indents were used and the approximate D = 100 μm indents were placed about 300 μm apart. X-ray beam was collimated with two right angle slits set at 600 μm apart such that the spot size was near 1mm x 1 mm to simultaneously irradiate at least 4 indents. The details of the SAXS measurements are described elsewhere [5] but for the current purpose it is pointed out that the 3x3-hole specimen holder could be rotated about a horizontal axis parallel to RD. The rotation up to 12 ° were performed remotely over the same volume. In this manner the probable source of DBS could be elucidated.

4. Results: *Observations of surface upheaval and rotated structures around the indent*

The slip pattern around the Vickers indentation are shown in Fig. 2. The indent with the diagonal parallel to the $[\overline{1}00]$ (referred as RD) shows single slip trace parallel to the interface and D/W is 1.53 whereas for D parallel to $[\overline{1}10]$ (referred as 45°) careful scrutiny indicate two fine slip traces, nearly 90° to each other with D/W = 1.73. These fine traces become undetectable near the minimum W location near the centre that correspond to {001} dislocation boundaries described in Fig. 1. Since for isotropic deformation, D/W is 1.414, the calculations shows that the interfaces parallel to RD (the 45° one), that is the {001} boundaries, is more resistant to deformation than that at 45° to RD. Examination of the standard [001] stereographic projection (not shown) indicate that for RD orientation the traces are due to $(\overline{1}11)$ on one side such that for equal operation of collinear slip the effective Burgers vector (BV) is

$[1\bar{1}2]$; on the other side for $(1\bar{1}1)$, BV is $[\bar{1}12]$. On the other hand for the 45° case, the combination set of $(\bar{1}\bar{1}1)$ and $(\bar{1}11)$ which upon equal strain on each, BV effectively becomes [101] and for the set of $(1\bar{1}1)$ and (111), $[\bar{1}01]$. Thus, possible wavy slip may be replaced by $\{101\} <\bar{1}01>$ slip [15] as indicated in Fig. 3. Note the Schmid factor for {011} slip is larger than that for {111} at [001].

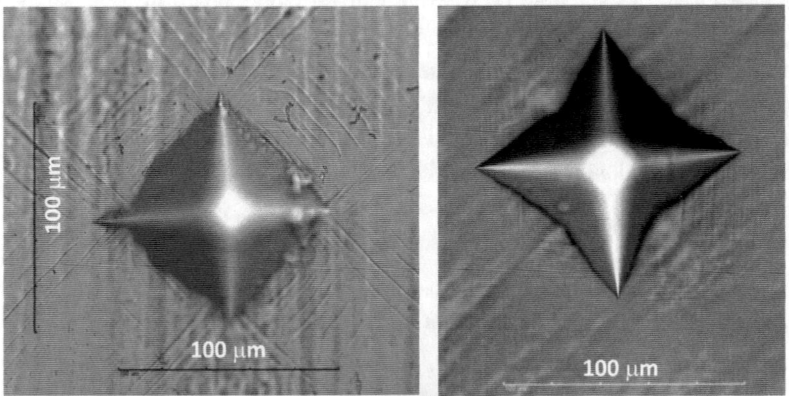

Figure 2. The indents formed at 295 K due to D parallel to RD (left) and 45° (right) are shown. Note the slip lines for RD are parallel to the trace of the interface whereas for the 45° one it is at 45 ° to it.

Figure 3. This indent occurred on part of sub-boundary such that the fourth interface region (lower right side) shows slip traces parallel to {011}. EBSD pattern [5] shows that the sub-boundary cannot be rotated by 45° to form {111} traces. Line markers are 100 µm.

Figure 4a shows the length measurements of D and minimum W and their ratios at various test temperatures for the 45° orientation. The slope of the D line is 0.283 and is larger than that for W at 0.131. However D/W is constant with temperature. These results indicate that the work-hardening mechanisms are temperature dependent but remain the same. On the other hand in Fig. 4b for RD orientation, D is linear with temperature with slope of 0.278 but that for W shows a break near 80°C (353 K) wherein the temperature at which annealing of stacking fault tetrahedral (SFT) occurs and ends near 120°C (393 K) [16] as denoted. The change of mechanism is attributed to the termination of SFT formation leading to primary-critical jog migration and their increase in length to form Stroh-Hirth locks. The work-hardening change is indicated by W becoming temperature independent.

In order to bypass the difficulty of quantifying the evolution of the indent using the observed slip traces, the Brown mechanism of circular flow [3] was invoked. In this model, lattice rotations of about 12° are predicted about the axis normal to the plane of V-shape indent. The circular flow are centred at the points of contact of the indenter interface to the specimen surface with the radius equal to W; that is, one semi-circle intersects the other at the mid-point below the V-shape indentation (*Fig. 2 in [3]*). The maximum lattice rotation which is independent of indent depth is found near the edges of the interfaces. In order to validate the crystal rotation, recent work [5] using SAXS indicated that DBS should occur if one of the possible diffraction poles coincided with (90° - θ_{Bragg}), a continuous circle within the stereographic projection. Thus SAXS images were taken with the rotation axis of the specimen holder parallel to the RD for the RD-specimen. The specimen was solid-body rotated up to +/- 12° to detect possible streaking. However, intrinsic to the Brown model, lattice rotation occurs to result in an asterism effect up to 12°. Thus for the RD specimen the possible axes are [$\bar{1}10$] and [$\bar{1}\bar{1}0$]. Subsequent to this deformation-induced lattice rotation, solid-body rotation up to +/- 12 ° were performed. Figure 5 schematically displays the observed streaks at the +12° position. Quantitative analysis using {420} and {422} pole figures indicate that the streaks can arise due to DBS from these planes. The details are not shown due to space limitations. The pole figures from EBSD determination [5] show that the cube texture is not ideally orientated and consists of two major closely-related orientations which are off cube axes by about 5°. For this reason, rotation about -12° did not show streaks; that is, the expected symmetry for ideal cube was not observed.

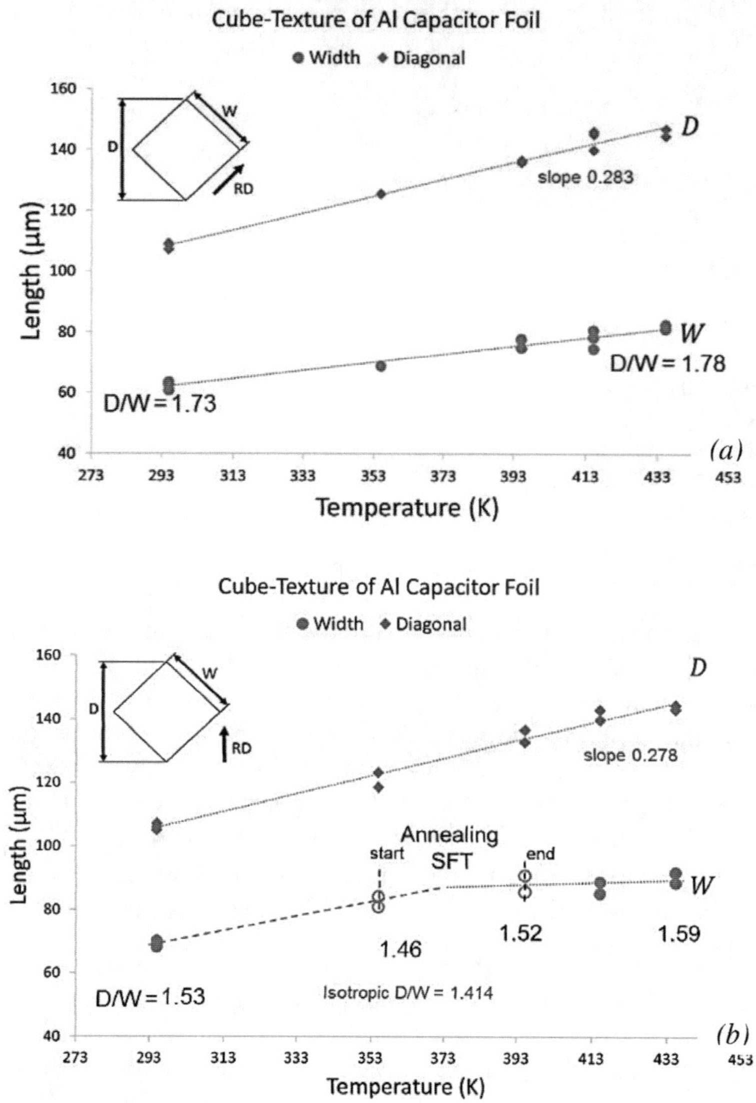

Figure 4. The parameter, length of D and W, and the ratio of these lengths versus temperature.
(a) For 45° indents the slope for D and W is constant.
(b) For RD, only that for D is constant and W shows temperature dependence.

5. Conclusion

The observation of slip patterns produced by Vickers diamond indenter on cube-textured Al foils showed the curious effect that for indents with the diagonal parallel to cube axis deformed with a more or less square impression whereas that for one rotated at 45° showed a concave shape with spearhead shape at the corners. The temperature dependence of the pattern showed that hardening mechanisms were temperature dependent but did not change for the 45° orientation. However for the RD one, the near-isotropic deformation behaviour at 80°C is taken to mean that SFT are no longer formed and are replaced by Stroh-Hirth locks. The lattice rotation due to the formation of indents can be detected by SAXS as it leads to streaking whenever the DBS condition is satisfied.

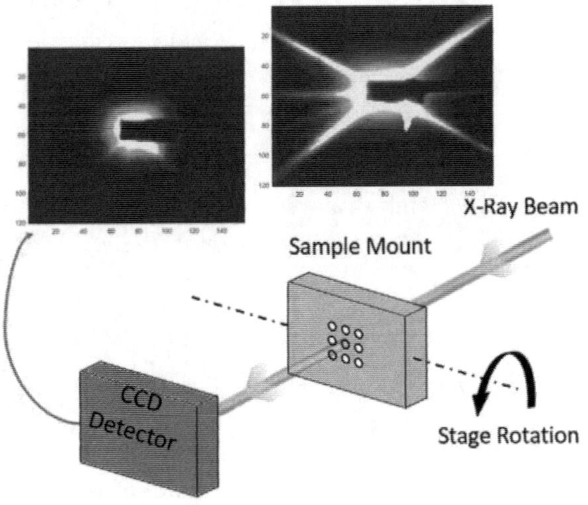

Figure 5. Schematic diagram of SAXS set-up with rotatable specimen holder. Left image shows 2-D SAXS image at 0° position and one with streaks at the 12° position.

References

[1] E. Demir, D. Raabe, N. Zaafarani, S. Zaefferer, Acta Mater. 57 (2009) 559-569.
[2] C.C. Tasan, M. Diehl, D.Yan, C. Zambaldi, P. Shamthraj, F. Roters, D. Raabe, Acta Mater. 81 (2014) 386-400.
[3] L.M. Brown, Mater. Sci. Forum, 550 (2007) 105-117
[4] K.K. McLaughlin, N.A. Stelmashenko, S.J. Lloyd, L.J. Vanderperre, W.J. Clegg, Mater. Res. Soc. Symp. Proc. 842 (2005) R1.3.1-R1.3.6.
[5] M.A. Singh, S. Saimoto, M.R. Langille, J. Lévesque, K. Inal, A.R. Woll, Philos. Mag. 97 (2017) 2496-2513.
[6] S. Saimoto, Low temperature deformation of copper single crystal oriented for multiple slip, Ph.D. thesis, Massachusetts Institute of Technology, Cambridge, MA. USA, (1964).
[7] B.H. Kear, Trans. Metall. Soc. AIME, 224 (1962) 674-677.
[8] L.E. Collins, Plane strain deformation and subsequent recovery of ideally oriented copper single crystals, M.Sc. Thesis, Queen's University, Kingston, Canada (1977).
[9] S. Saimoto, Philos. Mag. 86 (2006) 4213-4233.
[10] A. Akef, J.H. Driver, Mater.Sci. Engng. A 132 (1991) 245-255.
[11] C. Maurice, J. H. Driver, Acta metall. mater. 41 (1993) 1653-1664.
[12] C. Xu, Y. Zhang, F. Lin. G. Wu, Q. Liu, D. Juul-Jensen, Metall. Mater. Trans. 47A (2016) 5863-5870.
[13] Y. Nakata, Philos, Mag. 11 (1965) 251-261.
[14] G. Langelaan, S. Saimoto, Rev. Sci. Instrum. 70 (1999) 3413-3417.
[15] R. Le Hazif, J-P. Poirer, Acta Metall. 23 (1975) 865-871.
[16] H. Ohkubo, Y. Shimomura, I. Mukouda, K. Sugio, M. Kiritani, Mater. Sci. Eng. 350 (2003) 30-36.

Acknowledgements

The authors thank Cornell High Energy Synchrotron Source (CHESS) at which the SAXS studies were performed and the Natural Sciences and Engineering Council of Canada for support over many years.

IOP Conf. Series: Materials Science and Engineering **375** (2018) 012031 doi:10.1088/1757-899X/375/1/012031

The nucleation of cube grains during primary recrystallization of aluminum

M M Miszczyk[1,a] and H Paul[1,b*]

[1]Polish Academy of Sciences, Institute of Metallurgy and Materials Science, Krakow, Poland

[a]m.miszczyk@imim.pl, [b]h.paul@imim.pl

*Corresponding author: h.paul@imim.pl

Abstract. The formation of cube grains during recrystallization has been analyzed in commercial AA1050 alloy and pure Al single crystal of S{234}<524> orientation. Samples of AA1050 alloy were deformed along two deformation modes to form different as-deformed texture components and then lightly annealed. One group of the samples was plane strain compressed (PSC) in a channel-die. The second group was deformed by equal channel angular extrusion (ECAE). The results obtained on polycrystalline AA1050 alloy were compared with recrystallization behavior of PSC single crystal of S{234}<524> orientation. The textures were measured by SEM/EBSD and X-ray diffraction. It was found that a very weak residual cube{100}<011> component was observed in the samples of AA1050 alloy after both deformation modes. After annealing cube-oriented grains were extensively formed only in PSC samples. These grains were situated preferentially inside or in between the S-oriented areas with local disorientations about <111> axes. Cube-oriented grains were not formed during annealing of the ECAE samples, where main as-deformed texture components close to two complementary orientations of {124}<651>-type were transformed towards two components of ~{100}<011> and ~{221}<114>, essentially by <110> rotations. In single crystalline samples cube grains were extensively formed during annealing despite (near)cube-oriented areas haven't been observed in the as-deformed microstructure.

1. Introduction

Control of earing properties of thin sheets relies on a careful control of the texture. It is well-known that the minimal earing can only be obtained by well-balanced texture. In the case of Al and its alloys a key concern is to obtain a sufficient amount of cube{100}<001>-oriented grains. However, the origin of strong cube texture during annealing of cold-rolled face centered cubic (fcc) metals still constitutes an important scientific and technological problem, e.g. [1-13]. Most of the authors argue that {100}<001>-oriented grains evolve from 'cube bands', which are present in the deformed microstructure, often as thin (near)cube zones between high orientation gradients created by diverging rotations of unstable orientations [13] or fragmented leftovers of the original cube grains [8]. Two different cube grain nucleation models have been proposed. A *first 'nucleation' model* [2] proposes that the deformed cube-oriented bands recover faster due to the easy annihilation of two sets of pure screw dislocations with perpendicular Burger's vectors, without any requirement of orientation relation across recrystallization front. The *second 'micro-growth' model* [3, 4] is based on the strong growth rate dependence of ~40°<111>-type between areas of the deformed state of S{123}<634>

orientation and cube recrystallized nuclei. Both models, which may be concurrent or in competition, or even superseded by some other mechanisms, assume presence of initial cube nucleus in the deformed state.

However, the statement that cube-oriented nucleus originate from fragmented leftovers of the original cube grains is strongly disputable since the cube-oriented crystallites are very unstable during room temperature rolling or plane strain compression (PSC). At the beginning stages of deformation the grains of this orientation splits up by transverse and rolling direction rotations and evolve, in the range of very high deformations (> 90%), towards four variants of ~S orientation, e.g. [1,13]. To solve this problem some of the authors assume, e.g. [8] that despite large volumes of cube crystallite are unstable (and rotates towards different variants of S orientation) but small fragments could be metastable because they are subject to a different stress state. These less deformed areas conserve orientations near the original cube orientation and they are the source of cube grains nucleation during recrystallization. But this description of the phenomenon leads to another difficulty - *it is very difficult to accept the assumption that during annealing less deformed cube-oriented leftovers are more privileged sites for nucleation of new grains (areas of lower stored energy) as compared to highly deformed ones (areas of higher stored energy)*. Moreover, the experiments on deformed single crystals of fcc metals of stable orientations, deformed up to moderate strains clearly show that orientations of the as-deformed state are not retained in the new grains orientations during primary recrystallization [9, 10, 14, 15]. In these simple cases the orientation and the underlying microstructure of the as-deformed matrix are clearly defined due to the strong stability of such crystals. Since the orientations of new, recrystallized grains are not identified inside the as-deformed structure and the successful nuclei predominantly possess the ~25-55°(~<111>)-type orientation relationship with respect to the deformed matrix, e.g. [1-20], there must be valid a rotation mechanism which transforms the as-deformed state orientations into recrystallized one.

In the case of cold-rolled and annealed polycrystalline fcc metals the situation is ambiguous. As an example, cold rolled commercial purity Al after annealing often develops a mixed cube and 'R' texture, where: 'R' means Retained component, close to S. Since the as-deformed texture 'image' after typical industrial rolling up to 70 - 90% reductions is composed of strong S-orientation of 4 variants (supplemented by 2 variants of brass{110}<112> and copper{112}<111> orientations) and weak cube orientation, there is not clear whether the presence of cube and S-oriented grains after annealing results from validity of the oriented nucleation model or the rotation around selected the <111> axes of S orientation. (It is important to remember that ~40° rotation of ~S orientation around selected the <111> axes is able to transform the S-oriented areas of given variant into (near)S-oriented grain of another variant or into (near)cube orientation).

We believe that the main factor which influences cube grain nucleation are the specific texture components of the as-deformed state. Therefore, the samples of AA1050 alloy were deformed along two deformation modes to form different as-deformed texture 'images' and then lightly annealed. The results obtained on polycrystalline samples were compared with recrystallization behavior of PSC single crystal of S{234}<524> orientation. High stacking fault energy (SFE) of Al practically excludes mechanical and recrystallization twinning and it facilitates identifying the mechanisms responsible for the texture transformation during the initial stages of annealing. The basic research technique employed is scanning electron microscopy equipped with an electron backscattered diffraction (SEM/EBSD) facility.

2. Experimental

The materials were a commercial AA1050 alloy and Al(99.998%) single crystal of the (234)[52-4] orientation defined here as 'S' orientation. (In fact the orientation was few degree rotated away from 'classical' positions of S orientation usually defined as: {123}<634> or {124}<211>). One group of the samples of AA1050 alloy was deformed in a channel-die to the logarithmic strain of 0.52 (41%), whereas the second group was equal channel angular extruded (ECAE) up to six passes along route A. In order to limit the friction each PSC sample was wrapped in Teflon™ tape, whereas in the case of

ECAE samples the billet and tooling were well-lubricated with MnS_2 containing grease. Then the samples were annealed: AA1050 alloy (after PSC and ECAE) - for 1h at temperatures ranging between 100°C and 450°C with the purpose to obtain different stages of recrystallization, whereas single crystal of S orientation - for 30s at 420°C with the purpose to obtain final stage of primary recrystallization (95% of the sample surface is covered by recrystallized grains). The deformed and recrystallized microstructures were analyzed in the ND/ED section (where: ND and ED denote, the normal and the extension directions, respectively) with the use of SEM - QUANTA 3D FEG, equipped with a field emission gun and EBSD facility. At the sample scale, the textures were measured on the ED/TD plane (where: TD denote transverse direction) with the use of the Philips X Pert PW-1830 X-ray diffractometer.

3. Results and discussion

3.1. Microstructure and texture of the as-deformed state

3.1.1. PSC and ECAE of AA1050 alloy. The initial material was in the form of hot-rolled sheet taken from the medium stage of the technological process, i.e. after the reversible mill. At this initial stage, the material was characterized by a typical copper-type rolling texture composed of strong 4 variants of S{123}<634>, 2 variants of C{112}<111> (or copper) and brass{110}<112> orientations and minor cube{100}<001>. As expected, the texture of the initial material wasn't essentially changed during further PSC up to 41%. (After 41% PSC the texture image is quite similar to that after large deformations during cold-rolling of many fcc metals of medium to high SFE), e.g. [10].

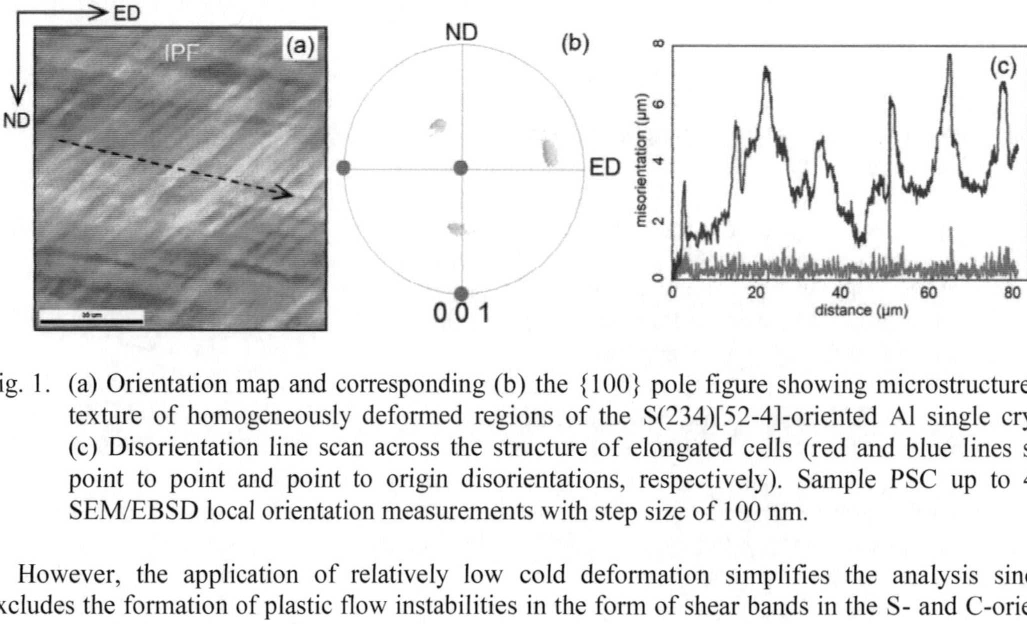

Fig. 1. (a) Orientation map and corresponding (b) the {100} pole figure showing microstructure and texture of homogeneously deformed regions of the S(234)[52-4]-oriented Al single crystal. (c) Disorientation line scan across the structure of elongated cells (red and blue lines show point to point and point to origin disorientations, respectively). Sample PSC up to 41%. SEM/EBSD local orientation measurements with step size of 100 nm.

However, the application of relatively low cold deformation simplifies the analysis since it excludes the formation of plastic flow instabilities in the form of shear bands in the S- and C-oriented grains [21]). It is clear that it is impossible to apply exactly the same amount of deformation *via* ECAE processing. After 1-2 passes along route A the texture is almost random, without clearly marked components. A 'clear texture image', with two strong components is obtained after 6 passes of ECAE processing (up to 5 passes one still detects additional weaker components of the initial state). The texture after 6 passes is 'stable' during further processing, i.e. is not essentially changed up to very high deformations (up to 12 passes). The texture 'image' is composed of two nearly complementarily oriented groups of orientations close to {124}<651> with slight scattering towards the shear{100}<011> orientation. Moreover, in this strongly sheared sample there is also a weak residual cube texture - a component that is in fact closer to exact cube than in the PSC sample.

3.1.2. PSC of aluminum single crystals of S(234)[52-4] orientation. The crystal undergoes a macroscopic $\varepsilon_{ED/TD}$ shear, typical of plane strain compressed S or near S-oriented crystals. Optical microscopy revealed significant changes of the sample shape in the ND/ED section with clearly marked traces of thin bands of localized strain. Generally, the aim was to generate a homogeneous microstructure throughout the samples but this turned out to somewhat over-ambitious since in S-oriented single crystals the regions of both homogeneous and heterogeneous deformation structures were found. After 40% deformation the heterogeneities were thin layers of localized strain crossing the samples at 15-20° to ED. The rest of the sample underwent homogeneous deformation in which the disorientation between neighboring areas is <15°.

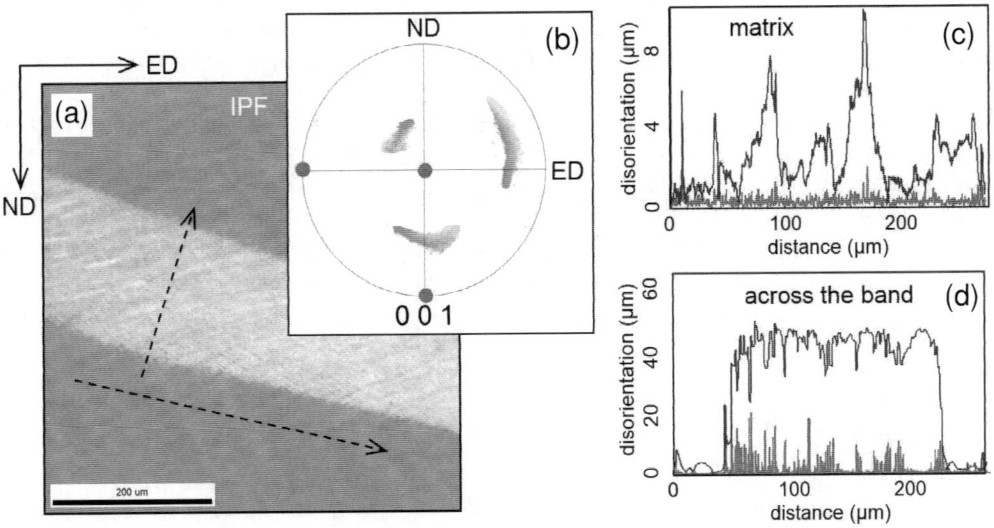

Fig. 2. Orientation map (a) and corresponding the {100} pole figure (b) showing microstructure and texture of heterogeneously deformed region of the S(234)[52-4]-oriented Al single crystal. Disorientation line scans across the matrix (c) and across the band (d). Red and blue lines show point to point and point to origin disorientations, respectively. Sample PSC up to 41%. SEM/EBSD local orientation measurements with step size of 100 nm.

In homogeneously deformed areas the SEM/EBSD local orientation measurements revealed microstructures composed of elongated microbands surrounded by low angle boundaries of one or two families (Fig. 1). The inclination of microbands with respect to the external directions depends on the analyzed area. However, strong coincidence with the traces of {111} planes is observed. The crystal lattice of the homogeneously deformed matrix (for larger distances) displayed rotation away from the initial crystal orientation around the axes lying close to TD. However, the disorientation angles in 'pixel to pixel' disorientation relation were below 15°. Consequently, the textures of the small areas corresponding to the homogeneously deformed microstructures are only slightly scattered, as shown by the {111} pole figures of the mapped areas. The crystal lattice of areas of the heterogeneous deformation undergoes strong ~TD rotation towards the orientation situated complementarily with respect to the initial orientation. The disorientation across the boundary separating the band and the less deformed matrix very often attains 40-50°. However, it is important to note that disorientation between particular cells inside the bands of localized strain is significantly lower and typically do not exceeds 15° (Fig. 2).

One of the most important conclusion which results from the detailed crystallographic characterization of the as-deformed sample is that the areas of (near)cube orientation are not present in the as-deformed samples PSC up to 40%.

IOP Conf. Series: Materials Science and Engineering **375** (2018) 012031 doi:10.1088/1757-899X/375/1/012031

3.2. Texture changes during annealing

Recrystallization behavior of commercial aluminum AA1050 alloy. The exact position of the main texture components of the PSC samples annealed at temperatures below 300°C were quite similar to those observed for the as-deformed state (Figs. 3a-c). At 300°C a radical textural changes occurs due to the development of a strong cube orientation, the retention of the 4 variants of the S orientation as recrystallized grains orientation (with lower intensity compared to the deformed state) and the disappearance of both C (copper) components in the fully recrystallized samples. It was observed that the volume fractions of all four S variants decreased, whereas the volume fraction of the cube component increased with annealing temperature. However, it was also noticed that similar area fractions of near cube-oriented areas, clearly observed in the as-deformed state were still-conserved during annealing up to 325°C. In the case of ECAE samples the most important textural changes were initiated at annealing temperature of 270°C (Figs. 3d-f). The intensity of the as-deformed components near the {124}<561> orientations strongly decreases with the progress of recrystallization and this process coincides with an increase of the {100}<011> and {221}<114> orientations which dominate in the recrystallized structure. Despite the presence of cube-orientation in the as-deformed state and over a wide range of annealing temperatures (of the same intensity) the cube texture development during annealing was not observed. This leads to the conclusion that cube-oriented bands in the as-deformed state are insufficient 'factor' to influence cube texture formation after annealing.

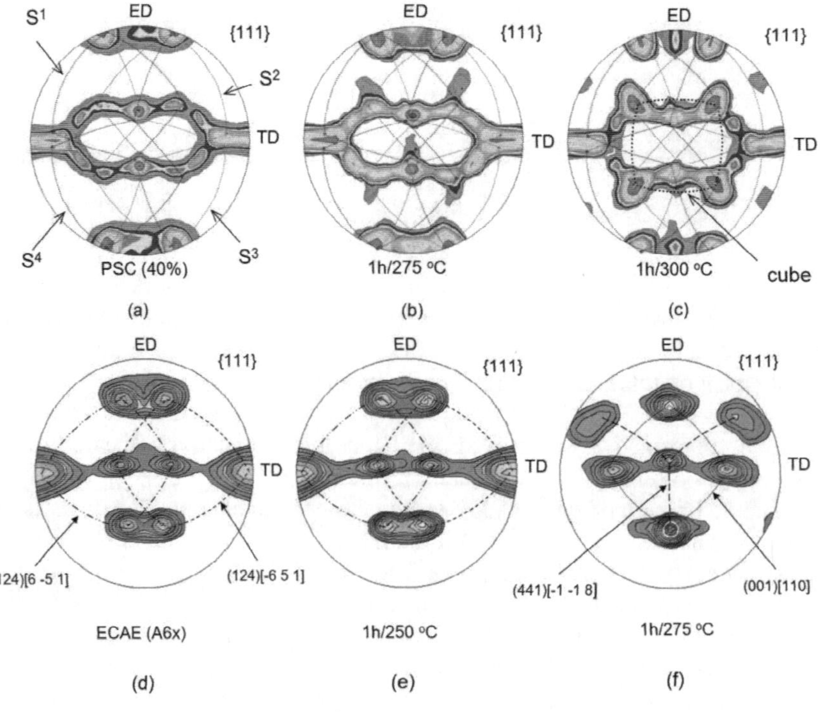

Fig. 3. Texture evolution during annealing: (a)-(c) PSC samples, (d)-(f) ECAE samples. The {111} pole figures showing essential texture changes. X-ray diffraction.

As for (near)cube-oriented grains there are observed two characteristic groups of places of their nucleation and further privileged growth to the inside of the S-oriented areas (of different variants) (Figs. 4). First group of cube grains is growing near the cube bands (Figs. 4a, c), whereas the second one is growing without any contact with cube bands (Figs. 4b, d), as visible on two dimensional orientation maps. Figure 4a shows a cube-oriented grain growing preferentially into S-oriented areas of the as-deformed state. The bottom part of the grain sticks with a layer (cube band) composed of the

IOP Conf. Series: Materials Science and Engineering **375** (2018) 012031 doi:10.1088/1757-899X/375/1/012031

cells~or (sub)grains of significantly larger diameter as those typically observed in the other as-deformed areas. In many cases the spread of the orientations of strongly recovered cells mark the transition from the near S to the (near)cube orientations and suggests that the rotation mechanism of ~40°<111>-type play a decisive role in the transformation of the as-deformed area orientations (~S) into the recrystallized one (~cube). Since the recrystallized grain is elongated along the boundaries of the as-deformed/recovered flat grains, therefore, it is likely that new grain nucleate inside ~S oriented areas of the as-deformed state and its preferred growth is connected with rapid migration of the high angle boundary. When the cube grain stick with cube band (marked as ~cube) the high angle grain boundary across recrystallization front disappears. This explanation of the situation observed in Fig. 4a could be confirmed by the cases showing the growth of cube grains to the inside of the as-deformed areas of S orientation, without necessity to be in contact with cube bands, as observed in Fig. 4b.

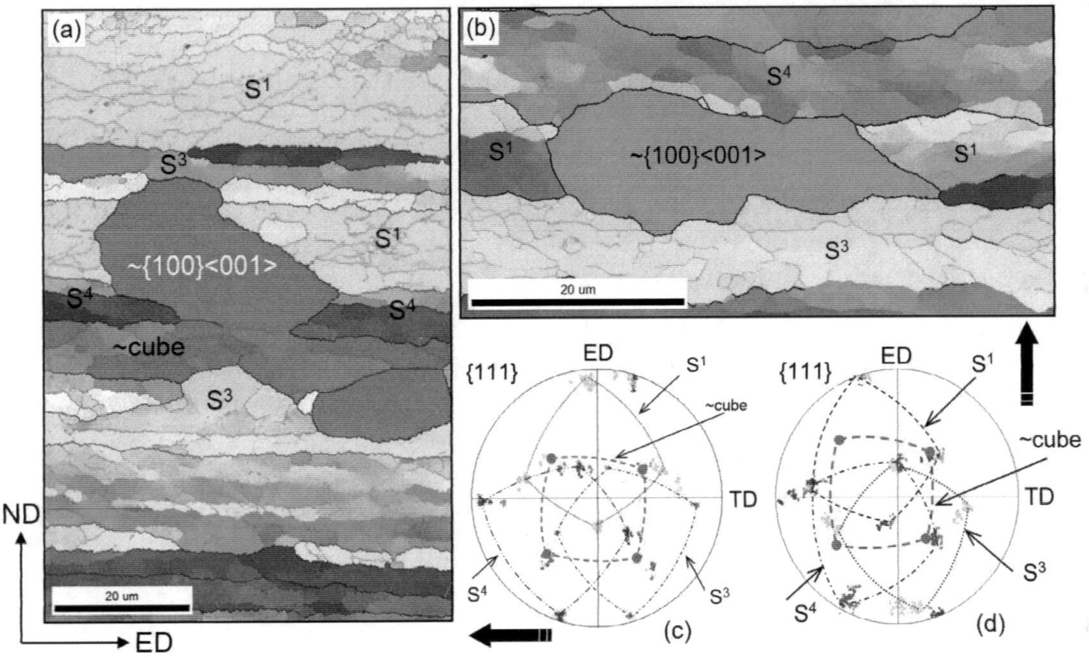

Fig. 4. Cube grains nucleation in the as-deformed state of ~S orientation: (a) cube grain growing to the inside of S-oriented areas (of different variants) without any contact with cube bands, and (b) cube grain growing near cube bands (marked as '~cube'). The new grain orientations and their nearest deformed/recovered neighborhood are presented in (c) and (d) as the {111} pole figures; they corresponds to the orientation maps presented in (a) and (b), respectively. PSC sample up to 40% and annealed for 1 h at 300 °C. SEM/EBSD local orientation measurements with step size of 100 nm.

However, in the case of polycrystalline metals the situation is always ambiguous, since '*there is a plenty room under bottom...*' and we never know if the cube bands are placed up/bottom observed section or not. The problem can be solved by the analysis of single crystalline samples of S orientation which is semi-stable during PSC up to moderate strains.

3.2.1. Recrystallization behavior of aluminum single crystals of S(234)[52-4] orientation. In the case of single crystalline samples the recrystallized grain orientation were analyzed inspects whole sample area. Figures 5a and b show the microstructure and orientations of recrystallized grains after annealing at 420 °C for 30 s. This recrystallized state can be described as final stage of primary recrystallization

(~95% of the recrystallized fraction) as analyzed on the sample section perpendicular to TD. Generally, the recrystallized grain orientations are symmetrically situated with respect to the external directions. This symmetry results from the positive and negative rotations of the recrystallized grain orientations around the axes grouped near all normals of the {111} planes. The disorientation angles are ranged between 25° and 55° with the maximum near the 40°, as presented many times in the past, e.g. [16, 18, 22-27].

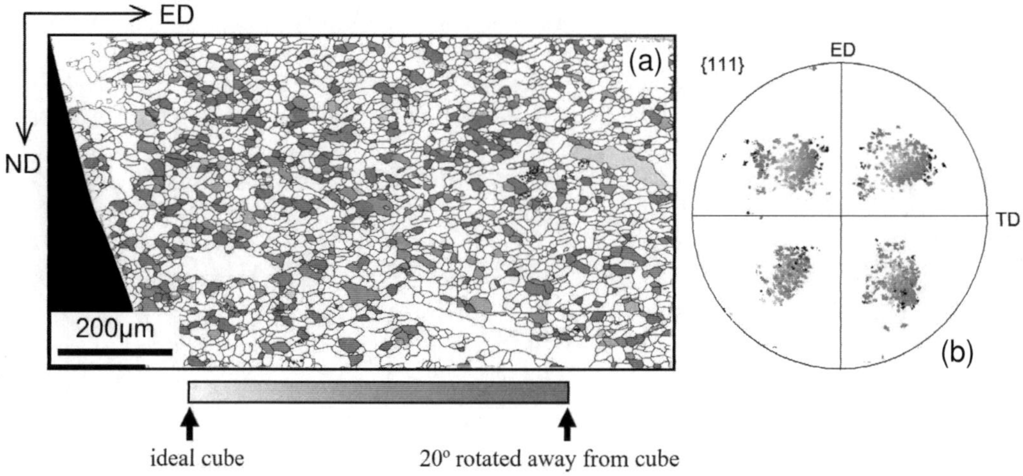

Fig. 5. Cube grains nucleation in the S(234)[52-4]-oriented Al single crystal. (a) Orientation map presented as a 'function' of grain boundaries (>15°) and texture component (20° scattering of ideal cube{100}<001> orientation, where: yellow – ideal cube, red – 20° away from cube) and corresponding (b) the {111} pole figure. Sample PSC up to 41% and then annealed at 420°C for 30s. SEM/EBSD local orientation measurements with step size of 5µm.

The high resolution orientation maps made in the deformed state well-documented that (near)cube-oriented areas were not identified in the as-deformed state and despite this the (near)cube grains nucleate intensively during annealing. This confirm the validity of the mechanism of the recrystallized grain orientations formation based on ~α<111>-type rotation. It is clear that anti-clockwise rotation of ~S-oriented areas of the as-deformed/recovered state around common for ~cube and ~S orientations the <111> axis directly leads to (near)cube grains. This also confirm the thesis *that presence of cube nuclei in the as-deformed structure is not necessary condition for (near)cube grains nucle*ation.

The calculation of disorientation relations between the recrystallized grains of cube orientation and the average orientation of the as-deformed/recovered state shows that most of the disorientation angles across the migrating recrystallization front are within the range of 25 and 55°, whereas the disorientation axes are concentrated around all the <111> poles 'describing' the orientation of the as-deformed/recovered state. However, the axes lying near the two highly stressed slip planes during deformation are preferred. It was also noticed that the scattering of the disorientation axes is larger than that of the orientations in the deformed state.

4. Conclusions

The present results provide detailed information about the development of cube texture in a commercial AA1050 alloy and a model Al single crystal of S{234}<524> orientation. Based on local orientation measurements and X-ray diffraction it was found that:

(i) A very weak residual cube texture component was observed in the samples of AA1050 alloy after both PSC and ECAE deformation modes. Despite this cube-oriented grains were extensively formed during annealing only in PSC samples. These grains were situated preferentially inside or in between the S-oriented as-deformed areas with local disorientations about <111> axes.

(ii) Cube-oriented grains were not formed during annealing of the ECAE samples. During annealing main as-deformed texture components of ECAE samples close to {124}<651> were transformed to two components of ~{100}<011> and ~{221}<114>-type, essentially by <110> rotations.

(iii) In the case of Al single crystal of S{234}<524> orientation, the cube grains were extensively formed during annealing despite no (near)cube-oriented areas were observed in the as-deformed microstructure. The strong cube texture after recrystallization results from the ~40°(~<111>)-type rotation of the as-deformed orientations.

(iv) Finally, it is concluded that during annealing of Al samples the cube-oriented grains can nucleate from S-oriented areas of the deformed state free of cube-oriented nuclei.

Acknowledgments:
This work was partially supported by the National Centre of Science and the Polish Ministry of Science and Higher Education, projects no: 2016/23/B/ST8/00439 and 2016/21/D/ST8/01700, respectively.

References

[1] Dillamore I L, Katoh H, 1974 *Met. Sci* **8** 73
[2] Ridha A A, Hutchinson W B, 1984 *Acta Metall* **30** 1929
[3] Duggan B J, Lücke K, Köhlhoff G D, Lee C S, 1993 *Acta Metall Mater* **41** 1921
[4] Samajdar I, Doherty R, 1996 *Scripta Mater* 32 845
[5] Vatne H E, Furu T, Nes E, 1996 *Mater Sci Techn* **12** 201
[6] Daaland O, Nes E, 1996 *Acta mater* **44** 1389.
[7] Engler O, *1996 Mater Sci Techn* **12** 859
[8] Albou A, Raveendra S, Karajagikar P, Samajdar I, Maurice C, Driver J H, 2010 *Scr Mater* **62** 469
[9] Miszczyk M M, Paul H, 2015 *IOP Conf Series: Mater Sci Engn* **89** 383
[10] Miszczyk M M, Paul H, Driver J H, Poplewska J, 2017 *Acta Mater* **129** 378
[11] Davenport S B, Higginson R L, Sellars C M, 1999 *Phil Trans Roy Soc* **357** 1654
[12] Nah J J, Kang H J, Huh M Y, Engler O, 2008 *Scr Mater* **58** 500
[13] Akef A, Driver J H, 1991 *Mat Sci Engn* **A 132** 245
[14] Miszczyk M M, Paul H, Driver J H, Maurice C, 2015 *Acta Mater* **83** 120
[15] Miszczyk M M, Paul H, Driver J H, Drzymała P 2017 *Acta Mater* **125** 109
[16] Doherty R D, Hughes D A, Humphreys F J, Jonas J J, Juul Jensen D, Kassner M E, King W E, McNelly T R, McQueen H J, Rollett A D, 1997 *Mater Sci Engn* **A 238** 219
[17] Miszczyk M M, Paul H, Driver J H, 2016 *Mater Character* **112** 68
[18] Ardakani M G, Humphreys F J, 1994 *Acta Metall Mater* **42** 763
[19] Juul Jensen D, 1995 in: Proc. of the 16th Risø International Symposium on Materials Science: Hansen N, Jensen D J, Lin Y L, Ralph B (Eds.), Risø Nat Lab, Roskilde, Denmark, 119
[20] Blicharski M, Liu J, Hsun Hu, 1995 *Acta Metall Mater* **43** 3125
[21] Paul H, Darrieulat M, Piątkowski A, 2001 *Zeitschrift fiir Metallkunde* **11** 1213
[22] Rios P R and Sandim H R Z 2005 *Mat Res* **8/3** 225
[23] Godfrey A, Juul Jensen D and Hansen N 1998 *Acta Mater* **46** 823
[24] West S S, Winther G and Juul Jensen D 2011 *Metall Mater Trans A* **42** 1400
[25] Fan G H, Zhang Y B, Driver J H and Juul Jensen D 2014 *Scripta Mater* **72-73**
[26] Paul H, 2003, *Materials Chemistry and Physics* **81** 531
[27] Inoko F, Kashihara K, Tagami M, Okada T, 2010 *Mater Trans* **51** 597

IOP Conf. Series: Materials Science and Engineering **375** (2018) 012032 doi:10.1088/1757-899X/375/1/012032

Relationship between initial hydrogen absorption properties and microstructures of Mg/Cu super-laminate composites with different accumulative roll bonding cycles

K Tanaka[1], D Nishino[2], R Kondo[3] and H T Takeshita[3]

[1]Research Institute of Electrochemical Energy, National Institute of Advanced Industrial Science and Technology (AIST), 1-8-31 Midoriagoka, Ikeda, Osaka 563-8577, Japan
[2]East Tower, Gate City Ohsaki, 1-11-2 Osaki, Shinagawa-ku, Tokyo 141-8643 Japan
[3]Faculty of Chemistry, Materials and Bioengineering, Kansai University, 3-3-35 Yamate-cho, Suita, Osaka 564-8680, Japan

E-mail: koji.tanaka@aist.go.jp

Abstract. The relationship between initial hydrogen absorption properties and microstructures of Mg/Cu super-laminate composites (SLCs) with different accumulative roll bonding (ARB) cycles was investigated. Hydrogenation of Mg was dominant in Mg/Cu SLCs with 5 and 10 cycles of ARB, whereas that of Mg_2Cu was dominant in Mg/Cu SLCs with 20 and 30 cycles of ARB. The growth rate constants did not show a significant difference among Mg/Cu SLCs with different ARB cycles. It is considered that the average thickness of Mg + Cu layers in as-rolled Mg/Cu SLCs as a function of ARB cycles has a large influence on initial hydrogenation properties and microstructure formation processes.

1. Introduction

Mg is still considered as a good candidate for hydrogen storage materials because of its high hydrogen storage capacity of 7.6 mass% in the form of reversible MgH_2 [1, 2], low cost, light weight and high abundance. Nevertheless, its hydrogen absorption/desorption kinetics is sluggish even at elevated temperatures (> 573 K) associated with its strong thermodynamic stability of standard formation enthalpy -74 kJ (mol $H_2)^{-1}$ and this effectively places a limit on its utility [3]. To solve these drawbacks, many efforts have been made recently, such as appropriate alloying of Mg [4-6], adding catalysts [7-11] and manufacturing nano-crystalline powder [12, 13].

Since it was reported that accumulative roll bonding (ARB) is effective for the preparation and hydrogenation of Mg-Ni based hydrogen storage alloys [14], super-laminate composites (SLCs) prepared by this method have been investigated on the hydrogenation properties of Mg-Cu[15, 16], Mg-Pd [17-19], Mg-Al [20], and Mg-Ti-Ni [21] systems.

In the case of Mg-Cu system, for example, Mg/Cu SLCs exhibit reversible hydrogen absorption/desorption at 473 K after initial activation at 573 K under the hydrogen atmosphere of 3.3 MPa [15], although Mg_2Cu alloys prepared by conventional melting method require 573 K [22]. From our previous research [16], it is found that the large surface area and the short diffusion distance

IOP Conf. Series: Materials Science and Engineering **375** (2018) 012032 doi:10.1088/1757-899X/375/1/012032

because of fine structures of Mg/Cu SLCs after initial activation at 573 K improve the dynamics of hydrogen absorption/desorption, and therefore lower the reaction temperature to 473 K.

Hydrogen absorption starts about at 453 K during the heating process of initial activation of Mg/Cu SLCs (hydrogenation under the condition of 573 K, in H_2 atmosphere of 3.3 MPa and dehydrogenation under the condition of 573 K in vacuum). During this, hydrogenation of Mg and alloying of Mg with Cu followed by hydrogenation of Mg_2Cu occur competitively [23-25]. However, the formation mechanism of microstructures is not fully understood yet. : For example, how much of Mg_2Cu can be formed in about ten minutes before temperature reaches 453 K during the heating process of initial activation.

It is reported that severe plastic deformation (SPD) such as high-pressure torsion (HPT) and equal-channel angular pressing (ECAP) can enhance atomic diffusion and promote solid-state reactions because of an increase in the density of lattice defects such as vacancies, dislocations and grain boundaries [26-31]. Therefore, it is important to know the effect of ARB. In this paper, we examined the relationship between initial hydrogen absorption properties and microstructures of Mg/Cu SLCs with different ARB cycles.

2. Experimental procedure

Commercial pure Mg (Takeuchi Metal Foil & Powder Co., LTD., 99.9 % of purity, 250 μm in thickness) and oxygen-free Cu (Fukuda Metal Foil & Powder Co., LTD, 99.98 % of purity, 10 μm in thickness) were used as starting materials. Mg foils were polished with #600 sandpaper in order to remove oxide layers and cold-rolled to 40 μm. These were cut into 20 mm x 40 mm in size and annealed at 673 K for 3.6 ks in Ar atmosphere of 0.1 MPa in order to remove internal stress and strain. Cu foils were cut into 20 mm x 40 mm in size and annealed at 873 K for 3.6 ks in Ar atmosphere of 0.1 MPa. Surfaces of Mg and Cu foils were cleaned with dilute hydrochloric and nitric acid, respectively. Twenty pairs of Mg and Cu foils were stacked in one. By combining Mg foil with Cu foil in different thickness, a stack with molar volume ratio Mg/Cu of 2 was prepared. Prior to ARB, this stack was wrapped up in a Cu plate with 1 mm in thickness and pressed under the condition of 735 MPa for 300 s in order to enhance adhesion and to prevent slip between Mg and Cu foils during ARB. The wrapped stack was roll-bonded at room temperature in air to get a Mg/Cu laminate using a conventional two high-roll mill until its length became about twice longer than the original one. Then the laminate was cut into half and piled. The piled laminate was wrapped up in a Cu plate again and then roll-bonded as described above. This procedure was repeated to fabricate a Mg/Cu super-laminate composite (SLC). Four types of specimens were prepared by changing the number of above procedure, which are 5, 10, 20 and 30 cycles of ARB.

Mg/Cu SLCs were hydrogenated under the conditions of 573 K, 86.4 ks in H_2 atmosphere of 3.3 MPa with a Sieverts' type instrument to measure the initial hydrogen absorption properties. Constituent phases were identified by powder X-ray diffraction (XRD) and energy dispersive X-ray spectroscopy (EDS). Microstructures of Mg/Cu SLCs were observed with a scanning electron microscope (SEM) and a scanning transmission electron microscope (STEM). SEM and STEM specimens were prepared by conventional mechanical grinding and polishing followed by ion milling.

The formation mechanism of microstructures in Mg/Cu SLCs during initial hydrogenation through competitive reactions was estimated by evaluating the growth rate constants of Mg_2Cu in Mg/Cu SLCs under 453 K with different ARB cycles. The measurement method of the growth rate constants of Mg_2Cu in Mg/Cu SLCs was reported in detail in Ref [30] and the experimental results of the growth rate constants of those with different cycles of ARB will be reported in detail elsewhere.

3. Results and discussions

Fig. 1 shows cross-sectional backscattered electron (BSE) SEM images of as-rolled Mg/Cu SLCs with different ARB cycles; (a) 5, (b) 10, (c) 20, (d) 30 cycles, respectively. Dark contrast is Mg and bright contrast is Cu. The microstructures of Mg/Cu SLCs became finer as the number of ARB cycles increased. For example, the average thickness of Mg and Cu layers in the case of 10 cycles of ARB

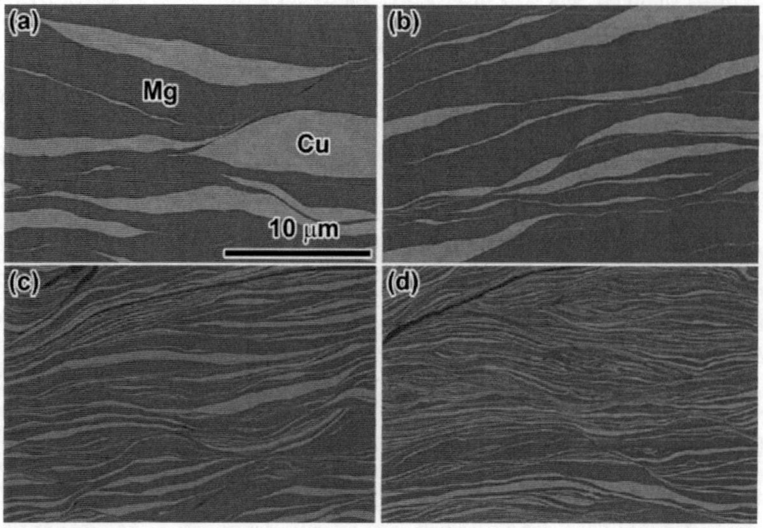

Figure 1. Cross-sectional BSE-SEM images of the as-rolled Mg/Cu SLCs with different ARB cycles; (a) 5, (b) 10, (c) 20, (d) 30 cycles, respectively.

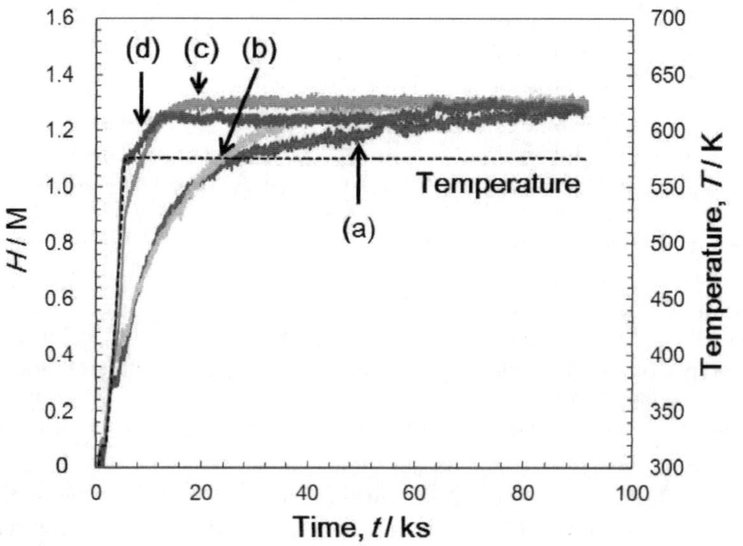

Figure 2. Initial hydrogenation properties of as-rolled Mg/Cu SLCs with different ARB cycles under the conditions of 573 K, 86.4 ks (24 h) in H_2 atmosphere of 3.3 MPa; (a) 5, (b) 10, (c) 20, (d) 30 cycles, respectively.

(fig. 1 (b)) was 443 and 162 nm, respectively, whereas that in the case of 30 cycles (fig. 1 (d)) was 123 and 64 nm, respectively. Fig. 1 (a) and (b) look similar, and fig. 1 (c) and (d) look similar. It seems that there are some differences between these two groups.

Fig. 2 shows the initial hydrogenation properties of as-rolled Mg/Cu SLCs with different ARB cycles under the conditions of 573 K, 86.4 ks (24 h) in H_2 atmosphere of 3.3 MPa. Mg/Cu SLCs with 5 and 10 cycles of ARB absorb hydrogen slowly, whereas those with 20 and 30 cycles of ARB absorb hydrogen fast. It also seems that there are some differences between these two groups.

Fig. 3 shows cross-sectional BSE-SEM images of Mg/Cu SLCs after initial hydrogenation under the conditions of 573 K, 86.4 ks (24 h) in H_2 atmosphere of 3.3 MPa with different ARB cycles; (a) 5, (b) 10, (c) 20, (d) 30 cycles, respectively. The size and the total area of MgH_2 layers were (a) > (b) > (c) > (d). Hydrogenation reaction of Mg_2Cu is as below.

$$2Mg_2Cu + 3H_2 = 3MgH_2 + MgCu_2$$

When Mg_2Cu absorbs hydrogen, it disproportionate to MgH_2 and $MgCu_2$ and show network structures of MgH_2 + $MgCu_2$ [15, 23, 25]. The size and the total area of network structures of MgH_2 + $MgCu_2$ were (d) > (c) > (b) > (a).

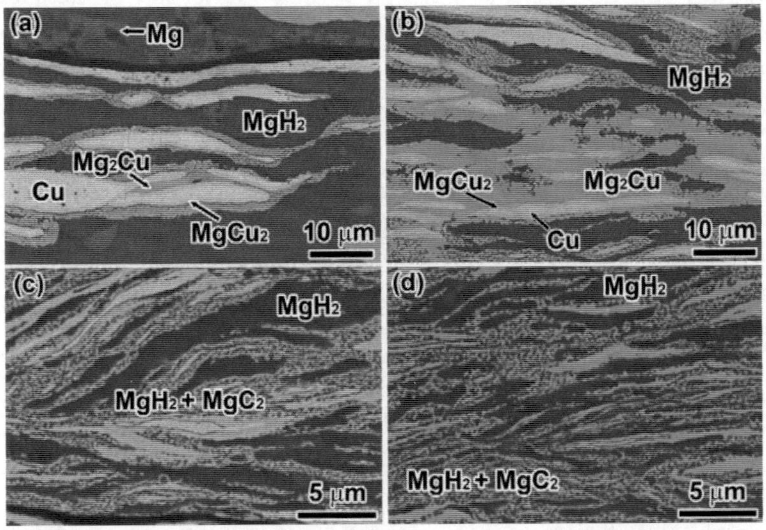

Figure 3. Cross-sectional BSE-SEM images of Mg/Cu SLCs after initial hydrogenation under the conditions of 573 K, 86.4 ks (24 h) in H_2 atmosphere of 3.3 MPa with different ARB cycles; (a) 5, (b) 10, (c) 20, (d) 30 cycles, respectively.

Table 1. Average thickness of Mg and Cu before initial hydrogenation, and approximate values of area fractions of MgH_2 and MgH_2 + $MgCu_2$ after initial hydrogenation in Mg/Cu SLCs with 10, 20 and 30 cycles of ARB.

Samples	Thickness of Mg and Cu		Approximate values of area fractions	
	Mg (nm)	Cu (nm)	MgH_2 (Hydrogenated Mg)	MgH_2 + $MgCu_2$ (Hydrogenated Mg_2Cu)
10 cycles of ARB	443	162	0.4	0.1
20 cycles of ARB	173	65	0.3	0.4
30 cycles of ARB	123	64	0.2	0.6

The average thickness of Mg and Cu before initial hydrogenation, and the approximate values of area fractions of MgH_2 (hydrogenated Mg) and MgH_2 + $MgCu_2$ (hydrogenated Mg_2Cu) were estimated from fig. 1 and 3, and summarized in table 1. It is seen that hydrogenation of Mg was dominant in Mg/Cu SLCs with less 10 ARB cycles, and that of Mg_2Cu was dominant in Mg/Cu SLCs with over 20 ARB cycles. This result can explain very well the features of fig. 2, since the hydrogen absorption/desorption kinetics of Mg is sluggish even at elevated temperatures (> 573 K) [3] and Mg_2Cu can absorb hydrogen fairy rapidly at 573 K [22]. From these results, it is expected that something happened in Mg/Cu SLCs between 10 cycles and 20 cycles of ARB, and showed different initial hydrogenation properties.

Fig. 4 shows schematic models of microstructures in Mg/Cu SLCs formed during (a) the heating process and (b) the hydrogenation process of initial hydrogenation. Since Mg/Cu SLCs were heated from room temperature to 573 K in H_2 atmosphere during initial hydrogenation, hydrogenation of Mg and alloying of Mg with Cu followed by hydrogenation of Mg_2Cu are competitive. From the experiments of Mg-Cu diffusion couples [32, 33], it is expected that core-shell like structures shown in fig. 4 (a) can be formed during the heating process [15, 25]. Constituent phase of the core-shell like structures are Cu, $MgCu_2$, Mg_2Cu, and Mg in the order from the core to the shell. A Mg shell becomes MgH_2 and a Mg_2Cu shell disproportionates to MgH_2 + $MgCu_2$ to form network structures after hydrogenation as shown in fig. 4 (b). It is considered that the difference of the amount of Mg_2Cu formed before hydrogenation causes the big differences of microstructures in Mg/Cu SLCs after hydrogenation. Therefore, it is expected that the growth rate constants of Mg_2Cu change depending on the number of ARB cycles.

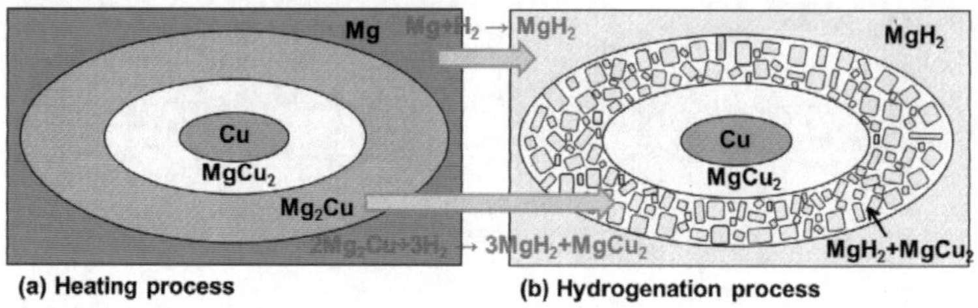

Figure 4. Schematic models of microstructures in Mg/Cu SLCs formed during (a) heating and (b) hydrogenation process of initial hydrogenation [15, 25].

Since $MgCu_2$ is not formed under 473 K [34, 35] and hydrogen absorption of Mg/Cu SLCs with 20 cycles of ARB starts at about 453 K, the growth rate constants of Mg_2Cu in Mg/Cu SLCs with different ARB were measured under 453 K. The measurements were performed following the method written in ref. [30] and the details of results will be reported elsewhere. Table 2 summarizes the growth rate constants of Mg_2Cu in Mg-Cu diffusion couples and Mg/Cu SLCs with 10, 20 and 30 cycles of ARB. The growth rate constants of Mg/Cu SLCs with 5 cycles of ARB could not be measured because Mg_2Cu did not grow enough owing to the insufficient contact between Mg and Cu foils.

Table 2. Growth rate constants of Mg_2Cu in Mg-Cu diffusion couples, and Mg/Cu SLCs with 10, 20 and 30 cycles of ARB.

Samples	Growth constant k / m^2 s^{-1}		
	373 K	413 K	453 K
Mg-Cu diffusion couples[32]	2.0×10^{-25}	2.7×10^{-23}	1.5×10^{-21}
10 cycles of ARB	1.2×10^{-21}	1.7×10^{-20}	4.1×10^{-19}
20 cycles of ARB	2.2×10^{-21}	9.5×10^{-20}	7.7×10^{-19}
30 cycles of ARB	1.7×10^{-21}	4.6×10^{-20}	7.6×10^{-19}

The growth rate constants of Mg_2Cu in Mg/Cu SLCs is approximately ten to the second - forth order larger than that of Mg-Cu diffusion couples below 453 K. It indicates that Mg_2Cu can grow at low temperature with sufficient rate in Mg/Cu SLCs. It can be said that the growth rate constants were

IOP Conf. Series: Materials Science and Engineering **375** (2018) 012032 doi:10.1088/1757-899X/375/1/012032

increased by ARB. However, there was no big difference among Mg/Cu SLCs with different ARB cycles contrary to our expectation.

As mentioned above, it is reported that SPD such as high-pressure torsion (HPT) and equal-channel angular pressing (ECAP) can enhance atomic diffusion and promote solid-state reactions because of an increase in the density of lattice defects such as vacancies, dislocations and grain boundaries [26-31]. It is reasonable to consider the introduction of a high density of vacancies and dislocation, and the grain refinement during ARB process.

Fig. 5 shows cross-sectional bright field STEM images of as-rolled Mg/Cu SLCs with 5 and 10 cycles of ARB. Mg_2Cu was formed during ARB process [30]. The grain size of Mg was 590 nm in Mg/Cu SLCs with 5 cycles of ARB (fig. 5 (a)) and 180 nm with 20 cycles of ARB (fig. 5 (c)). Since the grain size of annealed Mg is about 15 um, it can be said that Mg showed dynamic recrystallization even in Mg/Cu SLCs with 5 cycles of ARB. The average grain size of Mg in the case of 5 cycles of

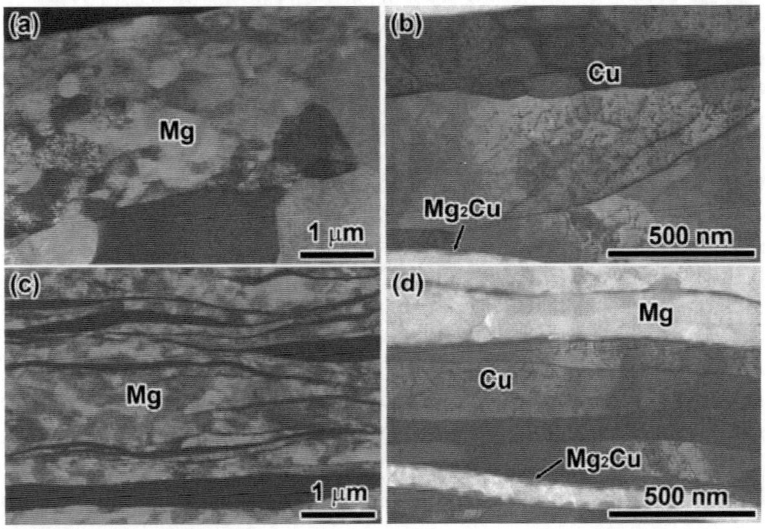

Figure 5. Bright field STEM cross-sectional images of Mg/Cu SLCs with different ARB cycles; (a) and (b) 5, and (c) and (d) 20 cycles of ARB, respectively.

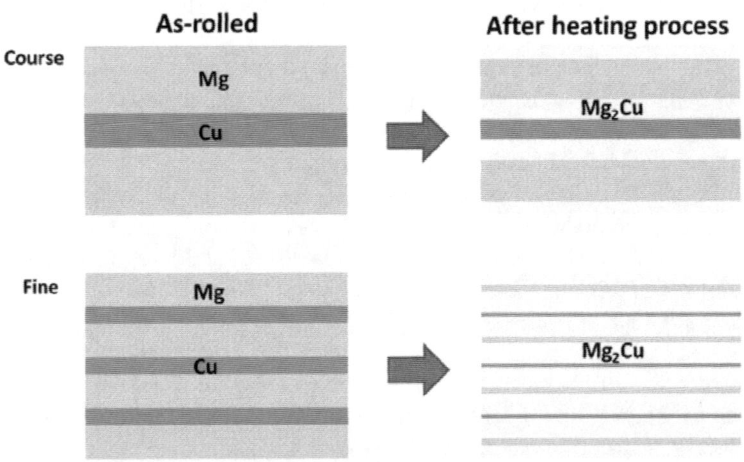

Figure 6. Schematic models representing change of microstructures in Mg/Cu SLCs during heating process.

ARB was larger than that of 30 cycles of ARB. However, that is because of inhomogeneity. The grain size of most grains in Mg/Cu SLCs with 5 cycles of ARB was almost as same as 20 cycles of ARB, although large grains in Mg/Cu SLCs with 5 cycles of ARB (lower area in fig. 5(a)) made the average grain size lager. The dislocation density of Mg was high in both specimens. As seen in fig. 5 (b) (5 cycles of ARB) and (d) (20 cycles of ARB), the same things can be said about Cu. That is the reason why the growth rate constants were almost same among Mg/Cu SLCs with 10, 20 and 30 cycles of ARB.

Fig. 6 shows schematic models representing a change of microstructures in Mg/Cu SLCs during the heating process. The upper row is a thick layer structure case and the lower row is a thin layer structure case. If the growth rate constants are the same, Mg_2Cu can grow the same length during the heating process. As can be seen in fig. 6, Mg is still remained in thick layer structure case, whereas Mg_2Cu is dominant in the thin layer structure case. Therefore, it can be concluded that this is the reason why the hydrogenation of Mg was dominant in the case of 5 and 10 cycles of ARB, and the hydrogenation of Mg_2Cu was dominant in the case of 20 and 30 cycles of ARB. This change of dominant hydrogen absorption phase seems to occur at the critical average thickness of Mg + Cu layers. From table 1, the critical average thickness of Mg + Cu layers can be estimated about 200 nm.

4. Conclusion

The growth rate constants of Mg_2Cu did not show a significant difference among Mg/Cu SLCs with over 10 cycles of ARB. However, the average thickness of Mg + Cu layers in as-rolled Mg/Cu SCLs became thinner as the cycle number of ARB increased.

The average thickness of Mg + Cu layers is the key factor which regulates the volume fraction of Mg_2Cu formed during the heating process of initial hydrogenation since the growth rate constants of Mg_2Cu is relatively constant in Mg/Cu SLCs with over 10 cycles of ARB.

Hydrogenation of Mg is dominant in Mg/Cu SLCs with over the critical average thickness and hydrogenation of Mg_2Cu is dominant in Mg/Cu SLCs with under the critical average thickness. The critical thickness is about 200 nm which is reached by ARB more than 10 cycles.

The volume fraction of Mg_2Cu formed during the heating process of initial hydrogenation determines initial hydrogen absorption properties and microstructures of Mg/Cu SLCs after initial hydrogenation.

Acknowledgement

This work was supported by JSPS KAKENHI Grant number 15K06519.

References

[1] Stampfer Jr J F, Holley Jr C E and Suttle J F The magnesium-hydrogen system 1960 *J. Am. Chem. Soc.* **82** 3504-8

[2] Danaie M, Mauer C, Mitlin D and Huot J Hydrogen storage in bulk Mg-Ti and Mg-stainless steel multilayer composites synthesized via accumulative roll-bonding (ARB) 2011 *Int. J. Hydrogen. Energ.* **36** 3022-36

[3] Lima G F, Triques M R M, Kiminami C S, Botta W J and Jorge Jr A M Hydrogen storage properties of pure Mg after the combined processes of ECAP and cold-rolling 2014 *J. Alloy. Compd.* **520** 287-94

[4] Vegge T, Hedergaad-Jansen L S, Bonde J, Munter T R and Nørskov J K Trends in hydride formation energies for magnesium-3d transition metal alloys 2005 *J. Alloy. Compd.* **386** 1-7

[5] Sato T, Blomqvist H and Noreus D Attempts to improve Mg_2Ni hydrogen storage by aluminum addition 2003 *J. Alloy. Compd.* **356-357** 494-6

[6] Selvan P, Viswanathan B, Swamy C S and Srinivasan V. Studies on the thermal characteristics of hydrides of Mg, Mg_2Ni, Mg_2Cu and $Mg_2Ni_{1-x}M_x$ (M = Fe, Co, Cu or Zn; $0 < x < 1$) alloys 1988 *Int. J. Hydrogen Energ.* **13** 87-94

[7] Hanada N, Ichikawa T, Hino S and Fujii H. Remarkable improvement of hydrogen sorption kinetics in magnesium catalysed with Nb_2O_5 2006 *J. Alloy. Compd.* **420** 46-9
[8] Isobe S, Ono A, Yao H, Wang Y, Hashimoto N and Ohnuki S. Study on reaction mechanism of dehydrogenation of magnesium hydride by in situ transmission electron microscopy 2010 *Appl. Phys. Lett.* **96** 223109-1-3
[9] Ma T, Isobe S, Morita E, Wang Y, Hashimoto N, Ohnuki S, Kimura T, Ichikawa T and Kojima Y Correlation between kinetics and chemical bonding state of catalyst surface in catalyzed magnesium hydride 2011 *Int. J. Hydrogen Energ.* **36** 12319-23
[10] Ma L-P, Wang P, Chen H-M. Hydrogen sorption kinetics of MgH_2 catalyzed with titanium compounds 2010 *Int. J. Hydrogen Energ.* **35** 3046-50
[11] Mao J, Guo Z, Yu X, Liu H, Wu Z and Ni J Enhanced hydrogen sorption properties of Ni and Co-catalyzed MgH_2 2010 *Int. J. Hydrogen Energ.* ***35*** 4569-75
[12] Shao H, Wang Y, Xu H and Li X. Preparation and hydrogen storage properties of nanostructured Mg_2Cu alloy 2005 *J. Solid State Chem.* **178** 2211-17
[13] Au M. Hydrogen storage properties of magnesium based nanostructured composites materials 2005 *Mater. Sci. Eng.* B **117** 37-44
[14] Ueda T T, Tsukahara M, Kamiya Y and Kikuchi S Preparation and hydrogen storage properties of Mg-Ni-Mg_2Ni laminate composites 2005 *J. Alloy. Compd.* **386** 253-7
[15] Tanaka K, Takeshita H T, Kurumatani K, Miyamura H and Kikuchi S. The effect of initial structures of Mg/Cu super-laminates on hydrogen absorption/desorption properties 2013 *J. Alloy. Compd.* **580** S222-5
[16] Tanaka K, Takeichi N, Tanaka H, Kuriyama N, Ueda T T, Tsukahara M, Miyamura H and Kikuchi S Investigation of micro-structural transition through disproportion and recombination during hydrogenation and dehydrogenation in Mg/Cu super-laminates 2008 *J. Mater. Sci.* **43** 3812-6
[17] Takeichi N, Tanaka K, Tanaka H, Ueda T T, Kamiya Y, Tsukahara M, Miyamura H and Kikuchi S Hydrogen storage properties of Mg/Cu and Mg/Pd laminate composites and metallographic structure 2007 *J. Alloy. Compd.* **446-447** 543-8
[18] Dufour J and Huot J Rapid activation, enhanced hydrogen sorption kinetics and air resistance in laminated Mg-Pd 2.5 at.% 2007 *J. Alloy. Compd.* **439** L5-7
[19] Dufour J and Huot J Study of Mg6Pd alloy synthesized by cold rolling 2007 *J. Alloy. Compd.* **446-447** 147-51
[20] Suganuma K, Miyamura H, Kikuchi S, Takeichi N, Tanaka K, Tanaka H, Kuriyama N, Ueda T T and Tsukahara M Hydrogen storage properties of Mg-Al alloy prepared by super lamination technique 2007 *Adv. Mater. Res.* **26-28** 857-60
[21] Mori R, Miyamura H, Kikuchi S, Tanaka K, Takeichi N, Tanaka H, Kuriyama N, Ueda T T, and Tsukahara M Hydrogenation characteristics of Mg based alloy prepared by super lamination technique 2007 *Mater. Sci. Forum* **561-565** 1609-12
[22] Reilly J J and Wiswall R H The reaction of hydrogen with alloys of magnesium and copper 1967 *Inorg. Chem.* **6** 2220-3
[23] Tanaka K, Shibata K, Kurumatani K, Ikeuchi S, Kikuchi S, Kondo R and Takeshita H T. Formation mechanism of micro/nano-strucutures through competitive reactions in Mg/Cu super-laminate composites during initial hydrogenation 2015 *J. Alloy. Compd.* **645** S72-5
[24] Shibata K, Tanaka K, Kurumatani K, Nishida Y, Kondo R and Takeshita H T Formation of $MgCu_2$ from MgH_2 and Cu in pressurized hydrogen atmosphere 2015 *Mater. Trans. JIM* **56** 785-9
[25] Tanaka K, Takeshita H T, Shin H, Kurumatani K, Kiyobayashi T, Takeichi N, Miyamura H, and Kikuchi S Micro/nano-structural transition and hydrogen absorption mechanism in Mg/Cu super-laminate composites 2014 *Mater. Trans. JIM* **55** 1122-8
[26] Straumal B B, Mazilkin A A, Baretzky B, Schütz G, Rabkin E and Valiev R Z Accelerated diffusion and phase transformation in Co-Cu alloys driven by the severe plastic deformation

2012 *Mater. Trans.* **63** 63-71

[27] Fujita T, Horita Z and Langdon T G Using grain boundary engineering to evaluate the diffusion characteristics in ultrafine-grained Al-Mg and Al-Zn alloys 2004 *Mater. Sci. Eng.* A. **371** 241-250

[28] Edalati K, Toh S, Watanabe M and Horita Z In situ production of bulk intermetallic-based nanocomposites and nanostructured intermetallics by high-pressure torsion 2012 *Scrpta Mater.* **66** 386-9

[29] Oh-ishi K, Edalati K, Kim HP, Hono K and Horita Z High-pressure torsion for enhanced atomic diffusion and promoting solid-state reactions in the aluminum-copper system 2013 *Acta Mater.* **61** 3482-9

[30] Tanaka K, Nishino D, Hayashi K, Ikeuchi S, Kondo R and Takeshita H T Formation of Mg_2Cu at low temperature in Mg/Cu super-laminate composites during initial hydrogenation 2017 *Int. J. Hydrogen Energ.* **42** 22502-10

[31] Divinski S V, Reglitz G, Rösner H, Estrin Y and Wilde G Ultra-fast diffusion channels in pure Ni severely deformed by equal-channel angular pressing 2011 *Acta Mater.* **59** 1974-85

[32] Nonaka K, Sakazawa T and Nakajima H Reaction diffusion in Mg-Cu system 1995 *Mater Trans JIM* **36** 1463-6

[33] Dai J, Jian B, Zhang J, Yang Q, Jian Z, Dong H and Pan F Diffusion kinetics in Mg-Cu binary system 2015 *J. Phase Equilib. and Diffus.* **36** 613-9

[34] Hong Q Z and d'Heurle F M The dominant diffusing species and initial phase formation in Al-Cu, Mg-Cu, and Mg-Ni systems 1992 *J. Appl. Phys.* **72** 4036-40

[35] Arcot B, Murarka S P, Clevenger L A, Hong Q Z, Ziegler W, and Harper J M E Intermetallic formation in copper/magnesium thin films – kinetics, nucleation and growth, and effect of interfacial oxygen 1994 *Appl. Phys.* **76** 5161-9

IOP Conf. Series: Materials Science and Engineering **375** (2018) 012033 doi:10.1088/1757-899X/375/1/012033

Three-Dimensional Characterization of Spalled Al5083 by Micro CT and In-Situ Femtosecond Laser - FIB

T Sano, J Ligda, T Walter, C Williams

U.S. Army Research Laboratory, Aberdeen Proving Ground, 21005-5066 USA

tomoko.sano.civ@mail.mil

Abstract. Spall cracks were formed in an equal channel angular extruded aluminum 5083 sample shock loaded to 1.45 GPa. A combined 3D characterization method using micro x-ray computed tomography (micro-CT) and in-situ femtosecond laser serial sectioning in the focused ion beam was used to characterize these spall cracks. The benefit of using both techniques lies in the ability to characterize the microstructure and spall cracks at multiple length-scales. The micro-CT characterization method is applied to non-destructively analyze the spall cracks at a higher targeted length-scale, and destructively explore the microstructure at a lower length-scale using the in-situ method. The femtosecond laser milling in the focused ion beam (FIB) in-situ method also allows for electron backscattered diffraction for crystallographic orientation determination. This paper describes the development of using both methods for 3D characterization of a single region of interest.

1. Introduction

1.1. Aluminum 5083

Aluminum (Al) 5083 is a lightweight alloy that has desirable properties including high strength to weight ratio, corrosion resistance, and weldability for numerous commercial and military applications. The strengthening mechanism of Al 5083 is by obstacles such as solutes, precipitates, and grain boundaries inhibiting dislocation movement [1]. Equal channel angular extrusion (ECAE), a technique for processing by severe plastic deformation, has been determined to increase strength as well as high strain rate properties by decreasing the grain size, increasing the number of grain boundaries, and increasing the number of dislocations [2 - 5]. Recently, the spall behavior of commercially obtained Al 5083-H321 processed by ECAE and subsequently cold rolled to 30% reduction was studied. The spall strength and the Hugoniot elastic limit (HEL) of the post processed Al 5083-H321 were determined by plate impact shock experiments conducted at a shock stress of 1.45 GPa [6, 7]. The results showed a 78% increase in the HEL, but a 37% decrease in the spall strength compared to as-received Al5083-H321 that was not processed by ECAE nor cold rolled. The authors correlated the re-alignment of the often observed manganese (Mn) rich precipitates [8] due to the ECAE and cold rolling processing, to the reduction of the spall strength. Preliminary micro CT analysis of the spall cracks confirmed the scanning electron microscopy (SEM) and energy dispersive spectroscopy analysis of the debonding of the Mn precipitates from the matrix magnesium, which became nucleation site for the cracks.

IOP Conf. Series: Materials Science and Engineering **375** (2018) 012033 doi:10.1088/1757-899X/375/1/012033

1.2. 3D Characterization Methods

X-ray computed tomography (CT) is a 2D and 3D imaging method that has been around for decades, first applied in the medical field [9]. In the last decade, micro CT techniques, in which "micro" indicates the pixel resolution being in the micrometer length scale, have been applied to non-destructive characterization of defects such as porosity, as well as density varying phase distributions in materials, which are distinguishable due to the differences in x-ray attenuation [10]. In typical CT systems the imaging magnification and resolution is dependent on the source to object and object to detector distances, often referred to as geometric magnification. Due to the finite spot size of the x-ray source a penumbra effect is created in the image. As the geometric magnification increases, the penumbra increases creating geometric unsharpness which eventually limits the maximum magnification achievable using this method. To achieve resolution beyond this limitation, the Zeiss Xradia 520 Versa uses optical magnification between the scintillator and detector. This allows the penumbra effect to be minimized while achieving submicron resolution. The accuracy of the measurements are dependent on image artifacts, noise, and thresholding methods [11]. Therefore, to obtain a high-resolution, 3D reconstructed microstructure, and also a multimodal dataset, a different characterization technique is necessary. The technique of serial sectioning with the FIB has been well developed [12-18] allowing for numerous 2D secondary electron micrographs, energy dispersive spectroscopy (EDS) elemental maps, as well as electron backscattered diffraction (EBSD) inverse pole figure maps to be reconstructed into 3D volumes. However, the milling of 2D layers of material with the gallium ion in the FIB is very slow, and limits the volume one can analyze. Hence a much faster 2D milling method that allows for 3D reconstruction, the femtosecond laser milling in the FIB technique, was applied to complement the micro CT results.

The technique of femtosecond laser milling in the FIB was developed at University of California, Santa Barbara. Numerous papers have been published by Echlin et al. on the technique and its application [19-22]. In this technique, the femtosecond laser beam is introduced into the chamber of the FIB through an objective lens threaded onto a micromanipulator attached to a port hole. The translation of the laser beam is controlled by a fast steering mirror. Inside the FIB chamber, the femtosecond laser beam geometry is such that it is parallel to the surface of the sample to be ablated. The laser beam ablates a pre-set layer at the region of interest from the surface of the sample, and the sample is rotated 180 ° by the movement of the stage, to obtain EBSD patterns from the ablated surface. If EDS maps or secondary electron images are desired, the stage would be tilted to the correct angle to capture those maps and images. After the collection of maps, images, and/or diffraction patterns, the stage is moved back to the laser milling position, and the process is continued for serial sectioning of the region of interest (ROI) of the sample. Similar to those of other serial sectioning methods, [23] the collected 2D layers are reconstructed into a 3D volume using a program such as DREAM.3D by BlueQuartz Software, LLC.

2. Experimental

A rolled Al5083-H321 plate was obtained and processed by ECAE at 250ºC following the 4E route. After ECAE processing, the plate was cold rolled to a reduction of 30% in thickness. Plate impact samples for the shock experiments were electro-discharge machined (EDM) from the processed plate. The shock experiments were conducted at 0.2 km/s shock velocities with the real-time free surface velocity-time histories measured with a Velocity Interferometry System for Any Reflector (VISAR). The post-shocked sample was recovered for 2 dimensional (2D) and 3 dimensional (3D) microstructural and spall crack analysis.

The shock loaded sample was first cross sectioned in the transverse direction to reveal the spall cracks running perpendicular to the loading direction. The 2D characterization of the cut sample was conducted with various techniques to obtain multimodal data. For higher millimeter scale analysis, Nikon Epiphot 300 optical microscope with the NIS Elements D software for image capture was used, and for lower micron scale analysis the FEI Nova NanoSEM 600 scanning electron microscope (SEM) was used.

Energy dispersive spectroscopy (EDS) for elemental analysis and electron backscattered diffraction (EBSD) for crystallographic orientation characterization were conducted with EDAX Inc.'s Pegasus XM4 analysis system attached to the FEI Nova NanoSEM. The 3D characterization was conducted with two different techniques. The first was by using the micro CT with the Zeiss Xradia 520 Versa, and the second was a femtosecond laser – FIB system, in which a Clark MXR femtosecond laser is incorporated into a FEI Nova Nanolab FIB with the TEAM Pegasus Integrated EDS-EBSD system from EDAX Inc. for collecting EBSD patterns.

The transverse plane was mechanically polished then ion polished on the Leica TIC-3X ion beam milling system at 6 kV, 2.2mA for 8 hours, and 2D characterization was conducted. To create micro CT samples, 5 mm thick short transverse plane samples were cut out by EDM in the longitudinal direction. Scans of a nominally 5.0 mm x 5.0 mm x 0.5 mm sample (now referred to as the 0.5 mm thick sample) were conducted at 80kV with 2.8x total magnification providing approximately 9.75 μm resolution. After identifying a specific spall crack ROI to examine, additional high resolution scans were performed. For these scans a 4x objective was used providing a total magnification of 9x and a resolution of 3 μm. Vertical stitching was used to generate a single large volume of interest which included 3 separate scans.

The 0.5 mm thick sample was loaded into the FIB, onto a 70° pre-tilted holder with a nanometer precision linear positioner substage (SmarAct Inc.) for fine and accurate Z movement. The FIB stage was then tilted 50° at a working distance of 8 mm, aligning the sample surface to be parallel to the femtosecond laser beam. This position was set as the "milling position". This position was also used for the secondary electron imaging at 20kV, 350 X magnification after each femtosecond laser milling of the surface.

An area by one of the cracks analyzed with the micro CT was chosen and the femtosecond laser milling parameters were set in LabView. LabView is the main program used to open and close the laser shutter, move the fast steering mirror, and control the SmarAct controllers which control the movement of the sample's substage and the shield that shields the laser beam from detectors, the pole pieces, and other components in the FIB chamber. LabView is also used to call FEI's control of the FIB stage movements, and the automated secondary electron image capture, directed by Runscript which uses the AutoIT scripting program to communicate with LabView. LabView through AutoIT is also used to run the EDAX software and capture EDS maps and/or EBSD patterns. More detailed descriptions of the femtosecond laser - FIB system setup, components, procedures, and capabilities can be found in the report by Ligda et al. [24]. The laser milling parameters for this dataset was 0.1 ms delay, 16 nm step size, speed of 160 μm/s, and 25 scans per layer. The laser power was set to 30 μJ, and the repetition rate was 1.0 kHz. After each layer was ablated, a secondary electron image was captured at the laser milling position, the stage was rotated 180° to the EBSD position (at the same stage tilt and working distance) and a 30 by 40 μm^2 area scan was collected at 0.3 μm step size at the 1200 x magnification. Once the EBSD scan was complete, the stage rotated back to the laser milling position, and the substage was moved 0.5 μm up to ablate the next layer. A total of 17 layers were laser milled, secondary electron micrographs captured, and EBSD scans obtained. The 3D microstructure was reconstructed by using DREAM.3D (version 6.1.77), and visualized with Paraview (version 4.3.1)

3. Results

The cross section of the target plate revealed an extended spall crack that was visible with the naked eye. However, upon evaluation with an optical microscope, it became clear that what was assumed to be a large crack was in fact numerous smaller cracks that were jagged as shown in Fig. 1. SEM analysis revealed that along many of the cracks were manganese (Mn) and iron (Fe) rich precipitates.

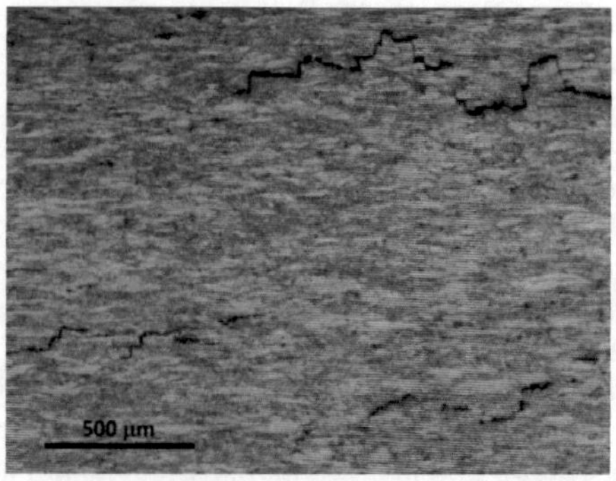

Figure 1. Optical micrograph of a cross section surface of the post-shock loaded sample.

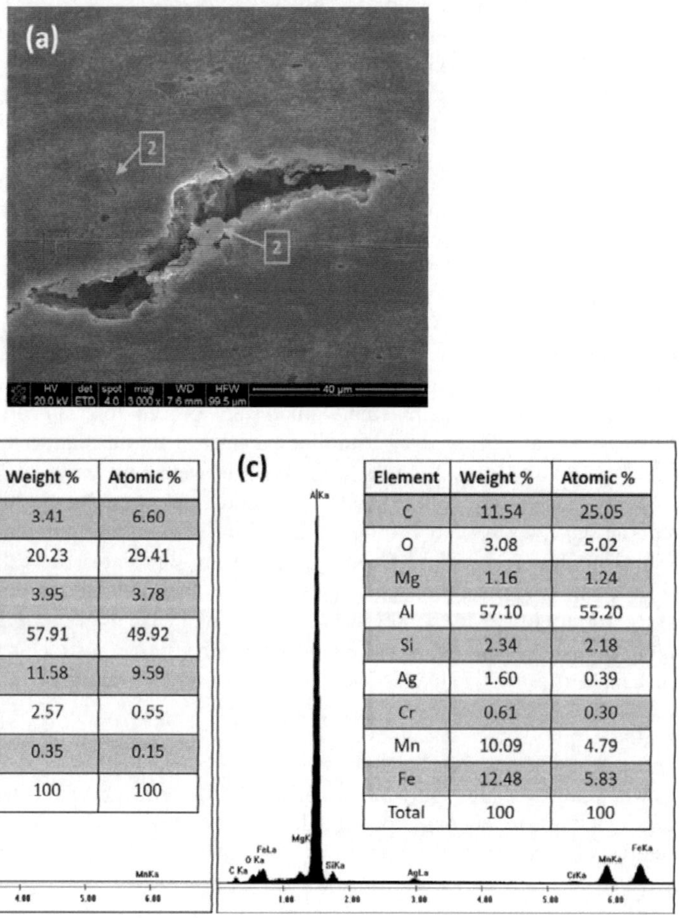

(b)

Element	Weight %	Atomic %
C	3.41	6.60
O	20.23	29.41
Mg	3.95	3.78
Al	57.91	49.92
Si	11.58	9.59
Ag	2.57	0.55
Mn	0.35	0.15
Total	100	100

(c)

Element	Weight %	Atomic %
C	11.54	25.05
O	3.08	5.02
Mg	1.16	1.24
Al	57.10	55.20
Si	2.34	2.18
Ag	1.60	0.39
Cr	0.61	0.30
Mn	10.09	4.79
Fe	12.48	5.83
Total	100	100

Fig. 2. (a) SEM micrograph of a spall crack identifying 2 precipitate compositions. (b) EDS spectra, weight and atomic percentages of the first (red) precipitate composition and (c) of the second (yellow) precipitate composition (the C and O are from surface contaminations).

It has been determined that the decohesion of these precipitates from the matrix resulted in voids that linked to create the spall cracks [6]. The EDS results in Fig. 2 (a) shows the location of these Mn-Fe rich precipitates, and Fig. 2(b) their spectrum and elemental composition. However the EDS spectra in Fig. 2(c) reveals that other precipitate compositions were also observed.

Micro CT scan of the cross sectioned sample was conducted. The results are shown in Fig. 3, where Fig. 3(a) is a scan of the bulk sample, and (b) is of an interior layer, revealing significant number of dark areas which are voids and planar spall cracks. In these scans, the resolution was not high enough to pick out the precipitates along the voids and spall cracks. However with the 0.5 mm thick sample, the micro CT scan was able to distinguish the precipitates due to the difference in densities. The lighter spots in Fig. 4 are the Mn-rich precipitates, and the area in the purple box is the crack ROI that was chosen for the femtosecond laser milling.

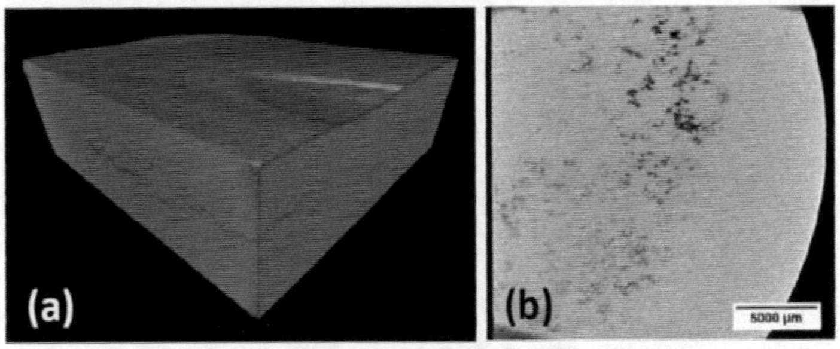

Fig. 3. Micro-CT scan of (a) the cross sectioned shock loaded sample, and (b) of an internal layer where the dark areas are the voids and spall cracks.

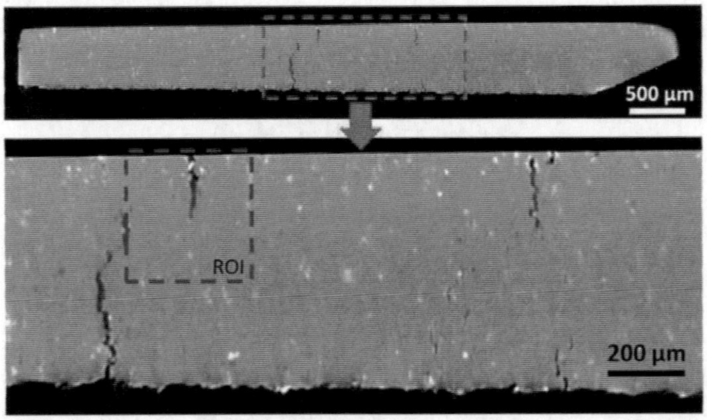

Fig. 4. Micro-CT scan of the 0.5 mm thick sample, revealing the light-colored Mn-rich precipitates and spall cracks. The crack ROI in purple was serial sectioned with the femtosecond laser.

With the femtosecond laser, serial sections of the ROI were milled. However the laser induced periodic surface structures (LIPSS) caused the surface to have a wavy structure that blocked the EBSD or diffracted poorly. To reduce the LIPSS structure to obtain better EBSD patterns, the stage was tiled

IOP Conf. Series: Materials Science and Engineering **375** (2018) 012033 doi:10.1088/1757-899X/375/1/012033

to -10 ° and the surface scanned with the Ga ion beam for a short time. The comparison of inverse pole figure (IPF) maps of an area that was scanned with the Ga ion beam versus one that was only femtosecond laser milled is shown in Fig. 5. The Ga ion beam reduction of the LIPSS was consequently used for each layer. The reconstruction of the 17 layers collected was conducted with DREAM.3D and visualized with Paraview and is shown in Fig. 6.

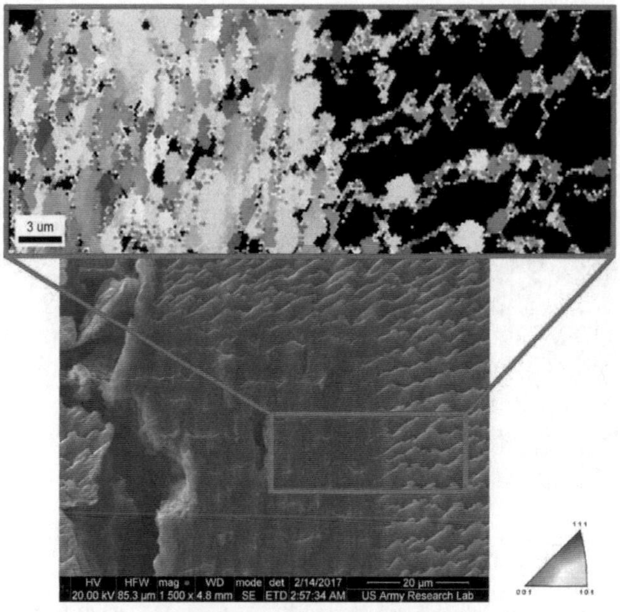

Fig. 5. EBSD IPF map of the smoother Ga ion beam exposed area and the LIPSS area, showing the effect of the surface on the EBSD pattern indexing.

Fig. 6. 3D reconstructed femtosecond laser milled EBSD dataset. Random colors were assigned to each grain. The dark areas are locations where the EBSD patterns were poor or lacking due to cracks or voids in the microstructure.

IOP Conf. Series: Materials Science and Engineering **375** (2018) 012033 doi:10.1088/1757-899X/375/1/012033

4. Discussions

A multi-dimensional, multi-lengthscale characterization was conducted on spalled Al 5083. The 2D microstructural characterization revealed 2 precipitates, one that was oxide-rich and one that was Mn-Fe rich. Analysis of the SEM micrographs and reconstructed 3D micro CT scans showed that the Mn-Fe rich precipitates were observed by the spall cracks, indicating that the possible nucleation site of spall cracks is the interface between the precipitate and the Al matrix. It is believed that the precipitates debonded from the Al matrix when the reflected compressive wave hit, causing the spall crack formation.

Serial sectioning with the femtosecond laser and capturing secondary electron images and EBSD scans were shown to supplement the 2D and 3D micro CT data. Because of the higher resolution, finer spall cracks and voids not detected with the micro CT were observed. With EBSD, observations on the finer grain sizes along the spall crack, effects of deformation at the grain boundaries near the spall cracks, crystallographic orientation data, and qualitative strain information were obtained.

The LIPSS formation on the as-femtosecond laser milled surface clearly makes obtaining large area EBSD difficult. The ridges on the LIPSS block the electron beam, creating region of poor diffraction. Scanning the surface with the Ga ion beam reduced the LIPSS and though not perfect, significantly improved the EBSD collection and indexing. The quality of the EBSD diffraction and indexing in the Ga ion treated surface shown in Fig. 5, is comparable to those of a mechanically then ion polished surface of the spalled Al 5083 H321 ECAE sample, shown in Fig. 7. However, it does add more time per layer, as it adds time to move the stage and up to 30 seconds of scanning with the Ga ion beam.

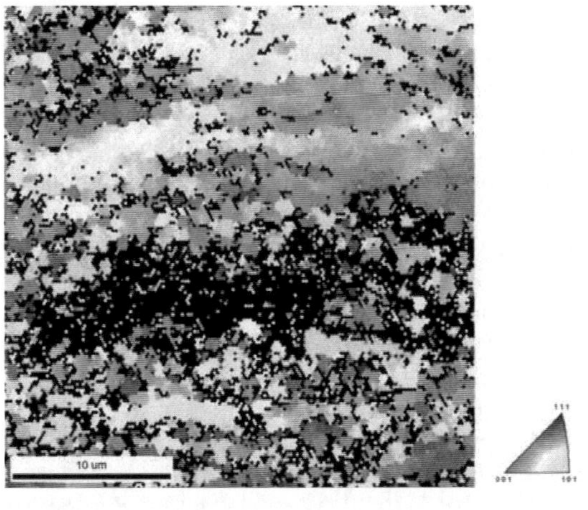

Fig. 7. EBSD IPF map collected from the spalled Al 5083 H321 ECAE sample mechanically and ion polished. The black area in the center is a cracked region.

Although there were numerous Mn-rich precipitates observed in the micro CT reconstructed volume, the small crack volume that was chosen to interrogate with the femtosecond laser-FIB serial sectioning method, did not reveal as many as expected. The one precipitate that was observed on the surface was ablated away while setting up the serial sectioning ROI. The author believe this is not a systematic problem with the femtosecond laser milling technique and that precipitates (above a few microns in size) can be characterized by the femtosecond laser – FIB serial sectioning method.

5. Conclusions

In this paper, the multi-dimensional and multi-lengthscale characterization of a spalled Al 5083 sample was described. Micro CT was used to first non-destructively characterize a larger volume of the sample, revealing precipitates, numerous spall cracks, and voids. A crack region of interest was chosen from the

IOP Conf. Series: Materials Science and Engineering **375** (2018) 012033 doi:10.1088/1757-899X/375/1/012033

reconstructed micro CT dataset to be serial sectioned with the in-situ femtosecond laser – FIB method. EBSD scans collected with this method were improved with the application of the Ga ion to the surface to reduce the effects of the LIPSS.

References

[1] Huskins E L, Cao B, and Ramesh K T 2010 *Mater. Sci. Eng.* A **527** 1292
[2] Dupuy L, Blandin J J, and E. F. Rauch E F 2000 *Mater.Sci. Tech.* **16** (11-12) 1256
[3] Malopheyev S and Kaibyshev R 2015 *Mater. Sci. Eng.* A **620** 246
[4] Thiyaneshwaran N and Sureshkumar, P 2013 *IJERT* **2** (5)
[5] Whelchel R L 2014 "Effects of Microstructure on the Spall Behavior of Aluminum-Magnesium Alloys," Ph.D. Thesis, Georgia Institute of Technology
[6] Williams C L, Sano T, Walter T R, Bradley J, and Kecskes L J, 2016 *JDBM* **2** 476
[7] Williams C L, Sano T, Walter T R, Kecskes L J 2017 *Behavior of Materials*, Vol 1. *Conference Proceedings of the Society for Experimental Mechanics Series*. eds D Casem, L Lamberson, J Kimberley, (Cham: Springer International Publishing) p 159.
[8] Goswami R, Spanos G, Pao P S, and Holtz R L 2010 *Mater. Sci. Eng.* A **527** 1089
[9] Goebbels J, 2013 "Computed Tomography," *Handbook of Technical Diagnostics*, ed H Czichos (Berlin Heidelberg : Springer-Verlag) p 249
[10] Patterson B M et al 2016 *Exp. Mech.* **56** 1585
[11] Miller J and Hornberger P 2016 *Precis. Eng.* **45** 18
[12] Uchic M D, Groeber M A, Dimiduk D M, and Simmons J P 2006 *Scripta Mater.* **55** 23
[13] Xu W, Ferry M, Mateescu N, Cairney J M, and Humphreys F J 2007 *Mater. Charact.* **58** 961
[14] Lee S-B, Rollett A D, and Rohrer G S 2007 *Mater. Sci. Forum* **558-559** 915-920.
[15] Dillon S J and Rohrer G S 2008 Three-Dimensional FIB-OIM of Ceramic Materials, *Applications of Texture Analysis* ed A D Rollett, (Hoboken, NJ: John Wiley & Sons, Inc.) Chapter 12, p 117
[16] Dillon S J and Rohrer G S 2009 *J. Am. Ceram. Soc.* **92** 7 1580
[17] Soldera F A, Lasagni F A and Mücklich F 2011 "Nano Characterization of Structures by Focused Ion Beam (FIB) Tomography" eds F A Lasagni and A F Lasagni, *Fabrication and Characterization in the Micro-Nano Range*, (Berlin Heidelberg: Springer-Verlag) p 171
[18] Zankel A, Wagner J, Poelt P 2014 *Micron* **62** 66
[19] Echlin M 2011 "A New Femtosecond Laser-Based 3D Tomography Technique," Ph.D. thesis, The University of Michigan
[20] Echlin M, Husseini N S, Nees J A, and Pollock T M 2011 *Adv. Mater.* **23** 2339
[21] Echlin M P, Mottura A, Torbet C J, and Pollock T M 2012 *Rev. Sci. Instrum.* **83** 023701
[22] Echlin M P, Straw M, Randolph S, Filevich J, and Pollock T M 2015 *Mater. Charact.* **100** 1
[23] Uchic M D 2011 *Computational Methods for Microstructure-Property Relationships* eds S Ghosh and D Dimiduk, (New York, NY: Springer Science+Business Media, LLC) p 31
[24] Ligda J, Jordan W, Lorenzo N, Sano T, and Schuster B, 2018 ARL-TR-8295

IOP Conf. Series: Materials Science and Engineering **375** (2018) 012034 doi:10.1088/1757-899X/375/1/012034

Texture, microstructure and mechanical properties evolution in Fe-x (x = 36 and 48 wt.%) Ni alloy after accumulative roll bonding

S Boudekhani-Abbas[1], K Tirsatine[1], H Azzeddine[1,2,*], B Alili[1], A L Helbert[3], F Brisset[3], T Baudin[3] and D Bradai[1]

[1] Faculty of Physics, University of Sciences and Technology Houari Boumediene, BP 32 El-Alia, 16111, Algiers, Algeria.
[2] Departments of Physics, University Mohamed Boudiaf, BP 166, 28000, M'Sila Algeria
[3] ICMMO, SP2M, Univ. Paris-Sud, Université Paris-Saclay, UMR CNRS 8182, 91405 Orsay Cedex, France
*Corresponding author: azehibou@yahoo.fr

Abstract. The texture, microstructure and Vickers microhardness evolution of Fe-36Ni and Fe-48Ni (wt. %) alloys processed by accumulative roll-bonding (ARB) up to 6 cycles were investigated using Electron Back Scatter Diffraction (EBSD) and microhardness measurements. Both alloys underwent a substantial grain refinement. Grains were of elongated shape parallel to the rolling direction and slightly smaller for Fe-48Ni alloy. Both alloys exhibit a different texture evolution upon straining influenced by the deformation conditions. The microhardness of both alloys also showed the same trends up to 6 cycles and Fe-48Ni alloy has higher strength than Fe-36Ni alloy. These differences were mainly ascribed to the deformation conditions, solute pinning and stacking fault energy.

1. Introduction

Severe plastic deformation (SPD) techniques such as equal channel angular pressing (ECAP) [1], high-pressure torsion (HPT) [2] and accumulative roll bonding (ARB) [3] have proven their ability to produce Ultra-Fine Grain (UFG) materials with high strengthening [4]. Up to now, among SPD techniques, only ARB processing has a great potential to be adapted to the industry in order to produce UFG materials in the form of large sheets due to its possibility as continuous process [5]. The evolution of crystallographic texture strongly depends on the SPD techniques [5–8]. Moreover, SPD techniques can lead to significant changes in the texture evolution compared to the conventional ones such as compression or rolling. Texture that develops after ARB processing is generally characterized by rolling-type components at the mid-thickness and shear-type components near the surface [5, 6]. However, there is a strong lack of studies on the effect of some special parameters such as deformation temperature, sample preparation or solute content on microstructure, texture and mechanical properties.

The aim of the present work is to investigate the effect of solute content on the texture, microstructure and mechanical properties evolution after severe plastic deformation of a Fe-x (x=36 and 48 wt.%) Ni alloy by ARB processing.

2. Experimental Procedure

The Fe-36%Ni (wt.%) and Fe-48%Ni (wt.%) alloys sheets were kindly provided by APERAM alloy society in France. ARB processing was carried out on a two-high mill of 2.2 kW power with rolls of 67 mm in diameter and rotating speed was 14 s^{-1}. Between cycles the Fe-36Ni and Fe-48Ni samples were preheated at 550 °C and 350°C for 10 min, respectively. The ARB processing was repeated up to 6 cycles.

Microstructure and microtexture were characterized using a scanning electron microscope FEG-SEM SUPRA 55 VP operating at 20 kV with OIM™ software. For EBSD measurements, the Fe-36Ni and Fe-48Ni samples were cut in the cross-section (RD-ND plan, rolling direction-normal direction) near the mid-thickness and to the surface of samples, respectively and were mechanically polished and then electropolished. The quantitative texture analysis was carried out by calculating the Orientation Distribution Function (ODF) using MTex software [9].

The Vickers microhardness of the samples was measured by SHIMADZU type HMV-2 tester using a load of 0.05 kg ($HV_{0.05}$) and indentation time of 10 seconds.

3. Results and discussion

Figure 1 presents the inverse pole figure (IPF) maps in the (RD-ND) plane showing the microstructure of the Fe-36Ni and Fe-48Ni alloys after ARB processing up 6 cycles. The EBSD maps reveal a substantial grain refinement and the grains are elongated parallel to the rolling direction. Figure 2 demonstrates the evolution of High angle grain boundaries (HAGB), the length (L along rolling direction) and the thickness (l along normal direction) of Fe-36Ni and Fe-48Ni ultrafine grains. The HAGB fraction decreases after 1 cycle leading to the formation of low angle grain boundaries (LAGB) due to the subdivision of primary grains and formation of subgrains.

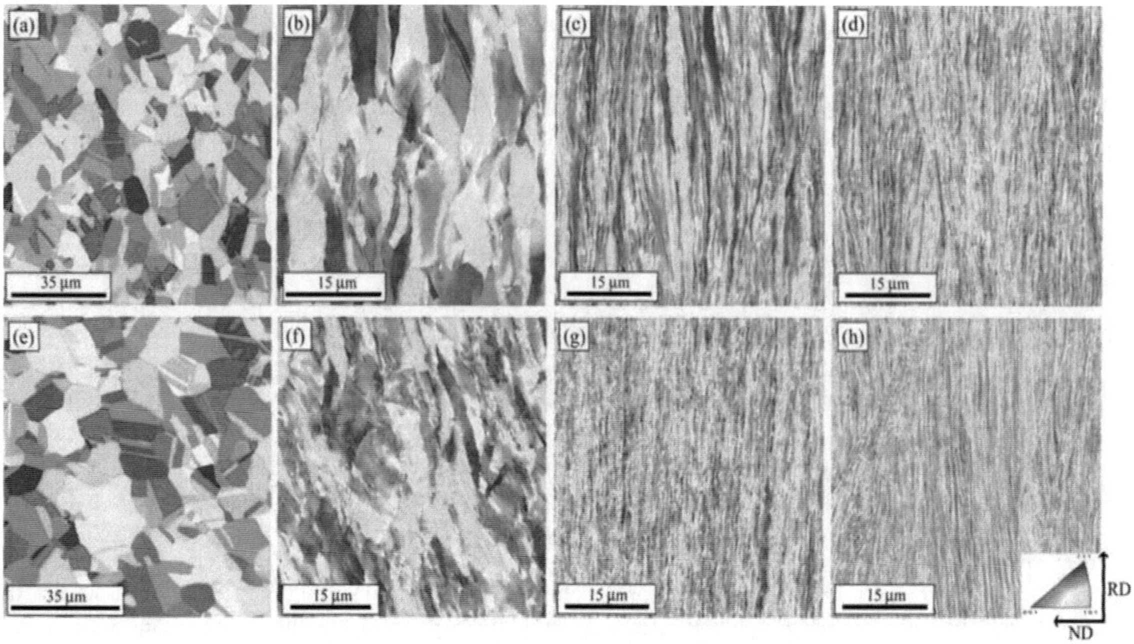

Figure 1: IPF maps showing the microstructure evolution of Fe-36Ni: a) 0 cycle, b) 1 cycle, c) 3 cycles, d) 6 cycles and of Fe-48Ni: e) 0 cycle, f) 1 cycle, g) 3 cycles and h) 6 cycles.

Then, the HAGB increases with increasing number of ARB cycles because of dislocation accumulation and the transformation of subgrain boundaries to HAGB to saturate near 68 % after 4 cycles for Fe-36Ni and 74 % after 5 cycles for Fe-48Ni. For both alloys the grain size parameters (L

IOP Conf. Series: Materials Science and Engineering **375** (2018) 012034 doi:10.1088/1757-899X/375/1/012034

and l) decrease with increasing number of ARB cycles and saturate after 4 cycles. It is interesting to note that Fe-48Ni alloy produces slightly smaller grain size compared to Fe-36Ni alloy due the effect of solute pinning on grain boundary migration [10].

The effect of heating treatment between cycles could also affect the final grain size by promoting dynamic recovery of dislocation [11]. In fact, during ARB processing the Fe-36Ni sample was heated at 550 °C for 10 min which is a suitable temperature for recrystallization (~ 600 °C) [12]. Although, the Fe-48Ni sample was treated at 350 °C which is far less than the effective temperature for recrystallization resulting hence in smaller grain size.

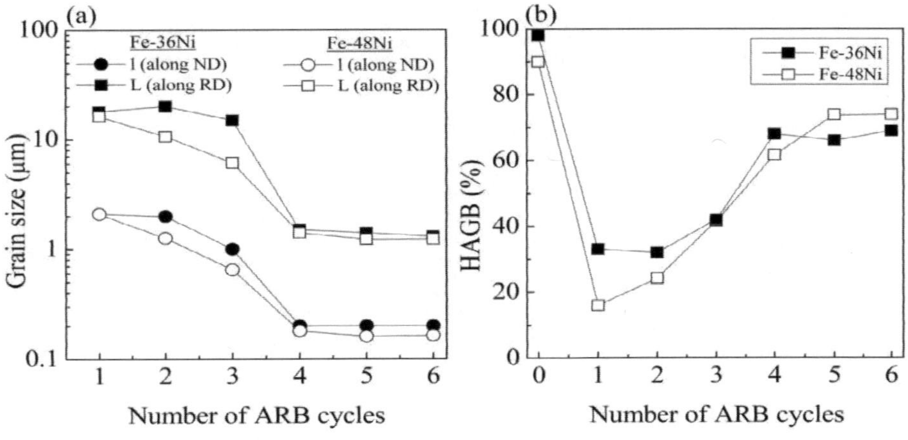

Figure 2: (a) Grain size parameters (length along RD (L) and ND (l)) and (b) HAGB fraction of Fe-36Ni and Fe-48Ni alloys as function of number of ARB cycles.

Figure 3 presents the texture evolution of Fe-36Ni and Fe-48Ni alloys upon straining via ODF sections (φ_2 = 0, 45 and 65°). The main ideal texture component position of FCC alloys is also presented and their descriptions are given in Table 1. The EBSD measurements were carried out in the cross section near to the mid-thickness of the Fe-36Ni sample where no shear components have been detected [5]. By contrast, the EBSD data of the Fe-48Ni samples were collected in the cross section near to the surface of the samples, where the shear texture is dominated by the R-Cube component [5].

Table 1: Main ideal rolling texture components of FCC alloys.

Component	{hkl}<uvw>	Euler Angle		
		φ_1	Φ	φ_2
□ Brass	{110}<112>	35°	35°	45°
◇ Goss	{110}<001>	0°	45°	0°
● Cube	{001}<100>	0°	0°	0°
O R-Cube	{001}<110>	45°	0°	0°
△ Copper	{112}<111>	90°	35°	45°
▽ S	{231}<346>	59°	29°	63°

Obviously, Fe-36Ni and Fe-48 alloys exhibit a different texture evolution. The Fe-36Ni alloy, with a strong Cube initial texture, developed a Brass component (after 1 cycle) and finally a Copper-type texture (Brass, S and Copper components) after 6 cycles. The Fe-48Ni alloy, with a quasi-random initial texture, developed a Rotated-Cube component after 1 cycle with is a typical shear texture component and subsequently a Copper component dominated after 6 cycles. This is in good agreement with results of Jamaati et al. [13] who assumed that the shear (R-Cube) components present in the surface regions of the sample could rapidly rotate towards rolling components as soon as they are moved to the quarter thickness regions. Indeed, a simulation with VPSC has confirmed this assumption [6]. It is interesting to note that Brass component has not been developed in Fe-48Ni alloy during ARB processing. Instead of Brass component a new weak (1.9 mrd) component $\{012\}\langle 2\bar{2}1\rangle$ (43°, 24°,0°) has been development (ODF section at φ_2 =0°) starting from 3 cycles. Such texture component was already reported in Fe-36Ni alloy after cross ARB [6].

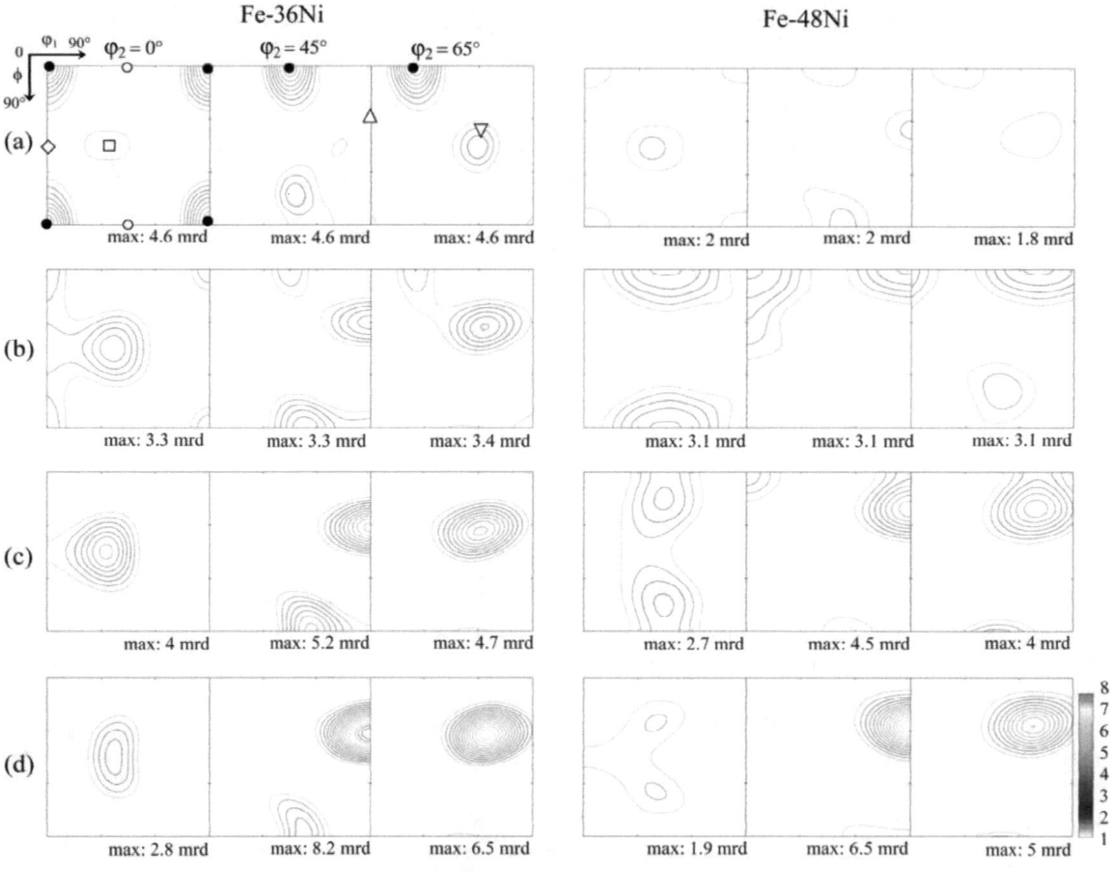

Figure 3: ODF sections at φ_2 =0, 45 and 65° of Fe-36Ni and Fe-48Ni alloys after ARB processing: a) 0 cycle, b) 1 cycle, c) 3 cycles and d) 6 cycles.

The evolution of microhardness of Fe-36Ni and Fe-48Ni alloys after nnnARB processing is shown in Figure 4. For both alloys, Hv increased up 3 cycles due to the combination of dislocation generation and grain refinement [14]. After 3 cycles the Hv values tends to stabilize well probably due to the occurrence of a concurrent dynamic recovery process [15]. It is interesting to note that the Hv values of Fe-48Ni alloy are almost 10 % higher that Fe-36Ni ones. This difference can be attributed to the effect of deformation temperature. It is known that the micorhardness decrease with increasing deformation temperature [16] mainly to due to the grain refinement Moreover, the difference in Hv values should probably arise from the stacking fault energy variation between the two alloys. An estimation of the reduced stacking fault energy γ_{SFE}/Gb has been done using an empirical correlation between it and the ratio of Brass (V_{Brass}), Copper (V_{Copper}) and S (V_S) volume fractions as proposed in the literature [17]:

$$\frac{\gamma_{SFE}}{Gb} \propto \frac{2V_{Brass}}{V_{Copper}+V_S} \tag{1}$$

where G is the shear modulus and b the Burgers vector.

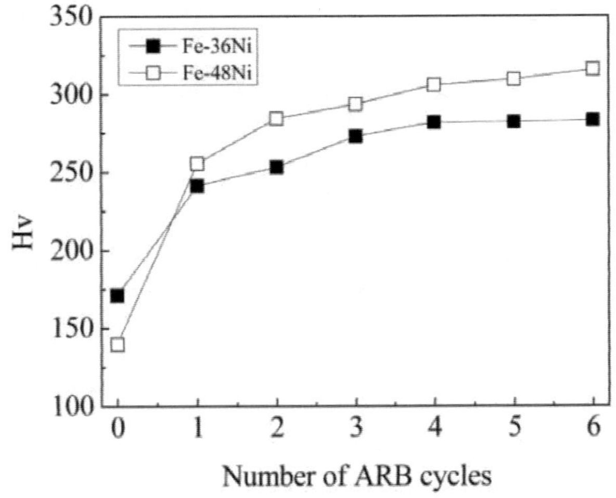

Figure 4: Microhardness evolution of Fe-36Ni and Fe-48Ni alloy as function of number of ARB cycles.

Table 2 listed the estimated γ_{SFE} for Fe-36Ni and Fe-48Ni alloy after ARB processing. The deduced values are surprisingly in good agreement with the literature data [18]. Indeed, the Fe-48Ni exhibits a lower γ_{SFE} compared to Fe-36Ni alloy. Many authors have demonstrated that a reduction in SFE leads to a smaller grain size and increased strength and ductility in FCC deformed alloys [19, 20]. Following these authors, the higher strength was mainly due to grain refinement and higher density of twins formed during deformation processing. In fact, as the SFE decreased, the stacking faults became wider, making cross-slip more difficult. Hence, mechanical twinning was favored. It is well known that low stacking fault energy promotes strain hardening due to the inhibition of dynamic recovery process by suppression of cross slipping and this leads to the fast build-up of the critical dislocation density required for initiating twinning [21].

Table 2. Estimation of the Stacking Fault Energy γ_{SFE}.

γ_{SFE} (mJ/m^2)	1 cycle	3 cycles	6 cycles	Reference
Fe-36Ni	124.7	127.6	128.6	122 [17]
Fe-48Ni	74.2	75.4	81.2	78 [17]

For both alloys, the γ_{SFE} values seem to increase with increasing strain but this increase is not significant and lies within the relative error estimated to be ± 8 mJ/m^2. In the author's opinion, no systematic evolution of the stacking energy versus amount of straining should exist but only versus solute content.

4. Conclusion

- The Fe-36Ni and Fe-48 Ni (wt. %) alloys after ARB processing at 550 and 350 °C up to 6 cycles respectively, underwent a substantial grain refinement. Grains are of elongated shape parallel to the rolling direction and slightly smaller for Fe-48Ni alloy.
- Both alloys exhibit a different texture evolution upon straining. The Fe-48Ni alloy developed a Rotated-cube component after 1 cycle and subsequently a domination of Copper component after 6 cycles. While, the Fe-36Ni alloy developed a Brass component after 3 cycles and finally a Copper-type texture after 6 cycles.
- The microhardness of both alloys showed the same trends up to 6 cycles. That of Fe-48Ni alloy was higher than that of Fe-36Ni owing to its smaller stacking fault energy.

Acknowledgements
This work was supported in part by the PHC Maghreb Program N°16MAG03.

References

[1] Valiev R Z, Islamgaliev R K and Alexandrov I V 2000 *Prog. Mater. Sci.* **45** 103–189.
[2] Smirnova N A, Levit V I, Pilyugin V I, Kuznetsov R I, Davidova L S and Sazonova V A 1986 *Phys. Met. Metall.* **61** 127-134.
[3] Saito Y, Tsuji N, Utsunomiya H, Sakai T and Hong R G 1998 *Scripta Mat.* **39** 1221-27.
[4] Valiev R Z, Estrin Y, Horita Z, Langdon T G, Zehetbauer M and Zhu Y T 2006 *JOM* **58** 33–39.
[5] Tirsatine K, Azzeddine H, Baudin T, Helbert A L, Alili B and Bradai D 2014 *J. Alloys Compd.* **610** 352-360.
[6] Azzeddine H, Tirsatine K, Baudin T, Helbert A L, Brisset F and Bradai D 2014 *Mater. Charact.* **97**140–149.
[7] Khereddine Y, Hadj Larbi F, Azzeddine H, Baudin T, Brisset F, Helbert A L, Mathon M, Kawasaki M, Bradai D and Langdon T G 2013 *J. Alloys Compd.* **574** 361–367.
[8] Hadj Larbi F, Azzeddine H, Baudin T, Mathon M H, Brisset F, Helbert A L, Kawasaki M, Bradai D and Langdon T G 2015 J. Alloys Compd. 638 88–94.
[9] Bachmann F, Hielscher R and Schaeben H 2010 *Solid. State. Phenom.* **160** 63–68.
[10] Hersent E, Marthinsen K and Nes E 2013 *Metall and Mat Trans* A **44** 3364–75.
[11] Pirgazi H, Akbarzadeh A, Petrov R, Sidor J and Kestens K 2008 *Mat. Sci. Eng.* A **492** 110-117.
[12] Zaefferer S, Baudin T and Penelle R 2001 *Acta Mater.* **49** 1105–22.
[13] Jamaati R and Toroghinejad M R 2014 *Mat. Sci. Eng.* A. **598** 263–276.
[14] Hansen N, Huang X and Hughes D A 2001 *Mat Sci Eng.* A **317** 3–11.
[15] Kumar S S S and Raghu T 2011 *Materials and Design* **32** 4650–57.
[16] Mungole T, Nadammal N, Dawra K, Kumar P, Kawasaki M and Langdon T G 2013 *J Mater Sci* **48** 4671–80.
[17] Witte M, "Texture Optimization of Ni-5at.%W for Coated Conductor Application", Diss. RWTH Aachen University, 2013.
[18] Schramm R E and Reed R P 1976 *Metall. Trans.* A **7** 359–363.
[19] Pei-Ling S., Zhao Y H, Cooley J C, Kassner M E, Horita Z, Langdon T G, Lavernia E J and Zhu Y T 2009 *Mat. Sci. Eng.* A. **525** 83–86.
[20] Zhao Y H, Zhu Y T, Liao X Z, Horita Z and Langdon T G 2006 *Appl. Phys. Lett.* **89** 121906.
[21] El-Danaf E A, Soliman M S and Al-Mutlaq A A 2015 *Adv. Mater. Sci. Eng.* **2015** 1–12.

IOP Conf. Series: Materials Science and Engineering **375** (2018) 012035 doi:10.1088/1757-899X/375/1/012035

Development and verification of simultaneous measurement system for texture and phase fraction by time-of-flight neutron diffraction at iMATERIA

Y Onuki[1], A Hoshikawa[1], S Sato[1,2], S Nishino[1,2], T Ishigaki[1] and T Tomida[3]

[1] Frontier Research Center for Applied Atomic Sciences, Ibaraki University, 162-1 Shirakata, Tokai, Ibaraki 319-1106, Japan.
[2] Graduate School of Science and Engineering, Ibaraki University, 4-12-1 Nakanarusawa, Hitachi, Ibaraki 316-8511, Japan.
[3] Ibaraki Prefectural Government, 162-1 Shirakata, Tokai, Ibaraki 319-1106, Japan.

yusuke.onuki.0@vc.ibaraki.ac.jp

Abstract. The authors have developed the simultaneous measurement scheme for texture and phase fraction by using the Time-Of-Flight (TOF) neutron diffractometer, iMATERIA at J-PARC MLF, Japan. The Rietveld texture analysis (RTA) was applied for the data analysis method, which has been developed mainly at HIPPO diffractometer at LANSCE, USA. Because of the differences of measurement scheme and machine specifications, the analysis for the data from iMATERIA requires some special treatments. In this paper, we provide instructional information for the measurement at iMATERIA and subsequent RTA to determine the textures and phase fraction in metallic materials. The instrumental features and measurement procedures are firstly introduced. Second, some unique data treatments, especially about the technique called "scale factor bounding" is explained, including how to set it. We also suggest some refinement strategies especially aiming to have accurate phase fraction determination.

1. Introduction

The textures and phase fractions in metallic materials are key features of the microstructure to control the performances. As discussed in many papers, consideration of texture is necessary to determine phase fractions accurately by diffraction techniques [1-3]. As reported in the previous papers, the authors have developed the measurement scheme of texture and phase fraction in steels by using time-of-flight (TOF) type neutron diffractometer, iMATERIA at J-PARC MLF, Japan [4-7]. By acquiring multiple diffractograms (TOF-intensity relationships) corresponding to various sample directions, Rietveld texture analysis (RTA) can simultaneously refine ODF, *i.e.*, texture, and various phase parameters including the volume fraction of each phase.

The RTA based on TOF neutron diffraction data was firstly proposed by Wenk *et al.* [8] and extensively utilized at HIPPO diffractometer at LANSCE, USA [9,10]. The data importing, and subsequent RTA can be conducted by using MAUD software, which has been developed by Lutterotti [11,12].

Although the Rietveld analysis is a computational approach, it does not "automatically" give users an appropriate result in all cases. Users should understand the procedures, have a strategy of refinement, and carefully review the result [13].

There are many practical instructions and examples for analysis using MAUD based on data measured at HIPPO [9,10]. Also in case of iMATERIA, most of those can be applied in the same manner. However, users may need some specific features and strategies for the data measured by iMATERIA because of the characteristics of instrument and measurement scheme. Additionally, the techniques at iMATERIA focus on the application for metallic materials used in industry while most of instructive literatures for HIPPO showed examples of minerals. The authors' previous papers somewhat focused on the verifications of the measurement and analysis but limited information was provided about the detailed procedures. In this paper, therefore, we introduce some features of iMATERIA and practical procedures of RTA by using MAUD to determine textures and phase fractions in metallic materials.

2. Instrumental features

2.1. *Fundamental specifications of iMATERIA*

iMATERIA locates at 20th beamline at J-PARC MLF, Japan. It accepts the pulse neutron beam from the poisoned moderator [6,7], resulting in an ideal balance between intensity and resolution. Numerous detectors are equipped around the incident beam axis as shown in figure 1, which enables the simultaneous detection at various 2θ. The names of "Banks", groups of the detectors, in figure 1 indicate approximate 2θ angles in degree. Although iMATERIA has more detectors below $2\theta = 30°$ and small angle neutron scattering (SANS) bank, they are not used in the current purpose. The detectors consist of the arrays of position-sensitive tube detectors (1D-PSDs). Hence, it is possible to set arbitrary positions and areas on the detectors as the independent 0D detectors. We call them "observation points" (OPs). As shown in figure 1, we set 132 OPs (small rectangles). Each OP acquires a diffractogram corresponding to different scattering vector direction. Further details *e.g.* the geometrical distribution of OPs, can be found in the previous papers [4].

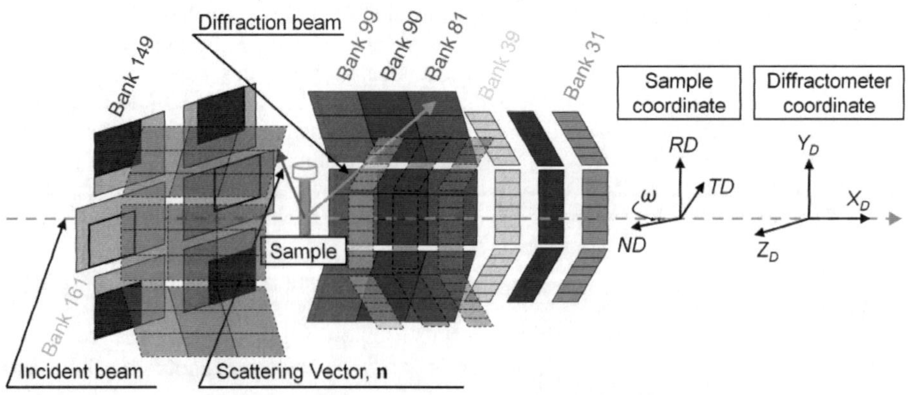

Figure 1. Schematic drawing of iMATERIA.

2.2. *Sample Setting*

Currently, 4 types of sample holders shown in figure 2 are available for the measurement in ambient environment. The sheet holder (figure 2 (a)) is mainly used for the samples from the industry. The standard length and width of the sample are 65 mm and 8 mm, respectively. There should be a hole for a screw pin. The diameter of the hole is 2.5 mm and the center of the hole is 5 mm away from the top end. It can hold the sample thickness up to 6 mm. Thin sheets are usually laminated so that the total thickness becomes 6 mm.

The cylindrical vanadium cells (figure 2 (b) and (c)) are used for smaller test pieces. Since the coherent scattering length of vanadium is very small, the diffraction from the cell is negligibly weak [14]. Therefore, vanadium is frequently used as the "transparent" material for various neutron diffraction experiments. In order to fix a small sample inside the cell, polyimide films are attached both on top and bottom of the sample as shown in figure 2 (e). These prevent the sample tilt or rotate during the transfer to the measurement position.

For the samples which cannot be inserted into the cells, the cramp holder shown in figure 2 (d) is applied. In the photograph, a quartz single crystal (approximately 15 mm×10 mm×10 mm) is put between the cut pieces of vanadium cells. Because of the side pillars (white bars in the photograph), available sample angle (ω in figure 1) is limited. However, the pillars do not shut the paths of neutron beams at all when ω is set at around 30°. Use of glue to hang the sample is discouraged at iMATERIA because it is often embrittled in vacuum and the sample falls from the height of around 1.5 m to the bottom of sample chamber.

All the sample holders are attached to the white plastic (polyoxymethylene) caps seen in figure 2 (a ~ c), which are held by the sample changer [15]. The cap has a radio-frequency identification (RFID) chip which records an individual sample ID number for the measurement/storage management. The sample changer can introduce ω rotation and vertical movement along Y_D. Therefore, the beam spot position can be adjusted by moving the sample if necessary. The beam spot size is also adjustable by changing the size of the incident beam slit. The squares shown in figure 2 indicate the typical spot sizes (20 mm × 20 mm or 15 mm × 15 mm) and positions.

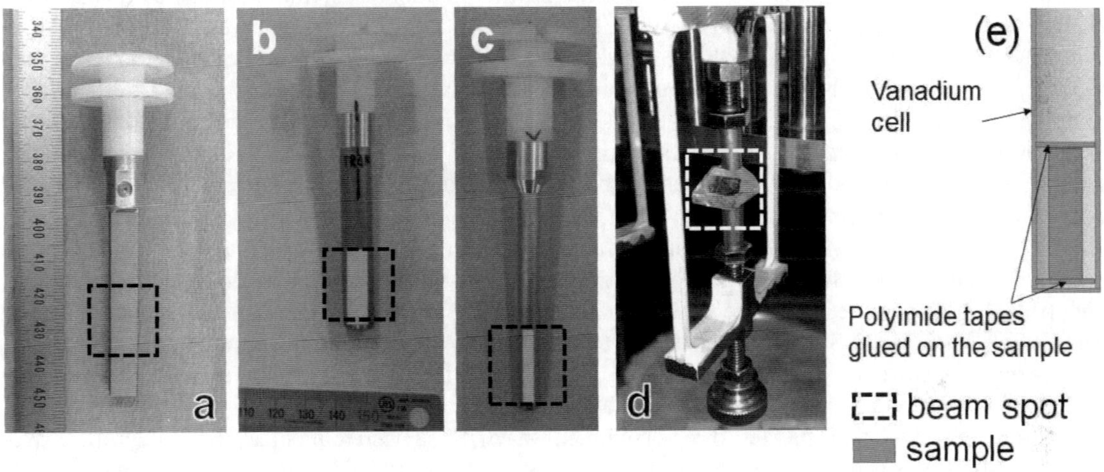

Figure 2. Sample holders used at iMATERIA. (a) sheet sample (65 mm × 8 mm) holder, (b) ϕ9.8 mm (inner diameter) cylindrical cell made of vanadium, ϕ5.8 mm vanadium cell, (d) cramp holder for irregular shape, and (e) schematic for the sample fixing in the vanadium cell.

2.3. Neutron diffraction measurement

The neutron pulse at J-PARC MLF is generated from the proton beam with the frequency of 25 Hz. For most of the metallic crystals, iMATERIA accepts every pulse (single frame mode). This results in the maximum d of approximately 0.25 nm for the backscattering banks (Banks 161 and 149 in figure 1), 0.35 nm for the side banks (Banks 99-81) and 0.9 nm for the low angle banks (Banks 39 and 31). By limiting the accepting frequency to 12.5 Hz (double frame mode), the maximum d becomes double but the scattering intensity per unit time becomes half. The double frame mode is advantageous for materials having large unit cells, *e.g.*, natural rocks and intermetallics [6].

The time for the measurement depends on the coherent scattering cross section, σ_c of the elements, the sample volume, and the proton beam power at J-PARC. In case of a steel sample held by the sheet holder, satisfactory data could be acquired in 5 minutes with the single frame mode when the proton beam power was 150 kW. The beam power is currently increasing, and it will reach to 1 MW in close future. It is confirmed that the incident neutron flux proportionally increases with the proton beam power. Therefore, the measurement time will become much shorter in future.

3. Analysis

3.1. Data format and importing

Since MAUD accepts the input data described in the format of GSAS, another Rietveld solver, details of the data format can be found in the GSAS manual [16]. Although there are several types of the input data formats, we use the "ESD" format and give the extension "gda". In the following, this type of data file is called a "GDA file". The other files are also referred by capitalized extensions. The appearance of a GDA file in the text editor is shown in figure 3. In this format, the intensity and its expected standard deviation are alternatively described with a constant binning width of TOF (usually 8 μs is applied for cubic and HCP metals). It should be noted that the values of intensity and standard deviation recorded in this file are not "raw" measured data but results of external normalization by the incident beam intensity [14]. Although MAUD/GSAS can deal with the incident spectrum as a continuous function, the incident spectrum profile at iMATERIA is too complex to be represented by the prepared functions.

The GDA file includes the 132 datasets (diffractograms) corresponding to the OPs. The sequential numbers are distributed as BANK 1, BANK 2, …, BANK 132. It should be noted that the terminology, "BANK" used here actually corresponds to the OP, a part of the detector. In this paper, we use "Bank" for the group of OPs having similar 2θ, as indicated in figure 1.

There is also a file describing the instrumental parameters of the 132 OPs (*.PRM). This file describes the positions of the OPs and instrumental profile functions describing the standard shapes of diffraction peaks. In addition, the conversion parameters, DIFC, DIFA, ZERO are also recorded. These parameters are used to convert TOF to d spacing in angstrom. These instrumental parameters are basically dealt as the constant (fixed) values in the following analysis. Therefore, regardless of the sample, the same PRM file can always be used. Strictly speaking, DIFC is often the value to be refined in multi-diffractogram analysis because this can be affected by the sample shape, kind of element and sample density [9]. However, the refinement of DIFC is usually very fine tuning, which makes the variation less than 1 percent of the value.

The "HIPPO wizard" can be used to import the GDA and PRM files (even though the instrument is not HIPPO but iMATERIA!). The sample setting angle ω is also asked in the dialog. After the data import, the whole parameters can be saved as the analysis file (*.PAR file). Subsequently, phase initial parameters are prepared. Since the crystal structures of metals are very simple, it may be possible to set them manually. However, it is easier to import CIF or MDB file. The former is written in the universal format [17] to describe the crystal structure of a certain phase. The latter is a set of CIF files. You may find "structures.mdb" in the MAUD directory and chose what you need to include from numerous phase data. The MDB files can be created by simply copy and paste the multiple CIF data into a single file. It is convenient to have one's own MDB library including only the phase data which you frequently refer to.

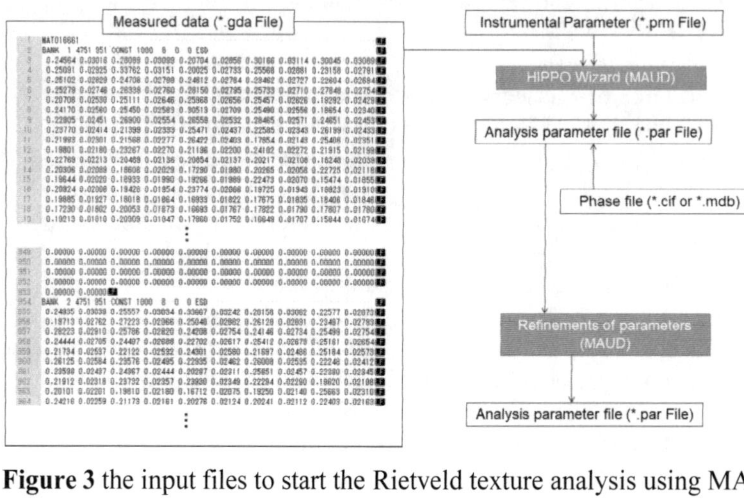

Figure 3 the input files to start the Rietveld texture analysis using MAUD.

IOP Conf. Series: Materials Science and Engineering **375** (2018) 012035 doi:10.1088/1757-899X/375/1/012035

3.2. Characteristic settings for iMATERIA

3.2.1. Bank grouping Tolerance. In order to have the Bank groups shown in figure 1, firstly, the tolerance angle of 2θ for the Bank grouping must be changed to 6°. By selecting "Analysis > Preferences" at the top menu bar, a table of values become visible. In this list, you can find "hippoWizard.maxDelta2ThetaForBankGrouping" and set the value as "6". Then click "save on disk", and "OK" to close the table, and restart MAUD. This is not mandatory, but the appropriate bank grouping makes the comparison between measured and computed diffractograms easier, especially in the 2D plotting. This should be done before starting "HIPPO wizard" and creating the PAR file.

3.2.2. Scale factor bounding. After creating the PAR file (either before or after importing phase information), we introduce the "scale factor bounding", which forces the scale factors belonging to the same Bank group to have the same value. The scale factor is the value to adjust the overall height of the computed diffractogram. This is basically a meaningless value but one of the most important values to have a good agreement between the computed and measured diffractograms.

When the authors developed the texture measurement scheme with only one neutron exposure, we encountered the problem that the scale factors belonging to the same Bank group varied in a certain range, resulting in inferior fitting. On the other hand, the scale factors in the same Bank group converged into an almost identical value for non-textured powder materials and a good fitting quality was obtained. The latter seemed to be more appropriate because the intensity normalization described above should compensate the differences of detecting efficiencies among the OPs. Therefore, we introduced the scale factor bounding, which improved the result and stability of the analysis for textured materials.

The scale factor variation that we observed was probably because that one OP measures only one diffractogram at iMATERIA. When we use 3~5 GDA files measured at different ω angles (as HIPPO does), the variation of the scale factors was suppressed. Since one OP measures multiple diffractograms for different sample directions in this case, the scale factor for this OP should be chosen to fit all the diffractogram. In other words, the effect of texture on the scale factor is somewhat cancelled or averaged. On the other hand, when one OP measures just one diffractogram, there arises the ambiguity; which should be applied to explain the diffractogram having large peaks, high pole densities or large scale factor?

When the scale factor bounding is introduced, a unique scale factor is determined by considering all the diffractograms belonging to the Bank group. As the complementary effect, this increases the calculation speed due to the reduction of parameters to be refined.

The scale factor bounding can be introduced by changing the refinement status of the scale factors. One of them in a certain Bank group needs to be "refined" but the others are set "equal to" the refined one. Since there are 132 scale factors and 7 Bank groups, 125 statuses of scale factors should be changed. Because it is very time-consuming to do this procedure manually on MAUD GUI, we developed a simple software to modify the PAR file so that the above procedure is conducted in a moment.

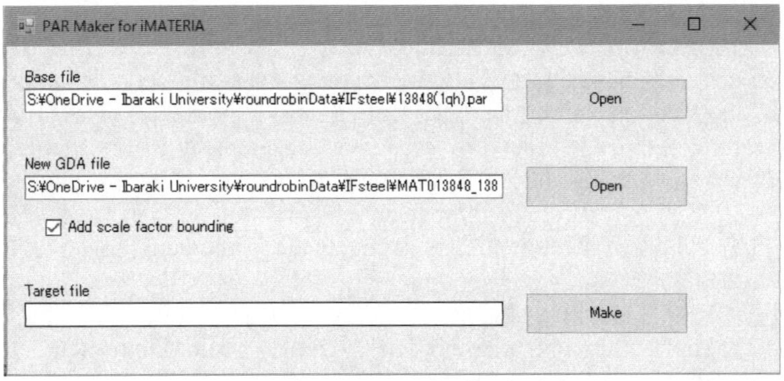

Figure 4. Screenshot of the "PAR Maker", which applies the scale factor bounding and/or makes a clone PAR file.

The screenshot of this software is shown in figure 4. This software can also be used to create a "clone" PAR file, which has the same parameters as the base but GDA file data is replaced. When the replacement of measured data is not needed, a user chooses the PAR file made by HIPPO wizard, specify the GDA file which was chosen in the wizard, and check the "Add scale factor bounding". Finally, by clicking "Make", the new PAR file with the scale factor bounding is created. The file name can be specified by the save-file dialog. The textbox of "Target file" shows the full-path name of the new PAR file after the job is completed.

By specifying different GDA file from that included in the base PAR file, the data is replaced by the new GDA file, *i.e.,* another measurement. This is useful to analyse the series of samples having similar conditions. It is unnecessary to repeat HIPPO wizard and set phases for every sample if a user uses this software. However, one should pay attention that this process may introduce a certain bias on the analysis. We recommend resetting ODF for every phase in the clone PAR file at the start of refinement. Based on our experience, the ODF calculation converges faster by stating from a random ODF than a certain "wrong" texture.

3.3. Refinement

3.3.1. General settings. Before starting the refinement, several settings are required. The detailed procedures can be found elsewhere [9]. In this paper, we simply introduce what we apply. We recommend using MAUD version 2.7.1 or higher.

As the ODF calculation method, E-WIMV method [11] is applied. The ODF resolution of 5° works the best for most of structural metallic materials. The default sample coordinate is fixed at the diffractometer coordinate, X_D-Y_D-Z_D, in MAUD. Therefore, artificial rotation of the sample is necessary so that X_D, Y_D, and Z_D respectively correspond to the RD, TD, and ND in case of the rolled sheet. figure 5 shows the values of rotation angles for the rolled sheet with three typical setting geometries.

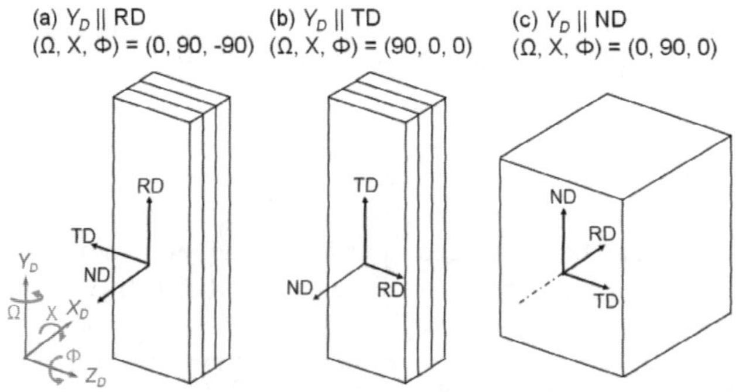

Figure 5. Sample rotations to fit the RD-TD-ND coordinate to the diffractometer coordinate,

3.3.2. Automatic refinement wizard. In the Rietveld analysis, a user can "refine" (deal as variables) almost all the parameters at the same time. However, having too many variables at the beginning quite often results in divergence, *i.e.* the calculated fitting curve goes far away from the measured diffractogram. Therefore, one needs to have a "strategy of refinement"; what should be fixed or to be refined at first, and what to do after that? The automatic refinement wizard provides some sets of the refinement strategies. In case of textured metallic materials, one can use "Texture analysis" in the wizard. This automated refinement process automatically conducts the refinements in the following four steps.

i. background (BG) and scale factor (SF).
ii. lattice parameter (only *a* for cubic phases), DIFC, Debye-Waller factors (B_{iso}) + BG and SF.
iii. Microstrain (ε) and crystallite size (D) + BG, SF, *a*, DIFC, B_{iso}.
iv. Mutual refinement of ODF and Rietveld parameters: ε, D, a, DIFC, B_{iso}, BG, SF.

At each step, several cycles of iterative calculation are conducted. We recommend conducting 4 cycles. Applying too few cycles may result in divergence.

If the main purpose of the analysis is the texture analysis of a single-phase steel, a satisfactory result can be achieved only by doing this. However, in some cases, additional iteration with more parameters are necessary as described below.

3.3.3. Individual refinement of Debye-Waller factor for each phase. The Debye-Waller factor, B_{iso}, also called the temperature factor, represents the effect of the thermal vibration of the atoms on the diffractogram. This introduces the gradual decrease of the diffraction intensities with decreasing d (or increasing Q, technically) [18]. By conducting the above automatic refinement, the Debye-Waller factors, B_{iso} for all the phases are bound in to one value. This is a good assumption at the initial stage not to reach divergence. However, For the accurate determination of the phase fraction, B_{iso} for each phase should be separately refined. The separate refinement can simply be done by changing the refinement status from "equal to" to "refine", and a few cycles of refinement. With decreasing B_{iso}, the overall diffraction intensity increases. Therefore, underestimation of B_{iso} causes the underestimation of the phase fraction.

Table 1 shows examples of analyses with separated and bounded B_{iso} refinement for the laminated steel sample used in the previous study [4]. The FCC (austenite) phase tends to have larger Biso than that of BCC (ferrite) phase. When B_{iso} for both phases are bounded into one value, the phase fraction of austenite is underestimated, as expected above. Peng et al. reported that B_{iso} for BCC and FCC iron at 293 K were 0.3250 and 0.5577, respectively [18]. Although the values for alloy steels can be different, it is always important to check if the value of B_{iso} is around these values. Sometimes, a negative value or extremely large value (>1) is obtained. In such case, some other parameters (most likely texture) are wrongly determined, or an absorption model should be considered.

To take the absorption into account, "Harmonic Coeff." sample shape model works very well. The sample symmetry should be "*2/mmm*" for rectangular samples. "*Cylinder*" symmetry is also used for cylindrical bars and wires. Since both the absorption and thermal vibration bring gradual change of scattering intensity depending on d, they seem to affect each other in the refinement. Based on our experiences, the steels containing Mn or Co often require the consideration of absorption while there seems to be negligible absorption for low carbon steels and Si steels.

When the fraction of the secondary phase is a few percent or smaller, the refinement of B_{iso} tends to be very unstable. In such case, it is better to have a fixed value, according to the table provided by Peng *et al.*, rather than applying an unreliable refined value.

Table 1. Refinement conditions and resulted values of Debye-Waller factors and Austenite fractions in the laminated SUS430-SUS316L composite sample. The prepared fraction of austenite was 19.4 vol%.

B_{iso} refinement	Separate	Bounded
B_{iso} (ferrite)	0.27	0.25
B_{iso} (austenite)	0.48	0.25
Austenite fraction (vol %)	19.1	17.4

3.3.4. Chemical composition. In most multiphase materials, each phase has a different chemical composition even if it is a solid solution. As the neutron scattering intensity depends on the elements, the chemical compositions should be considered for the accurate phase fraction determination. In the previous paper [4], we have reported a verification using composite samples consisting of SUS 316L austenitic stainless steel (Fe-18Cr-12Ni-1Mo, mol %) and SUS430 ferritic stainless steel (Fe-18Cr). At the beginning of the study, we obtained smaller analysed fractions of austenite than the prepared fractions in all the samples because both FCC and BCC phases were set as "pure Fe". Since SUS 316L contains a certain amount of Ni, the average scattering length was larger than that of SUS 430. By setting alloy elements (as the atoms partially occupies the Fe site), we have reached the satisfactory results reported in that paper. If possible, one should estimate the chemical compositions of the phases by the thermodynamic calculation or measure them by EDS, which is beneficial to obtain more accurate phase fraction. Especially the effect of chemical composition would be more critical when the sample contains high amount of Ti and/or V. Since they have negative values of coherent scattering lengths, the average scattering length can largely change by adding them.

4. Summary

The simultaneous measurement system for texture and phase fraction was developed at the TOF neutron diffractometer, iMATERIA at J-PARC MLF, Japan. Most of technical backgrounds and procedures of measurement/analysis at iMATERIA are common with the pioneer, HIPPO at LANSCE, USA. However, some specific considerations and procedures are required due to the machine specification and characteristics of the measurement. Especially, the scale factor bounding, and separate Debye-Waller factor refinement are keys to obtain accurate textures and phase fractions. In some cases, consideration of the absorption relating to the sample shape and/or chemical composition must be carefully considered, which depends the elements consisting the materials.

References

[1] Xu P G, Tomata Y, Arakaki Y, Harjo S and Sueyoshi H 2017 *Mater. Charact.* **127** 104
[2] Jacques P J et al. (14 more autors) 2009 *Mater. Sci. Tech.* **25** 567
[3] Scarlett N V Y, Madsen I C, Cranswick L M D, Lwin T, Groleau E, Stephenson G, Aylmoree M, and Agron-Olshina N 2002 *J Appl. Cryst.* **35** 383.
[4] Onuki Y, Hoshikawa A,Sato S, Ishigaki T and Tomida T 2017 J. Mater. Sci. **52** 11643
[5] Onuki Y, Hoshikawa A, Sato S, Xu P G, Ishigaki T, Saito Y, Todoroki H and Hayashi M 2016 *J. Appl. Cryst.* **49** 1579
[6] Ishigaki T et al. (16 more authors) 2009 *Nucl. Instrum. Methods Phys. Res. A* **600** 189
[7] Nakajima K et al (63 more authors) 2017 *Quantum Beam Sci.* **1** 9
[8] Wenk H R, Matthies S and Lutterotti L 1994 *Mater. Sci. Forum* **157-162** 473
[9] Vogel S C, Hartig C, Lutterotti L, Von Dreele R B, Wenk H R and Williams D J 2004 *Adv. X-Ray Anal.* **47** 431
[10] Wenk H R, Lutterotti L and Vogel S C 2010 *Powder Diffr.* **25** 283
[11] Lutterotti L, Matthies S, Wenk H R, Schultz A S and Richardson Jr J W 1997 *J. Appl. Phys.* **81** 594
[12] Lutterotti L, Chateigner D, Ferrari S and Ricote J 2004 *Thin Solid Films* **450** 34
[13] Toby B H 2006 *Powder Diffr.* **21** 67
[14] Izumi F, Asano H, Murata H and Watanabe N 1987 *J. Appl Cryst.* **20** 411
[15] Hoshikawa A et al. (14 more authors) 2009 *Nucl. Inst. Methods Phys. Res. A* **600** 203
[16] Larson A C and Von Dreele R B, 2004 *Los Alamos National Laboratory Report LAUR* 86-748
[17] Hall S R, Allen F H and Brown I D 1991 *Acta Cryst. A* **47** 655
[18] Peng L M, Ren G, Dudarev S L and Whelan M J 1996 *Acta Cryst. A* **52** 456

Acknowledgement

The neutron diffraction experiments were conducted at the Materials and Life Science Experimental Facility of J-PARC as part of proposal 2016PM0001. We appreciate Dr. Sven Vogel and Dr. Luca Lutterotti for their professional advices to the development of measurement and analysis schemes. We are also grateful to Ibaraki local government for the financial assistance.

IOP Conf. Series: Materials Science and Engineering **375** (2018) 012036 doi:10.1088/1757-899X/375/1/012036

Utilization of spatial resolved qualitative texture assessment method on an object of the Seuso Treasure

M Sepsi, V Mertinger and M Benke

Institute of Physical Metallurgy, Metalforming & Nanotechnology, University of Miskolc, Miskolc-Egyetemváros H3515, Hungary

femsepsi@uni-miskolc.hu

Abstract. Centerless X ray diffractometers are specially designed for nondestructive residual stress measurements. The similarity between the X-ray diffraction based stress and texture measurement methods has led us to introduce measurement method for nondestructive (sample cutting-free) stress measuring devices, to describe the texture. After the determination of identical lattice plane series distribution with centerless and conventional diffractometers, advanced anisotropy measurements techniques were invented. The new examination methods were applied for a Late Roman silver artefact, the Perfume Box from the Seuso treasure. The spatial resolved qualitative assessment of the texture provided valuable insight on the processing of the material and manufacturing of the different parts of the artefact. The developed techniques are completely nondestructive, therefore the examined artefacts suffer no harm. The methods inherit the privileges of centerless diffractometers such as no object size limits.

1. Introduction

For archeological finds, especially high-value objects, only nondestructive testing modes are permitted as test options. This is true even though destructive testing methods give answers to many questions that cannot be achieved by nondestructive means [1]. For example, in the case of metal sampling, a simple light or an electron microscope examination can be used to give a clear evidence of the production and manufacturing technology based on microstructural characteristics [2]. Since the conditions of metal forming are clearly imprinted into the objects, it appears in the state of residual stress and texture [3,4]. Therefore, nondestructive mapping of these properties can provide important information. The two characteristics can be measured very well with X-ray diffraction technique. Classical powder diffractometers require Bragg-Brentano layout with a limited sample size and geometry satisfying the focusing conditions. It provides some degree of freedom when the diffractometer can function with a Göbel mirror and parallel beam operation [5]. Then, curved surfaces can be tested, centered on the pattern plane, which requires a careful and time-consuming measurement setup. The sample size is also limited here, but we get information about the entire diffraction angle range.

Centerless diffractometers [6] do not require sample cutting and virtually have no sample size limit. Such diffractometers are designed to determine residual stress in a nondestructive manner of machine parts, even in field conditions. The similar actions, tilting and rotating movements during residual stress and pole figure measurement led to the idea of developing the texture test method for centerless diffractometers. A single limitation of the device is that, for a given setting, a narrower diffraction angle range, namely, a higher Bragg angle range can be mapped in contrast with standard powder

diffractometers. The development of the test and evaluation methods are continuing to extend the features of the equipment and to extract information beyond the residual stress.
The Seuso treasure is one of the most famous silver collection from the Late Roman period. It is composed of 14 silver vessels and copper cauldron in which they were found. The name comes from the original owner, Seuso to whom the treasure is addressed in a couplet inscribed on the niello-inhalid Hunting Plate. The objects were probably manufactured in the Roman Empire at various points between A.D. 350 and 450. The Cascet- Perfume Box- having a cylindrical body a conical lid and an interior disk pierced with 7 holes for toilet flasks. The body and lid of the Cascet were both manufractured from single sheet of silver. No evidence of any joints can be detected. Ultrasonic measurements indicate that the Cascet is made of thin silver measuring only 0.5 to 1.5 mm in thickness. Because of the areas of high relief are not thinner than the background, suggesting extensive working of the silver from both outside and inside [7,8]. However, no other evidence about the deformation processes during shaping had been revealed so far. Because of the very high value of the objects, no destructive investigation could have been performed to reveal the microstructure. In this research work, the aim is to extract as much information as possible about the material, the production and the afterlife of the objects, from the large amount of diffraction data obtained during standard residual stress measurement carried out without destruction. In the present case, the focus is to determine the texture characteristics from the residual stress measurement data by visualizing the characteristic sections of pole figures.

2. Experimental

The standardized modified $\sin^2\psi$ residual stress measurement method (the beam path is realized according to Figure 1) was used to determine the interferential function in different tilting position. The measuring parameters of the Stresstech G3R centerless diffractometer were the followings: exposure time is 5 seconds; accelerating voltage 28 kV; current 8 mA; collimator 4 mm in diameter; detector arc 50 mm; X ray source Cr, tilting range -45 / + 45 °; number of tilting made 5/5. The intensity distribution of {222} and {311} reflections of silver were measured for the texture analysis. To interpret the intensity data of residual stress measurements in terms of texture, it must be transformed to be comparable with the standard representation method. The effect of tilting on the intensity value (different absorption due to the different path length) must be also corrected. A measurement with the same tilting was done on the silver powder and a correction function was calculated and applied on the stress measured data.

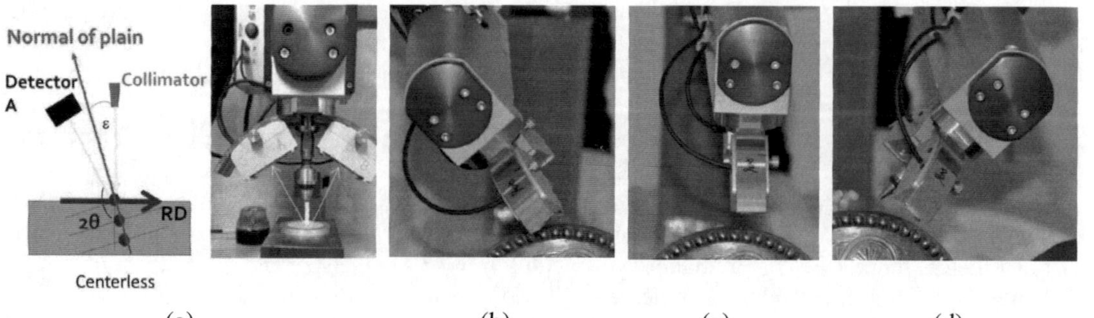

(a) (b) (c) (d)

Figure 1. The beam path and operating mode of the centerless diffractometer (a) schematic and photo in Ω mode, A is the position sensitive detector. Different ψ tilting positions in ψ mode (b) -45°, (c) 0°, (d)-45° during stress measurement.

Furthermore, due to the performed standard residual stress data acquisition, only a few intensity data point is available from one measurement location, therefore, the standard χ-sections of the pole figure cannot be reconstructed. Another problem is that in the classical representation mode, the intensity distribution is indicated by colors or contour lines, which is not feasible if only a fraction of the measured

IOP Conf. Series: Materials Science and Engineering **375** (2018) 012036 doi:10.1088/1757-899X/375/1/012036

pole figure is available. Therefore, it is necessary to leave the characterization with the standard polar coordinate system (χ,ϕ) and represent measured data, in the 3D Cartesian coordinate system.

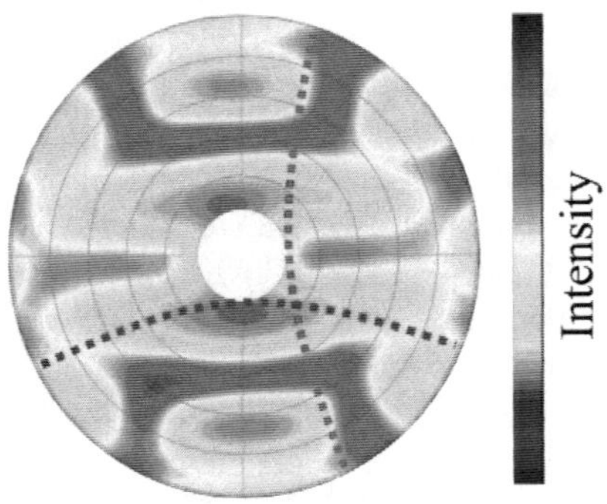

(a) {311} pole figure of rolled aluminum sheet, dashed lines represent data obtained from stress measurements

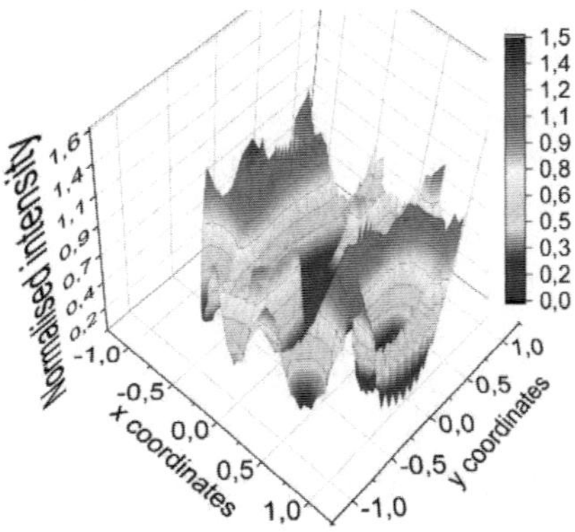

(b) The same {311} pole figure than (a) in Cartesian coordinate system

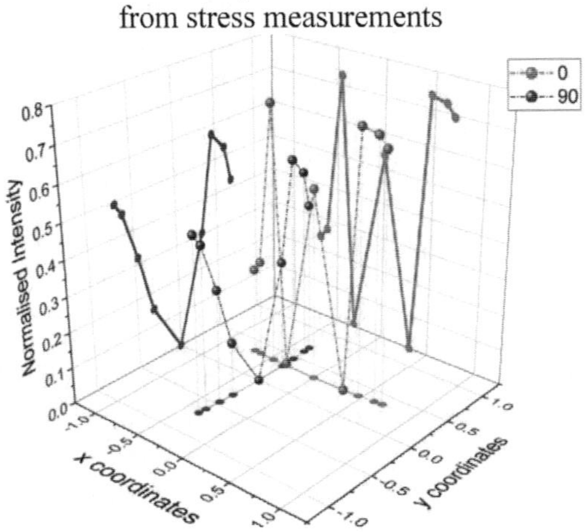

(c) Set of one colored data originated one stress measurement point on the {311} plane in the same Cartesian system than (b). 0 - in horizontal, 90 - in vertical direction

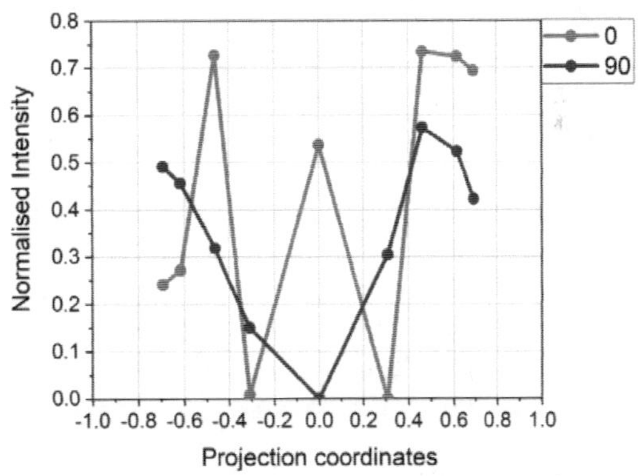

(d) Compare the projection curves of (c), representing the section of {311} pole figure in horizontal (0-red) and in the vertical (90-blue) direction

Figure 2. The procedure of visualization for qualitative texture description of a rolled aluminium sheet

The method of the calculation and visualization of the specific section of the pole figure is introduced through an example of a cold rolled aluminum sheet shown in Figure 2. After the determination of the identical lattice plane series distribution in centerless and in the conventional diffractometers, the {311} pole figure was measured with the conventional diffractometer Figure 2. a), while two data series (dashed line) of the {311} pole figure were obtained with the centerless diffractometer during stress measurements in modified ψ mode by tilting about the horizontal direction (Figure 2. a, dashed red line,0°) and vertical direction (90°, blue dashed line). The pole figure measured data was transformed from the polar coordinate system (χ,ϕ) to the 3D Cartesian coordinate system (x,y) using the (1); (2); (3) equations (Figure 2(b)):

$$x_i = \chi' \cos \phi_i \quad (1)$$

$$y_i = \chi' \sin \phi_i \quad (2)$$

$$\chi'_{.} = \frac{\chi_i}{\chi_{max}} \quad (3)$$

Due to this transformation, the value of χ is between -1 and +1. The value of z intensity coordinate was normalized within one measurement. The data obtained with the centerless X-ray diffractometer are visualized in the Cartesian system Figure 2(c) dashed lines. Because of the low information of the 3D visualization, the projection into the x-z and y-z plane were introduced (blue and red solid line curves, Figure 2. d). The two projection curves clearly represented the different texture features. If the measurement points of an object that were determined during stress measurement were measured in the same measuring direction related to the coordinate system of the object, the following analysis can be performed. Refer the selected measuring direction to a particular direction of the pole figure. Calculate and transform the measured stress point into the Cartesian coordinate system. Perform the correction for the intensity value based on the powder data was determined at the same tilting position. Visualize and compare the projection curves for each measurement point.

3. Results and discussion

Measurements have been carried out at many locations on the Perfume Box. In the gridded photo (Figure 3), the measurement locations and the direction are symbolically marked according to the tilting position, which is very important when evaluating the texture results. The Perfume Box consists of a cylindrical body (B) and a conical cover (C) (Figure 3(b)). Axial measured direction was applied on the body and the cover while tangential direction was used on the bottom of the body (Figure 3(a),(b)). Since the measured directions were identical at every measurement points, the calculated data are comparable.

The measurement points numbered from 205 to 275 can be divided into six groups depending on their locations on the artefact. From the machining point of view, (i) the bottom of the cylindrical body, (ii) the cylindrical body and the envelope of cone, (iii) the edge of the body and the cover, (iv) and the embossment pattern on the body and covers can be separated. The edges of the body and cover are overlapped on the gridded photos. The points 224-228 on the bottom of the box (B-i), the points 206-213 on the edge of the body (B-iii), the points 214-223 on the cylindrical body (B-ii), while the 270, 271, 272, 273, 274 points on the cone cover (C-ii), the 267,269,275 points on the cover embossment (C-iv) and the 247-255 points on the edge of the cover (C-iii) were measured. The position of the measurements 201 and 254 in the edge of body and the cover respectively are shown in Figure 3(c),(d). Figure 4 and 5 show the results of measurement points. The bottom plate data (Figure 4(a)) are distinctly different from the results obtained at the other measurement sites, both plain series show a non-characteristic, a quasi-isotropic structure. There are two reasons for this. One is that the deformation did not have a distinct orientation, which is a logical consequence if thinning was done with hammering. The other possibility is that heat treatment was applied after the deformation. Similarly, it has an uncoordinated texture of the data measured on the embossed areas of the cover (Figure 4(b)). This obviously comes from a non-preferred shaping direction which is also logic if any kind of shaping methods likes chasing, vibration embossing or die hammering were applied. The data on the body, however, have a clear deformation feature for each of the plane series in each measured point (Figure 4.(c)). Data from the plain series {222} shows larger sensitivity to the degree of formation on a more fragmented surface, such as the 214 measurement point, which is surrounded by embossed sections. In the case of the cover, this difference is even more pronounced (Figure 4(d)). The strongest forming characteristics are shown by both edge of body and the cover (Figure 5(a), (b)). The "earing" appearance is clearly visible as it can be seen in the measurement points 210 and 254 on the body and the cover respectively (Figure 3(c), (d)). The maximum normalized intensity values were also measured on the edge of the cover and the body. Since silver is a very well-formed metal, it is suitable for many shaping operations. The shaping technology of these archaic objects is radically different from today's industrial metal technology, so the practice in those industrial problems does not help in understanding the resulting texture image. However, it is clear from the results that the bottom is in an isotropic shaped or partially softened state. While in contrast, a texture-like pattern of unidirectional formation was visible on the body and cover, which increased even more strongly along the edges, that is likely to be the result of a metal spinning operation.

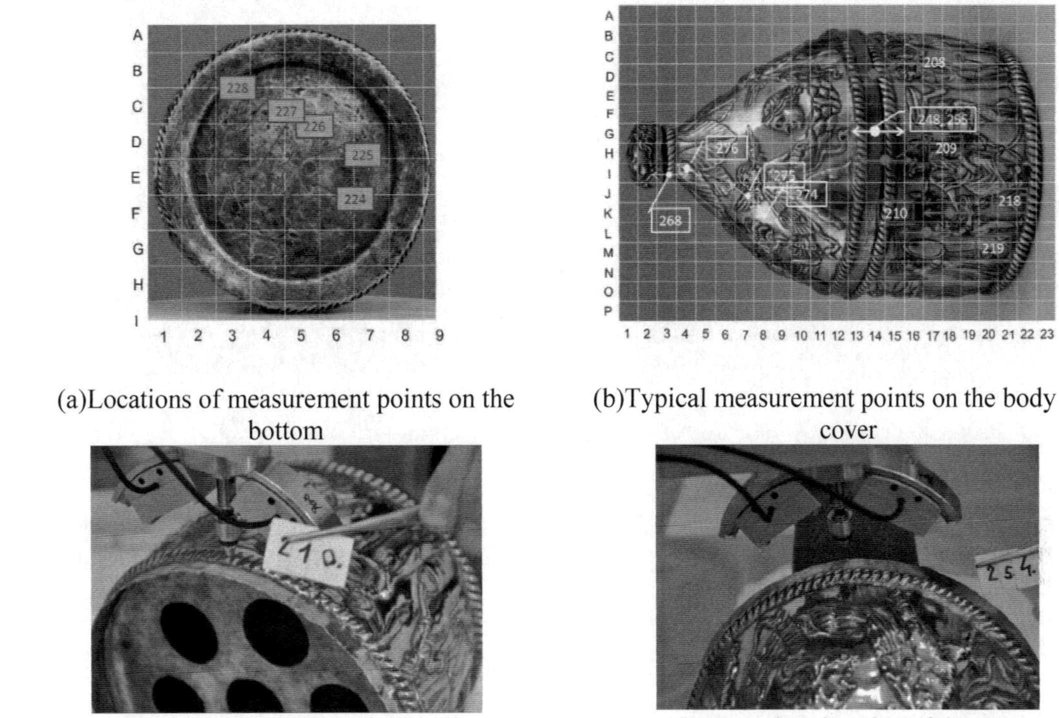

(a)Locations of measurement points on the bottom

(b)Typical measurement points on the body and cover

(c)Measuring the edge of the body

(d)Measuring the edge of the cover

Figure 3. The measurement points on the Perfume Box

4. Summary

A new texture characterisation method was developed based on the data of nondestructive (sample cutting-free) residual stress measurement performed with a centerless diffractometer. The method was used to describe the texture feature of a valuable Late Roman silver artefacts, the Perfume Box, the one part of the Seuso treasure. The intensity distribution of the silver {222} and {311} plain series were used for the texture analysis. Despite the few measuring data in one location which were determined during residual stress measurements it can be concluded that the method provided consistent and relevant texture information on the different part of the Perfume Box. Based on the evaluations it is possible to distinguish clearly the metallic state of the bottom, the side wall, the embossment and the edges of the box. The bottom plate and embossment data are distinctly different from the results obtained at the other measurement sites, both plain series show a non-characteristic, a quasi-isotropic structure. The body and the cover of the box have a clear deformation feature for each of the plane series. The strongest forming characteristics are shown by both edge of body and the cover, earing was also visible. These results were confirmed by several parallel measurements on the similar part of the Perfume Box. The results of these evaluations give guidance not only to the production technology, but also to the possible future target texture measurements. Future plans can also help in the technological reconstruction by determining the texture images that can be linked to the targeted shaping.

Acknowledgements

The described article was carried out as part of the GINOP-2.3.2-15-2016-00027 "Sustainable operation of the workshop of excellence for the research and development of crystalline and amorphous nanostructured materials" project implemented in the framework of the Szechenyi 2020 program. The realization of this project is supported by the European Union. The described research was also supported through the National Research, Development and Innovation Office – NKFIH K119566 project. M Sepsi's research was supported through the new national excellence program of the ministry of human capacities (Hungary).

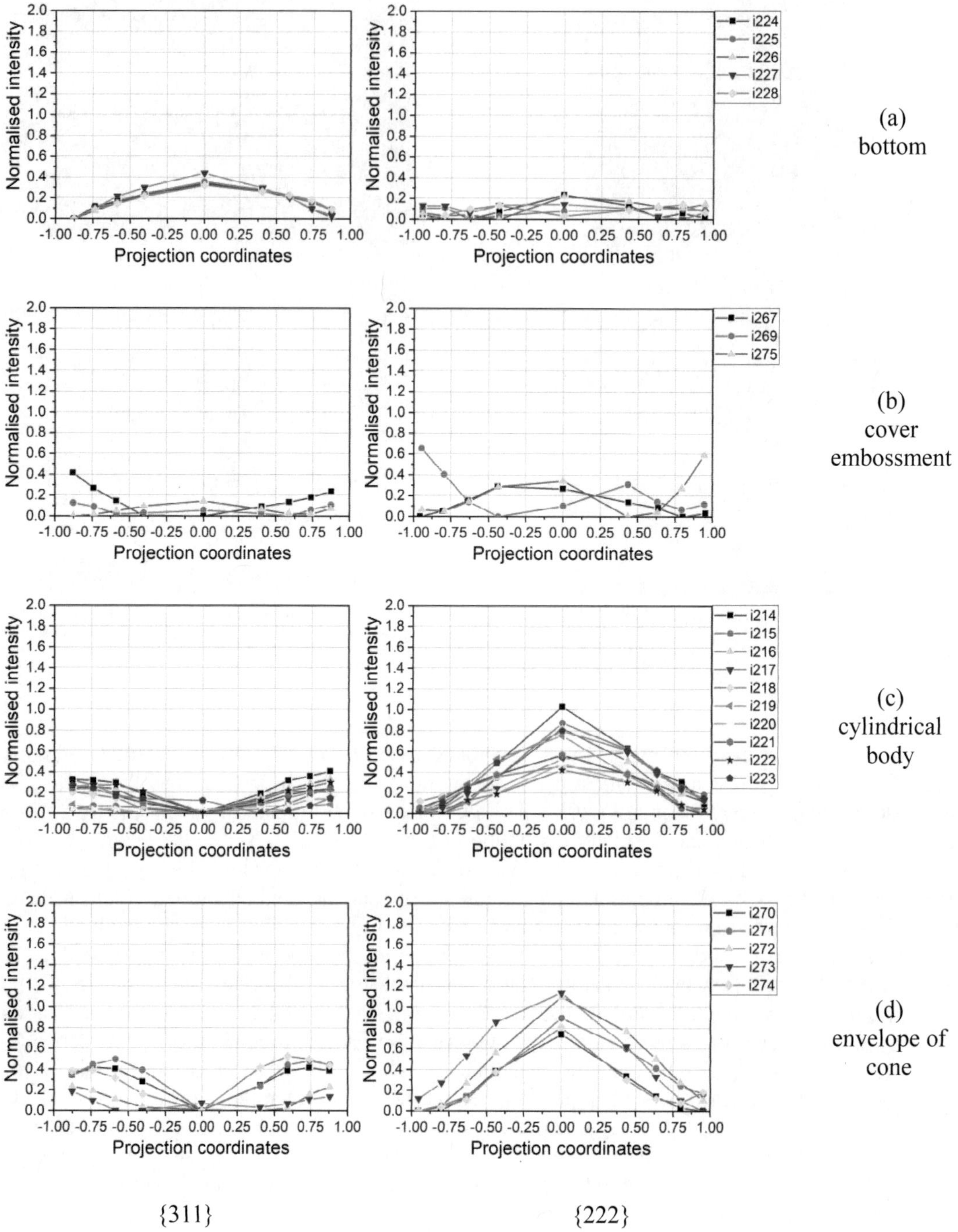

Figure 4. The calculated texture feature (projection curves) of {311} and {222} on the different part of the Perfume Box

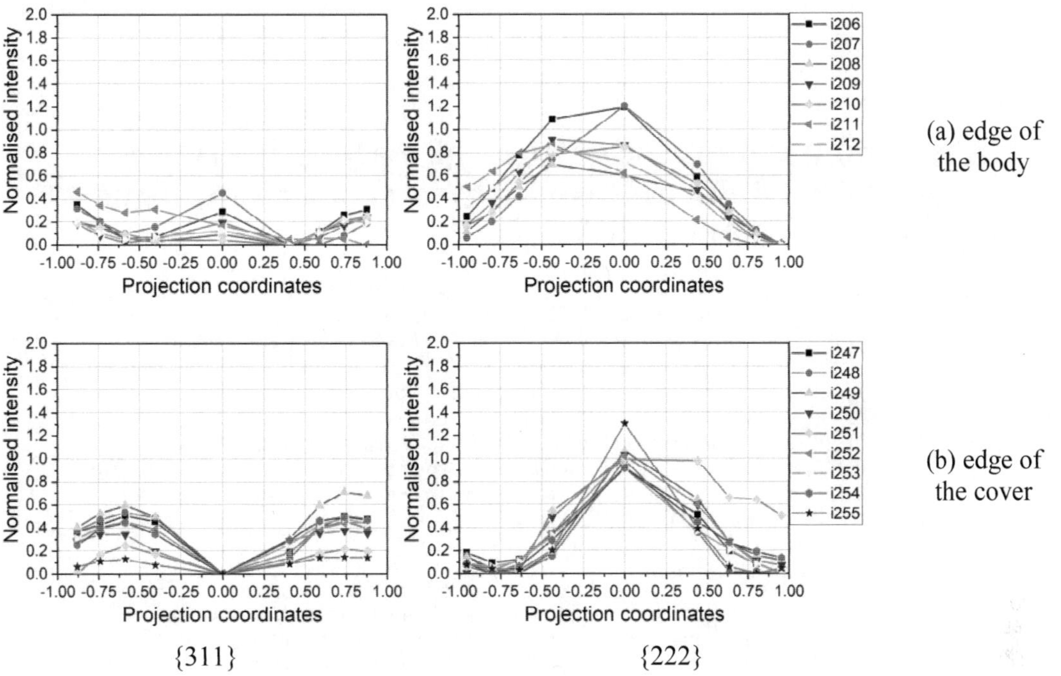

Figure 5. The calculated texture feature of {311} and {222} on the different part of the Perfume Box

Authors thank the Hungarian National Museum and Zsolt Mráv for the object and the photographs of it and Viktoria Mozgai to provide the photos with a square grid. Archaeological findings were investigated within the framework of contractual cooperation with the Institute of Astronomy and Earth Sciences, Geological and Geochemistry of the Hungarian Academy of Sciences, as part of the Seuso Research Project.

References

[1] Light J 2000 A Field Guide to the Identification of Metal *Studies in Material Culture Research* ed K Karklins (California, PA: Society for Historical Archaeology) pp. 3–19

[2] Adriaens A 2005 Nondestructive analysis and testing of museum objects: An overview of 5 years of research *Spectrochimica Acta Part B* **60** pp 1503 – 1516

[3] Kocks F, Tomé C and Wenk H 1998 *Texture and Anisotropy- Preferred Orientations in Polycrystals and their Effect on Materials Properties* (Cambridge: Cambridge University)

[4] ed Totten G, Howes M, and Inoue T 2008 *Handbook of Residual Stress and Deformation of Steel* (USA ASM International)

[5] Duran A, Herrera L, Jimenez de Haro M, Justo A and Perez-Rodriguez J 2008 Nondestructive analysis of cultural heritage artefacts from Andalusia, Spain, by X-ray diffraction with Göbel mirrors, *Talanta* **76**, Issue 1, pp 183-188

[6] Fitzpatrick M, Fry A, Holdway P, Kandil F, Shackleton J, and Suominen L. 2005. A National Measurement Good Practice Guide, Determination of Residual Stresses by X-ray Diffraction - Issue 2 No. 52 (Teddington: Crown)

[7] Mango M and Bennett A 1994, *The Sevso Treasure. Part One* (USA Rhode Island: Journal of Roman Archaeology Supplementary) Series 12.1.

[8] ed Visy Zs 2012 *The Sevso treasure and Panonia, Scientific contributions to the Sevso treasure from Hungary*, Vil.1, (Pécs:GeniaNet)

Microstructural investigations of materials after severe plastic deformation by means of orientations mapping in TEM and SEM

M. Bieda[1], A. Jarzębska[1], P.Koprowski[1,2], J.Kawałko[1,4], K. Kudłacz[1], S. Boczkal[2], M. Faryna[1], M. Kulczyk[3],W. Pachla[3], K. Sztwiertnia[1]

[1]Institute of Metallurgy and Materials Science, Polish Academy of Sciences, Krakow, Poland
[2]Institute of Non-Ferrous Metals, Light Metals Division, Skawina, Poland.
[3]Institute of High Pressure Physics, Polish Academy of Sciences, Warszawa, Poland
[4]AGH University of Science and Technology Academic Centre for Materials and Nanotechnology, Krakow, Poland

E-mail:m.bieda@imim.pl

Abstract. Qualitative and quantitative microstructure descriptions of crystalline materials are crucial for characterization and better understanding of the mechanisms of deformation and recrystallization processes. In this study orientation mapping in SEM and TEM were applied to investigations of materials after plastic deformations by means of cold rolling and hydrostatic extrusion (HE) methods. Statistical analysis of local orientation, texture and grain boundary characterization were done for chosen materials of different symmetry like cubic aluminium and hexagonal titanium with high resolution capability down to a few square nanometres. It was proved that TEM/ Orientation Mapping (OM) investigations can be successfully applied for complementary analysis to conventional SEM/EBSD.
.

1. Introduction

Properties of the crystalline materials depend on their microstructure. Materials after plastic deformation are characterised by reduction of grain size and high density of dislocations. Microstructure descriptions need to be performed in a unified manner with resolution adjusted to grain size dimensions. Introducing the SEM/EBSD technique to microstructural analysis, made the process more versatile and comparable. Thanks to that many characteristics of the materials can be obtained and the influence of processing on the properties can be better described. However, the Orientation Mapping (OM) technique in TEM has also been developed lately, in addition to EBSD studies in SEM. While TEM mapping is not so meaningful from a statistical point of view (such as the measurement of texture), it is well suited for local detailed analysis critical to particular problems.

In this paper, two metals of different crystal structure fcc (aluminium 6013) and hcp (titanium grade 2) were investigated after plastic deformation by cold-rolling and hydrostatic extrusion (HE), respectively. Both materials were characterised by large strain applied and strong enhancement of mechanical properties.

2. Material and methods

Materials after severe deformation are, because of their very good properties, especially interesting for researchers. The classical EBSD methods are generally insufficient for the observation and understanding of mechanisms of deformation and recrystallization in highly deformed materials. Investigations performed by means of transmission electron microscopy (TEM) in cases when the high resolution facilitated this method is necessary are useful supplements to SEM/ESBD. However, in order to obtain statistical information, a combination with EBSD analysis is required.

For several years, many attempts were made to create TEM systems (using different kinds of diffraction patterns or dark field images), including those that were commercialised [1-5]. Also for the SEM, the new technique of Transmission Kikuchi Diffraction (TKD) was developed [6-7]. Yet none of the considered systems are as common as conventional EBSD. In this paper, the results from investigations performed by means of conventional EBSD together with OM in TEM are presented.

Samples for investigations were chosen from the two groups of different symmetries and different deformation method. Aluminium alloy 6013, after cold-rolling up to 90 %, and pure Titanium grade 2 after HE were investigated. In both cases the material was prepared for observation in TEM and SEM by standard polishing and electropolishing methods.

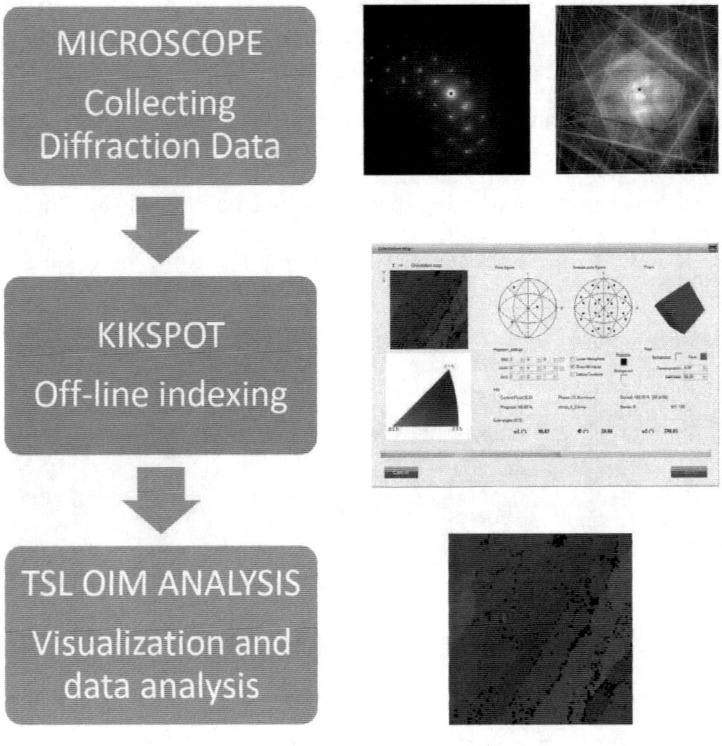

Fig. 1. Schematic of the Orientation Mapping system in the TEM

For SEM/EBSD investigations, the FEI Quanta 3D FEG SEM was used, equipped with the EDAX TSL OIM system for data collection and analysis.

For orientation mapping in the TEM, the system established in OML Skawina was used [8]. This system is based on a FEI TECNAI G20 transmission electron microscope with Lab6 cathode and equipped with a GATAN ORIUS SC200 camera. It operates on the same idea as the previously built systems presented in [3,8] with Digital Micrograph scripts to control the microscope activities. The new commercial software KikSpot was used to index sets of diffraction patterns [9]. Unlike in previous investigations [10-12] both spot and Kikuchi line type diffraction patterns (fig.1), obtained in different TEM modes, were indexed. In [8], results from the application of convergent beam Kikuchi type electron diffraction patterns for the precise analysis of changing orientations, with angular resolution about 0.1 ° are presented. In this research, spot patterns in microdiffraction mode were used to obtain better spatial resolution (less than 10 nm) than in EBSD systems.

For statistical analysis OIM Analysis 7.2 software was used in both cases to obtain good compatibility of the statistics.

3. Results and discussion

After deformation aluminium alloy 6013 presents grains elongated in the rolling direction (RD) and a bimodal distribution of precipitation. In the laminar microstructure of the matrix, small (<<1 μm) and large (1-3 μm) precipitates of other phases are distributed. In the region surrounding large particles, deformation zones (DZ) formed, and these play important roles in the recrystallization process [11]. From this point of view, it is most interesting to analyse the orientations and grain size in these areas. Conventional methods of EBSD were applied with a step size below 100 nm (fig. 2a) to very high quality prepared surfaces of those samples. The presented map shows that good quality patterns were not possible to obtain from the DZ areas. Only matrix grains could be characterized using the SEM/EBSD method. From the same material, thin foils were prepared to perform TEM investigation. The microdiffraction spot patterns were acquired from the particular area with a step size of 5 nm, and indexed using KIKSpot software. From the orientation map obtained in that way, particular features of the microstructure are visible. It is observed that RD elongated grains become curled around the second phase particles, and are narrower. Their dimensions in the directions perpendicular to the RD are less than 100 nm. The grain misorientations between neighbouring bands are characterized by High Angle Grain Boundaries (HAGB). Apart from that, very small equiaxed grains are visible in the closest vicinity of precipitations with diameter less than 50 nm (fig.2.b). These could be responsible for nucleation of new grains in that area which are presented in [11].

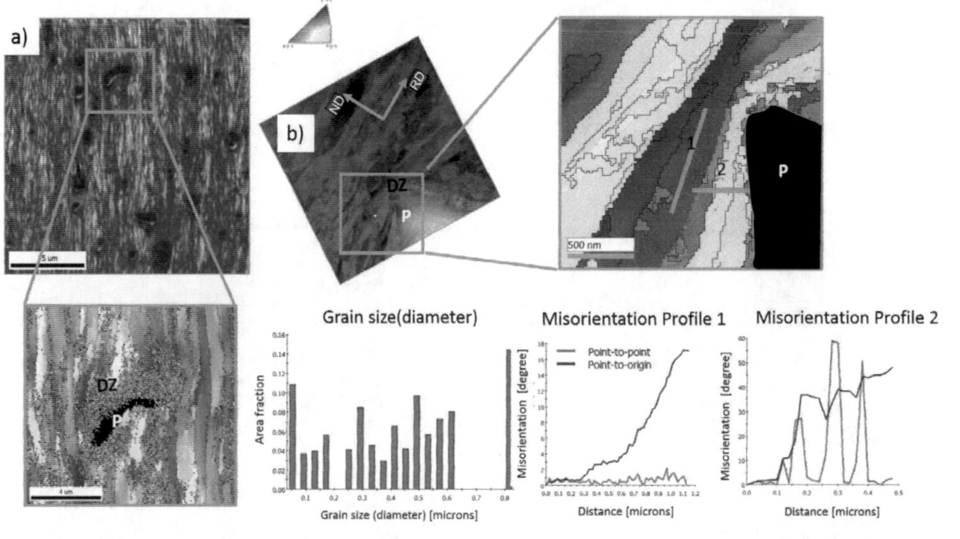

Fig. 2. Aluminium alloy 6013. (a) Orientation map (IPF coloring) from SEM/EBSD with IQ (Image Quality) map. (b) TEM Bright Field image with orientation map, grain size distribution and misorientation changes across the marked lines 1 and 2, DZ – Deformation Zone, P-Particle

The second investigated example was pure titanium grade 2, examined after HE. It possesses a strong fiber texture with grains elongated in the direction of extrusion (ED). Observation from SEM/EBSD maps reveals that average grain size is about 11 μm, and shows the existence of local areas with very high dislocation density, which cannot be resolved by

IOP Conf. Series: Materials Science and Engineering **375** (2018) 012037 doi:10.1088/1757-899X/375/1/012037

conventional SEM/EBSD (fig. 3c). These "black spaces" on EBSD maps correspond to the low Confidence Index of the pixels which do not form the grains. Observations of the thin foils cut from the bulk samples were performed using TEM. On the bright field images it could be observed that, among the elongated grains in the ED, other, smaller ones were also visible. It was not possible to index the acquired Kikuchi diffraction pattern in those areas of smaller grains, because of strong diffusion of the lines on diffraction (Fig. 3c). However, with the help of microdiffraction analysis, it was possible to obtain orientation maps (Fig. 3b). The average grain size diameter for this map was about 380 nm and HAGB fraction was on the level of 0.8. In Fig. 3. c) grains with diameters about 100 nm are coloured black. Additionally, the orientation of the grains in those areas deviates strongly from <10-10> direction. This means that local TEM characterisation of the microstructure brings completely new information about the investigated material which was not resolved by the SEM/EBSD mapping.

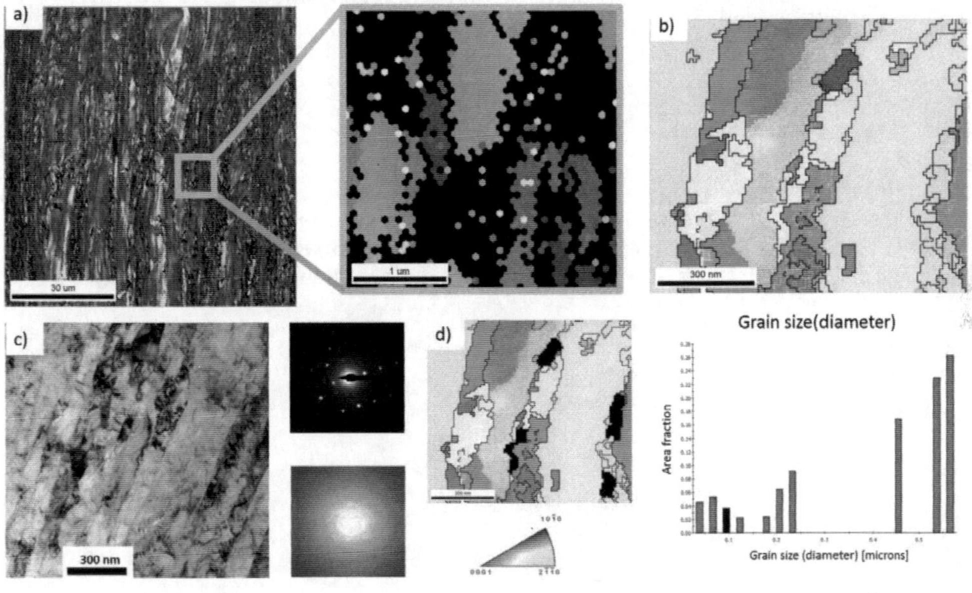

Fig. 3. Titanium grade 2. (a) Orientation map (IPF coloring) from SEM/EBSD enlarged part with low CI (black), (b) TEM orientation map, (c) TEM Bright Field image with diffractions patterns of two types (d), grain size distribution with marked grains with diameter about 100 nm

4. Conclusions

The orientation mapping with the TEM is an appropriate tool for qualitative and quantitative characterisation of microstructure at nanoscale for materials after severe plastic deformation. It closes the gap between EBSD/ SEM and BF/DF TEM with spatial and angular resolution on the nanolevel.

A combination of both diffraction techniques provides a complementary characterization of material microstructure, especially with high density of dislocations produced after plastic deformations.

Acknowledgements

This work has been partly supported by the National Centre for Research and Development in Poland through the Project No PBS2/A6/18/2014.

The CCD camera was purchased with funding from National Science Centre (NCN) project NR 2012/07/B/ST8/04025.

References

[1]Dingley, D 2006 Microchim Acta **155** pp 19- 29
[2]Wu G, Zaefferer S. 2009 Ultramicroscopy **109** pp 1317-1325,
[3]Fundenberger JJ,Morawiec A, Bouzy E, Lecomte JS 2003 Ultramicroscopy **96** pp 127-137
[4]Rauch EF, Dupuy L 2005 Arch Metall Mater **50** pp 87 -99
[5]Rauch EF, Véron M, Portillo J, Bultreys D, Maniette Y, Nicolopoulos S 2008 Microsc Anal **22** pp S5-S8
[6] Trimby PW, Cao Y, Chen Z, Han S, Hemker KJ, Lian J, Liao X, Rottmann P, Samudrala S, Sun S, Wang JT, Wheeler J, Cairney JM 2014 Acta Mater **62** pp 69-80
[7] Snedon G, Trimby PW, Cairney JM 2016 Mat Sci Eng R **110** pp 1-12
[8] Bieda M, Boczkal S, Koprowski P, Sztwiertnia K, Pieła K 2016 Arch Metall Mater **61** pp 461-468
[9] KIKSPOT www.crysorient.com
[10]Bieda M 2012 Sol St Phen **186** pp 53-57
[11] Bieda M, Sztwiertnia K, Korneva A, Czeppe T 2010 Solid State Phenom. **163** pp 13-18
[12] Bogucka J, Paul H, Bieda M, Baudin T 2012 Solid State Phenom **186** pp 112-115.

IOP Conf. Series: Materials Science and Engineering **375** (2018) 012038 doi:10.1088/1757-899X/375/1/012038

Neutron and X-ray Diffraction Texture Analysis of Novel Al-Si-Mg Alloy

M Kucerakova[1], S Vratislav[1], L Kalvoda[1], J Capek[1], K Dragounova[1] and J Cech[2]
[1] Department of Solid State Engineering, Faculty of Nuclear Sciences and Physical Engineering, Czech Technical University in Prague, Trojanova 13, 120 00 Prague 2, Czech Republic
[2] Department of Materials, Faculty of Nuclear Sciences and Physical Engineering, Czech Technical University in Prague, Trojanova 13, 120 00 Prague 2, Czech Republic
E-mail: monika.kucerakova@fjfi.cvut.cz

Abstract. Crystallographic texture and microstructure of as-casted specimen of a novel aluminum-based alloy is analyzed. Pole figures recorded by means of neutron and X-ray diffraction methods are used as the primary data in texture analysis. The data collection is performed on the SmartLab diffractometer (Rigaku) diffractometer with Cu rotating anode X-ray tube, and the KSN-2 neutron diffractometer located at the research reactor LVR-15 in the Nuclear Research Institute, plc. Rez, Czech Republic. Software tools GSAS, X'Pert Texture, and MTEX are applied in the subsequent data processing. Additional characterization methods applied include elemental composition analysis by the instrumental neutron activation analysis, phase analysis by X-ray fluorescence and X-ray diffraction, and microstructure analysis by scanning electron microscopy and metallographic optical microscopy. Elemental composition of the alloy is dominated by Al (79.91 wt.%), Si (9.90 wt. %) and Mg (4.19 wt. Aluminum (cubic, Fm3m), silicon (cubic, Fd3m) and magnesium silicide (Mg2Si, cubic, Fm3m) phases are then identified as the main crystallographic phases. Crystallographic preferential orientation of the main Al phase is only analyzed and discussed; the other two phases do not provide well-resolved diffraction maxima necessary for a reliable texture analysis.

1. Introduction

Depending on the inherent single crystalline grain anisotropy, textured material as a whole can exhibit some anisotropy of mechanical, physical and even chemical properties. In principle, the preferred orientation of grains can be found even in simple powder samples. In this case, the texture is caused by an irregular habitus of the constituting grains, and the subsequent packing of these. Such shape texture can be of importance in, e.g., sintered materials or sediment rocks. However, vast majority of textured materials studied until recently has a form of compact polycrystalline aggregates, the observed crystallographic texture of which is invoked by their mechanical deformation, heat treatment, or combination of both [1].

Looking on the diffraction techniques used to characterize crystallographic textures of polycrystalline aggregates, the three radiation types are in use, including X-rays (provided by a conventional X-ray tube or a synchrotron source), electrons, and thermal neutrons (obtained from a stationary or a spallation source). Different interaction of the mentioned radiations with matter results then in different specifics of the related diffraction methods [2-4].

The Al-Mg-Si alloys show generally excellent casting properties needed for intended use of these e.g. in automotive and aerospace industries [5]. In order to fulfil the applications requirements, further research is recently in progress focused on further improvement of mechanical properties at elevated temperatures, and a higher corrosion resistance [6, 7]. There are two main ways of accomplishing the mentioned goals: modification/refinement of the alloying elements composition, and thermal treatment of the 'as-cast' raw alloy [7]. Composition of the alloy studied in this paper (Al80 Si10 Mg4 Pb1 Fe0.7 Ca0.2) is selected with aim to improve the target properties. With exception of Pb, all the main constituting elements show a negative mixing enthalpy with the Al matrix [8]. Addition of a small amount of lead is intended to positively contribute to the corrosion resistance of the resulting alloy [9, 10]. The neutron (ND) and X-ray diffraction (XRD) methods are applied in order to investigate crystallographic texture of an as-casted Al-Mg-Si alloy sample; scanning electron microscopy (SEM), energy dispersive X-rays analysis (EDX), instrumental neutron activation analysis (INAA), metallographic optical microscopy and XRD methods are then used to elucidate microstructure and phase composition of the sample. Mechanical properties are tested by instrumental indentation method.

2. Experimental

Specimen of a circular cross-section 50 mm in diameter and thickness 5 mm was cut from the bottom part of the cylindrical ingot of the casted studied alloy sample and used in the ND and XRD measurements. No further processing of the as-casted specimen was performed.

Elemental composition of the sample was analyzed by INAA performed on the research reactor LVR-15 in the Nuclear Research Institute, plc. (NRI) Rez, Czech Republic. For quality control purposes, 75 – 100 mg aliquots of NIST standard reference material (SRM) 1633b Constituent Elements in Coal Fly Ash were prepared for irradiation. The amount of lead was determined by X-ray fluorescence analysis.

Microstructure of the samples was characterized by two main methods. On the larger scale, the sample structure was observed with metallographic microscope Neophot 32. The reflection micrographs were taken on the surface with the grain structure enhanced by electro-polishing method. The microstructure details were then revealed by SEM (JEOL JSM840 A) providing the back-scattered electron mode and EDX analysis.

Instrumental indentation measurements were performed with NHT2 system (Anton Paar). The Vickers indentor pressed into the sample surface under the maximal load of 3 N was used. The obtained data were processed by the Oliver-Pharr method [11].

The theta/theta X'Pert PRO diffractometer with a Co X-ray tube was used for the XRD phase analysis. The X'Pert HighScore (PANalytical) software was used to process the obtained diffraction patterns.

The XRD pole figures (PFs) were collected on the SmartLab diffractometer (Rigaku) with a Cu rotating anode X-ray generator. Incomplete PFs {222}, {400} and {311} of the Al-phase were recorded with the maximum polar angle 75 degrees (measured from the specimen normal). The experimental PFs were then used to calculate the orientation distribution function (ODF) of grains which then served as a source for reconstruction of the complete PFs {200}, {220}, {111} and {311}. The calculations were performed with aid of the SW tool MTEX [12].

The ND measurements took place at KSN-2 neutron diffractometer (equipped with a Huber Eulerian Cradle 511.5 goniometer) located at the horizontal channel of the research reactor LVR-15 in NRI Rez. The KSN-2 diffraction device offers a good sample irradiation intensity with the best inter-planar distance (d) resolution value of $\Delta d/d = 0.007$ in the range of d ~ 1.0÷0.1 nm. A beam of monochromatic neutrons with the mean wavelength $\lambda = 0.1362$ nm obtained by reflection from the single-crystal Cu (200) monochromator was used to record the complete experimental PFs of the plane sets {111}, {200}, {220} and {311} of the Al-phase. The same procedure as for XRD PFs was then used to obtain the re-calculated PFs {200}, {220}, {111} and {311}.

3. Results and Discussion

3.1 Sample composition and morphology

The elemental composition obtained by INAA is in Tab. 1. The coarse-grained Al phase dominates the metallographic micrographs, accompanied by precipitates of other phases (Fig. 1). The estimated mean size of Al grains is ca 40 μm.

Table 1. Elemental composition of the studied as-casted sample. Only the main constituting elements with the concentration exceeding 500 ppm are included.

Al	Si	Mg	Pb	Fe	Ca	Cr	Mn
[wt%]	[wt%]	[wt%]	[wt%]	[wt%]	[wt%]	[mg/kg]	[mg/kg]
79.91	9.90	4.19	1.10	0.69	0.18	644	1644

The Vickers hardness ($HV = 95.5 \pm 5.5\, MPa$) and Young modulus ($E = 81.8 \pm 2.9\, GPa$) values determined by instrumental indentation method exceeds in ca 20 per-cent the values reported for pure aluminum. Such enhancement is well known and occurs due to the precipitation hardening effect.

IOP Conf. Series: Materials Science and Engineering **375** (2018) 012038 doi:10.1088/1757-899X/375/1/012038

Morphology and phase composition obtained from SEM is shown in Fig. 2. In agreement with the optical micrograph (Fig. 1), the matrix consists of Al grains and contains further segregated phases, including mono-elemental particles (Cu, Pb) and grains of intermetallic phases (Mg_2Si, $Al_8Si_6FeMg_3$).

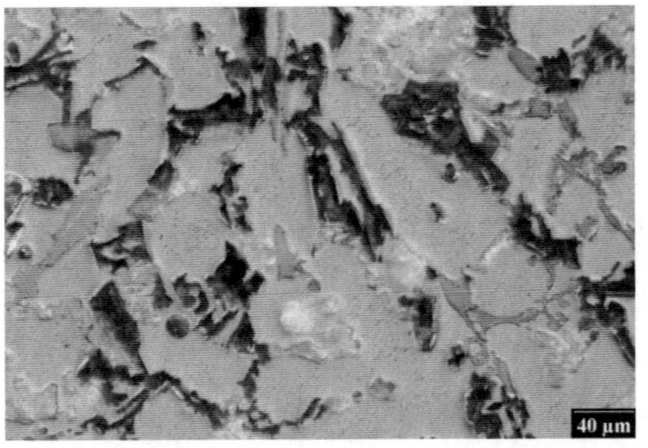

Figure 1: Typical micrograph obtained by metallographic microscopy. Electrolytic polishing is used to enhance the grain boundaries contrast. The pink background corresponds to aluminum grains; precipitates of other phases appear at the Al grains boundaries.

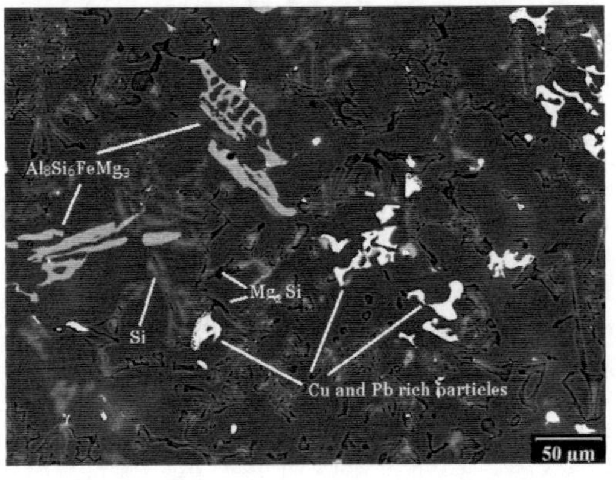

Figure 2: SEM micrograph recorded in the backscattered electrons mode. The matrix is formed by aluminum (black grains). Other phases identified by SEM/EDX are marked.

The XRD patterns used in phase composition analysis (Fig. 3) revealed presence of several main phases including aluminum (cubic, Fm3m), silicon (cubic, Fd3m), aluminum iron magnesium silicide $Al_8Si_6FeMg_3$ (hexagonal, P-62m), lead (cubic, Fm-3m) and magnesium silicide Mg2Si (cubic, Fm3m) (Tab. 2).

Table 2. Concentration of the main phases as identified by the quantitative X-ray phase analysis.

Al [% wt.]	Si [% wt.]	Mg_2Si [% wt.]	Pb [% wt.]	$Al_8Si_6FeMg_3$ [% wt.]
79.91	9.90	4.19	1.10	0.69

IOP Conf. Series: Materials Science and Engineering **375** (2018) 012038 doi:10.1088/1757-899X/375/1/012038

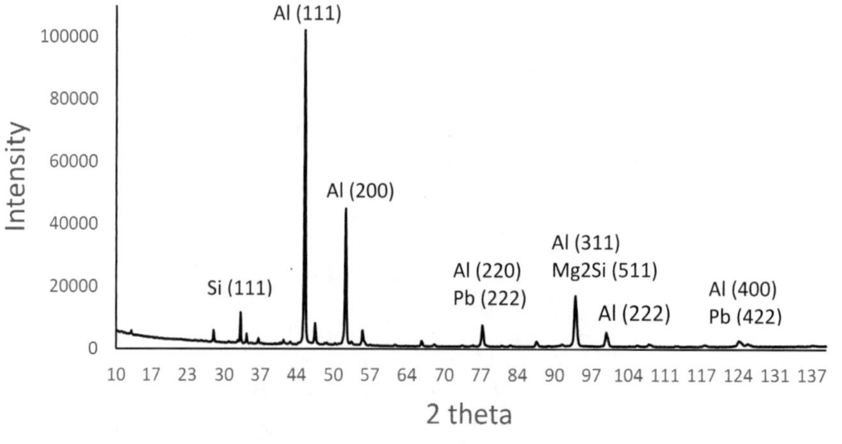

Figure 3: XRD patterns used in phase analysis. The phases corresponding to the main diffraction maxima are indicated.

3.2 Crystallographic texture

The PFs re-calculated for the Al phase from the primary XRD and ND data are given in Fig. 4 and Fig. 5, respectively. The directions marked "X" and "Y" are arbitrarily selected and "physically indistinguishable" from the point of the cylindrical symmetry of the casting cuvette. The recalculated PFs agree well with the primary experimental PFs (not shown here). The observed isolated maxima reflect the coarse-grained microstructure of the Al phase. Taking in account the large size of grains, it is worth to note that the PFs obtained by ND should be considered statistically more significant than the data obtained by XRD.

The PF {200} contains, for both applied methods (although more distinctly for XRD), some extremes located in the central part of the projection, surrounding the normal direction identical with the axis of the casting cuvette. In an ideal and sharp form, such alignment of the {100} poles parallel to the normal direction is well known to appear in cast aluminum alloys [1]. However, contrary to the XRD PFs, the ND PFs show maxima near the outer rim of the PFs, indicating some preference in the radial direction. The apparent difference of the XRD PFs from the ND results is likely due to additional surface deformation introduced during the specimen cutting and thus, the ND results are more reflective of the texture in the bulk. The XRDs PFs have maxima just below 5 m. r. d (mean random distribution); whereas, the maxima observed in the ND PFs are below 3 m. r. d. reflecting a relatively random texture in the cast material.

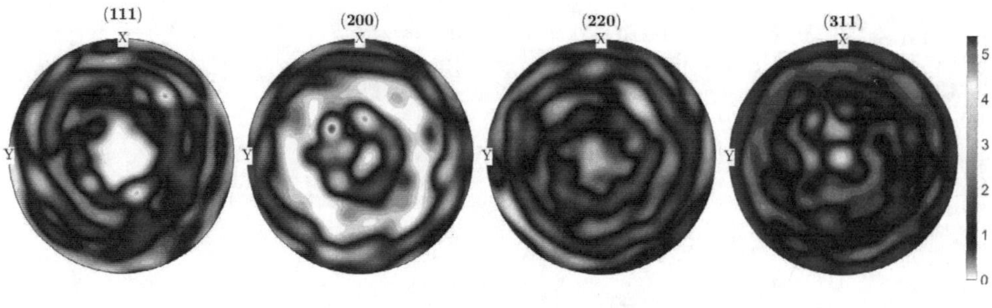

Figure 4. PFs {111}, {200}, {220} and {311} of the Al phase recalculated from the XRD data. The scale given in m. r. d. units

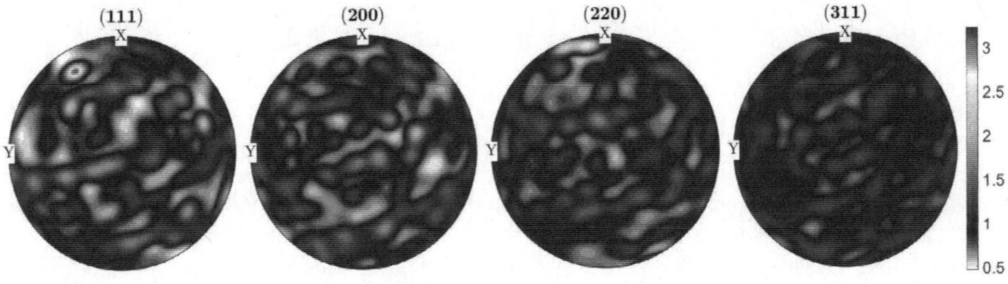

Figure 5. PFs {111}, {200}, {220} and {311} of the Al phase recalculated from the ND data. The scale given in m. r. d. units

4. Conclusions

The performed research showed that the microstructure of the investigated as-cast sample of Al-Si-Mg alloy is composed of a coarse-grained aluminum matrix filled with precipitates of other phases localized at the Al grain boundaries. The main minor phases include silicon, aluminum iron magnesium silicide, lead, and magnesium silicide. Presence of the phases was confirmed by EDX and quantitative XRD phase analysis. The Vickers hardness and Young modulus show enhancement following from the precipitation hardening process amounting ca 20 percent increase of the values observed for pure aluminum.

The aluminum matrix shows crystallographic texture of low-to-medium sharpness, with isolated maxima that are in line with the coarse-grain character of the aluminum phase. The pole orientations are declined from the positions known to appear in casted Al and Al-Si alloys characterized by the texture component with prevailing orientation of {100} poles along the casting cuvette axis. Randomization of the aluminum matrix texture and reduction of its sharpness can be likely attributed to the coarse-grained microstructure of the Al matrix interpenetrated by other segregated phases, and influence of the alloying elements.

Acknowledgement
This work was supported by the Czech Science Foundation, grant 14-36566G.

References

[1] Hsun Hu 1974 *Texture of metals*, Technical report (United States: Steel Corporation Research Laboratory).
[2] Bacon G E 1975 *Neutron Diffraction* (Oxford: Clarendon Press).
[3] Kocks U F, Tomé C N, Wenk H-R 2000 *Texture and Anisotropy: Preferred Orientations in Polycrystals and their Effect on Materials Properties* (Cambridge University Press).
[4] Engler O, Randle V 2009 *Introduction to Texture Analysis: Macrotexture, Microtexture, and Orientation Mapping* 2nd Ed. (CRC Press).
[5] Kumar K C H, Chakraborti N., Lukas H L, Bodak O, Rokhlin L 2005 *Light Metal Systems Part 3* **11A3** pp 1-13.
[6] Liang W J, Rometsch P A, Cao L F, Birbilis N 2013 *Corrosion Science*, **76** pp 119–128.
[7] Chong L, Jiayue S, Zedi L, Zhiming G, Yongchang L, Liming Y, Huijun L 2016 *Materials Characterization*, **122** pp 142–147.
[8] Takeuchi A, Inoue A 2005 *Materials Transactions*, **46(12)** pp 2817 – 2829.
[9] Zeng Feng-li , Wei Zhong-ling, Li Jin-feng, LiChao-xing, Tan Xing, Zhang Zhao, Zheng Zi-qiao 2011 *Trans. Nonferrous Met. Soc. China* **21** pp 2559–2567.
[10] Linqian Wang, Richu Wang, Yan Feng,z Min Deng, Naiguang Wang 2017 *Journal of The Electrochemical Society*, **164 (2)** pp A438-A446.
[11] Qianhua K, Wenyi Y, Guozheng K, Qingping S 2013 Journal of the Mechanics and Physics of Solids, **61** pp 2015-2033.
[12] Bachmann F, Hielscher R, Schaeben H 2010 Solid State Phenomena, **60** pp 63-68.